ASTRONOMY AND ASTROPHYSICS LIBRARY

Series Editors:
I. Appenzeller, Heidelberg, Germany
G. Börner, Garching, Germany
M.A. Dopita, Canberra, ACT, Australia
M. Harwit, Washington, DC, USA
R. Kippenhahn, Göttingen, Germany
J. Lequeux, Paris, France
A. Maeder, Sauverny, Switzerland
P.A. Strittmatter, Tucson, AZ, USA
V. Trimble, College Park, MD, and Irvine, CA, USA

Springer
New York
Berlin
Heidelberg
Hong Kong
London
Milan
Paris
Tokyo

 # ASTRONOMY AND ASTROPHYSICS LIBRARY

Series Editors: I. Appenzeller · G. Börner · M.A. Dopita · M. Harwit · R. Kippenhahn · J. Lequeux · A. Maeder · P.A. Strittmatter · V. Trimble

The Design and Construction of Large Optical Telescopes
By P.Y. Bely

Stellar Physics (2 volumes)
Volume 1: Fundamental Concepts and Stellar Equilibrium
Volume 2: Stellar Evolution and Stability
By G.S. Bisnovatyi-Kogan

Theory of Orbits (2 volumes)
Volume 1: Integrable Systems and Non-perturbative Methods
Volume 2: Perturbative and Geometrical Methods
By D. Boccaletti and G. Pucacco

Galaxies and Cosmology
By F. Combes, P. Boissé, A. Mazure, and A. Blanchard

The Solar System 2nd Edition
By T. Encrenaz and J.-P. Bibring

Physics of Planetary Rings
Celestial Mechanics of Continuous Media
By A.M Fridman and N.N. Gorkavyi

Compact Stars
Nuclear Physics, Particle Physics, and General Relativity
2nd Edition
By N. K. Glendenning

The Physics and Dynamics of Planetary Nebulae
By G.A. Gurzadyan

Stellar Interiors
Physical Principles, Structure, and Evolution
By C.J. Hansen and S.D. Kawaler

Astrophysical Concepts 3rd Edition
By M. Harwit

Physics and Chemistry of Comets
Editor: W.F. Huebner

Stellar Structure and Evolution
By R. Kippenhahn and A. Weigert

Continued after Index

Pierre Y. Bely
Editor

The Design and Construction of Large Optical Telescopes

With 327 Illustrations

 Springer

Pierre Y. Bely, Space Telescope Science Institute, Science and Engineering Systems
Department, Baltimore, MD 21218, USA

Series Editors

Immo Appenzeller
Landessternwarte, Königstuhl
D-69117 Heidelberg
Germany

Gerhard Börner
MPI für Physik und Astrophysik
Institut für Astrophysik
Karl-Schwarzschild-Str. 1
D-85748 Garching
Germany

Michael A. Dopita
The Australian National University
Institute of Advanced Studies
Research School of Astronomy
 and Astrophysics
Cotter Road, Weston Creek
Mount Stromlo Observatory
Canberra, ACT 2611
Australia

Martin Harwit
Department of Astronomy
Space Sciences Building
Cornell University
Ithaca, NY 14853-6801
USA

Rudolf Kippenhahn
Rautenbreite 2
D-37077 Göttingen
Germany

James Lequeux
Observatoire de Paris
61, Avenue de l'Observatoire
75014 Paris
France

André Maeder
Observatoire de Genève
CH-1290 Sauverny
Switzerland

Peter A. Strittmatter
Steward Observatory
The University of Arizona
Tuscon, AZ 85721
USA

Virginia Trimble
Astronomy Program
University of Maryland
College Park, MD 20742
and Department of Physics
University of California
Irvine, CA 92717
USA

Library of Congress Cataloging-in-Publication Data
The design and construction of large optical telescopes / editor Pierre Y. Bely.
    p. cm. — (Astronomy and astrophysics library)
  Includes bibliographical references and index.
  ISBN 0-387-95512-7 (alk. paper)
    1. Large astronomical telescopes—Design and construction. I. Bely, Pierre-Yves.
  II. Series
  QB90 .D48 2002
  522'.29—dc21   2002070552

ISBN 0-387-95512-7   Printed on acid-free paper.

© 2003 Springer-Verlag New York, Inc.
All rights reserved. This work may not be translated or copied in whole or in part without the
written permission of the publisher (Springer-Verlag New York, Inc., 175 Fifth Avenue, New
York, NY 10010, USA), except for brief excerpts in connection with reviews or scholarly analysis.
Use in connection with any form of information storage and retrieval, electronic adaptation,
computer software, or by similar or dissimilar methodology now known or hereafter developed is
forbidden.
The use in this publication of trade names, trademarks, service marks, and similar terms, even if
they are not identified as such, is not to be taken as an expression of opinion as to whether or
not they are subject to proprietary rights.

Printed in the United States of America.

9 8 7 6 5 4 3 2 1   SPIN 10881149

Typesetting: Pages created by the author in LaTeX2e.

www.springer-ny.com

Springer-Verlag   New York Berlin Heidelberg
A member of BertelsmannSpringer Science+Business Media GmbH

To dreamers, then, now, and always

George W. Ritchey's proposed 8-meter telescope at the Grand Canyon, 1929.

Reproduced from *L'évolution de l'astrophotographie et les grands télescopes de l'avenir*, by permission of the Société Astronomique de France.

Preface

There is no dearth of books on telescope optics and, indeed, optics is clearly a key element in the design and construction of telescopes. But it is by no means the only important element. As telescopes become larger and more costly, other aspects such as structures, pointing, wavefront control, enclosures, and project management become just as critical.

Although most of the technical knowledge required for all these fields is available in various specialized books, journal articles, and technical reports, they are not necessarily written with application to telescopes in mind. This book is a first attempt at assembling in a single text the basic astronomical and engineering principles used in the design and construction of large telescopes. Its aim is to broadly cover all major aspects of the field, from the fundamentals of astronomical observation to optics, control systems, structural, mechanical, and thermal engineering, as well as specialized topics such as site selection and program management.

This subject is so vast that an in-depth treatment is obviously impractical. Our intent is therefore only to provide a comprehensive introduction to the essential aspects of telescope design and construction. This book will not replace specialized scientific and technical texts. But we hope that it will be useful for astronomers, managers, and systems engineers who seek a basic understanding of the underlying principles of telescope making, and for specialists who wish to acquaint themselves with the fundamental requirements and approaches of their colleagues in other disciplines.

We have deliberately chosen to treat ground and space telescopes with a common perspective. Scientific institutes and industrial companies working on such observatories have historically been compartmentalized, so that the design and fabrication of ground and space telescopes have mostly been carried out by scientists, engineers, and industries of different "cultures." In practice, however, many of the problems are similar and we feel that there is actually a great advantage in understanding how each of these cultures solves them.

Since our subject is so broad, it has been our approach to invite contributions from a number of scientists, engineers, and managers. However, rather than using the traditional one section/one author format, these contributions were then edited so as to adhere to a common structure in the interest of consistency of approach and treatment. Finally, to ensure objectivity and completeness, the manuscript was then reviewed and sometimes expanded by yet other specialists. Overall, this book is therefore the product of a large number of individuals currently active in the field. Their names are listed in the following pages.

As the editor of this work, I am grateful to the Space Telescope Science Institute and the European Southern Observatory for their support and, in particular, to Ann Feild of the Space Telescope Science Institute for the preparation of the graphics. I must also thank Louise Farkas, senior editor at Springer-Verlag, and her staff for their valuable assistance in the manuscript preparation. Above all, I wish to express my gratitude to my colleagues at many institutions and in industry who have generously contributed their time to the making of this book, and to my wife Sally for much help with the text.

Baltimore, Maryland
Pierre Y. Bely
October 2002

Corrections: Although this text has passed through the hands of many reviewers, some errors undoubtedly persist. Readers are requested to bring such errors or possible misinterpretations that they may note to the attention of Pierre Y. Bely care of Springer-Verlag, or via e-mail to bely@stsci.edu.

Contents

Preface   vii

Contributors and Reviewers   xix

Credits for Figures and Tables   xxiii

Introduction   1

1 Astronomical Observations   5
- 1.1 Role of astronomical telescopes 5
- 1.2 Source characteristics 5
  - 1.2.1 Intensity 5
  - 1.2.2 Distribution of sources of interest in the sky 7
- 1.3 Observing through the atmosphere 9
  - 1.3.1 Atmospheric extinction 9
  - 1.3.2 Atmospheric emission 11
  - 1.3.3 Atmospheric refraction 12
  - 1.3.4 Atmospheric turbulence: basic notions 13
  - 1.3.5 Atmospheric turbulence: the physics of seeing 17
- 1.4 Background sources 19
  - 1.4.1 Celestial backgrounds 19
  - 1.4.2 Atmospheric background 22
  - 1.4.3 Stray light and detector background 22
  - 1.4.4 Coping with atmospheric and telescope thermal emission   22

1.5	Signal-to-noise ratio	24
1.6	Time	28
	1.6.1	Sidereal time	28
	1.6.2	Julian date	28
1.7	Coordinate systems	29
1.8	Pointing corrections	31
	1.8.1	Precession and nutation	32
	1.8.2	Proper motion	32
	1.8.3	Parallax	32
	1.8.4	Aberration of starlight	33
	1.8.5	Atmospheric refraction	35
1.9	Telescope pointing and tracking procedure	35
	1.9.1	Target acquisition	35
	1.9.2	Guiding	36
	1.9.3	Guide star catalogs	36
1.10	Telescopes and interferometers	37
	References	39
	Bibliography	40

2 Instruments 41
2.1	Main types of instrument	41
	2.1.1	Cameras	42
	2.1.2	Photometer	43
	2.1.3	Polarimeters	44
	2.1.4	Dispersing spectrometers	44
	2.1.5	Fabry-Perot spectrometer	47
	2.1.6	Fourier transform spectrometer	48
2.2	Optical through mid-infrared detectors	49
	2.2.1	Photon detection in semiconductors	50
	2.2.2	CCD detectors	52
	2.2.3	Infrared array detectors	53
	2.2.4	Specific detector characteristics	55
2.3	Relay optics	59
2.4	Cryogenic systems	60
	References	61
	Bibliography	61

3 Design Methods and Project Management 62
3.1	The project life cycle	63
3.2	The tools of systems engineering	66
	3.2.1	Design reference program	67
	3.2.2	Requirements "flowdown"	69
	3.2.3	Error budgets	72
	3.2.4	End-to-end computer simulations	74
	3.2.5	Design testability and forgiveness	75

	3.2.6	Scaling laws	76
	3.2.7	Cost models	78
	3.2.8	Cost as a design variable	79
	3.2.9	Observatory performance metrics	81
3.3	Project management	86
	3.3.1	General principles	87
	3.3.2	Project organization	88
	3.3.3	Work breakdown structure	89
	3.3.4	Project data base	90
	3.3.5	Procurement strategy	90
	3.3.6	Technology development	91
	3.3.7	Reliability	93
	3.3.8	Quality assurance, verification, and validation	94
	3.3.9	Interface documents	95
	3.3.10	Configuration management	95
3.4	Project scheduling	96
3.5	Risk analysis	99
3.6	Cost estimates and budgeting	100
	3.6.1	Approaches to cost estimating	100
	3.6.2	Budgets of main funding agencies	101
	3.6.3	Cost estimate accuracy	101
	3.6.4	Construction of multiple units	102
	3.6.5	Budgeting and resource planning	103
	References	104
	Bibliography	105

4 Telescope Optics 106

4.1	Optical design fundamentals	106
	4.1.1 Fundamental principles	106
	4.1.2 Equations of conic surfaces	108
	4.1.3 Stops and pupils	108
	4.1.4 Primary aberrations	110
	4.1.5 Wavefront errors	111
	4.1.6 Diffraction effects	114
	4.1.7 Image formation	115
4.2	Telescope optical configurations	121
	4.2.1 Single-mirror systems	121
	4.2.2 Two-mirror systems	122
	4.2.3 Three- and four-mirror systems	124
	4.2.4 Systems with spherical mirrors	126
	4.2.5 Auxiliary optics	126
4.3	Optical error budget	127
4.4	Criteria for image quality	129
4.5	System issues	134
	4.5.1 Focus selection	134

xii   Contents

   4.5.2 Selection of f-ratio 136
   4.5.3 Matching plate scale to the detector resolution 137
 4.6 Mirror blank materials 139
   4.6.1 Generalities 139
   4.6.2 Borosilicate glass 142
   4.6.3 ULE fused silica 143
   4.6.4 Low-thermal-expansion glass ceramic 143
   4.6.5 Silicon Carbide 144
   4.6.6 Beryllium 145
   4.6.7 Aluminum 146
 4.7 Mirror structural design 146
   4.7.1 Lightweighted mirrors 150
   4.7.2 Segmented mirror systems 151
   4.7.3 Thermal effects 155
 4.8 Mirror production 157
   4.8.1 Computer-controlled lapping 160
   4.8.2 Stressed mirror figuring 161
   4.8.3 Active lap figuring 162
   4.8.4 Ultraprecision machining 162
   4.8.5 Ion beam figuring 163
   4.8.6 Postfiguring mechanical deforming 164
 4.9 Optical surface testing during manufacture 165
   4.9.1 Testing philosophy 165
   4.9.2 Main testing techniques 167
   4.9.3 Testing the figure of primary mirrors 171
   4.9.4 Testing secondary mirrors 172
   4.9.5 Measuring the radius of curvature 173
   4.9.6 Eliminating the effect of gravity 173
   4.9.7 Testing cryogenic mirrors 173
 4.10 Mirror coatings and washing 174
   4.10.1 Mirror cleaning 174
   4.10.2 Coating plant 176
   References 177
   Bibliography 181

5 Stray Light Control   **183**
 5.1 Causes of stray light 183
 5.2 Finding and fixing stray light problems 184
 5.3 Baffles and stops 185
   5.3.1 Aperture stop 185
   5.3.2 Field stop 186
   5.3.3 Lyot stop 186
   5.3.4 Baffles 187
   5.3.5 Baffles for Cassegrain systems 187
 5.4 Scattering processes 188

5.5		Stray light analysis	189
5.6		Surface scattering properties	193
	5.6.1	Scatter from mirrors	193
	5.6.2	Scatter from diffuse black surfaces	195
5.7		An example of protection against off-axis sources: HST	197
5.8		An example of minimizing stray light from self-emission: NGST	198
5.9		Minimizing thermal background in ground-based telescopes	199
		References	200
		Bibliography	201

6 Telescope Structure and Mechanisms — 202

6.1		General principles	203
	6.1.1	Kinematic mounting	203
	6.1.2	Minimizing decollimation	205
	6.1.3	Use of preload	207
	6.1.4	Load paths	208
	6.1.5	Designing out "stick-slip" and "microlurches"	208
	6.1.6	Choice of materials	210
	6.1.7	Athermalization	212
	6.1.8	Structural design	213
6.2		Design requirements	214
	6.2.1	Operational requirements	214
	6.2.2	Survival conditions	215
6.3		Mirror mounts	219
	6.3.1	Mounts for single mirrors	219
	6.3.2	Mounts for segmented-mirror systems	223
6.4		Telescope "tube"	224
	6.4.1	Tube truss	225
	6.4.2	Tripod and tower-type supports for secondary mirrors	228
	6.4.3	Thermal effects	229
	6.4.4	Cassegrain mirror "spider"	230
	6.4.5	Primary mirror cell	233
6.5		Mounts for ground-based telescopes	233
	6.5.1	Equatorial mount	233
	6.5.2	Altitude-azimuth mount	235
	6.5.3	Altitude-altitude mount	236
	6.5.4	Fixed-altitude and fixed-primary-mirror mounts	236
6.6		Bearings for ground telescopes	237
	6.6.1	Rolling bearings	238
	6.6.2	Hydrostatic bearings	238
6.7		Miscellaneous mechanisms	242
	6.7.1	Overall telescope alignment	242
	6.7.2	Optics alignment and focusing devices	242
	6.7.3	Active secondary mirror for infrared chopping and field stabilization	243

xiv Contents

      6.7.4   Balancing systems 245
      6.7.5   Cable wrap and cable twist 245
      6.7.6   Mirror cover 246
  6.8 Safety devices. 247
      6.8.1   Brakes. 247
      6.8.2   End stops 248
      6.8.3   Locking devices 248
      6.8.4   Earthquake restraints 249
    References. 250
    Bibliography 251

7 Pointing and Control 252
  7.1 Pointing requirements 253
  7.2 System modeling 253
      7.2.1   First-order lumped-mass models 255
      7.2.2   Medium-size lumped-mass optomechanical models . . . 257
      7.2.3   Integrated models 257
  7.3 Pointing servo system 263
      7.3.1   Fundamentals of servo systems 263
      7.3.2   Telescope control system implementation 265
      7.3.3   Disturbance rejection 271
  7.4 Attitude actuators 273
      7.4.1   Drives for ground-based telescopes 273
      7.4.2   Space telescope attitude actuators 279
  7.5 Attitude sensors and guiding system 280
      7.5.1   Position encoders 280
      7.5.2   Tachometers 282
      7.5.3   Gyroscopes 283
      7.5.4   Star trackers and Sun sensors 284
      7.5.5   Guiding system 285
  7.6 Ground-based telescope disturbances 288
      7.6.1   Effects of wind: Generalities 289
      7.6.2   Effects of wind on telescope structure 291
      7.6.3   Effect of wind on primary mirror 293
      7.6.4   Effect of wind on telescope pier 293
  7.7 Disturbances in space 294
      7.7.1   Gravity gradient torque 295
      7.7.2   Aerodynamic torque 296
      7.7.3   Solar radiation torque 296
      7.7.4   Magnetic torque 297
      7.7.5   Reaction wheel disturbances. 298
      7.7.6   Other internally generated disturbances 300
  7.8 Active and passive vibration control 302
      7.8.1   Passive isolation of the vibration source 303
      7.8.2   Active isolation 305

Contents   xv

    7.9   Observatory control software 306
            References 308
            Bibliography 310

8   Active and Adaptive Optics   311
    8.1   Fundamental principles 311
            8.1.1   Respective roles of active and adaptive optics 311
            8.1.2   Active and adaptive optics architectures 313
    8.2   Wavefront sensors 315
            8.2.1   Shack-Hartmann sensor 315
            8.2.2   Curvature sensing 316
            8.2.3   Phase retrieval techniques 318
    8.3   Internal metrology devices 319
            8.3.1   Edge sensors 319
            8.3.2   Holographic grating patches and retroreflector systems   322
            8.3.3   Laser metrology systems 324
            8.3.4   IPSRU 324
    8.4   Wavefront correction systems 325
            8.4.1   Fine steering mirrors 325
            8.4.2   Deforming the main optics 327
            8.4.3   Dedicated deformable mirror 330
    8.5   Control techniques 331
    8.6   Typical active optics system implementations 332
            8.6.1   The VLT active optics system 332
            8.6.2   Coaligning, cofocusing, and cophasing segmented systems 333
    8.7   Correction of seeing 338
            8.7.1   Historical developments 339
            8.7.2   Adaptive optics using natural guide stars 340
            8.7.3   Adaptive optics with laser stars 340
            References 342
            Bibliography 344

9   Thermal Control   345
    9.1   General requirements 345
    9.2   Thermal environmental conditions 346
    9.3   Temperature control techniques 346
    9.4   Thermal control for dimensional control 348
            9.4.1   Mirror figure control 348
            9.4.2   Controlling optics separation and alignment 349
    9.5   Avoiding locally induced seeing 351
            9.5.1   Thermal control of the enclosure during the day 352
            9.5.2   Seeing caused by a warmer floor 353
            9.5.3   Seeing due to heat sources or sinks in the telescope chamber 354

Contents

    9.5.4 Seeing due to telescope structure cold areas 356
    9.5.5 Mirror seeing 356
  References 360
  Bibliography 360

10 Integration and Verification   361
10.1 Integration and verification program, methods, and techniques 362
    10.1.1 Verification methods 363
    10.1.2 Incremental verification 363
    10.1.3 Verification requirements matrix 364
    10.1.4 Verification based on end-to-end computer modeling . . 365
10.2 Observatory validation 366
    10.2.1 Engineering verification 366
    10.2.2 Science verification 366
  References 367
  Bibliography 367

11 Observatory Enclosure   368
11.1 Enclosure functions and requirements 369
11.2 Overall enclosure configuration 369
11.3 Height of telescope chamber above the ground 372
11.4 Wind protection and flushing 372
    11.4.1 Basic principles 372
    11.4.2 Windscreens and louvers 374
    11.4.3 Wind- and water-tunnel studies and numerical modeling 375
    11.4.4 Acoustic modes in the enclosure 376
11.5 Thermal design 376
    11.5.1 Basic principles 376
    11.5.2 Enclosure external skin emissivity 380
11.6 Structural and mechanical design 380
    11.6.1 Loading cases 380
    11.6.2 Enclosure shape 381
    11.6.3 Shutter 382
    11.6.4 Bogies and drive 383
    11.6.5 Weather seals 384
11.7 Telescope pier 385
11.8 Handling equipment 386
  References 387

12 Observatory Sites   389
12.1 Ground versus space 389
    12.1.1 Advantages of ground-based facilities 389
    12.1.2 Advantages of space-based facilities 390
    12.1.3 Aircraft and balloons 392

		12.1.4 Capabilities of various observatory platforms	392
	12.2	Desirable characteristics for ground-based sites	393
		12.2.1 Seeing	394
		12.2.2 Criteria for extremely large telescopes of the future	397
	12.3	Location and characteristics of the best observing sites	398
		12.3.1 Characteristics of the major observatory sites	400
	12.4	Evaluation methods for ground-based sites	401
		12.4.1 Methods for testing image quality	401
		12.4.2 Microthermal sensors	402
		12.4.3 Acoustic sounder	402
		12.4.4 Site flow visualization	404
		12.4.5 Radiosondes	404
		12.4.6 Numerical modeling of the atmosphere	405
		12.4.7 Optical seeing monitors	406
	12.5	Space orbits and the moon	408
		12.5.1 Low-inclination low Earth orbit	408
		12.5.2 Sun-synchronous orbits	410
		12.5.3 Geostationary and geosynchronous orbits	411
		12.5.4 High Earth orbits	412
		12.5.5 Sun-Earth Lagrangian point 2	412
		12.5.6 Drift-away orbit	414
		12.5.7 Heliocentric elliptical orbit	415
		12.5.8 Moon	415
		12.5.9 Sun-Jupiter Lagrangian point 2	416
	12.6	Radiation in the space environment	417
		12.6.1 Sources of radiation	417
		12.6.2 Radiation effects	418
		12.6.3 Dependence of radiation levels on observatory location	419
	12.7	Launchers	422
		References	423
		Bibliography	425
A	**Commonly Used Symbols**		**426**
B	**Basic Data and Unit Conversions**		**427**
C	**The Largest Telescopes**		**429**
D	**Sharpness**		**433**
E	**Derivation of the Equation of Motions**		**436**
F	**Glossary**		**439**
	Index		**486**

Contributors and Reviewers

This book is the result of a team effort. A great many scientists and engineers active in the field and possessing a vast and diverse experience have participated in its making. The list below gives the names and institution of the contributors and reviewers together with the main topics to which each has contributed. The main contributors are identified with an asterisk.

Greg Andersen, Lockheed Martin: *space systems pointing control*

Roger Angel, Steward Observatory: *mirror fabrication, general telescope concepts*

Charles Atkinson, TRW: *optical design and fabrication*

Janet Barth,* NASA, Goddard Space Flight Center: *space environment*

Christopher Benn,* Isaac Newton Group: *telescope performance metrics*

Pierre Bely,* Space Telescope Science Institute (retired): *contributions to all topics and general editor*

Daniel Blanco,* EOS Technologies, Inc.: *mechanical design*

Allen Bronowicki,* TRW Space & Electronics Group: *space system dynamics, isolation, attitude control*

Richard Burg*, NASA, Goddard Space Flight Center: *astronomical observations, general review*

James Burge, Optical Science Center, University of Arizona: *mirror fabrication, optical shop testing*

Robert Burke, TRW Space & Defense: *project management*

Christopher Burrows, Consultant, previously at the Space Telescope Science Institute: *image quality*

Marvin (Tim) Campbell,* Vertex RSI: *control systems*

Stefano Casertano, Space Telescope Science Institute: *astronomical observations*

Marc Cayrel,* REOSC, France: *mirror material and mirror manufacture*

Gary Chanan, University of California, Irvine: *phasing of segmented systems*

Jingquan Cheng, National Radio Astronomy Observatory: *general telescope design*

Martin Cullum,* European Southern Observatory: *thermal control of ground-based telescopes*

Larry Daggert, NOAO, Gemini Observatory: *systems engineering and project management*

Phillipe Dierickx,* European Southern Observatory: *optical design, mirror manufacture*

Rodger Doxsey, Space Telescope Science Institute: *project management*

Scott Ellington, Space Science and Engineering Center, University of Wisconsin: *control systems*

Toomas Erm, European Southern Observatory: *control systems*

Fred Forbes, NOAO (retired): *atmospheric seeing and seeing control*

George Frederick, Meteorological Systems URS Radian: *acoustic sounding*

Paul Gillett, NOAO, Gemini Observatory: *enclosures*

Paul Giordano, European Southern Observatory: *mirror washing and coating*

Gary Golnik, Schafer Corporation: *systems engineering and project management*

Peter Gray, European Southern Observatory: *assembly and integration*

Hashima Hasan, NASA Headquarters: *image quality*

Thomas Hawarden, UK Astronomy Technology Centre, Royal Observatory, UK: *astronomical observations, thermal issues, infrared telescopes*

John Hill,* Large Binocular Project, Steward Observatory: *telescope concepts, mirror manufacture*

James Janesick, Advanced Sensors Group, Sarnoff Corporation: *solid-state detectors*

Helmutt Jenkner, Space Telescope Science Institute: *astronomical observations, guide star catalog*

Debora Katz, U.S. Naval Academy: *astronomical observations*

Philip Kelton, Hobby Eberly Telescope, University of Texas: *telescope design*

Michael Krim, Raytheon (retired): *space structures, mirror fabrication and support*

John Krist, Space Telescope Science Institute: *image modeling*

Mark Lake, Jet Propulsion Laboratory: *microdynamics*

Marie Levine, Jet Propulsion Laboratory: *microdynamics, damping*

Richard Lyon, University of Maryland: *diffraction-limited optics*

Barney Magrath, Canada France Hawaii Telescope Corp.: *mirror washing and coating*

Jean-Pierre Maillard, Institut d'Astrophysique, France: *astronomical observations, instruments*

Terry Mast,* University of California, Santa Cruz: *segmented optics, optical fabrication and testing, active optics*

Rebecca Masterson, TRW Space & Electronics Group: *disturbances and isolation*

John Mather, NGST Project Scientist, Goddard Space Flight Center: *general review*

Craig McCreight, NASA, Ames: *detectors, instruments*

Stefan Medwadowski, Consulting structural engineer: *telescope structural design, dome and building design*

Aden Meinel,* Optical Science Center, University of Arizona (retired): *optical design, telescope concepts*

Marjorie Meinel,* Optical Science Center, University of Arizona (retired): *optical design, telescope concepts*

Mike Menzel, Lockheed Martin: *project management, systems engineering*

Luciano Miglietta, Observatorio Astrofisico di Arcetri, Italy: *mechanical systems*

Gary Mosier,* NASA, Goddard Space Flight Center: *space telescope pointing systems*

Jerry Nelson,* Keck Observatory: *mirror design and fabrication, telescope systems, observatory concepts*

Lothar Noethe,* European Southern Observatory: *mirror support systems, wind action*

James Oschmann, NOAO, Gemini Observatory: *project management*

Mette Owner-Petersen,* Lundt Observatory, Sweden: *optical design*

Roger Paquin, Advanced materials consultant: *material properties and fabrication*

Thomas Parsonage, Brush Wellmann: *beryllium materials and fabrication*

Earl Pearson, Kitt Peak National Observatory: *mechanical design*

Charles Perrygo,* Swales Associates: *mechanical and structural design*

Gary Peterson,* Breault Engineering: *stray light*

Larry Petro, Space Telescope Science Institute: *astronomical observations*

Joe Pitman,* Lockheed Martin: *space systems structures*

Marco Quattri,* European Southern Observatory: *telescope and dome structures*

Bernard Rauscher,* Space Telescope Science Institute: *detectors*

Martin Ravensbergen,* AMSL, Netherlands, previously at the European Southern Observatory: *control systems*

David Redding, Jet Propulsion Laboratory: *active optics, phase retrieval techniques*

François Rigault, Gemini Observatory: *adaptive optics*

Massimo Robberto,* Space Telescope Science Institute: *infrared telescopes and instruments*

François Roddier,* Institute for Astronomy, University of Hawaii (retired): *atmospheric turbulence, image quality, optical testing, adaptive optics*

Joseph Rothenberg, NASA Headquarters (retired): *management of large space projects*

Marc Sarazin,* European Southern Observatory: *site selection, seeing*

Daniel Schroeder,* Beloit College (retired): *optical design*

Bernard Seery, Goddard Space Flight Center: *system issues, project management*

Michael Schneermann, European Southern Observatory: *telescope mechanics, domes*

David Shuckstes, TRW Space & Defense: *project management*

Walter Siegmund,* University of Washington: *telescope structure and mechanisms, dome design*

Mark Sirota, Corning, Inc.: *pointing control system*

Alessandro Spagna, Osservatorio Astronomico di Torino, Italy: *guide star counts*

Philip Stahl, NASA, Marshall Space Flight Center: *optics fabrication and testing*

Larry Stepp,* NOAO, Gemini Observatory: *optical design and fabrication*

Mark Stier, Goodrich: *space telescopes, optics fabrication*

Conrad Sturch Space Telescope Science Institute: *astronomical observations*

Marco Venturini, Phase Motion Control S.r.l., Italy: *direct drives, control systems*

Merle Walker, Lick Observatories, University of California (retired): *site testing*

Mark Warner, National Solar Observatory: *telescope structure and mechanics*

Robert Williams, Space Telescope Science Institute: *project management, science metrics*

Krister Wirenstrand European Southern Observatory: *observatory control software*

Eve Woolridge, NASA, Goddard Space Flight Center: *space optics contamination*

James Wyant, Optical Science Center, University of Arizona: *optical testing*

Lorenzo Zago,* Swiss Federal Institute of Technology, previously at European Southern Observatory: *atmospheric turbulence, dome and mirror seeing*

Credits for Figures and Tables

Acknowledgment and grateful thanks are due to the following publishers for permission to use figure material from their books. In all cases, the relevant figures have been redrawn and sometimes modified to correspond to the content of the text.

Academic Press: Fig. 4.10 (right), adapted from Schroeder, D.J., *Astronomical Optics*, figure 10.5.

Dover Publications: Figure 12.13 from Bate, R.B., Mueller, D.D., and White, J.E., *Fundamentals of Astrodynamics*, 1971, p. 157.

International Society for Optical Engineering (SPIE): Figures 3.12, 4.52, 5.10, 7.37, 8.24, 9.2, 11.6, 11.8, and 11.16 (references cited in figure captions).

John Wiley & Sons, Inc. and *Praxis Publishing Ltd.:* Figure 2.9 adapted from McLean, I. S., *Electronic Imaging in Astronomy: Detectors and Instrumentation*, 1997, figure 6.3.

Nature: Figure 3.16 adapted from Leverington, D., "Star-gazing funds should come down to Earth," Nature, Vol. 387, p. 12, figure 1, 1997.

Publications of the Astronomical Society of the Pacific: Figure 4.12 adapted from Hasan, H. and Burrows, C.J., "Telescope image modeling (TIM)," PASP, Vol. 107, p. 291, 1995, figure 1.

Scientific American, Inc.: Figure 2.8 adapted from Kristian, J. and Blourke, M., "Charge-coupled devices in astronomy," Scientific American, Vol. 247, No. 4, second figure on p. 70, 1982.

We also thank the following corporations and institutions for providing or permitting the use of drawings, data, and plots. Once again, the original drawings or plots have been redone and, at times, modified to better illustrate the text.

xxiv   Credits for Figures and Tables

Canada-France-Hawaii Telescope Corporation: Figures 6.20, 6.35 (left), 6.49, and 6.51 (left), and 9.7, and Table 3.5.

Draper Laboratory: Figure 8.13.

European Southern Observatory: Figures 4.18, 6.15, 6.33 (left), 6.35 (right), 6.43 (left), 7.13, 7.27 (left), 7.28, 7.29, 8.19, 8.22, 11.5, 11.7, and 12.4, also Table 3.5 and part of the cover illustration.

Gemini Observatory: Figures 3.8, 3.21, 3.22, 4.20, 6.27 (right), 6.48 (left), 9.11, 11.2, and 11.16, and Tables 3.5 and 7.3.

Goodrich Corporation: Figure 7.20 (right).

Heidenhain Corporation: Figure 7.22.

W.M. Keck Observatory: Figures 4.40, 6.18, 6.23, 6.24, 6.27 (center), 6.33 (right), 6.43 (right), 7.15, 8.9 (right), 8.10, 8.24, and 8.25, and Table 3.5.

Jet Propulsion Laboratory: Figures 7.4 and 7.41.

McDonald Observatory: Figure 6.39.

Multiple Mirror Telescope Observatory: Figure 6.4 (center).

Next Generation Space Telescope Project Office: Figures 3.1, 3.18, 4.17, 4.42 (right), 5.17, 6.30, 7.12, 7.25, 7.33, 7.35, 8.18, 8.26, and 10.3, and part of the cover illustration.

Observatoire Midi-Pyrénées: Figure 11.9 (right).

Phase Motion Control, Inc.: Figure 7.16.

REOSC: Figures 4.39, 4.42 (left), and 4.20.

SKF USA, Inc.: Figure 6.42.

Societé Astronomique de France: Illustration on page v.

Space Telescope Science Institute: Figures 1.1, 1.2, 4.11, 4.24, 4.27 (left), 5.7, 5.16, 6.27 (left), 7.26, 7.36, 7.37, and 7.42, and Table 3.5.

Subaru Telescope National Astronomical Observatory of Japan: Figure 6.45 and Table 6.2.1.

University of Arizona Mirror Lab: Figure 4.41.

Isaac Newton Group, La Palma: Figure 4.51 and Table 3.5.

Introduction

Since the inception of research in astronomy some 400 years ago, the need to study fainter and fainter objects has naturally led to telescopes of ever larger diameter (Fig. 1). Early in the nineteenth century, George Hale recognized the significant advantage to be gained from locating to better sites (e.g., California), but found that instruments at even the best sites were still limited by flux. He therefore began championing the use of large mirrors, a concept which culminated with the Mount Palomar 5-meter telescope, conceived in the 1930s and completed in 1949. For the next 40 years, 4- to 6-meter class telescopes were to remain the norm, on one hand because telescope technology had reached a plateau, and on the other because alternative means of increasing sensitivity without increasing mirror size were available.

Indeed, existing and new telescopes of this size saw a manyfold increase in *sensitivity* thanks to the following advances in understanding and technology:

- Observatory sites were found (Chile, Hawaii) where seeing was approximately twice as good as before, affording a gain in sensitivity comparable to that obtained with telescopes twice as large.
- The importance of dome and mirror seeing became understood and eliminating most of it led to improvement in sensitivity of the same order of magnitude as that obtained from going to better sites.
- Fast automatic guiding replaced the inherently slow visual guiding, thus eliminating most of the tilt component in the image blur and increasing sensitivity accordingly.

2   Introduction

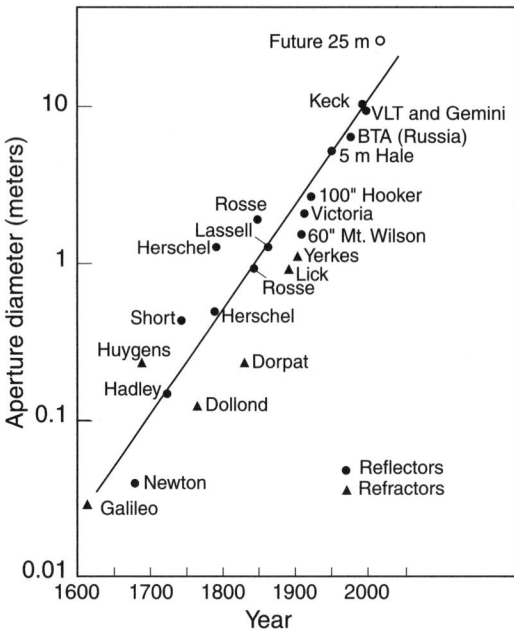

Fig. 1. Evolution of telescope aperture diameter over the last four centuries. According to the trend line shown, the diameter of the largest telescopes doubles about every 40 years. The 20- to 30-meter class telescopes planned for the 2015 time frame display a somewhat faster growth rate than the historical trend.

– Finally and most importantly, photoelectric detectors replaced the photographic plate, creating a dramatic improvement (with a quantum efficiency of up to 80% compared to 4% for the photographic plate, roughly equivalent to a fourfold gain in telescope diameter).

Eventually though, a new barrier in sensitivity was reached in the mid-1980s, and with photon-hungry cosmology being the most active field of astronomy at the time, there was no escape from going to larger telescope diameters or eliminating the atmospheric limitations altogether by going to space. This led to the current crop of 8- and 10-meter telescopes and to the immensely successful Hubble Space Telescope that, although quite small by today's standards, benefits from quasi-perfect imaging unaffected by the atmosphere.

This increase in telescope size was made affordable by a series of technological advances that substantially reduced costs and schedule. These included computerized design, faster and improved optical figuring techniques, the use of the altitude-azimuth configuration to reduce the mass and cost of telescope mounts, and faster f-ratios for smaller domes and buildings.

Next to sensitivity, *angular resolution* is arguably the most important factor in astronomical observations, and many important discoveries have indeed been made as a result of improvements in this capability. Theoretically, angular resolution is proportional to telescope size but, unfortunately, increasing aperture size has not led directly to better angular resolution because of atmospheric turbulence limitations (Fig. 2). Still, slow gains in resolution have been made by employing better and larger optics, by moving to better sites and, more recently, by compensating for atmospheric turbulence and going into space.

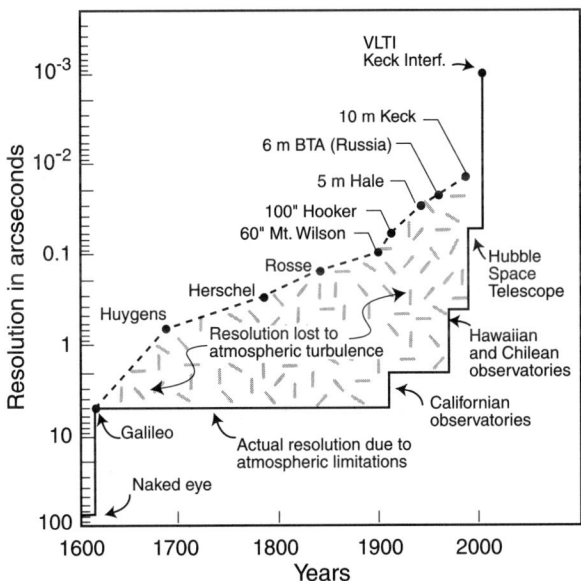

Fig. 2. The evolution of angular resolution in imaging optical astronomy. Ground-based telescopes never achieved their angular resolution potential because of atmospheric turbulence, but gains were progressively made by going to better sites. Now, atmospheric turbulence compensation techniques promise to approach the theoretical limit over at least part of the sky. Interferometry techniques, which consist of combining the light of several telescopes, make it possible to reach much higher resolution than that afforded by a single telescope, albeit with limited sensitivity. When completed, the Very Large Telescope Interferometric array (VLTI) and the Keck interferometer will reach milliarcsecond resolution.

In this book, we present the state of the art in astronomical optical telescope design and construction as it stands at the beginning of the new century. We have limited our treatment to *optical* telescopes, that is to say, those covering the optical wavelength domain, defined not just as the visible region but also including the adjoining spectral regions: the ultraviolet and the infrared up to about 500 μm. In the X-ray domain, optical systems are driven only by geometric effects (diffraction is negligible), whereas in the radio domain, diffraction is dominant (antenna beam theory applies). But from 100 nm to the submillimetric, the laws of geometric optics (reflection, refraction) apply *and* diffraction effects are neither negligible nor dominant. This results in telescope design principles that are essentially identical.

In the first two chapters, we review the notions of astronomy and principles of instruments needed to understand the function of telescopes and the conditions they have to satisfy.

Chapter 3 presents the methods used in the design and management of a large telescope project. Chapters 4 to 9 then cover the various engineering disciplines involved in telescope design and construction: optical, structural, mechanical, control, and thermal. Because of its growing importance, an entire chapter, Chapter 8, is devoted to active and adaptive optics.

The approaches followed for assembly and verification of the telescope system during manufacture and for commissioning are described in Chapter 10. The remaining two chapters address environmental issues. The design and construction of enclosures for ground-based telescopes is covered in Chapter 11, and site or orbit selection and environmental conditions are presented in Chapter 12.

A list of basic reference books and journal articles is supplied at the end of each chapter for those who wish to pursue their study further. Finally, basic astronomical and engineering data, a list of the major telescopes now extant, and an extensive glossary are provided in the Appendixes.

It is our hope that this text will serve as a foundation for the astronomers and engineers who face the challenge of building the ever larger telescopes, both in space and on the ground, that are needed to work at the forefront of knowledge.

1
Astronomical Observations

1.1 Role of astronomical telescopes

Unlike all other branches of science, astronomy is limited to *observations*. Aside from the analysis of meteorites, and perhaps the use of space probes, no *experimentation* is possible; the astronomer on Earth is a passive observer. Except for specific particles (cosmic rays, neutrinos), the only carrier of cosmic information is the electromagnetic radiation received on or near Earth, and the purpose of telescopes is to collect as much of this radiation as possible and measure it with ever greater sensitivity and accuracy.

In this chapter, we examine the main characteristics of astronomical sources and the complex background radiation that must be dealt with. We also cover the basic astronomical concepts with which the telescope designer needs to be familiar.

1.2 Source characteristics

1.2.1 Intensity

Astronomical sources cover an extremely wide range of brightness. To quantify this range, it is conventional to use a scale which, in astronomy, is called "magnitude." The magnitude system was established in the second century B.C.E. by Hipparchus, who classified the stars visible to the naked eye into six categories, with stars in one category appearing to be about twice as

1. Astronomical Observations

bright as those in the next. Since the response of the eye to brightness is roughly logarithmic, Hipparchus's categories constituted a logarithmic scale. The magnitude scale in use today was formalized in the nineteenth century using precise intensity measurements and was adjusted so that its first six levels would correspond to Hipparchus's categories. Because the ancient system attributed the first category to the brightest stars, the magnitude scale follows a counterintuitive progression, with the larger numbers representing fainter brightness. In the magnitude system, two objects with apparent flux density ϕ_1 and ϕ_2 have magnitudes m_1 and m_2 such that

$$m_1 - m_2 = 2.5 \log \frac{\phi_2}{\phi_1} \,. \tag{1.1}$$

Conversely, one has

$$\frac{\phi_2}{\phi_1} = 10^{0.4(m_1 - m_2)} \,. \tag{1.2}$$

Table 1.1 illustrates the correspondence between magnitude differences and brightness ratios.

Table 1.1. Magnitude and brightness

Magnitude difference	0.5	0.75	1.	2.	2.5	5	10
Brightness ratio	1.58	2.	2.51	6.31	10.	100	10 000

By convention, at all wavelengths, magnitude 0 has been attributed to the bright star Vega (a blue main-sequence star of spectral type A0). Objects brighter than Vega (Sun, bright planets) have negative magnitudes.

Accurate photometry is accomplished with photoelectric and solid-state devices and filters which accept only certain wavelength bands. One widely used photometric system is the UBV system, which has been extended to cover bands in the red and infrared (see Section 1.3.1). The characteristics of these bands and the flux of a magnitude zero source in each of these bands are listed in Table 1.2.1. It should be noted that several photometric systems are in use which differ in central wavelength and bandwidth and which also depend on instrumental responses particular to each observatory. The data supplied here are for quick approximations, not for actual observational work.

A flux-density unit less esoteric than the magnitude system has been imported from radioastronomy and is becoming widely accepted. It is the Jansky, which is defined as

$$1 \text{ Jansky(Jy)} = 10^{-26} \,\text{Wm}^{-2}\text{Hz}^{-1} \,. \tag{1.3}$$

For those astronomers who prefer to think in magnitudes but want to use measurements in Janskys, the "AB magnitude" has been devised. It is based on the Jansky, but expresses the result in magnitude format. It is defined as

$$\text{ABmag} = -2.5 \log(\text{Jansky}) + 8.90 \,, \tag{1.4}$$

1.2 Source characteristics   7

with the constant defined so as to correspond to the normal magnitude in the V ("visual") band.

Table 1.2. Photometric wavelength bands and flux densities for a magnitude zero object. Approximate values – see text

Band	λ μm	ν 10^{14} Hz	$\Delta\lambda$ μm	F_λ photons/(m^{2}μm s)	F_ν Jy
U (ultraviolet)	0.365	8.3	0.068	$7.9\cdot 10^{10}$	1810
B (blue)	0.440	7.0	0.098	$1.6\cdot 10^{11}$	4260
V (visible)	0.550	5.6	0.089	$9.6\cdot 10^{10}$	3540
R (red)	0.700	4.3	0.22	$6.2\cdot 10^{10}$	2870
I (near infrared)	0.880	3.7	0.24	$4.9\cdot 10^{10}$	2250
J (near infrared)	1.25	2.4	0.26	$2.02\cdot 10^{10}$	1520
H (near infrared)	1.65	1.8	0.29	$9.56\cdot 10^{9}$	1050
K (near infrared)	2.20	1.4	0.41	$4.53\cdot 10^{9}$	655
L (near infrared)	3.40	0.86	0.57	$1.17\cdot 10^{9}$	276
M (near infrared)	5.0	0.63	0.45	$5.06\cdot 10^{8}$	160
N (mid-infrared)	10.4	0.30	5.19	$5.07\cdot 10^{7}$	35.2
Q (mid-infrared)	20.1	0.14	7.8	$7.26\cdot 10^{6}$	9.70

Source: Refs. [1] and [2]

Table 1.3 gives apparent magnitudes and flux densities outside the atmosphere in the V-band (visible) for a few typical sources.

Table 1.3. Apparent magnitude and flux density of typical objects in V

Object	Magnitude	Flux in photons/(m^{2}μm s)	Flux in Janskys
Sun	-26.5	$3.8\cdot 10^{21}$	$1.4\cdot 10^{14}$
Full Moon	-12.7	$1.1\cdot 10^{16}$	$4.2\cdot 10^{8}$
Jupiter	-2.6	$1.0\cdot 10^{12}$	$3.9\cdot 10^{4}$
Sirius	-1.5	$3.8\cdot 10^{11}$	$1.4\cdot 10^{4}$
Faintest galaxies	~ 30	~ 0.1	$\sim 3\cdot 10^{-9}$

1.2.2 Distribution of sources of interest in the sky

A number of factors must be considered when selecting targets for a given scientific program. Certain targets are unique or nearly so and leave little leeway for optimizing observations. But in the case of "generic" objects that may be found in many locations in the sky, observations gain from being optimized by the proper choice of the time of year (so that the source appears high enough in the sky) and Moon phase (e.g., new Moon for a darker sky) and by selecting regions with reduced background from zodiacal light and galactic dust. Figure 1.1 shows a near-infrared map of the whole sky, illustrating the regions of high zodiacal background and the band of galactic emission from

8   1. Astronomical Observations

stars and nebulas in the Milky Way. Both of these regions must be avoided if sensitivity is to be maximized. The map also shows the location of those regions which are especially important for extragalactic research.

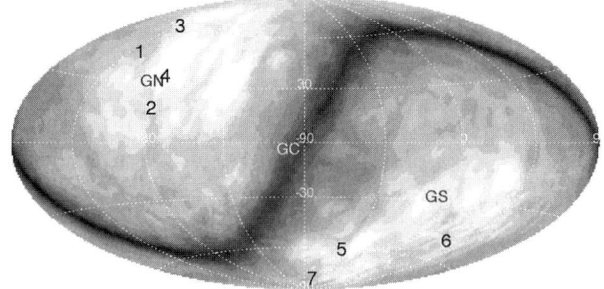

Fig. 1.1. Sky map in ecliptic coordinates showing the region of high zodiacal background and galactic emission in the near infrared, as well as several selected regions of interest. The numbers correspond to (1) Lockman hole (a region of especially low far-infrared galactic emission), (2) Virgo cluster of galaxies, (3) Hubble Deep Field (an HST long-exposure target area), (4) Coma cluster of galaxies, (5) Small Magellanic Cloud (a satellite of the Milky Way galaxy), (6) Fornax cluster of galaxies, (7) Large Magellanic Cloud. GN and GS are the north and south galactic poles, respectively; GC is the galactic center.

It is interesting to note that when constraints related to observing from the surface of the Earth are eliminated, as in the case of space telescopes, the distribution on the sky of targets selected by observers is surprisingly random (Fig. 1.2), except for those specific regions of high interest referred to above.

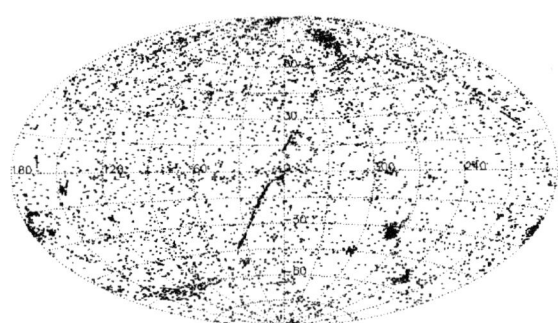

Fig. 1.2. Distribution of the targets observed by the Hubble Space Telescope (HST) over a period of 11 years. (Data from the Multimission Archive at the Space Telescope Science Institute.)

1.3 Observing through the atmosphere

The atmosphere affects observations in several ways: (1) extinction, which reduces the flux of the source, (2) line and thermal emission, which creates unwanted background, especially in the infrared, (3) refraction, which alters the apparent position of the source and disperses its image spectrally, and (4) turbulence, which blurs the image of the observed object. These effects are quantified and described in more detail below.

1.3.1 Atmospheric extinction

Atmospheric extinction results from the absorption and scattering of incoming photons by collision with air molecules or particles. In the absorption process, the photon is destroyed and its energy transferred to the molecule, which may lead to subsequent emission. The primary absorbers are H_2O, CO_2, O_2, and O_3. In the scattering process, the photon is not destroyed, but its direction and energy are changed. Scattering by air molecules having a typical size much smaller than the wavelength of light, λ, is roughly proportional to λ^{-4} and is called *Rayleigh scattering*. Scattering by small solid particles with sizes close to λ is proportional to λ^{-1} and is referred to as *Mie scattering*.

The combination of absorption and scattering essentially prevents the detection of electromagnetic radiation from extraterrestrial sources, except for a few spectral regions called "windows," the most important of which are (1) the optical window, which includes the visible range, the near ultraviolet, and the infrared up to $\simeq 25$ μm, and (2) the radio window (Fig. 1.3).

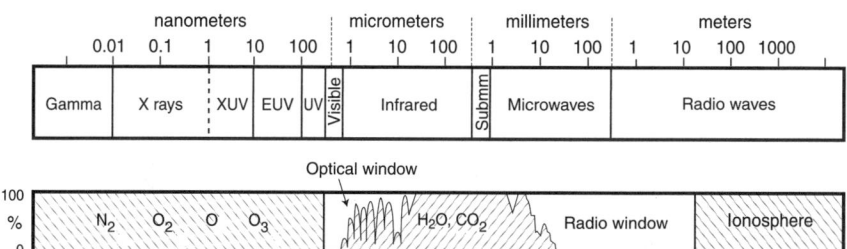

Fig. 1.3. Electromagnetic spectrum (top) and absorption of the atmosphere as a function of wavelength (bottom) with an indication of absorbing molecules.

In the visible, extinction is only about 10–15%, but the atmosphere becomes opaque below 300 nm due to the ozone layer, which is at an altitude of about 20 to 30 km. In the near infrared, between 0.8 and 1.35 μm, there are some absorption bands caused by water vapor and oxygen, but the atmosphere is never completely opaque. Beyond 1.3 μm, there begin to occur absorption bands where the atmosphere is completely opaque, especially at low-altitude sites. The transparent wavelength regions (windows), which cor-

10   1. Astronomical Observations

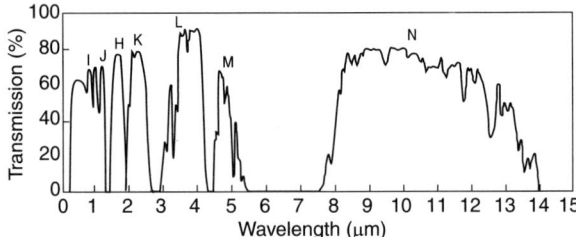

Fig. 1.4. Atmospheric transmission in the visible and infrared as a function of wavelength. The letters identify the infrared windows.

respond to the photometric bands listed in Table 1.2.1, are shown in Fig. 1.4. Beyond ∼ 25 μm, the atmosphere at low-altitude sites is totally opaque up to a wavelength of about 1 mm.

The particle number density for most absorbers falls off almost exponentially with altitude. For H_2O, the dominant absorber in the near infrared, the scale height is 2 km, hence the enormous advantage afforded by high altitude sites. For example, the top of the Hawaiian mountain Mauna Kea (4200 m) is above 95% of the atmospheric water vapor, with a remaining H_2O column depth (the equivalent thickness of a layer containing all precipitable water in the upper atmosphere) of only 1.5 mm. Much lower values can be found in the Antarctica plateau, where precipitable water vapor is typically in the 0.1–0.3 mm range. At both of these locations, markedly wider wavelength ranges are usable for astronomy. The very low amount of precipitable water above 10 km in altitude is also a major incentive for observing from high-flying platforms such as balloons and airplanes (Fig. 1.5).

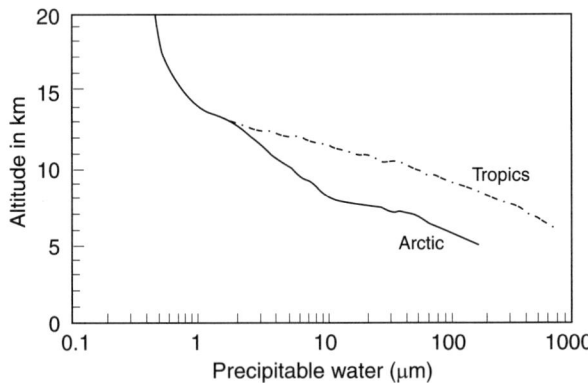

Fig. 1.5. Precipitable water as a function of altitude.

Obviously, extinction also depends on the zenith angle, since the path through the atmosphere increases with that angle. This effect is traditionally expressed in terms of "air mass," which is the ratio of the quantity of air along the observed direction to that in the zenith direction. For zenith angles of less than 60°, the atmosphere may be considered a flat slab, and the air mass is then simply proportional to the inverse of the cosine of the zenith angle (i.e., $\sec z$) [3].

1.3.2 Atmospheric emission

During daytime, atmospheric radiation is dominated by scattering of sunlight, which prevents observations in the visible and near infrared. At night, aside from the possible contribution of moonlight scattering, the major source of atmospheric emission at these wavelengths is fluorescence ("airglow"). Atoms and radicals in the upper atmosphere ($\simeq$ 100 km) undergo radiative deexcitation, emitting characteristic spectral lines. This phenomenon is most important in the near infrared due to the strong intensity of the OH^- spectrum. The spatial and temporal fluctuations of the airglow lines limit the photometric accuracy of ground-based near-infrared observations.

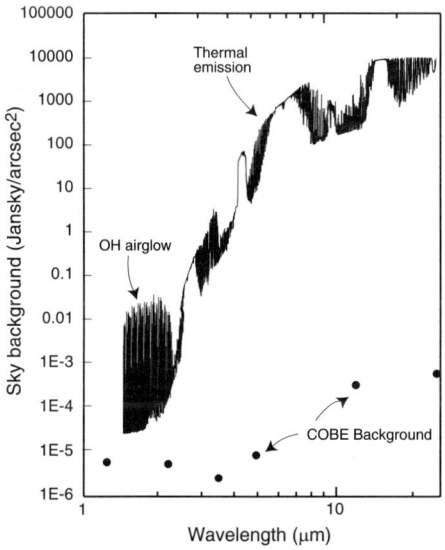

Fig. 1.6. Typical infrared background emission for a ground-based telescope at a good, high-altitude site (Mauna Kea). Thermal emission from the telescope and atmosphere dominates the background beyond 2.3 μm, whereas OH airglow lines dominate at shorter infrared wavelengths. Also shown is the minimum sky background from space as measured by COBE (dots). (From Ref. [4].)

Beyond about 2.3 μm, day or night, atmospheric radiation is dominated by its thermal emission. The effective temperature of the various atmospheric components is in the 230 – 280 K range, but the atmosphere actually radiates less than a corresponding blackbody because of its gaseous nature. The emission will approach that of a blackbody, which peaks at about 12 µm, only in those bands which have strong absorption and thus strong emission. A typical background flux measured by an infrared-optimized telescope is plotted in Fig. 1.6. It shows low emission compared to that of a blackbody except in the strong bands of CO_2 at 15 µm and H_2O at 6.3 µm. This has the fortunate result that thermal emission will be low in those bands where the atmosphere is relatively transparent. On the other hand, beyond 2 µm, observations from the ground become increasingly difficult because of thermal emission by the telescope itself. It is clear, in any case, that the exponential rise of background flux with wavelength dramatically reduces sensitivity at those wavelengths.

1.3.3 Atmospheric refraction

Atmospheric refraction is the bending of incoming light due to variable atmospheric density along the light path, making the source appear higher in the sky than it actually is (Fig. 1.7). The effect is a strong function of the zenith angle, being 0 at the zenith and close to half a degree at the horizon (Fig. 1.7, right), and also varies with altitude, humidity, and wavelength. The overall error in pointing direction can be corrected in the pointing control system, but the *differential refraction* across the field induces field rotation and can be significant for wide fields [5].

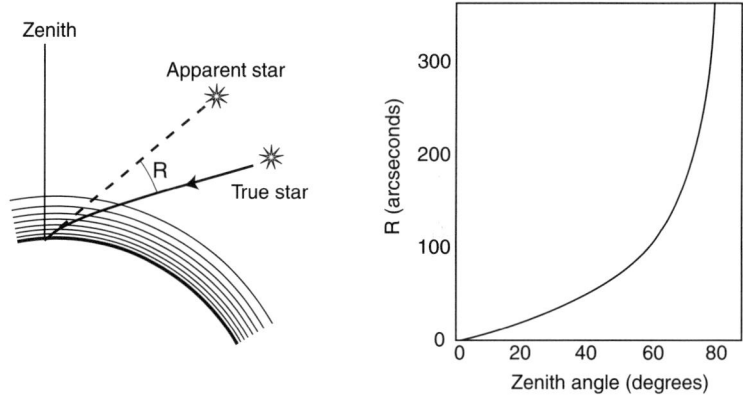

Fig. 1.7. Refraction in the atmosphere of the Earth (left). The variation of the atmospheric refraction with the zenith distance at Mauna Kea is shown on the right.

One secondary effect of atmospheric refraction results from the variation of the index of refraction of air with wavelength, with shorter wavelengths being more refracted than longer ones. At large zenith angles, the differential refraction between red and blue can be as much as several arcseconds. This effect can be corrected by introducing a dispersing element in the instruments. Since this dispersion varies with the zenith angle, correction is usually implemented by installing two rotating prisms to adjust the total refraction angle.

1.3.4 Atmospheric turbulence: basic notions

The atmosphere is never totally calm. Wind and convection induce turbulence which can mix layers with slightly different refraction indexes, causing changes in the direction of the light passing through. As a result, the amount of light reaching the aperture of a telescope varies constantly, both in intensity and direction. This phenomenon is referred to as "seeing."

The index of refraction of air depends on its density, which is proportionally much more affected by the *temperature* fluctuations likely to occur in the free atmosphere or near a telescope than by the aerodynamic *pressure* variations associated with wind. Thus, "seeing" is strongly dependent on temperature fluctuations but negligibly on wind effects. Such temperature fluctuations result from turbulent mixing of air layers at different temperatures caused by natural convection or mechanical turbulence. Convection is essentially limited to the ground layer and to the troposphere below the inversion layer, but mechanical turbulence exists throughout the lower and upper atmosphere. Mechanical turbulence is most pronounced in the weakly stratified troposphere, especially in the regions of high wind shear just above and below the jet streams. The stratosphere, the layer above the troposphere, is, as its name implies, much more stratified and is generally very stable.

During turbulent mixing, the temperature of an air parcel will change adiabatically as the parcel rises or descends. If the local temperature gradient is equal to the adiabatic lapse rate[1] ($\gamma_d = -9.8$ °C/km), the parcels of air displaced by mechanical turbulence will always be at the same temperature as the surrounding air, and no optical distortion will occur. But the greater the difference between the actual temperature gradient and the adiabatic lapse rate, the greater the risk of optical distortion due to mechanical turbulence. This situation is common at the tropopause in the mid-latitudes because of the temperature profile upturn and the wind shear created by jet streams.

In general, turbulence occurs in very thin layers just a few meters deep. A typical profile of the intensity of turbulence contributing to seeing as a function of altitude is shown in Fig. 1.8.

The effect of turbulence on optical distortion naturally decreases with the index of refraction of air, which is proportional to density, which itself is proportional to pressure and inversely proportional to absolute temperature. In practice then, turbulence-generated optical disturbance above 20 km altitude is negligible because the index of refraction has become very small.

---

[1] The adiabatic lapse rate is the rate of change of temperature with altitude of a particle of dry air which is raised or lowered in the atmosphere without exchanging heat.

14   1. Astronomical Observations

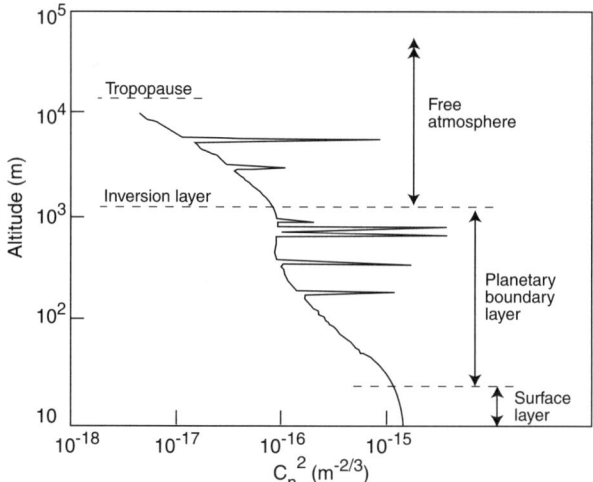

Fig. 1.8. Representative profile of the contribution to seeing as a function of altitude. The intensity of the fluctuations is expressed in terms of the index-of-refraction structure coefficient as defined in Section 1.3.5. Most of the fluctuations occur near the ground and in relatively thin turbulent layers generated by wind shear. (From Ref. [6].)

The "Fried length," also called "Fried parameter" or "coherence length," is a statistical parameter which permits a simple characterization of seeing. Simply stated, r_0 is the diameter of the bundle of rays issuing from a source at infinity which travel together through the various turbulent atmospheric layers and arrive, still parallel and in phase, at the telescope entrance.

A telescope with an aperture equal to r_0 would primarily suffer from image motion (as the tilt of the ray bundle changes), but not much from image blur. To reach diffraction-limited performance, that is to say the imaging performance of a quasi-perfect system limited only by diffraction (see Chapter 4), r_0 must be somewhat larger than the telescope diameter, about 1.6 times. Then, with an adequate guiding system to remove wavefront tilt, the telescope would essentially be free of atmospheric turbulence effects, as if it were in space. For a telescope with an aperture which is large compared to r_0, the full width at half-maximum (FWHM) of the image is given by [7]

$$\text{FWHM} = 0.98 \frac{\lambda}{r_0}, \tag{1.5}$$

where λ is the wavelength. Note, however, that r_0 is itself a function of λ with $r_0 \propto \lambda^{6/5}$ (see equation 1.12), so that seeing varies as $\lambda^{-1/5}$ and is thus most pronounced at the lower end of the optical range. In the visible, r_0 varies from a typical value of 10 cm to 30 cm at the best sites, which results in seeing of $1''$ to $0.35''$, respectively. Under the same conditions, seeing would be between $0.75''$ and $0.25''$ in the near infrared at around 2 μm.

1.3 Observing through the atmosphere

In general, seeing will degrade an image in two ways: image motion and image blur. At any one time, the apparent direction of an observed object is determined by the average direction of the wavefront entering the telescope. Small-aperture telescopes experience greater image motion than larger telescopes because wavefront distortions tend to have larger slope changes over small scales (Fig. 1.9). The reverse is true for image blur: larger telescopes suffer from a larger image spread than smaller ones.

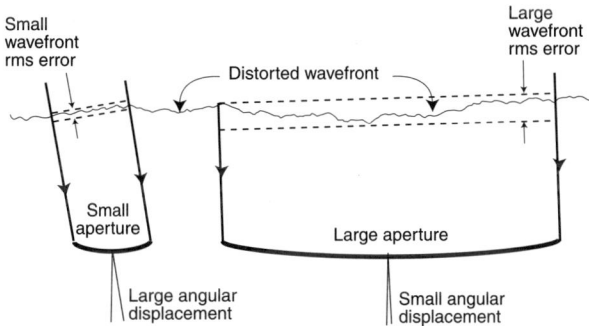

Fig. 1.9. Image motion decreases as telescope size increases, whereas the reverse is true for image blur. (Adapted from Ref. [8].)

Scintillation is the variation in intensity of the image. It is due to the curvature of the wavefront over the surface of the aperture, which tends to focus or defocus the image and results in brightness variations. Scintillation affects only small telescopes in which the aperture size is r_0 or less.[2] Large apertures average out the effect and the image brightness does not vary much with time.

The characteristic time of optical turbulence, τ_0, called "coherence time,"[3] is the transit time of the statistical coherence region of diameter r_0 over the line of sight. To the first order, it is determined by the wind speed, v, at the level where the main turbulence occurs (Fig. 1.10, left) and is thus given by

$$\tau_0 \simeq \frac{r_0}{v}. \tag{1.6}$$

Another characteristic of seeing is the angle on the sky over which the incoming beam remains coherent (i.e., within which the effects of turbulence are correlated). This angle, called the "isoplanatic angle" (Fig. 1.10, right), is given by

$$\theta_0 \simeq 0.6 \frac{r_0}{h}, \tag{1.7}$$

where h is the altitude of the main turbulence layer above the telescope.

---

[2] The pupil of the eye being much smaller than r_0, stars seen with the naked eye "twinkle" noticeably under almost all conditions.

[3] The inverse of τ_0 is known as the Greenwood frequency.

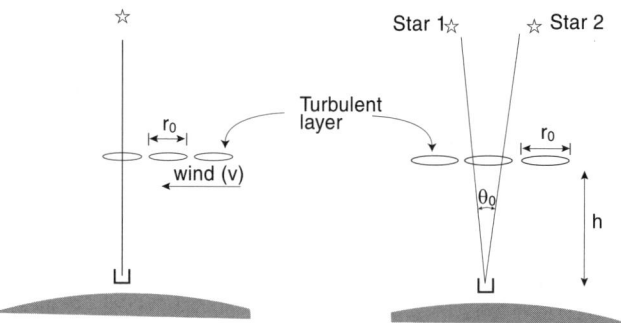

Fig. 1.10. Assuming that the turbulence is occuring in a single layer, seeing effects can be viewed as resulting from the passage of atmospheric coherent cells of diameter r_0. The characteristic time of optical turbulence is a function of atmospheric coherence cell size and the speed of wind carrying these cells (left). The isoplanatic angle is a function of the size of the atmospheric cells and the height of the turbulent layer above the telescope (right).

The typical distribution of seeing at an excellent observatory site (Mauna Kea, Hawaii) is shown in Fig. 1.11, and typical values for the seeing characteristic parameters at that same site are given in Table 1.4.

Table 1.4. Typical seeing parameters at an excellent site

Parameter	Visible 0.5 μm	IR 2.2 μm
Fried's parameter r_0 (m)	0.20	1.35
Seeing disk (arcsecond)	0.5	0.33
Coherence time τ_0 (milliseconds)	10	50
Isoplanatic angle θ_0 (arcseconds)	2	10

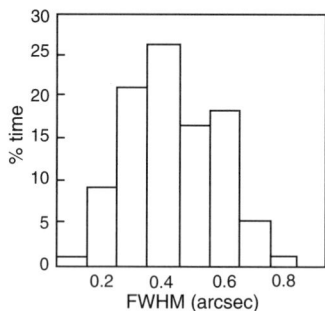

Fig. 1.11. Histogram (0.1″ bins) of natural seeing on Mauna Kea at 0.55μm. (From Ref. [9].)

1.3.5 Atmospheric turbulence: the physics of seeing

As indicated in the preceeding subsection, seeing is due to turbulent fluctuations of the index of refraction. The index of refraction, n, varies with the density and composition of the medium. For air, this may be expressed by Cauchy's formula (extended by Lorenz for humidity)

$$n - 1 = \frac{77.6 \cdot 10^{-6}}{T} \left(1 + 7.52 \cdot 10^{-3} \lambda^{-2}\right) \left(p + 4810 \frac{v}{T}\right). \quad (1.8)$$

where λ is the wavelength, p is the atmospheric pressure (mb), T is the absolute temperature (K), and v is the water vapor pressure (mb).

Fluctuations in humidity are only significant in extreme cases, such as in fog or in proximity to the sea surface, and are not normally relevant to seeing at astronomical sites, which generally experience very low humidity values (< 20%). Large pressure fluctuations also have negligible effects on the index of refraction. Thus, for all practical purposes, the refractive index is affected only by air temperature, so that fluctuations in the index of refraction are most intimately linked to the structure of thermal turbulence in the atmosphere.

Turbulent energy in the atmosphere is produced by buoyancy and wind shear over relatively large dimensions, on the order of several tens of meters. These large turbulent eddies set up wind shears on a smaller scale, and these give rise to still smaller eddies. As the process is repeated, smaller and smaller eddies are created. Finally, eddies with linear dimensions on the order of a few millimeters are produced. The shears in such eddies are so large that the air viscosity, although small, is sufficient to transform their kinetic energy into heat. There, the process of turbulence decay stops. The spectrum of turbulence can thus be divided into three ranges:

(1) the energy-producing range, the characteristics of which are controlled by the energy-producing process (this range is called the "outer scale of turbulence," L_0, which has values between 1 and 100 m);

(2) the "inertial" subrange, in which energy is neither created or destroyed, simply transferred from larger to smaller scales; in this range local isotropy exists;

(3) the dissipation subrange, l, in which energy is destroyed by viscosity (l being on the order of millimeters).

It is in the inertial subrange that thermal fluctuations occur. In this isotropically turbulent region, the spatial variation of temperature has a spectrum proportional to $k^{-5/3}$, where k is the wave number. This very general law is referred to as the "Kolmogorov spectrum" [10]. The law is conveniently expressed by means of a statistical "structure function," which is a measure of the mean squared fluctuation (i.e., the variance) over a span r. For the temperature field, for one dimension, it is defined as

$$D_T(r) = <(T(x+r) - T(x))^2>, \quad (1.9)$$

where $T(x)$ is the temperature at a given point in the field and $T(x+r)$ is the temperature at another point at a distance r from the first one; angle brackets denote an average. For a turbulent medium with a Kolmogorov spectrum, the temperature structure function, $D_T(r)$, has the form

$$D_T(r) = C_T^2 r^{2/3}, \qquad (1.10)$$

where C_T^2 is called the "temperature structure coefficient."

From equation 1.8, ignoring the very minor effects of humidity and pressure, the structure coefficient for the index of refraction, C_n^2, is related to C_T^2 by

$$C_n^2 = C_T^2 \left[77.6 \cdot 10^{-6} \left(1 + 7.52\, 10^{-3} \lambda^{-2}\right) \frac{P}{T^2} \right]^2. \qquad (1.11)$$

From this law, the photometric profile of the so-called seeing disk (long-exposure stellar image) can be predicted, and the theory is found to be in excellent agreement with observations [11]. The seeing profile can be fully described with a single parameter, the Fried parameter, r_0, introduced in the previous subsection, which is related to the index-of-refraction structure coefficient as a function of altitude by

$$r_0 = \left[1.67\, \lambda^{-2} (\cos \gamma)^{-1} \int C_n^2(z) dz \right]^{-3/5}, \qquad (1.12)$$

where z is the altitude and γ is the zenith angle. Thus, image quality depends only on the integral of C_n^2 over the light path. The full width at half-maximum (FWHM) of the seeing disk, θ, can be derived from r_0 and is given in arcseconds by

$$\theta = 2.59 \cdot 10^{-5} \lambda^{-1/5} \left[(\cos \gamma)^{-1} \int C_n^2(z) dz \right]^{3/5}. \qquad (1.13)$$

For a vertical direction and $\lambda = 500$ nm, the FWHM angle in typical conditions of astronomical mountain sites (pressure 770 mb, temperature 10 °C) is expressed as

$$\theta = 0.94 \left[\int C_T^2(z) dz \right]^{3/5}. \qquad (1.14)$$

The theory of atmospheric seeing summarized above was developed in the 1960–1980s and is now well proven: measuring the profiles of C_T^2 or C_n^2 does yield reliable estimates of image quality. Detailed treatment of the subject can be found in the works listed in the bibliography at the end of this chapter.

1.4 Background sources

"Background sources" are those that affect an observation but do not originate in the source being observed. They include natural sources in the sky, atmospheric emission, thermal emission from the telescope, and side effects in the detector itself. These sources are schematically illustrated in Fig. 1.12, and the brightness of the celestial and atmospheric backgrounds is shown in Fig. 1.13.

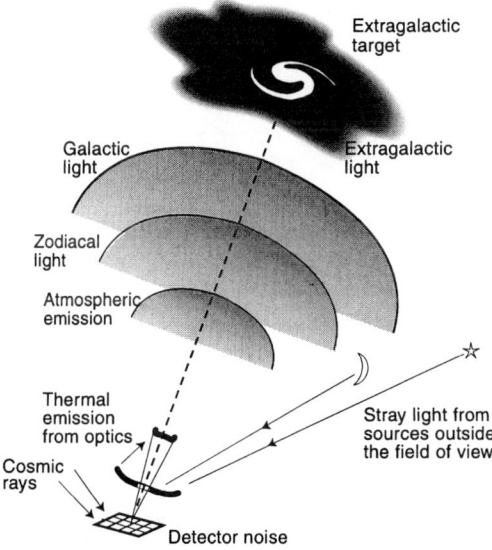

Fig. 1.12. The various sources of background.

1.4.1 Celestial backgrounds

Galactic

The galactic background is due to faint stars and dust. Down to the smallest resolvable scale the galactic background due to dust is characterized by highly irregular patches of emission, and for this reason, this component is commonly known as *galactic cirrus*.

Zodiacal

The zodiacal light is due to dust grains orbiting the Sun and concentrated in the ecliptic plane. It is the result of two effects: scattering of sunlight and thermal emission by the dust grains heated by the Sun (Fig. 1.14). The scatter component has a spectrum close to that of the Sun, whereas the thermal emis-

20   1. Astronomical Observations

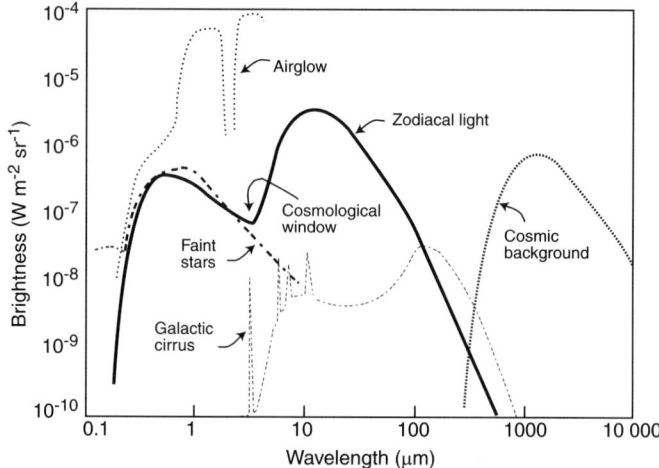

Fig. 1.13. Brightness of the celestial background sources and of the atmosphere as a function of wavelength. (After Leinert et al. [12].)

sion component approximates that of a blackbody. Between these two regimes, there is a minimum at ~ 3.5 µm that defines a *cosmological window* permitting observation from space with the lowest possible celestial background.[4]

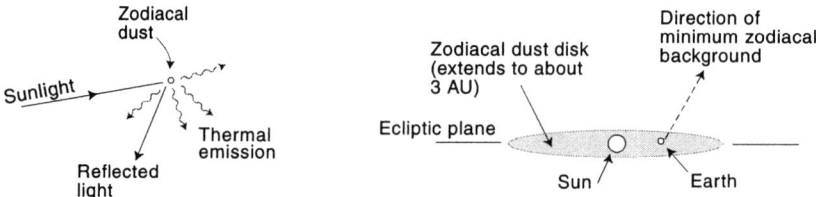

Fig. 1.14. Scatter and thermal emission by dust particles in the zodiacal disk (left). Schematic view of the zodiacal disk (right).

Zodiacal background is not uniform. It is at a maximum toward the Sun and also, at short wavelengths, directly away from the Sun due to backscattering (the "gegenschein"). The minimum occurs away from the Sun at around 60° of ecliptic latitude, due to the combination of minimum thickness of the zodiacal cloud and cooler dust temperature.

The zodiacal light has been well characterized by measurements from space by the DIRBE experiment on board the COBE spacecraft and a detailed

---

[4]The zodiacal cloud extends to about 3 astronomical units from the Sun, and a space telescope would have to be located that far away to be essentially free of the zodiacal light background.

model is available [13]. A simple model, valid above 1 μm in wavelength, is given by

$$Z(\lambda) = \frac{15.1}{\lambda^2} + 6.0 \cdot 10^{-8} \frac{2c}{\lambda^3} \frac{1}{e^{hc/(265k\lambda)} - 1} , \quad (1.15)$$

where Z is the zodiacal light flux from the antisun direction at about 45° above the ecliptic expressed in photons per m^2 per second per steradian and for a bandpass $(\Delta\lambda/\lambda)$ of 1, λ is the wavelength in meters, c is the velocity of light in m/s, h is the Planck constant in J·s, and k is the Boltzman constant in J/K [14]. One recognizes the first term as being the fall-off of the scatter component, whereas the second term corresponds to the thermal emission of a diluted blackbody at a temperature of 265 K.

Cosmic rays

Cosmic rays are atomic nuclei (mostly protons) and electrons that have been accelerated to extremely high energies and permeate space and the Earth's environment (technically they are not rays, but particles). They travel at about 0.9 the velocity of light and have energies up to 10^{21} eV. Some cosmic rays come from the surface of the Sun, but most originate in our galaxy or other galaxies, with the most energetic ones thought to come from supernovas. Some of these particles are trapped by the magnetic field of the Earth forming zones of high radiation called the "Van Allen radiation belts."

Cosmic rays are attenuated by the atmosphere, but reactions take place and generate secondary particles. The cosmic rays at ground level (0 to 4 km) consist almost entirely of secondary particles (mostly muons). The rates of these secondary particles depend slightly on latitude and strongly on altitude.

Cosmic rays are an important source of degradation of astronomical observations, especially in space where they produce spurious charges in detectors. They affect single pixels primarily, but, at times, cosmic rays with grazing incidences can affect several adjacent pixels. Spurious counts can also be generated within the instruments themselves, either by the electronics or by Cerenkov radiation in refractive optics.

On the ground, the rate of these events is approximately 50 per cm^2 per hour at sea level and twice that at 4000 m altitude. In space, the rate is much higher, about 1 per cm^2 per second. Cosmic rays can be subtracted from the data by splitting the observation into subexposures and comparing the frames.

A dip in the lower Van Allen belt caused by a reduced magnetic field above Brazil increases the cosmic ray rate there to such an extent that low-Earth-orbit space telescopes passing through that zone essentially have to shut down (see Chapter 12). This zone is known as the South Atlantic Anomaly (SAA).

1.4.2 Atmospheric background

As discussed in Section 1.3.2, the background created by the atmosphere falls into three regimes: the "optical regime" below about 1 μm, which is dominated by Moon scatter, the "nonthermal infrared regime," from 1 to 2.5 μm, which is dominated by narrow OH emission lines, and the "thermal regime," 3.0 μm and above.

Typical values for the brightness of the night sky at a dark, high altitude site are given in Table 1.5. Moonlight scatter results in nights being classified according to the phase of the Moon. "Dark time" is when the Moon is less than a quarter full, "bright time" is when the Moon is more than half full, and other nights are classified as "gray time." Dark time is reserved for the most demanding observations in the visible, while bright time is generally used for high resolution spectrographic or infrared observations, which are almost unaffected by Moon scatter.

Table 1.5. Typical brightness of the night sky for each photometric band at a high-altitude site, in magnitude per arcsecond square

Days from New Moon	U	B	V	R	I	J	H	K	L	M
0	21.3	22.1	21.3	20.4	19.1	15.7	14.0	12.0	3.4	0.5
7	19.2	20.9	20.7	19.9	18.9	15.7	14.0	12.0	3.4	0.5
14	15.0	17.5	18.0	17.9	18.3	15.7	14.0	12.0	3.4	0.5

1.4.3 Stray light and detector background

Stray light affecting observations has two origin: light from celestial sources outside the field of view and thermal emission from the telescope and instruments. These effects will be studied in detail in Chapter 5.

Although not a true source of background, detectors can produce effects with similar characteristics to those of natural background. They are due to unwanted photoelectrons generated by the detector itself or by the readout process. These sources will be discussed in detail in Chapter 2, Section 2.2.4.

1.4.4 Coping with atmospheric and telescope thermal emission

Ground-based observing in the infrared differs from observing in the optical because of the very large atmospheric and telescope thermal background flux that peaks near 10 μm. To observe sources which can be several orders of magnitude fainter than the background per square arcsecond, one must subtract the background.

Up to about 2.5 μm, the background is still manageable, and if the object is small compared to the field of view, it is not necessary to observe the sky with a separate exposure. One can "dither" around the source and use the field surrounding the source to subtract the sky background. Dithering is simply the operation of placing the object in different positions on the detector by moving the telescope. Beyond 2.5 μm, on the other hand, the sky brightness is so high and so variable with position and time that it cannot be modeled and subtracted in this straightforward manner to better than 1 part in 10 000. Special modulation techniques must be used.

The background fluctuates on a time scale of minutes or less due to turbulent motion of the atmosphere and temperature drift in the telescope. Since most sources are much fainter than the range of the associated sky fluctuations, any small error in the estimate of the background will dramatically affect extraction of the signal. The solution is to repeatedly point at the source and then at a nearby empty sky area, switching back and forth at a rate commensurate with the temporal variations in the sky background.

The choice of the pointing shift frequency depends on various factors, such as observing wavelength, weather conditions, telescope location, but is typically between 3 and 10 Hz. Since it is virtually impossible to move a telescope at these frequencies, the solution is to rapidly modulate a single optical element between two slightly different positions. To minimize pupil misalignment at the cold stop (which thermal infrared instruments all have, see Chapter 4), it is usually the secondary mirror of the telescope that is modulated. This classic technique is called "chopping" (Fig. 1.15).

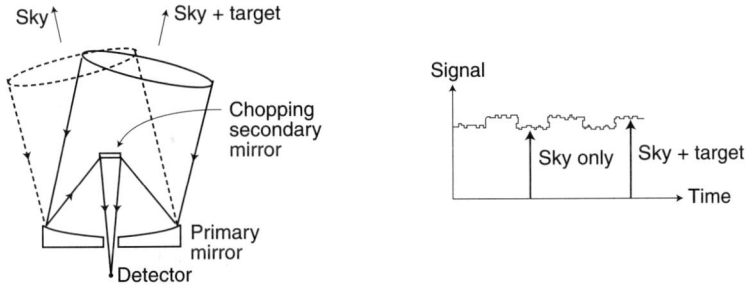

Fig. 1.15. Chopping with the secondary mirror for background subtraction.

Chopping is often complicated by the presence of astronomical sources in the "sky beam," so that a judicious choice of chopping amplitude and angle is generally necessary to avoid background-subtraction problems in crowded fields. The chopping system must thus allow for changes in the direction of chopping and for variable amplitude, usually up to 30 arcseconds. The chopping profile should be as close to a square wave as possible. In general, the maximum chopping frequency is constrained by the settling time of the secondary mirror, which is typically in the 20–50 ms range.

The problem with moving an internal optical element is that the detector sees the high-emissivity surfaces in the telescope, such as the central obstruction, mirror edges and spider, from a slightly different angle on the two sides of the chop. This results in a residual background variation which limits the accuracy of the background subtraction.

To remove this residual background difference, which is stable over time scales of minutes, one must repoint the telescope as a whole. This is the so-called "beam-switching" or "nodding" technique. This operation needs to be done often enough, usually on the order of 60 seconds, to eliminate telescope thermal variations.

1.5 Signal-to-noise ratio

Modern astronomical detectors are "linear," or almost linear, meaning that the recorded signal will be proportional to the number of photons received. But, as we have seen above, there are several "background" sources of radiation, besides the object of interest, that affect the detector. The problem, then, is to extract the true signal coming from the source from this additional flow of unwanted electrons.

At first thought, one might assume that a signal should be stronger than the background and that this would be the condition for detecting it. This is not the case, however, because the average value of the background can be subtracted from the signal and is, thus, irrelevant. What matters are the *fluctuations* around the mean value of the background, called "background noise," by analogy with radiobroadcast "static" (Fig. 1.16).

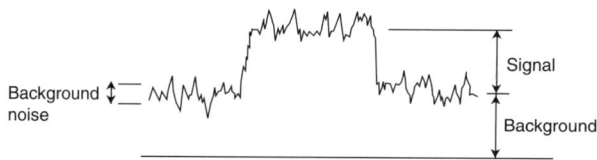

Fig. 1.16. Detection of a signal in the presence of background.

All background noise processes, being the result of independent events (i.e., photon arrivals) occuring at a constant rate, can be described by Poisson statistics. If N is the average rate (electrons/s) at which photoelectrons are collected, the probability for n photoelectrons to be collected in a given area of the detector in a time t is then given by

$$p(n,t) = \frac{(Nt)^n e^{-Nt}}{n}. \tag{1.16}$$

An important property of the Poisson distribution is that its standard deviation (rms of the fluctuation) is simply the square root of the average number of events:

$$\sigma = \sqrt{Nt}. \tag{1.17}$$

A common way to express the reality of a detected signal and the precision of its measurement is to compare the signal S to the fluctuation of the background (i.e., the "noise" of the background). Traditionally, this is done by taking the ratio of the two quantities forming the "signal-to-noise ratio," usually written S/N, which is the strength of the measured result expressed in units of its standard deviation. If the errors inherent in a measurement have a Gaussian distribution, which is generally the case, then if the resulting measurement has S/N = 1, there is a 68% probability that the signal is real (often referred to as the "significance" of the result). Conversely, there is about one chance in three that it is *not* real. Hence, an S/N of 1, also referred to as "1-σ" detection, is not regarded as a credible detection of a signal.

At a S/N of 3 ("3-σ" detection), however, the probability that the signal is not real (i.e., that an unfortunate combination of the natural statistical scatter in the measured quantities has combined to simulate a true signal) is reduced to 0.003. This means that a spurious result of this strength should only occur thrice in a 1000 measurements. This is often used as a reasonably safe level at which to believe the result to be real. Because there are often unquantified or unexpected error sources in even the most carefully studied experiments, however, many workers prefer to achieve a S/N of 5 (i.e., a 5-σ result), at which stage the formal statistical likelihood that the result is spurious is less than 1 in 10^5.

The reason for wanting to go beyond an S/N of 3 or 5 is dynamic range. As a rule of thumb, the signal-to-noise ratio should be equal to the dynamic range desired in the object being observed. With an S/N of 10, for example, one would be able to measure on the order of 10 levels in the intensity of an extended object, with the faintest level having an S/N of only 1.

Taking into account the common sources of noise, and using equation 1.17 to calculate the fluctuation of the background noise, the signal-to-noise ratio for a point source or an extended source covering n_{pix} pixels on the detector can be expressed as

$$\text{S/N} = \frac{S\,t}{\sqrt{(S + B n_{\text{pix}} + I_d\, n_{\text{pix}})\,t + R_{\text{n}}^2\, n_{\text{pix}} + \text{var}(B_t n_{\text{pix}}\, t)}}, \quad (1.18)$$

where S is the total number of photoelectrons received from the source per unit time, t is the integration time, B is the number of photoelectrons received from the background (zodiacal light, atmospheric emission, and telescope thermal emission) per pixel and unit time, I_d is the dark current of the detector expressed in electrons per pixel and unit time, R_{n} is the readout noise per pixel (i.e., the standard deviation of the readout electrons collected per pixel for each read), and $\text{var}(B_t n_{\text{pix}} t)$ is the variance of the estimate of the *total* background, B_t ($B_t = B + I_d + R_{\text{n}}/t$), per pixel per unit time. This last term reflects the uncertainty in the estimate of the background which does *not* arise from photon statistics; in other words, this term accounts for *true* variations in the measured background level generated by changes in that level itself or by inadequacies in the method by which it is measured.

One notes that the signal also appears in the denominator. This is because the source signal itself has a statistical variation which is indistinguishable from other fluctuations. This noise in the signal is referred to as "source noise." One also notes that the background can generally be estimated over a large number of pixels, or long integration times, so that the term $\mathrm{var}(B_t n_{\mathrm{pix}} t)$ is often negligible. But this term must be included when using detectors with a small number of pixels and when the background is variable (e.g., infrared observations on the ground). Estimation of this term can be very difficult and it is generally better to attempt to eliminate it by stabilizing the signal from all background sources.

The S/N formulation defined above is fundamental to all astronomical observations. It incorporates all "instrumental features" of the observation. It includes the detector's efficiency and noise characteristics as well as throughput of the telescope and instrument optics, via the fact that S and B are the number of photoelectrons received at the detector, not those impinging on the telescope aperture. It is this formulation that permits the determination of the limiting attainable magnitude and the exposure time required for a given observation.

Depending on the relative importance of each of the terms, one can distinguish three types of observation:

- **source-photon-noise limited** where the source of interest is bright and its photon noise dominates all other fluctuations. In this case, S/N simplifies to

$$\mathrm{S/N} = \sqrt{St}, \qquad (1.19)$$

and increased exposure time will bring a proportional increase in sensitivity. Since S is proportional to the telescope's collecting area (i.e., to D^2, D being the telescope diameter), increasing the telescope diameter will bring large gains in exposure time ($t \propto 1/D^2$). This is a rare case for large telescopes, as such "easy" observations are generally more cost-effective on smaller telescopes.

- **detector-noise limited** where the source and background signals are faint and the noise of the detector dominates. In this case, the signal-to-noise ratio simplifies to

$$\mathrm{S/N} = \frac{St}{\sqrt{I_d n_{\mathrm{pix}} t + R_{\mathrm{n}}^2 n_{\mathrm{pix}}}}. \qquad (1.20)$$

This case is typical of mid- to high-resolution spectroscopy because the fraction of the background noise per spectral element diminishes as the spectral resolution increases. Note that, for this case (and the following background-limited case), S/N is proportional to the total number of source photons detected. This gives rise to two important considerations.

First, the individual integrations (between reads) should be as long as possible: S/N will improve *linearly* with time until the accumulated background signal is large enough so that its fluctuations are larger than R_n (but other factors such as cosmic ray hits on the detector set limits to the practicable length of individual exposures). Second, the limiting sensitivity of a telescope used under these conditions will improve with the collecting area (i.e., as the square of the diameter D), and the time required to carry out a given observation will scale as the inverse fourth power of D.

- **background limited**, also called "sky limited" when observing from the ground. This occurs when the source is faint and the natural background (zodiacal light and, if applicable, atmospheric emission) dominates the noise. In this case, the signal-to-noise ratio simplifies to

$$\mathrm{S/N} = \frac{S\sqrt{t}}{\sqrt{Bn_{\mathrm{pix}}}}, \qquad (1.21)$$

and the accuracy of the measurement scales as the square root of the exposure time and inversely as the square root of the background. This is the ultimate case, where all instrumental effects (detector, instrumental thermal emission) have been minimized and the only remedies are to increase the diameter of the telescope, improve the image quality (via adaptive optics, for example), or reduce the background (e.g., by avoiding airglow emission or reducing thermal emission of the optics).

To the first order, the background term, B, is given by

$$B \sim \Phi_{\mathrm{bkgd}}\, \sigma^2 A t, \qquad (1.22)$$

where Φ_{bkgd} is the background flux per arcsecond square, σ is the angular diameter on the sky of the image of the source, A is the area of the telescope aperture (proportional to D^2), and t is the exposure time.

For a large telescope on the ground, the angular size of the image of a point source, σ, is driven by seeing and does not depend on the aperture size. The solid angle from which source photons arrive, and from which background photons come and must be coped with, is constant: increasing the telescope aperture increases both source and background signals at the same rate. Since noise is proportional to the square root of their sum, S/N increases only linearly with aperture, as in the source-photon-noise-limited case.

If, on the other hand, the telescope is diffraction limited and the detector pixels are matched to the point source image size, $\sigma \sim \lambda/D$ (see Chapter 4, Section 4.5.3), D cancels out and

$$\mathrm{S/N} \sim D^2 \sqrt{t}. \qquad (1.23)$$

In this latter regime (background- and diffraction-limited), the time needed to reach a given S/N scales with the fourth power of the telescope diameter. A 10-meter telescope is ∼ 10 000 times faster than a 1-meter telescope. In practice, this means that large telescopes will have ready access to limiting magnitudes that are essentially unattainable with a small telescope. But both conditions, celestial background-limited and diffraction-limited optics, must be satisfied. This is the case of space-based telescopes with diffraction-limited optics as long as detector noise is negligible. This is also the case on the ground in the mid-infrared (e.g., 10 μm) because detectors are almost perfect (at least with respect to the huge background), and imaging is quasi-diffraction-limited (the image size produced by the optics increases with wavelength, whereas seeing goes down).

1.6 Time

1.6.1 Sidereal time

Common time is determined by the position of the Sun with respect to the local meridian. A day has elapsed when the Sun returns to the local meridian. Since the Earth is rotating around the Sun, however, a distant celestial object which was on the meridian will have returned to the meridian slightly less than a day later (Fig. 1.17), by roughly 1/365 of a day or about 4 minutes. A sidereal day is the time interval between successive passages at the meridian of a given star, and is equal to 23 h 56 min 4 s, or 86 164 s.

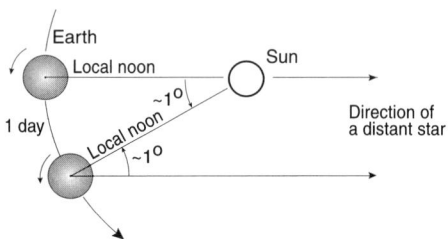

Fig. 1.17. Sidereal time (see text).

1.6.2 Julian date

It is sometimes convenient to use a system in which days are numbered consecutively rather than being measured in months and years. This is the case with research on variable stars, and for space observatories, which are in continuous operation and do not depend on a human-based schedule. In such cases, it is common practice to use "Julian dates," counting in days and fractions of days from an arbitrary day which has been set as 1 January, 4713 B.C.E., with each day beginning at noon rather than midnight. Julian day numbers (JD) are listed in the Astronomical Almanac or can be calculated from the following formula:

$$JD = 2451544 + 365(Y - 2000) + N + L - 0.5, \qquad (1.24)$$

where Y is the current year C.E., N is the number of days elapsed since the beginning of that year, and L is the number of leap years which have occurred between 2000 and the current year.

1.7 Coordinate systems

The most common coordinate systems used to locate the position of celestial objects are listed in Table 1.6.

Table 1.6. Coordinate systems

System	Reference plane	Reference direction	Latitude coordinate	Longitude coordinate
Horizontal	Horizon	North	Altitude (h)	Azimuth (A)
Equatorial	Celestial equator	Vernal equinox	Declination (δ or DEC)	Right ascension (α or RA)
Ecliptic	Earth orbit	Vernal equinox	Ecliptic latitude (β)	Ecliptic longitude (λ)
Galactic	Galaxy plane	Galaxy center	Galactic latitude (b_{II})	Galactic longitude (l_{II})

Observers may select the most convenient system for their field of research (e.g., galactic coordinates for galactic studies). But when it comes to defining the location of an object to be observed, only the *equatorial* system is used because it is defined with respect to Earth, yet is independent of the time of day and exact location of the telescope on Earth. The system is centered on the Earth and uses the celestial equator, the plane perpendicular to the rotation axis of Earth, as a reference plane (Fig. 1.18). The latitude angle is called "declination" and abbreviated as DEC or δ. It is measured in degrees, starting from the celestial equator, and is positive for objects in the northern hemisphere. The longitude angle is called "right ascension," abbreviated as RA or α, and is measured in hours, minutes and seconds, with eastward being the positive direction. The reference direction for the right ascension is arbitrary and has been selected as the vernal equinox (γ), the point on the sky where the Sun crosses the celestial equator at the spring equinox.

30   1. Astronomical Observations

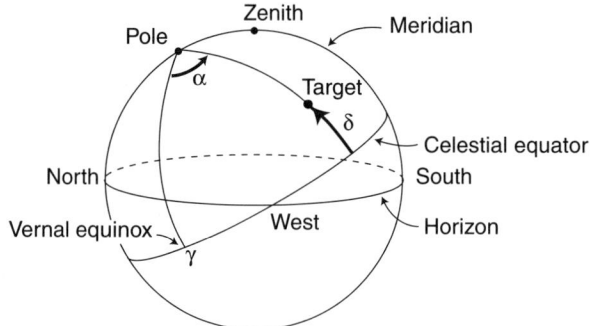

Fig. 1.18. Equatorial coordinates.

For an observer on Earth, the position of a given celestial object is found by determining its *hour angle* (HA) at that time. The hour angle is the angle measured westward along the celestial equator from the local meridian to the hour circle passing through the object. HA is expressed in hours, minutes and seconds and is given by

$$\text{HA} = \text{ST} - \text{RA}, \tag{1.25}$$

where ST is the sidereal time at the moment of observation and RA is the right ascension of the object. If the object has a positive hour angle, it is in the western part of the sky, and if it is negative, it is in the eastern part.

In the case of an equatorial telescope, which can be rotated around an axis parallel to Earth's rotation axis, the hour angle and declination can be used directly to point to the object. If the telescope is an "alt-az," with its rotation axes vertical and horizontal, one needs to convert the hour angle and declination of the object to be observed to altitude and azimuth. In addition, because the field rotates in this type of mount, one needs to know how the orientation of the field varies as the telescope tracks. This is defined by the "parallactic angle," q, which is the position angle (measured north through east at the target) of the arc that connects the target to the zenith, or loosely speaking, the position angle of "straight-up." The parallactic angle is zero for an object on the meridian (Fig. 1.19). The conversion is given by

$$\sin h = \sin\varphi \sin\delta + \cos\varphi \cos\delta \cos\text{HA}, \tag{1.26}$$

$$\tan A = \frac{\sin\text{HA}}{\sin\varphi \cos\text{HA} - \cos\varphi \tan\delta}, \tag{1.27}$$

$$\tan q = \frac{\sin\text{HA}}{\tan\varphi \cos\delta - \sin\delta \cos\text{HA}}, \tag{1.28}$$

where h is the altitude, A is the azimuth measured eastward from due north, HA is the local hour angle measured westward from the south, δ is the declination, and φ is the observatory's latitude [15]. The inverse transformation for the target coordinates is

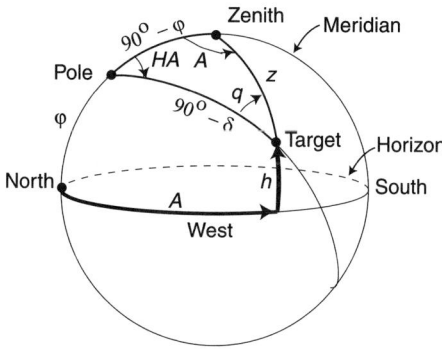

Fig. 1.19. Alt-az coordinates. Angles are defined in the text. The zenith angle, z, is the complement of altitude, h.

$$\sin \delta = \sin \varphi \sin h + \cos \varphi \cos h \cos A, \quad (1.29)$$

$$\cos \text{HA} = \frac{\sin h - \sin \varphi \sin \delta}{\cos \delta \cos \varphi}. \quad (1.30)$$

1.8 Pointing corrections

The coordinates given in catalogs for celestial bodies need to be corrected for the time at which the observation is actually made and for various other effects. These corrections used to be the observer's responsibility, but are now performed automatically by telescope pointing software. They must be applied for both the target to be observed and the guide stars used. The origin and order of magnitude of the effects to be corrected are briefly described below and summarized in Table 1.7. The formulas for making the corresponding corrections can be found in textbooks and astronomical almanacs.

Table 1.7. Order of magnitude of pointing corrections

Atmospheric refraction	2′ at 60° zenith distance
Precession	50″ per year
Annual aberration	20″
Velocity aberration (spacecraft only)	5″
Stellar parallax	<1″
Proper motion	< 1″
Diurnal aberration	0.3″
Differential velocity aberration (spacecraft only)	20 mas (LEO)

Note: mas = milliarcsecond; LEO=low-Earth orbit

1.8.1 Precession and nutation

The positions of celestial objects are normally referenced to the barycenter of the solar system and are given in RA and DEC coordinates, with RA having its origin at the vernal equinox. But the rotation axis of Earth is not fixed in space. The Sun's gravity field interacts with the Earth's equatorial bulge to generate a torque which causes the axis to precess around the normal to the orbit (to which it is currently inclined by 23°27′) with a period of 26 000 years. Consequently, the vernal equinox moves around the celestial equator with the same period, advancing by 50.25″ per year. To specify a celestial coordinate system, it is therefore necessary to specify its date. This date is referred to as the "equinox" and is always given along with the RA and DEC (as, e.g., 1900, 1950, or 2000).

The *equinox* is quite distinct from the *epoch* of an observation, which is the absolute time at which it takes place. Because stars (and other objects) move, even relative to an inertial coordinate system, the full definition of a measured position (or a specified position) must include both equinox and epoch; a position may therefore be given as "equinox 1950, epoch 2001.456," the latter referring to the true date of observation. It should be stressed that epoch and equinox are rarely the same, and much confusion in the location of fast-moving objects such as dwarf stars is caused by a widespread habit of conflating the two terms as *epoch*.

Superimposed on the precession is a much smaller "nodding" motion of the rotational axis, caused by the Moon's gravity pulling on the equatorial bulge. This effect is variable in amplitude and has a period of 18.7 years. It is referred to a the "nutation" of the axis.

1.8.2 Proper motion

The proper motion of stars results from their intrinsic motion through space with respect to the Sun. Several hundred stars have proper motions greater than 1″ per year.

1.8.3 Parallax

The orbital motion of Earth around the Sun creates a parallax (Fig. 1.20) that is negligible for extragalactic objects and distant stars, but needs to be corrected for the closest stars, particularly potential guide stars, since the effect can be a significant fraction of an arcsecond. When known, star catalogs give the heliocentric distance to the star expressed in "parsecs" (a star at 1 parsec has an annual parallax of 1″), from which the annual parallax can be calculated.

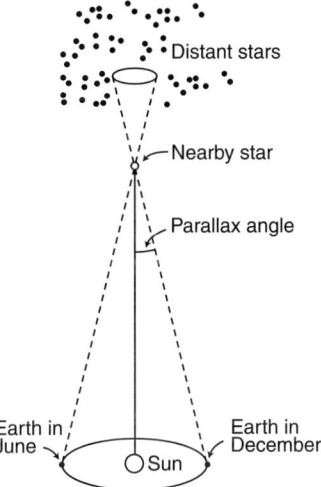

Fig. 1.20. Parallax due to the rotation of the Earth. The position of the nearby star moves with respect to background stars as the Earth rotates around the Sun.

1.8.4 Aberration of starlight

The aberration of starlight is the apparent displacement of a celestial object from its true position on the sky due to the finite velocity of light. As shown in Fig. 1.21 (left), a telescope moving through space with a velocity v has to be tilted forward by the angle v/c radian, where c is the velocity of light, in order to point to the target's apparent position. The main component of a telescope's velocity through space is due to the velocity of the Earth around the Sun (annual aberration), which is about 30 km/s. The effect is greatest when the target is in the direction perpendicular to the Earth's motion and reaches 20.5″. Aberration for any point on the celestial sphere forms an ellipse throughout the year. This ellipse is a circle at the ecliptic poles and collapses to a line along the ecliptic (Fig. 1.21, right).

For ground-based telescopes, a secondary component is due to the rotation of Earth around its axis (diurnal aberration). The effect is greatest at the equator (velocity of 0.46 km/s) and leads to a maximum aberration of 0.3″.

For space-based telescopes in low Earth orbit, there is an additional component due to the orbital velocity of the spacecraft. This orbital velocity is about 7 km/s and produces an aberration of up to 5″.

Since starlight aberration is a function of field angle, the aberration will not be the same for all points in the field, especially for the primary target vis-à-vis guide stars which may be several arcminutes away. This must be taken into account if guide stars are used to refine the pointing of the telescope.

Over the typical duration of an observation (a few hours), the velocity component due to Earth's motion in space remains essentially unchanged, so that once the pointing has been corrected for the overall effect as well as

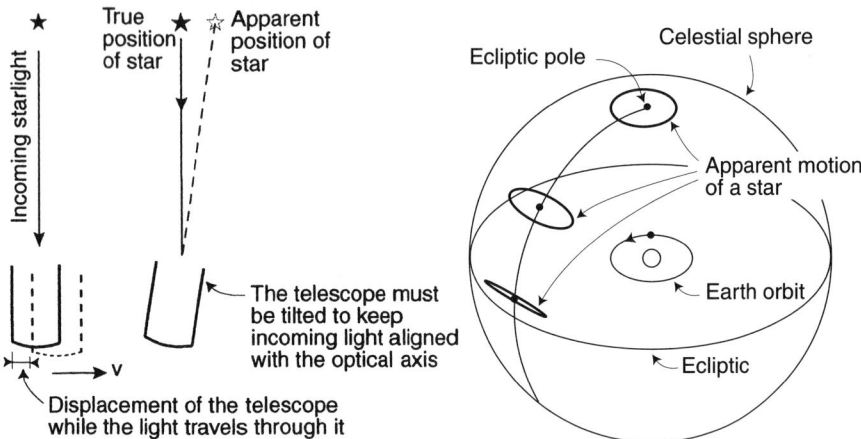

Fig. 1.21. At left, aberration of starlight due to the motion of the observer. At right, velocity aberration due to Earth's rotation around the Sun. The corresponding apparent motion of a star is a circle at the ecliptic poles and an ellipse elsewhere.

for any difference between target and guide stars, the apparent positions of the target and guide stars do not change. This is not the case, however, for space observatories in low Earth orbit, since the spacecraft's orbital period is of the same order of magnitude as that of observation durations. This effect, referred to as "differential velocity aberration," requires that the position of the guide stars in the field be continuously adjusted in order to maintain the target's position in the focal plane (Fig. 1.22). It must be emphasized that this effect is significant only for observatories in close orbit around the Earth. It is negligible for observatories in drift orbit or at the second Lagrange point of the Sun–Earth system.

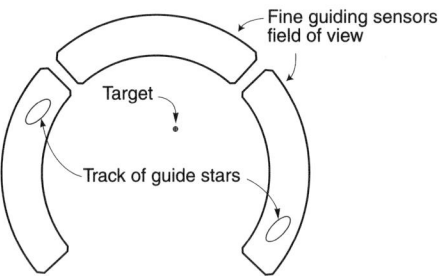

Fig. 1.22. Example of differential velocity aberration in the focal plane of HST. The long axis of the ellipse is about 20 milliarcseconds and is grossly exaggerated in this diagram for clarity. The guiding system must correct for this effect in order to keep a target at the same location in an instrument aperture.

1.8.5 Atmospheric refraction

As explained in Section 1.3.3, atmospheric refraction raises targets above their true position by as much as several arcminutes. The pointing system must correct for this absolute effect as well as for the differential effect in the field when guide stars are used as references.

1.9 Telescope pointing and tracking procedure

Once calibrated, telescopes can typically point to a new field with an absolute accuracy of about $1''$ for ground-based telescopes and about $10''$ for space telescopes. This may be enough for wide-field imaging, but small-field imaging and spectrographs need more precise pointing. This is accomplished by various techniques referred to as "target acquisition." Once pointed as desired, the telescope must "track" to compensate for inertial drifts in the case of space telescopes, or for the Earth's rotation in the case of ground-based telescopes. Tracking is an "open-loop" procedure, however, meaning that it relies on information provided by attitude sensors or encoders. For improved pointing accuracy and the compensation of optical, thermal, or gravity effects that vary with time, it is necessary to close the pointing system loop on the observed field itself. This is generally done with the use of relatively bright stars in the field of view, referred to as "guide stars." We briefly examine these procedures next.

1.9.1 Target acquisition

There are many possible acquisition schemes depending on the precision required, the brightness of the target, the availability of precise target coordinates, whether or not the instrument has an imaging mode, and so forth, but a common approach is as follows.

When the accuracy needed for locating the target is not demanding (e.g., several arcseconds, as in the case of wide-field imaging), the telescope is simply pointed using its attitude sensors or encoders. This is referred to as "blind pointing."

If the accuracy required is better than the pointing system is capable of, but still not too great (e.g., a fraction of an arcsecond), all that is needed is to refine the pointing of the telescope by identifying an object in the field. If the target itself can be observed, one can simply move the telescope so that the target falls on the desired fiducial position. If the target is extended or too faint, one must identify a bright star of known coordinates in the field. Such a star is called a "reference star." The exact position of the reference star in the focal plane is then measured and the pointing is corrected by the difference between the star's expected and actual position. This procedure, referred to as "offsetting," requires that the telescope's small open-loop maneuvers be accurate enough (e.g., 100 mas over a 10 arcminute offset).

When a higher precision is required, as for spectrography using very small slits (e.g., 10 mas), then it is necessary to "view" the science target using the imaging mode of the instrument if there is one, or with a "peak-up" observation. Peak-up consists of making slight pointing changes in a prescribed pattern, recording the flux received in the instrument's aperture, and determining the pointing direction that maximizes the flux, hence centers the target in the aperture.

If the target is too faint for the imaging mode or the peak-up procedure to work, the only remaining possibility consists of determining the position of the target with respect to a nearby reference star with high accuracy (e.g., by measurement of a previously taken long-exposure image of the field). The telescope is then pointed to that reference star, peaked-up on it, and then offset by the target/reference-star vector. This procedure is called "blind offset."

In the days of visual acquisition, the target would be acquired first and guiding would be turned on afterward. But with modern automation and the need for increased pointing accuracy, it is more efficient, sometimes indispensable, to have the telescope tracking and guiding before initiating target acquisition. This ensures that the above procedures are not defeated by pointing drifts during the acquisition phase.

1.9.2 Guiding

Although it is sometimes possible to guide on the target itself, the general procedure consists of using a dedicated focal plane instrument, called a "guider," to image one or possibly two bright stars in the field and correct the pointing accordingly. The technique will be studied in detail in Chapter 7, Section 7.5.5.

1.9.3 Guide star catalogs

Until recently, the selection of guide stars for guiding was "opportunistic"; that is, the stars to be used for guiding would be determined *after* pointing the telescope to the intended field. When planning the Hubble Space Telescope in the 1980s, it was realized that efficient operation would require the automation of the entire process. This led to the creation of a complete, all-sky catalog of stars up to a magnitude of about 14.5, called the "Hubble Space Telescope Guide Star Catalog" [16]. This catalog is based on a photographic survey performed with Schmidt telescopes at Mt. Palomar in California for the northern hemisphere and in Siding Spring in Australia for the southern hemisphere. The plates were digitized by scanning microdensitometers and the resulting digital images were processed to determine the location and brightness of all stars in the 7 to 16 magnitude range in a computerized form. This catalog, which contains about 19 million objects, gives the right ascension and declination of each one with an absolute accuracy of about 1 arcsecond and a relative accuracy with respect to neighboring objects of about 0.3″. Brightness is given in eight spectral bandpasses with an accuracy of 0.01 magnitude.

A second catalog, Guide Star Catalog–II, is currently in the making and will provide positions, proper motions, and colors for stars up to the 18th magnitude in V, based on multicolor and multiepoch Schmidt surveys.

Other catalogs are available, such as the Tycho and Hipparcos catalogs, which provide more accurate positions (in the 1 to 10 mas range), but for brighter stars (V < 11). These two catalogs are based on data obtained by the Hipparcos astrometric satellite.

1.10 Telescopes and interferometers

This book deals essentially with single-aperture, single-mount telescopes. This means that telescopes with a complete or almost complete primary mirror, that primary mirror having a single optical figure, whether segmented or not, and using a single mount for pointing.

There is another means of collecting celestial light, emphasizing angular resolution at the expense of sensitivity, which consists of dispersing the collecting area into two or more widely separated apertures. These astronomical instruments are referred to as "interferometers."

There is no fundamental difference between the two types of instrument. As shown conceptually in Fig. 1.23, an interferometer can be viewed as an incomplete traditional telescope. The information provided by an interferometer is referred to as "interference fringes" as opposed to an "image," but the physics is the same. An image is nothing other than the cumulation of interference fringes that would be produced by a series of subapertures. The difference,

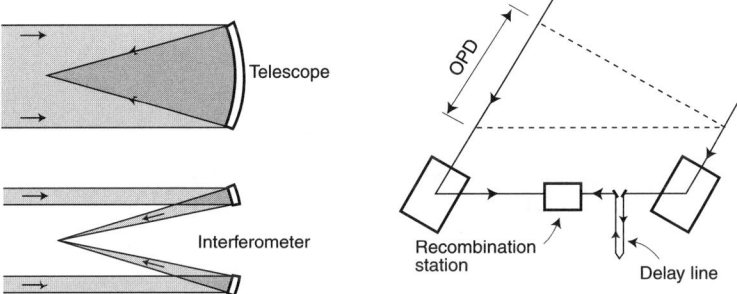

Fig. 1.23. Interferometers can be viewed as telescopes with incomplete apertures (left). As shown on the right, they are generally implemented as separate instruments the light of which is recombined. Delay lines are used to keep the optical paths of the individual telescopes the same: the length of the path in the delay line compensates for the optical path difference (OPD) in the incoming beams.

38   1. Astronomical Observations

however, is that the telescope does provide a true and direct image of the field observed, whereas the information produced by an interferometer is encoded in the fringes and needs to be processed.

To the first order, the two types of instruments have the same angular resolution, which is a function of the aperture diameter (the diameter of the circumscribed circle, in the case of the interferometer). But the interferometer achieves that resolution with a much smaller collecting surface. Conversely, for a given total primary mirror area, the interferometer can have a resolution several orders of magnitude higher than that of a conventional, full-aperture telescope.

Interferometers with only two apertures are very limited in their astronomical application because they provide information only in the direction of the vector joining the two apertures. But true imaging (i.e., two-dimensional imaging), can be obtained by using several apertures (typically a minimum of six) and making *sequential* measurements after spatial rearrangement of the individual apertures. The rearrangement can be made by physically changing the geometry of the aperture distribution, by rotating the entire set of subapertures around the line of sight (on the ground, this rotation is obtained "gratis," thanks to the Earth's rotation), or by a combination of both. Such a procedure is called "aperture synthesis," because one essentially reconstitutes in a *sequential* fashion the full aperture of the equivalent single-aperture telescope.

The notion of "aperture dilution" is used to quantify the fullness of the aperture. Aperture dilution is the ratio of the collecting area to the area of the circle circumscribing the individual apertures. A traditional telescope has a dilution close to 1. Typical interferometers have a dilution on the order of 1%.

The drawback of interferometers is a loss of sensitivity. This can be appreciated by comparing the image formed by a full aperture compared to that of an unfilled aperture with the same collecting area (Fig. 1.24). The core of the diluted aperture is narrower (hence, a better angular resolution), but the peak

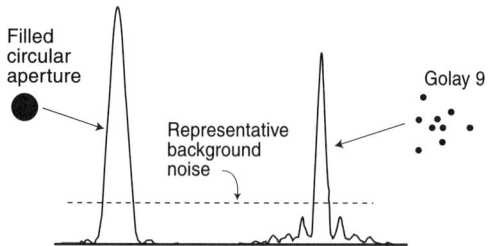

Fig. 1.24. Comparison of the image of a point source formed by a filled aperture and a diluted aperture with the same collecting area (here in a nonredundant configuration referred as a "Golay," with nine subapertures [17]). For faint objects, the light in the wings of the diluted aperture image is likely to be lost in the background.

intensity is lower and the nonnegligible amount of light in the wings may be lost in the background noise and is difficult to recover by image processing.

Interferometers composed of separate telescopes impose particularly difficult optical and mechanical requirements: the optical path of the individual telescopes must be kept the same to a fraction of wavelength. This calls for optical and mechanical techniques that will not be covered in this book, but the basic principles and techniques of interferometry can be found in the references listed in the bibliography.

References

[1] Cox, A.N., ed., *Allen's Astrophysical Quantities*, Springer-Verlag, 2000.

[2] Zombek, M.V., *Handbook of Space Astronomy and Astrophysics*, Cambridge Univ. Press, 1990.

[3] Hardie, R.B., in *Astronomical techniques*, Vol. II of *Star and Stellar Systems*, Hiltner, W.A., ed., Univ. of Chicago Press, 1962, p. 180.

[4] Gillett, F.C. and Mountain, M., *On the comparative performance of an 8 m NGST and a ground-based 8 m optical/IR telescope*, in *Science with the NGST*, ASP Conf. Series, Vol. 133, p. 42, 1998.

[5] Smart, W.M., *Textbook on Spherical Astronomy*, Cambridge Univ. Press, 1977, pp. 36 and 291.

[6] Coulman, C.E., The physics of seeing, in Proceedings of the Flagstaff Conference on *Identification, Optimization, and Protection of Optical Telescope Sites*, 1986, p. 2.

[7] Dierickx, P., Optical performances of large ground-based telescopes, J. Mod. Opt., Vol. 39, No. 3, p. 569, 1992.

[8] Young, A.T., Seeing and scintillation, *Sky and Telescope*, Sept. 1971, p. 139.

[9] Racine, R., Salmon, D., Cowley, D., and Sovka, J., Mirror, dome and natural seeing at CFHT, PASP, Vol. 103, p. 1020, 1991.

[10] Léna, P., Lebrun., F., and Mignard, F., *Observational Astrophysics*, Springer-Verlag, 1998, p. 42.

[11] Roddier, F., The effects of atmospheric turbulence in astronomy, in *Progress in Optics*, Wolf, E., ed., North-Holland, Vol. 19, 1981, p. 281.

[12] Leinert, C., et al., The 1997 reference of diffuse night sky brightness, Astron. Astrophys. Suppl. Ser., Vol. 127, No. 1, p. 1, 1998.

[13] Kelsall, T. et al., The COBE Diffuse Infrared Background Experiment search for the cosmic infrared background: II. Model of the interplanetary dust cloud, Ap. J., Vol. 508, p. 44, 1998.

[14] Thompson, R.I., Infrared detectors for a 10 m space or lunar telescope, in Proc. of *The Next Generation Space Telescope*, Bely, P.Y., Burrows, C.J. and Illingworth, G.D., eds., STScI, p. 310, 1989.

[15] Smart, W.M., *Textbook on Spherical Astronomy*, Cambridge Univ. Press, 1977, pp. 35 and 49.

[16] Lasker, B., Sturch, C., McLean, B., Russel, J., Jenkner, H., and Shara, M., The guide star catalog I, astronomical foundations and image processing, Astron. J., Vol. 99, p. 2019, 1990.

[17] Golay, M.J.E., Point arrays having compact nonredundant autocorrelations, J.O.S.A., Vol. 61, p. 272, 1971.

Bibliography

Astronomical observations

Baum W., *Astrophysical Techniques*, Stars and Stellar Systems, Vol. II, Univ. of Chicago Press, Chicago, Illinois, 1962, Chap.1.

Birney, D.S., *Observational Astronomy*, Cambridge Univ. Press, 1991.

Cox, A.N., ed., *Allen's Astrophysical Quantities*, Springer-Verlag, 2000.

Léna, P, Lebrun, F., and Mignard, F., *Observational Astrophysics*, Springer-Verlag, 1998.

Nicklas, H., Optical telescopes and instrumentation, in *Compendium of Practical Astronomy*, Vol. I, Instrumentation and Reduction Techniques, Roth, G.D., ed., Springer-Verlag, 1994.

Smart, W.M., *Textbook on Spherical Astronomy*, Cambridge Univ. Press, 1977.

Sterken, C. and Manfroid, J., *Astronomical Photometry — A Guide*, Kluwer Academic Publishers, 1992.

Walker G., *Astronomical Observations*, Cambridge Univ. Press, Cambridge, 1987.

Wolfe, W.L. and Zissis, G.J., eds., *The Infrared Handbook*, Office of Naval Research, Department of the Navy, Washington D.C., 1989.

Zombek, M.V., *Handbook of Space Astronomy and Astrophysics*, Cambridge Univ. Press, 1990.

Atmospheric turbulence

Hardy, J.W., *Adaptive Optics for Astronomical Telescopes*, Oxford books, 1998. (Contains a good introduction to atmospheric seeing).

Roddier, F., *The effects of Atmospheric Turbulence in Astronomy*, Progr. Opt., Wolf, E, ed., North-Holland, 1981, Vol. 19, p. 281.

Roddier, F., ed., *Adaptive Optics for Astronomy*, Cambridge University Press, 1999. (Contains a good introduction to atmospheric seeing).

Smith, F.G., ed., *Atmospheric Propagation of Radiation*, The Infrared and Electro-optical Systems Handbook, Vol. 2, SPIE Opt. Eng. Press, 1993.

Tatarskii, V.I., *The effects of the turbulent atmosphere on wave propagation*, Israel Program for Scientific Translations, 1971.

Interferometers

Kilometric Baseline Space Interferometry, European Space Agency Report SCI(96)7, 1996.

The VLT Interferometer Implementation Plan, European Southern Observatory Report No. 59b, 1999.

Roddier, F., Interferometric Imaging in Optical Astronomy, Phys. Rep. 170 (2), p. 97, 1988.

2
Instruments

In the broadest sense, astronomical observations consist of gathering light emitted or reflected from a distant source and making an image of it that is then analyzed for intensity, size, morphology, or spectral content. Collecting the light and forming the image is the role of the *telescope*. The analysis of the image is carried out by the *instruments*. But the telescope/instruments combination forms a tightly coupled system, and a telescope designer must understand the overall picture to properly optimize the system for which he is responsible. To that end, and although this is a book about telescopes, we give below a brief overview of the role and nature of instruments and of detection principles. An exhaustive treatment will be found in the voluminous literature on the subject, with a good introduction in the books and articles listed in the bibliography at the end of this chapter.

2.1   Main types of instrument

Astronomical observations fall into two broad classes: *imaging*, where one records the image of one or more celestial objects in order to measure their shape and relative brightness, and *spectroscopy*, where one disperses incoming light in order to measure the intensity of the received light as a function of wavelength. These two classes are not always clearly distinct, however, as cameras can be used for crude analysis of intensity as a function of wavelength by taking a series of images in various spectral bandpasses, and some spectrometers can be used to reconstruct an image in a narrow spectral bandpass.

2. Instruments

Generally, however, imaging refers to direct imaging onto a detector, using filters to control the spectral bandpass, whereas spectroscopy refers to the production of continuous spectra by means of a dispersing element or interferences. The corresponding instruments are called *cameras* and *spectrometers* (or *spectrographs*), respectively. A special case in the spectrometer class is the photometer, which is used to measure the light intensity of a single object in a given spectral bandpass. The various types of corresponding instruments are briefly described below.

2.1.1 Cameras

The simplest camera is a detector placed directly at the focal plane of the telescope (Fig. 2.1, left). If telescope aberrations, especially field curvature, are negligibly small, this avoids additional optics and thus benefits from high throughput. The main disadvantage is that the filters are in the converging beam and their optical thickness modifies the focus. This can be mitigated either by refocusing when a different filter is installed or by choosing the thickness of a filter as a function of its refractive index, so that the optical pathlength of every filter is the same. Other disadvantages are that the filters can be very large, especially in the case of wide-field cameras, and need to be extremely good optically, since any defect will directly affect image quality at the focus.

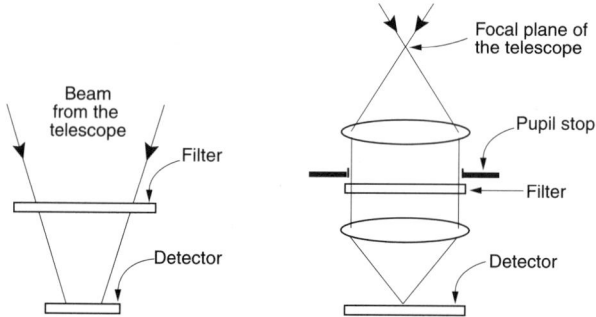

Fig. 2.1. Basic layout of cameras: direct imaging at the telescope focal plane (left) and with reimaging, where filters are placed in a collimated beam (right). (Although a refractive layout is shown for clarity, cameras often use mirrors to avoid chromatic aberration or provide improved throughput, especially in the infrared.)

A better solution consists of collimating the input beam so that the filters are in a parallel beam (Fig. 2.1, right). The focus is then unaffected by differences in filters, and the need for high-optical-quality ones is less stringent. This configuration also permits the creation of a physically real exit pupil. By positioning the filters at a pupil, their size can be minimized and this also prevents the beam from "walking" across the filter as a function of the field angle, which could affect transmission. A stop can also be placed at that exit pupil to control stray light or, in the case of an infrared camera, to prevent the surrounding infrared radiation from reaching the detector ("cold stop", see Chapter 5). Another advantage of this solution is that, by adjusting the relative magnification of the two camera lenses (or mirrors), one can change the plate scale delivered by the telescope to optimize it for the particular pixel size of the detector used (see "pixel matching" in Chapter 4, Section 4.5.3).

2.1.2 Photometer

A photometer is an instrument that measures the brightness of a single source within a given spectral bandpass. A single-cell detector (e.g., a photomultiplier tube) is all that is required. With the advent of high-efficiency two-dimensional detectors of high photometric quality, both in the visible and the infrared, precision photometry is now possible with modern photoelectric cameras. But photometers still have the edge in very-high-precision or high-speed photometry of bright objects and for inexpensive systems.

The main difference between a camera and a photometer is that, in a photometer, the detector is not placed at the focus. This is because the sensitivity of a single cell may not be uniform over its entire surface. If the image were to be at the focus, the ratio of detected photoelectrons to incoming photons could vary depending on where the image was actually formed, thus degrading the photometric accuracy of the system. This is avoided by inserting a lens, called a *Fabry lens*, directly in front of the detector so as to reimage the primary mirror of the telescope onto the detector, in other words, by placing the detector at the exit pupil (Fig. 2.2). A diaphragm (also called an aperture) is placed at the telescope focal plane to block out unwanted radiation from the sky surrounding the source and thus reduce background.

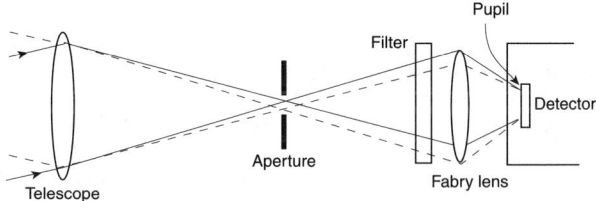

Fig. 2.2. Basic layout of a photometer. A lens immediately in front of the detector places the exit pupil on the detector so as to average out illumination of the detector and make it insensitive to the exact direction of the source and line-of-sight jitter. Illumination of the detector remains the same regardless of the direction of the source (e.g., solid or dotted line).

2.1.3 Polarimeters

Light can become polarized under a variety of conditions and this may reveal important characteristics of the emitting source. For example, polarization accompanies scatter by dust particles such as those surrounding stars or found in the interstellar medium. The effect is generally small (a few percent) and difficult to measure. The simplest way to measure polarization would be to place a birefringent material in the incoming beam and rotate it to determine the maximum intensity and, thus, the direction of polarization. But this measurement could be affected by polarization caused by the optics in the instrument itself. A better solution is therefore to introduce a calibrated phase shift (retarder) in the beam as far upstream as possible in the instrument, and then use a fixed polarizer downstream to measure polarization (Fig. 2.3). The phase-shift variation can be obtained by rotating a retardation plate or by using fixed retard plates of various values mounted on a wheel. Such a system can be incorporated in front of a photometer, camera, or spectrometer. The polarization created within the telescope itself due to coatings or nonnormal incidence mirrors must be calibrated out by observing standard sources. Care must be taken to avoid polarization effects due to nonnormal incidence in the optical train. For this reason, polarimeters are not placed at a Nasmyth or coudé focus, both of which involve folding mirrors.

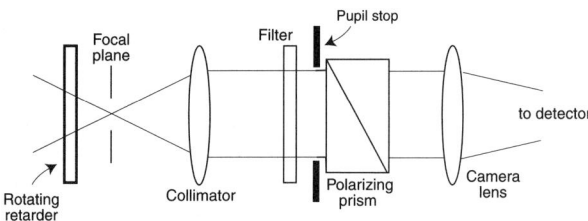

Fig. 2.3. Basic layout of a polarimeter with a rotatable retarder and fixed polarizer.

2.1.4 Dispersing spectrometers

Dispersing spectrometers rely on the dispersion of white light into its constituent wavelengths via a prism or a diffraction grating. Diffraction gratings are generally of the reflection type. A grating is a glass plate ruled with fine, parallel, equally spaced linear grooves[1] so that light can only be reflected between the grooves. These "Young's slits" diffract the incoming light, producing destructive interferences, except for specific directions which are a function of wavelength. Prisms are generally used for low spectral dispersion and gratings for high dispersion. In both cases, the dispersing elements must be fed by a

---

[1] A typical grating has about 1200 lines per millimeter and can have as many as 6000.

parallel beam to avoid mixing wavelengths. A hybrid device is the so-called "grism" (a contraction of grating + prism), in which a transmission grating is ruled or glued onto the surface of a prism; the prism deviation compensates for the grating dispersion angle, such that the output beam remains aligned with the input beam. Grisms are typically placed into the filter wheel of an imaging system to add spectroscopic capabilities. Grisms can be used in a converging beam.

The basic arrangements for the two types of spectrometer are shown in Fig. 2.4. In both cases, light from the observed source enters the spectrometer through a slit and is collimated to illuminate the dispersing element. The collimating lens or mirror is also used to form a real pupil on the dispersing element so that dispersion will be the same for all field angles.

Light emerging from the dispersing element is then captured by a lens or mirror, which forms images of the slit on the detector, one for each dispersed wavelength, resulting in a "spectrum." This last part of the spectrometer is referred to as the "camera." The camera's focal length is selected so that the image of the slit on the detector is properly sampled (i.e., two pixels per slit angular size on the sky).

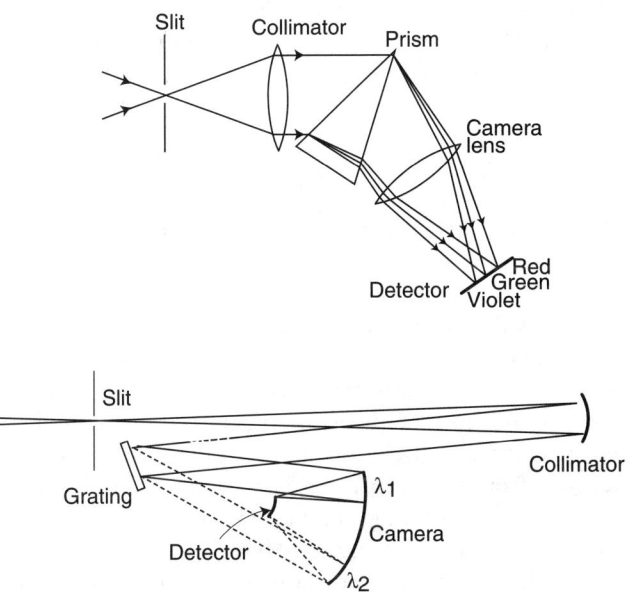

Fig. 2.4. Basic layouts of a prism (top) and diffraction grating (bottom) spectrometers.

The spectral resolving power of a spectrometer, $\mathcal{R}$, which measures the capacity to distinguish between two wavelengths $\Delta\lambda$ apart, is defined as $\mathcal{R} = \Delta\lambda/\lambda$, where λ is the mean wavelength. $\mathcal{R} < 100$ is generally considered low spectral resolution, $100 < \mathcal{R} < 1000$ mid-resolution, and $\mathcal{R} > 10000$ high resolution.

The resolving power of a grating spectrometer is given by

$$\mathcal{R} = \frac{m\lambda W}{\theta D d}, \qquad (2.1)$$

where m is an integer representing the grating's interference "order," W is the width of the grating, d is the spacing between adjacent lines on the grating, θ is the angular size of the entrance slit projected on the sky, λ is the mean wavelength, and D is the diameter of the telescope.

In a diffraction-limited system such as a space telescope, the slit width is optimal when its angular size on the sky is equal to the angular size of the image λ/D. A larger slit would reduce resolution as well as letting in sky background; a smaller slit would reduce the flux of the source entering the spectrometer, hence reducing sensitivity. In this case, equation 2.1 becomes

$$\mathcal{R} = \frac{mW}{d}, \qquad (2.2)$$

and it is noted that $\mathcal{R}$ is independent of the telescope diameter.

This is not the case for large ground-based telescopes, which are generally limited by atmospheric seeing. The image size being much larger than λ/D, the slit has to be widened to admit more light. For a slit width equal to the size of the seeing disk, σ, the resolving power is given by

$$\mathcal{R} = \frac{m\lambda W}{\sigma D d}. \qquad (2.3)$$

One notes that, in this latter case, the spectral resolving power is a function of telescope diameter and that, as D increases, the size of the grating (W) has to increase in the same proportion in order to maintain the same resolution.

When λ is not negligible with respect to the slit width, as is often the case for infrared spectrographs, diffraction effects become important. The resulting blur of the slit image introduces extra background that, at thermal infrared wavelengths, can significantly reduce instrument performance. For this reason, good infrared spectrograph design includes a fore-optics system, producing a cold pupil image in front of the slit. In this way, the slit sees a low-temperature environment and the extra background admitted is minimized.

Mid- to high-resolution spectrographs produce long narrow spectra. This was not an issue when photographic plates were in use because they could be produced in arbitrary lengths, but it is a problem with photoelectric detectors, which typically have square formats. The solution is to use "échelle gratings," which allow small portions of the spectrum to be stacked one on top of the other. An échelle is a grating with steps rather than rulings that is used in high order to produce high dispersion. This results in overlapping orders and a limited spectral range in each order, but these orders can be separated with a low-resolution cross-disperser (Fig. 2.5).

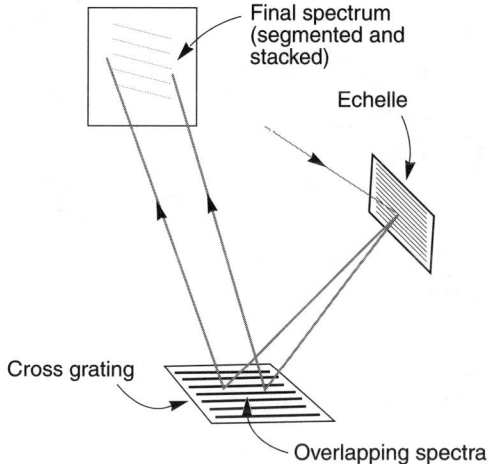

Fig. 2.5. Schematic view of an échelle spectrometer.

Whatever type of spectrometer is used, it is generally advantageous to artificially "widen" the spectra in order to make spectral features more visible. This is accomplished by moving the image along the slit, either by controlling the telescope pointing or by "wobbling" a glass plate in front of the slit.

A comparison spectrum is also generally added to the detector to calibrate the spectrum of the astronomical source. This is generated by a lamp filled with a gas that produces a large number of lines of well-known wavelengths (e.g., thorium). Infrared ground-based spectrometers simply use the airglow lines (OH) as reference.

2.1.5 Fabry-Perot spectrometer

Instead of using the dispersion of light by refraction or diffraction as dispersing spectrometers do, Fabry-Perot spectrometers use interferences to create the spectrum. A Fabry-Perot spectrometer behaves like a tunable narrow-band filter. Two highly reflective, very close plates (called an "étalon") are placed in a collimated beam. Multiple reflections are created in the gap between the two plates, resulting in destructive interference except for a specific wavelength which is a function of the gap width and the incidence angle of the incoming light. This light emerges from the étalon in a circular pattern and is imaged onto a detector. The spectrum is explored by changing the gap width of the étalon (Fig. 2.6). Fabry-Perot spectrometers work on extended sources and have extremely high spectral resolutions of 10^4 or larger. Their spectral coverage is very narrow and their use is generally limited to the study of emission line profiles.

48   2. Instruments

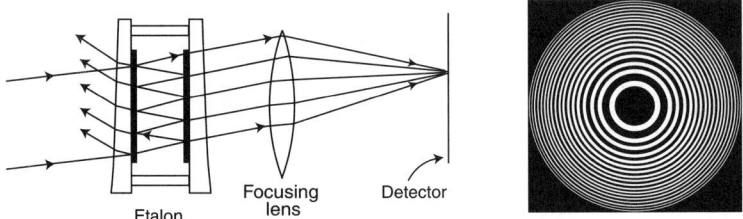

Fig. 2.6. The basic layout of a Fabry-Perot spectrometer is shown at left. A typical interference fringe pattern is shown at right.

2.1.6 Fourier transform spectrometer

As in the Fabry-Perot spectrometer, the Fourier transform spectrometer (FTS) uses interferences to create the spectrum. Light from a source is fed to a Michelson interferometer, and the output signal is recorded as one of the mirrors is scanned (Fig. 2.7). For monochromatic light, the intensity recorded will vary as a cosine law of the scanning distance due to successive constructive and destructive interferences, that is, as $\cos(4\pi x/\lambda)$, where x is the distance the scanning mirror is moved. For a polychromatic beam, the recorded intensity is the sum of all these cosine terms, and its spectrum is extracted by an inverse Fourier transform.

Fourier transform spectrometers can reach very high spectral resolutions ($\mathcal{R} > 100\,000$) and still have a wide spectral coverage thanks to their "multiplex" advantage. Their primary disadvantage is that their signal-to-noise ratio suffers from the photon noise of the full spectral range covered, rather than just that of the band analyzed.

The fact that they rely on continuous motion of a mirror is also an obstacle for space applications.

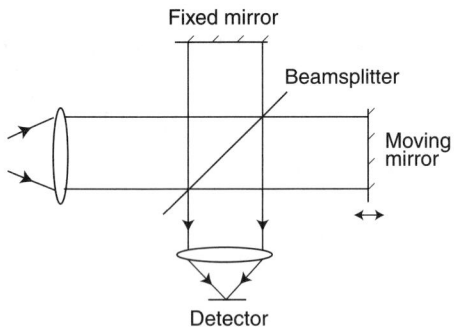

Fig. 2.7. Basic layout of a Fourier transform spectrometer.

2.2 Optical through mid-infrared detectors

The detector's role is to detect, as noiselessly and efficiently as possible, each precious photon collected by the telescope and instrument. In the optical through mid-infrared range (∼0.3–30 µm), most astronomical detectors are semiconductor based.

Until the late 1970s, optical astronomers relied essentially on photographic plates, photocathode devices, and single-pixel detectors in the infrared. Photographic plates were not particularly efficient, having quantum efficiencies (number of photons detected per incident photon) of a few percent at best. They were intrinsically noisy, affected by fog due to the natural formation of silver grains even in the absence of light, and were linear for only a limited range of exposure. Moreover, any quantitative analysis required that the plate be digitized (i.e., scanned with a microdensitometer), to turn the photographic record into computerized data, thus increasing the duration and cost of the process.

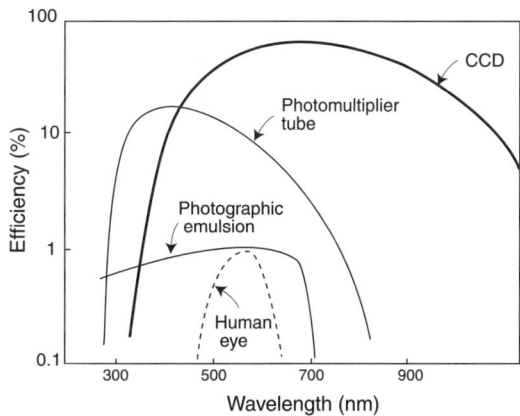

Fig. 2.8. Quantum efficiency of optical and near-infrared detectors. (After Kristian and Blourke [1].)

The photocathodes used in photomultipliers and electron-beam detectors (video type) had better quantum efficiency, but were still limited to about 20%, and were also limited in size.

All of this changed dramatically with the advent of solid state imaging detectors which are close to ideal devices: they are linear, intrinsically digitized, and have high quantum efficiency (Fig. 2.8). This breakthrough occurred first in the optical range in the mid-1970s, when charge-coupled devices (CCDs) were developed for astronomical application. And about a decade later, infrared array detectors which had been developed by the military became available for astronomy.

In the following sections, we will discuss the underlying physics of these detectors, then review the specific characteristics of the most common types. Readers who seek more detail should consult the standard texts by Janesick, McLean, and Rieke listed in the bibliography at the end of this chapter.

2.2.1 Photon detection in semiconductors

It is well known from quantum mechanics that the electrons in isolated atoms occupy discrete, well-defined energy levels. But when atoms come together to form a crystal, the outer energy levels distort and overlap, blending to create bands (Fig. 2.9). The continuum of blended energy levels within each band can allow electrons to move from one atom to another within the crystal. This sharing of electrons in bands gives rise to the covalent bonds that hold the crystal together.

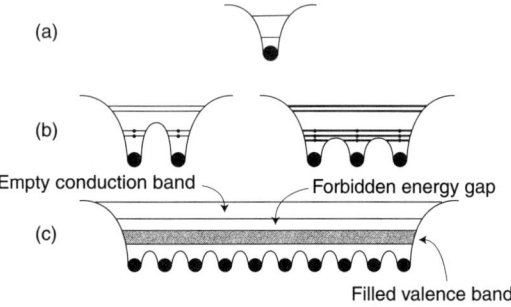

Fig. 2.9. As individual atoms (a) come together, the outer energy levels blend and overlap to create bands (b). The outermost filled band (c) is called the "valence band." (After McLean [2].)

In semiconductors, the lowest band that would be completely filled at a temperature of absolute zero is called the valence band. Above the valence band is a forbidden energy range called the band gap, E_g, and, at higher energies, partially filled conduction bands that can be populated by thermally excited electrons. In metals, the valence and conduction bands overlap, resulting in high conductivity. In insulators, the band gap is much wider, resulting in no appreciable conduction. In semiconductors, the band gap is sufficiently narrow, $0 < E_g < 3.5$ eV, that significant numbers of electrons can be thermally excited into conduction even at room temperature or by the absorption of individual optical-infrared photons. As such, semiconductors are intermediate in conductivity between conductors and insulators.

The elemental semiconductors are silicon (from which CCDs are made) and germanium. In principle, one could fabricate germanium CCDs, but because silicon semiconductor technology is much more mature, all astronomical CCDs to date have been silicon. Silicon and germanium form crystals with a diamond-lattice structure by sharing electrons with four neighbor atoms.

These elements, which appear in column IV of the periodic table, each have four valence electrons per atom.

Compounds that include elements from neighboring columns can also be formed, and these alloys can have semiconductor properties as well. Common examples include HgCdTe (mercury–cadmium–telluride, also called mercad-telluride or MCT) and InSb (indium–antimonide). HgCdTe and InSb are the bases today of the dominant detector technologies for astronomical applications in the near infrared (1–5 µm).

Silicon, germanium, HgCdTe, and InSb are all "intrinsic photoconductors," which means that single optical-infrared photons are sufficiently energetic to promote their electrons into conduction. The red wavelength limit of intrinsic photoconductors is therefore set by the wavelength of photons having energy equal to the band gap. With photon energy being equal to hc/λ, where h is the Planck constant and c is the velocity of light, the red wavelength limit is

$$\lambda_c = \frac{hc}{E_g}. \tag{2.4}$$

Intrinsic photoconduction works well for visible and near-infrared wavelengths, but at longer wavelengths, some other process must be used. Fortunately, by adding small amounts of impurities to a semiconductor (a process known as "doping"), its properties can be altered. Charge carriers are then created by promoting electrons from the doping atoms into conduction, rather than by promoting electrons from the semiconductor atoms. These devices are called *extrinsic photoconductors* and are described by the notation *semiconductor:dopant*. For example, Si:As designates silicon with arsenic as the major impurity. Si:As technology is currently the most mature one for mid-infrared (5–30 µm) arrays.

Doping can also be used to alter the properties of a semiconductor in less radical ways. For example, by adding small amounts of an impurity having a greater number of valence electrons than the semiconductor, one can create an "n-type" semiconductor, so called because the dominant charge carriers are electrons donated into the conduction band by the dopant. Likewise, one can add elements having fewer valence electrons than the semiconductor. This will create positively charged "holes" in the valence band that permit conduction. In this case, the material is called "p-type" because the dominant carrier is positively charged holes.

When an n-type semiconductor is butted against a p-type semiconductor, a p/n junction (or diode) is formed. In p/n junctions, electric fields are created by the diffusion of positively charged holes into the n-type material and by the diffusion of negatively charged electrons into the p-type material. This diffusion is halted by the electric field arising from the charge distribution.

As can be seen from Fig. 2.10, the resulting charge distribution is analogous to that of a parallel-plate capacitor. If a positive voltage is applied to the n-type material, and a negative voltage to the p-type material, the diode will conduct when the difference is strong enough to overcome the voltage

52   2. Instruments

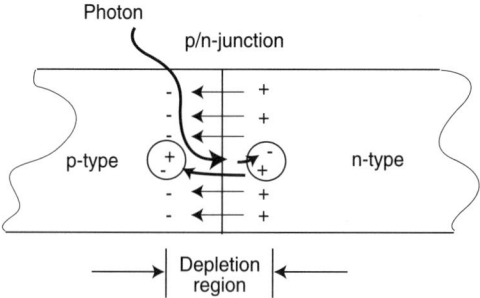

Fig. 2.10. A p-type semiconductor can be butted against an n-type semiconductor to create a "p/n junction," resulting in a diode. Charge diffusion at the junction creates an electric field that sweeps up photoexcited charge. The region affected by this charge diffusion is known as a "depletion region."

established by the charge diffusion. In this case, the diode is said to be forward biased. If, instead, we reverse the voltages, even more positive charge will pile up in the n-type material, with correspondingly more charge accumulating in the p-type material. In this case, the diode is said to be reverse biased. Such a reverse-biased diode is the basic photosensitive element in modern CCD and infrared array detectors.

As light enters the diode, it is absorbed and creates an electron/hole pair. Because such a charge is mobile, it may migrate to the depletion region near the p/n junction and there remove one unit of charge from the capacitor. This process of photoexcited charges bleeding off the bias is the physical mechanism of charge collection.

2.2.2 CCD detectors

A CCD is a two-dimensional array of p/n junctions made of silicon. Figure 2.11 (left) shows the basic construction of the popular "three-phase" CCD. In such a CCD, a giant p/n junction is formed where the p-type silicon meets the n-type silicon. This junction is divided into individual pixels by nonconducting "channel stops," which separate rows, and by voltages on control electrodes, which define columns. In any CCD, charges are physically shuffled around on the surface by changing these control voltages. Figure 2.11 (right) shows how this is done in the case of the three-phase CCD. Because charge is carried to the output amplifier, CCDs are intrinsically very quiet.

The front side of the CCD is partially obscured by metal electrodes. For this reason, although CCDs can be illuminated from either side, for many astronomical applications it is preferable to illuminate them from the rear. This is known as "backside illumination." Unfortunately, this can result in poor sensitivity to blue light due to the blue photons being absorbed far from the depletion region. To improve blue wavelength sensitivity, backside-illuminated CCDs can be thinned to shorten the path to the photosensitive

2.2 Optical through mid-infrared detectors

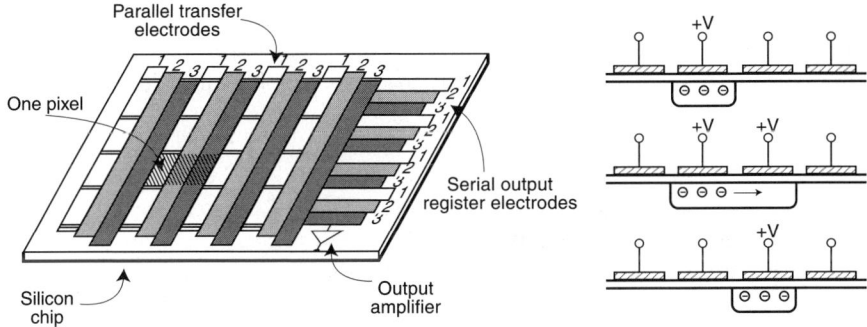

Fig. 2.11. Schematic view of a typical three-phase CCD, left. At right, in sequence from top to bottom, the principle of charge transfer in a CCD: charges are moved to the output amplifier by changing the electrode voltages.

depletion region. Figure 2.12 shows the spectral response of common front and backside illuminated CCDs. Antireflection coatings can be used to modify, to some extent, the wavelength uniformity and coverage of these curves. In the ultraviolet ($\lambda < 300$ nm), a photon may randomly create more than one carrier, and a correction factor must be applied to estimate the quantum efficiency in this regime.

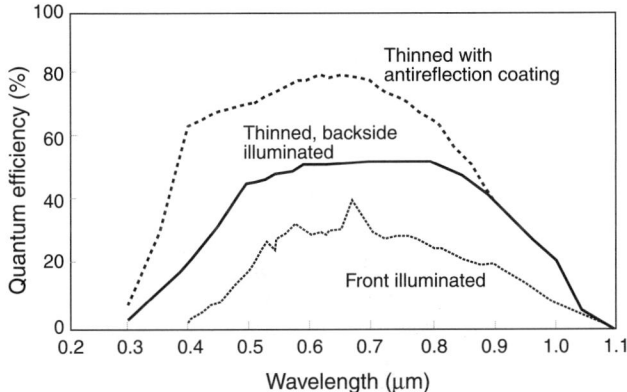

Fig. 2.12. Spectral response of common CCD types.

2.2.3 Infrared array detectors

Although CCDs have provided optical astronomers with nearly perfect quantum-limited photon detection, the same approach cannot be used in the infrared. The underlying problem is silicon's 1.12 eV band gap, which sets the material's red wavelength limit at about 1.1 µm. Although one might envision making CCDs out of infrared-sensitive materials such as InSb or HgCdTe, this is not currently possible, as the microelectronics technology for these materials is not yet well developed.

The main detector materials for the near infrared are HgCdTe (with a variable cutoff between 0.4 and 12 µm depending on the relative concentration of Hg and Cd) and InSb (1–5 µm). In both HgCdTe and InSb, the short-wavelength limit is at least partially determined by the substrate on which the detector material is grown. Once this substrate is removed, either by chemical etching or mechanical machining, their responses extend into the optical.

In the mid-infrared (5–28 µm), arsenic-doped silicon (Si:As) is the leading technology. Several other materials have been tried, but all currently either have lower performance or face implementation problems.

Infrared arrays are *hybrid* devices in which a silicon multiplexer, often referred to as a MUX, is bonded to a photosensitive substrate. This is done because the technology is not mature enough to provide a complex low-noise readout circuit on materials other than silicon. To transfer out the signals, multiplexers, which provide a direct electrical connection between each pixel and the detector output, are preferred to CCDs because CCDs exhibit poorer performance at the low temperatures needed by infrared detectors. The photosensitive slab is bonded to this MUX by an array of pixel-sized indium bumps that are cold-soldered under pressure (Fig. 2.13).

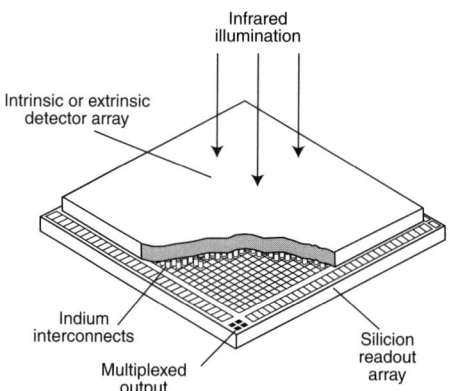

Fig. 2.13. Diagram of the basic "hybrid" structure of infrared array devices. The two slabs are bonded together by tiny indium bumps of the size of each pixel

In modern infrared arrays, unlike CCDs, charge is sensed *in situ*, usually using one source follower per pixel. Although this allows the same pixel to be read out nondestructively many times, each readout is subject to a variety of noise mechanisms that are absent in CCDs.

The current state of the art for these three types of detectors is shown in Table 2.1, and the typical spectral response of InSb and Si:As detectors is shown in Fig. 2.14.

Table 2.1. State-of-the-art infrared detector performance

Item	InSb	HgCdTe	Si:As
Representative array	"ALADDIN"	"HAWAII"	SIRTF
Manufacturer	SBRC	Rockwell	SBRC and Rockwell
Wavelength range (μm)	0.6–5.5	1–2.5	5–28
Format	1024^2	1024^2	256^2
Single sample read noise (e^-)	50	34	50
Dark current (e^-/s)	< 0.1	< 0.1	< 10
Well depth ($10^5\ e^-$)	3	0.9	1
Pixel size (μm)	27	18.5	30
Operating temperature (K)	35	80	6
Mean quantum efficiency	85% (0.9–5μm)	66% (K band)	40% (5–25 μm)
Readout time (μs/pixel)	3	3	3
Power dissipation (mW)	<3	<3	<1

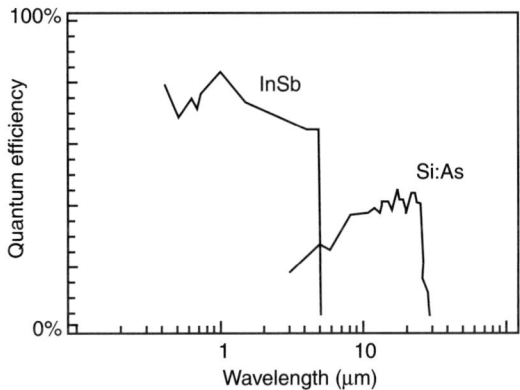

Fig. 2.14. Quantum efficiency of InSb and Si:As detectors.

2.2.4 Specific detector characteristics

We now examine some of the special requirements for detectors used in astronomical observations and ways of implementing them.

Pixel size

The physical size of the pixels is not, in itself, a factor as far as observations are concerned. Only the angle subtended on the sky by each pixel matters, and this can be adjusted by changing the optics on the telescope and camera. However, detector parameters such as dark current, readout noise, cross-talk,[2] dynamic range, and sensitivity to cosmic rays can vary with pixel size. Overall, performance improves with smaller pixels since (1) dynamic range decreases exponentially with decreasing pixel size, (2) readout noise for CCDs is independent of pixel size, and (3) dark current increases by pixel area. The main drawback is that cross-talk increases exponentially with decreasing pixel size.

From the telescope optics point of view, too, smaller pixels are preferable, since the magnification of the telescope/camera combination is reduced. For approximate Nyquist sampling (see Chapter 4, Section 4.5.3), the optimal focal ratio at the detector is given by

$$\frac{f}{D} = \frac{2p}{\lambda}, \qquad (2.5)$$

where f is the final system focal length, D is the diameter of the primary mirror, p is the pixel size, and λ is the operating wavelength. At a given wavelength, the focal ratio required is proportional to the pixel size. Hence, the smaller the pixel size, the easier it is to package the optics. In practice, a pixel size on the order of 20 to 30 μm is adequate from this point of view, but the smaller the better.

Elemental exposure time

With large telescopes, typically pushed to the limit, observations can be long, lasting from several hours to several days. In practice, in order to detect and eliminate cosmic rays effects, observations are split into shorter "elemental exposures." If Φ is the proton flux, p the pixel side dimension, and $f_{\rm cr}$ is the allowable fraction of "hit pixels," the maximum integration time, t, is set by

$$t \leq \frac{f_{\rm cr}}{\Phi p^2}. \qquad (2.6)$$

In deep space, for example at the L2 orbit planned for NGST, the cosmic flux is about 1 proton/(cm^2 s). For a pixel size of about 20 to 30 μm and an allowable fraction of hit pixels of a few percent, the elemental exposure time will have to be on the order of 1000 s. But for bright objects or for deep exposures in the thermal infrared, where the zodiacal foreground is much higher than in the near infrared, the elemental exposure will have to be even shorter due to the "full-well" limitation (i.e., the maximum number of electrons that the array can store in each pixel).

Dark current

Noise introduced by the detector is primarily of two kinds: dark current and readout noise. The term "dark current" refers to the current measured when

---

[2] Cross-talk is the leaking of charges between neighboring pixels due to diffusion in the silicon.

2.2 Optical through mid-infrared detectors   57

no light is falling on the detector. The dark current signal increases linearly with time and can be calibrated and subtracted. It has, however, an intrinsic uncertainty due the statistical nature of the charge generation process. The residual error, usually equal to the square root of the dark current signal, is the so-called "dark current shot noise."

In CCDs, dark current is thermally generated at the silicon–silicon dioxide interface and in the depletion and bulk regions of the device. These effects are strongly temperature dependent, and the dark current, d_c, expressed in electrons per pixel per second, follows the general equation

$$d_c = C T^{1.5} e^{-E_g/2kT}, \qquad (2.7)$$

where T is the absolute temperature, E_g is the band-gap energy, k is the Boltzmann constant, and C is a constant. Figure 2.15 gives an example of dark current in CCDs as a function of the operating temperature. CCD dark current can be essentially eliminated by cooling the detector, typically to about $-70\ ^\circ$C.

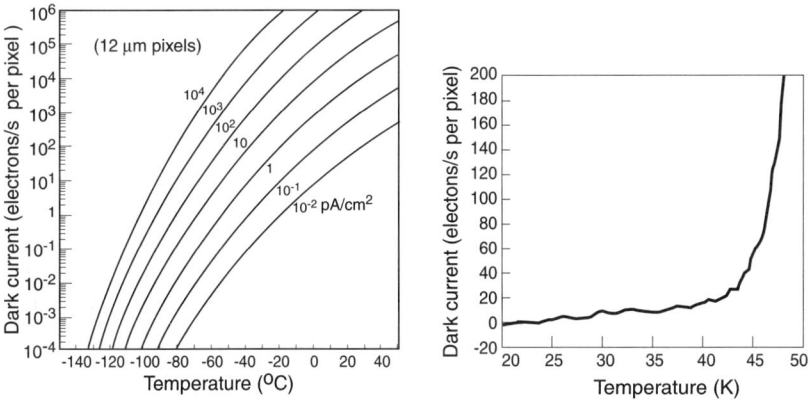

Fig. 2.15. Dark current dependence on temperature. At left, CCD dark current for 12 μm pixels as a function of temperature, and the value of the dark current in pA/cm² at a reference temperature of 293 K. At right, typical dark current of InSb.

The dark current of infrared detectors is also strongly dependent on temperature. InSb detectors must be cooled to about 30 K for the dark current to be negligible. HgCdTe detectors with cutoff at 2.5 μm need to be cooled to about 70 K. Mid-infrared detectors are even more sensitive and must be cooled to about 8 K. Typical dark current values for these detectors are shown in Table 2.1.

Readout noise

When the signal collected on CCD pixels is transfered, amplified, and converted to a digital value, noise is introduced at each step of the process. The noise added by reading the signal in each pixel is called the readout noise. This can be dominant for short exposure times and when dark current has been reduced to negligible levels. Typical values for infrared detectors are shown in Table 2.1. The readout noise for infrared detectors can be large compared to dark current noise over a typical subexposure of 1000 s. But one can take advantage of the fact that it is possible to read infrared arrays nondestructively by reading out the array several times during the integration ("up-the-ramp sampling") or by making multiple readouts at the beginning and end of each frame ("Fowler sampling").[3] These two methods are illustrated in Fig. 2.16.

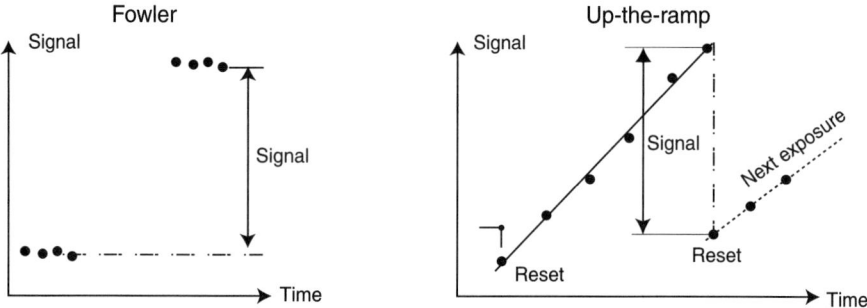

Fig. 2.16. Reducing readout noise in infrared detectors by "Fowler sampling" (left), or by "up-the-ramp sampling" (right).

Because the defining points at each end of the integration flux line have maximum leverage in the line-fitting process, Fowler sampling is better by a factor of $\sqrt{2}$ for white noise over linearly spread sampling. With Fowler sampling, readout noise can be substantially reduced and follows the expected inverse square law up to about 30 samples (Fig. 2.17).

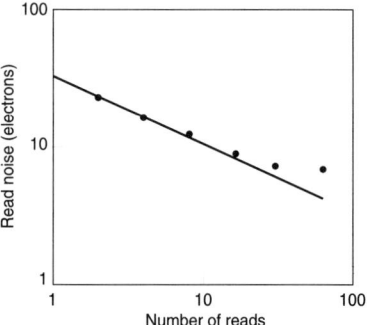

Fig. 2.17. Reduction of readout noise by Fowler sampling. Up to about 30 readouts, the gain follows a $\sqrt{n}$ law.

---

[3] Named after Al Fowler from NOAO who first proposed it.

Beyond 30 samples, systematic effects become significant and the gain levels out [3]. On the other hand, the advantage of up-the-ramp sampling is that it allows for the recovery of some of the pixels affected by cosmic rays or by other transients occurring during the integration.

2.3 Relay optics

In ground-based telescopes, instruments are generally mounted directly at a given focus and are removed and replaced to suit the needs of the observing program. In some cases, however, it is advantageous to keep all of the instruments in place and feed them by redirecting the beam with a rotating fold mirror. In the case of infrared telescopes, the mirror is often replaced by a *dichroic*, which reflects the infrared wavelengths to the science instruments and transmits the visible light to the guiding system. In the case of space-based telescopes, where such a moving mirror would be a "single point of failure," it is advantageous to have all of the instruments share the field. This requires some type of relay optics to avoid congestion at the focal plane. These relay optics can also be used to change the focal ratio in order to match the plate scale to the detector pixel size, remove residual aberrations (e.g., astigmatism and field curvature in a Ritchey-Chrétien system – see Chapter 4), and create real pupils where stops and filters can be placed.

One convenient relay optics system is the "Offner relay" [4], a 1-to-1 relay originally designed for copy machines. An Offner system is simple to fabricate and free of all third-order aberrations. It is a two-mirror, three-reflection system composed of two spherical mirrors, one concave and the other convex, with the same center of curvature, the concave one having a radius of curvature twice that of the convex one (Fig. 2.18). This results in cancelation of the spherical aberration and of all third-order aberrations when the system is fully symmetric and the input beam pupil is at infinity. If this is not the case, the aberrations can still be kept low. If the input beam pupil is relatively close by, the Offner creates a second pupil close to its secondary mirror, which can

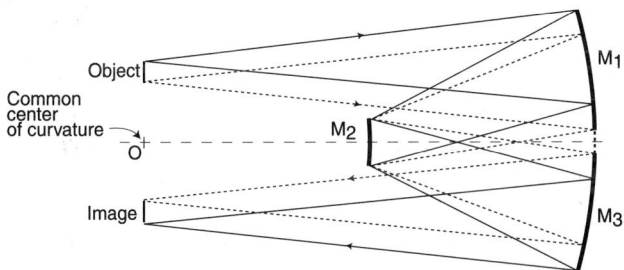

Fig. 2.18. Offner relay layout. The reflecting areas M1 and M3 are part of the same mirror. The object and image surfaces are on a plane containing the common center of curvature of the mirrors.

be used to locate filters, cold stops, or stray light control. By slightly changing the radius of curvature of the main mirror compared to its theoretical value, it is possible to correct for field curvature in the incoming beam. Finally, individual focusing of the downstream instrument is possible by actuating the secondary mirror of the Offner combination.

An alternate solution consists of using an inverted Cassegrain configuration (see Chapter 4), the drawback being that the secondary mirror then creates a large central obstruction.

2.4 Cryogenic systems

As indicated earlier, solid-state photoelectric detectors must be cooled to reduce dark current. In addition, for work in the infrared, the entire telescope and instrument should ideally be cooled to reduce thermal emission. This is only possible in space, however. On the ground, frost may form on any surface cooled to below ambient temperature unless it is placed in vacuum, and as this is clearly not possible for the main optics, only the detector and the optics near it can be cooled. Equipment can best be cooled by placing it inside a *cryostat* which is a dewar filled with a cryogen. A dewar is basically a thermos bottle, that is a vessel with a vacuum jacket to minimize thermal load due to the surrounding air.

For work in the visible, CCD cryostats typically use liquid nitrogen (LN), a fairly inexpensive and easy-to-manage fluid with a boiling temperature of 77 K.[4] For work in the infrared beyond 2.5 μm, detectors need to operate at temperatures lower than LN can provide. The fluid of choice is then liquid helium (LHe$_4$), which boils at 4.2 K.

An increasing number of infrared instruments now use closed-cycle coolers instead of, or together with, liquid coolants. Closed-cycle coolers do not require routine replenishment as traditional cryostats do, and they are particularly suited for remote locations where the delivery of cryogenic fluid can be problematic and bad weather can occasionally cut off supply. Closed-cycle coolers exploit one of the several thermodynamic cycles (Stirling, Gifford-McMahon, Joule-Thomson, etc.) to perform a refrigerator cycle, wherein a moving "piston" in a "cold head" causes an expansion that subtracts heat from the cryostat. The major issue with closed-cycle coolers is vibration induced by their moving parts. With proper isolation, however, vibration can be reduced to a negligible level and, in general, does not present a problem either for the instrument or for the telescope.

---

[4]When very low dark currents are not an issue, simple thermoelectric coolers can provide a reliable, inexpensive alternative to LN.

References

[1] Kristian, J. and Blourke, M., Kristian, J and M Blourke, *Charge-coupled devices in astronomy*, Scientific American, Vol. 247, No. 4, p. 66, 1982.

[2] McLean I. S., *Electronic Imaging in Astronomy: Detectors and Instrumentation*, John Wiley & Sons, 1997, p. 134.

[3] Fowler, A. et al., Aladdin, the 1024×1024 InSb array: design, description and results, SPIE Proc. Vol. 2816, p. 150, 1996.

[4] Offner, A., *Unit power imaging catoptric anastigmat*, U.S. Patent 3748015, 24 July 1973.

Bibliography

Amico, P. and Beletic, J.W., eds., *Optical Detectors for Astronomy II*, Kluwer Academic Publishers, 2000.

Eccles, M.J., Sim, E.M., and Tritton, K.P., *Low Light Level Detectors in Astronomy*, Cambridge Univ. Press, 1983.

Hudson, R.D., *Infrared Engineering*, John Wiley & Sons, 1969, p. 104.

Janesick, J., *Scientific Charge Coupled Devices*, SPIE Press, 2001, Vol. PM83.

Mackay, C.D.,Charge-coupled Devices in Astronomy, Annu. Rev. Astron. Astrophys., Vol. 24, p. 255, 1986.

McLean, I. S., *Electronic Imaging in Astronomy: Detectors and Instrumentation*, John Wiley & Sons, 1997.

Rieke, G.H., *Detection of Light: from Ultraviolet to the Submillimeter*, Visnovky K. ed., Cambridge Univ. Press, 1994.

Rodriguez-Espinosa, J.M., Herrero, A., and Sanchez, F., eds., *Instrumentation for Large Telescopes*, Cambridge Univ. Press, 1997.

3
Design Methods and Project Management

Design is primarily an engineering undertaking, but to be successful it must be accomplished in concert with astronomers every step of the way. Just as in the case of designing the perfect house, intimate cooperation between the client (the astronomer) and the architect/builder (the engineers) is essential.

The key to this cooperation is to make the process an *iterative* one. Astronomers should define their goals in broad, nonintransigent terms and, in collaboration with the engineers, arrive at a set of concrete requirements and priorities. Engineers should then try to formulate a realistic architecture to meet these requirements and evaluate the corresponding cost and schedule. The results are unlikely to meet the budgetary constraints. It will then be up to the astronomers and engineers to arrive at a revised set of requirements. This iterative process should continue until a satisfactory solution is found, leaving enough margin and "descope options" to alleviate the inevitable future surprises and unforeseen difficulties (Fig. 3.1).

This iterative design process is formally embodied in what is called "systems engineering." Systems engineering differs from pure *design* in that it deals with the nontechnical constraints (cost, schedule, risk) in addition to how the system accomplishes its strictly scientific objectives. Although systems engineering is the key to a successful design, it has a role in many subsequent phases of the project, particularly in the implementation phase, where it is often necessary to revisit choices made during design in order to cope with technical or scheduling difficulties.

Enveloping systems engineering is "project management." Project management, like systems engineering, deals with the observatory's scientific performance, cost, and schedule, but it is the decision-making function, whereas sys-

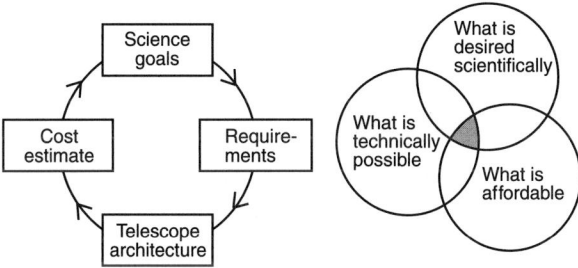

Fig. 3.1. The optimal solution is arrived at iteratively (left) and is at the intersection of what is needed scientifically and what is technically possible and affordable (right).

tems engineering is the analytical and advisory function. In addition, project management deals with the issues of human resources, contracts/procurement, and communication among the various groups involved.

In short, the role of systems engineering and project management is to coordinate the technical definition, development, test, and operation of the observatory so as to optimize the science return within program constraints, such as budget and schedule, while minimizing technical and programmatic risk. This is visually expressed by the classic diagram shown in Fig. 3.2.

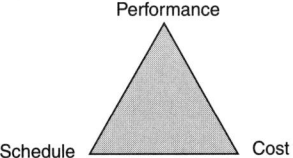

Fig. 3.2. The three poles of systems engineering and project management. The fundamental objective is to find the best balance among performance, cost, and schedule while keeping risks under control. The analyses are performed by the systems engineering team, whereas the decision making is the responsibility of project management.

In this chapter, we will explore the various tools and approaches used in systems engineering and project management. But to help organize these ideas, we first discuss an important notion, the "project life cycle."

3.1 The project life cycle

A fundamental concept introduced by NASA and which is applicable to the management of many major systems, including large ground-based observatories, is the decomposition of a project into sequential *phases*. This provides a means of organizing the project into manageable elements with phase boundaries chosen so as to provide natural points for go/no-go decisions for each

64   3. Design Methods and Project Management

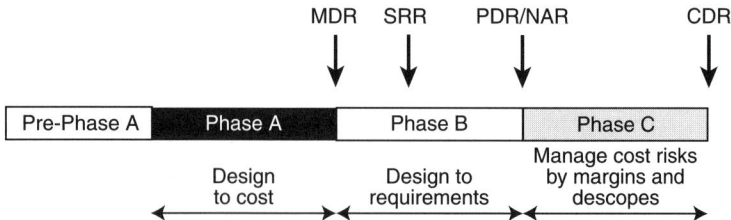

Fig. 3.3. The canonical project phases and reviews during the design phase (see text).

subsequent phase (Fig. 3.3). If authorization to proceed is not obtained, the project may be allowed to "go back to the drawing board," be "descoped" (i.e., its goals are reduced), or it may be terminated. The decomposition covers all stages of the project commencing with concept formulation and extending through operation and eventual retirement of the system. Collectively, the project phases are called the *project life cycle*.

Following NASA's terminology,[1] the project life cycle can be decomposed into the following phases.

- **Pre-Phase A – Concept studies.** The purpose of this phase is to explore both science and implementation ideas. The product of this phase is a set of science goals and design concepts warranting further study.
- **Phase A – Preliminary analysis.** The purpose of Phase A is to further examine the scientific desirability and technical feasibility of the proposed concept and demonstrate its value. It is during this phase that various architectures should be explored, and trade-off studies should be made to maximize performance within the expected cost envelope. The product of this phase is a formal set of science requirements and one or more credible designs and operation concepts. Phase A terminates with an "observatory design review" (called Mission Design Review or MDR in the case of space telescopes).
- **Phase B – Project definition.** The purpose here is to define the project in enough detail to establish a feasible and credible baseline design. A "feasible" design is one that meets the science goals within schedule and financial constraints. To be "credible," the design must not depend on breakthroughs in the state of the art. To meet these two conditions, it is necessary to produce a robust preliminary design and demonstrate that all the required technological developments have been mastered. This is an important phase because small amounts of

---

[1] This terminology has recently changed and projects are now decomposed into "formulation phase" and "implementation phase." We have retained the traditional phase definition in this work because of the extensive references to it in the literature.

effort or money invested here can bring large benefits later on; this is the time to use innovation and out-of-the-box thinking. It is during this design phase that 70% of the ultimate project costs are determined and that proper sizing of the project scope can avoid future grave and costly problems.

A few months after the start of this phase, there is a "system requirements review" (SRR) to establish the project's scientific and technical requirements. Near the end of the phase, the project is subjected to a "nonadvocate review" (NAR) to assess the clarity of the objectives and the thoroughness of the management plan and trade studies performed. Phase B culminates in a "preliminary design review" (PDR). Design issues that may be uncovered during the PDR must be resolved before the next phase. To arrive at an impartial evaluation, both the NAR and PDR must be attended by experts and managers not related to the project and preferably not belonging to the agency in charge of the project. After PDR and agency approval, the project baseline, budgetary envelope, and schedule are, in principle, assumed to be fixed. The preliminary design must therefore be robust and the project must include sufficient technical, schedule, and financial margins to cope with unexpected problems and technology shortfalls during the following phases. Most overruns in large projects can be traced either to insufficiently funded Phases A and B (Fig. 3.4), or to undue programmatic pressure.

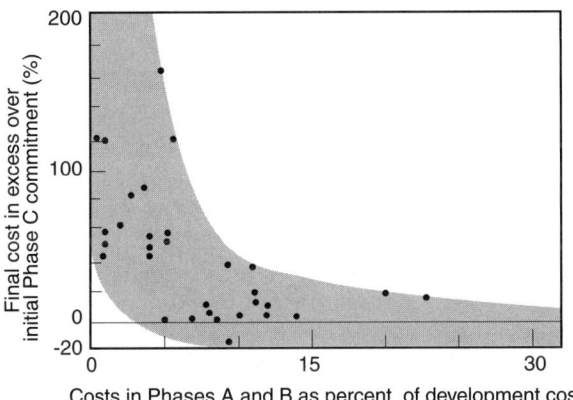

Fig. 3.4. The likelihood of project overruns increases when Phases A and B are underfunded. (From Ref. [1].)

- **Phase C – Design.** The purpose of this phase is to establish a complete design that is ready to be built, integrated, and verified. Trade studies may continue but only at the subsystem level, and engineering prototypes are fabricated to establish confidence in the design. Changes to the baseline should represent successive refinements, not fundamental changes. In a fixed budgetary environment, cost risks are managed by utilizing design margins, schedule margins, and descopes. Phase C culminates in a "critical design review" (CDR).

- **Phase D – Development.** In this phase, the observatory is built, launched and deployed in the case of a space system, and commissioned. Commissioning is the process of verifying that the observatory has been successfully activated, meets all its science requirements, and is ready for operation. In parallel with these main activities, operation and maintenance procedures and manuals are prepared and the initial operating and maintenance staff is trained. As in Phase C, cost overruns during fabrication are managed by utilizing design margins, schedule margins, and descopes.

- **Phase E – Operations** This is the observation phase for which the observatory was built, but it also encompasses all upgrades to the observatory and dealing with the system once the operations are completed. For a space mission, this last issue may mean a planned return to Earth or jettisoning.

3.2 The tools of systems engineering

As indicated earlier, observatory design is an iterative process. The sequence of steps and feedback loops vary according to the degree of novelty of the design, funding-agency regulations, and the ways in which design and development are competed in private industry. Very generally, however, the process follows the sequence shown schematically in Fig. 3.5.

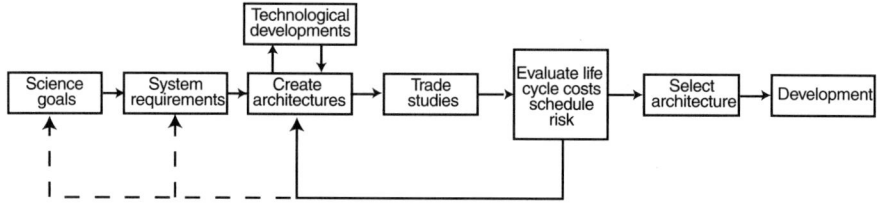

Fig. 3.5. Systems engineering process for observatory design.

More specifically, the procedure should progress through the following steps:

- Recognize the scientific opportunity and establish the general scientific goals based on the desires of the scientific community and institutions. This step triggers the activity but is not really part of systems engineering. However, systems engineering feedback will help delineate the exact set of science goals that are both feasible and affordable.

- Identify a set of realistic, well-defined scientific requirements in concert with the science advisory group.
- Convert these scientific requirements into first-order system and subsystem requirements.
- Create a basic architecture that is responsive to the requirements.
- Define a trade-off metric tightly linked to the system requirements, and promote the study of alternative architectures and subsystem trade-offs. Evaluate cost, schedule of completion, and risks for each alternative.
- Use the results of these alternative concept and trade-off studies to refine system and subsystem requirements. If necessary, adjust the science goals themselves in concert with the science advisory group.
- On completion of this iterative process, select the optimum architecture in concert with the project manager and the science team.
- Identify reductions or modifications of the science requirements (e.g., reduction of wavelength coverage, field of view, Strehl ratio) which could lead to significant cost savings should the project encounter budget overruns.
- Use these results to derive the "scientific requirement document" and "requirement budget allocation," which will guide the engineering studies.

In the following subsections, we explore the various approaches and tools that help carry out this procedure.

3.2.1 Design reference program

Scientific goals in astronomy can be translated into five basic requirement categories: sensitivity, angular resolution, wavelength coverage, spectral resolution, and temporal resolution. Unfortunately, these requirements are often conflicting. For example, requirements for the study of high-redshift supernovae are different from those for studying high-redshift galaxies. For the supernovae study, the rarity of the events requires a large field of view, whereas moderate angular resolution is sufficient since these are point sources. High-redshift galaxies, on the other hand, are numerous and extended yet small (a few tenths of an arcsecond), and therefore require exquisite angular resolution but only a small field of view. Another example might be the conflicting demands placed on an observatory by the need to observe nearby, relatively bright, moving targets in the Solar System as well as to conduct deep exposures of the distant universe. Observatories are "general purpose" by nature, obliged to serve the different needs of a large number of observers. To find a practical compromise, one must find a way of "weighting" the various scientific goals and their corresponding observational requirements.

An essential tool for accomplishing this optimization is the "design reference program." This concept has its origin in space programs, where a "design reference mission" (DRM) serves as a basis for simulations to validate the hardware and operational software before launch. The idea behind the DRM is to define a strawman observation program to exercise all the functions of the observatory in a manner as similar as possible to real observations, then to run this program against the completed hardware in simulated on-orbit conditions. The concept of the DRM can be extended from serving as a check for the validity of a particular implementation to providing a metric for comparing various possible observatory architectures. This trade-off tool is referred to as the design reference program (DRP), the process of which is as follows.

(1) Establish a strawman observation program as close as possible in its scientific goals and content to the one that the completed observatory is expected to execute, at least during its first few years. This strawman program should be determined after exhaustive consultation with the scientific community and be under the control of a scientific advisory committee. Each program component should be given a scientific "weight," expressed in terms of fractions of the total available observing time.

(2) Define the observing requirements for each program component (e.g., source flux, type of instrument, bandpass, and signal-to-noise ratio). As an example, Table 3.1 gives a simplified excerpt of the programs and weights used for an early study of NGST [2].

Table 3.1. Excerpt from the NGST design reference program

Program	Flux (nJy)	S/N	$\lambda/\Delta\lambda$	Mission fraction
Primordial galaxies – survey	0.4	2	0.5	6%
Primordial galaxies – spectr.	4	10	100	11%
Distant supernovae	1.4	5	5	14%
etc.				

(3) Determine the set of principal observatory parameters to be explored and their realistic range based on available funding, current technology, and potential technological development. These main parameters may include, for example, the telescope aperture diameter, instrument field of view, optics temperature, spectral coverage, optics throughput, parallel use of instruments, detector readout, and dark current noise.

(4) Develop a mathematical end-to-end model of the telescope and instrument to calculate exposure times for observations.

(5) For a given set of observatory parameters, calculate the exposure times required for each program, including all necessary overheads.

(6) Compare the total calculated exposure time to the expected mission or observatory lifetime, and the fraction of the time attributed to each program to the desired allocation.

(7) Repeat the process, adjusting the expected science goals, science weights, and observatory parameters until a satisfactory combination is found.

As an example, Fig. 3.6 shows the results of such a parametric study which helped pinpoint the main characteristics of NGST [2]. Although the DRP is most beneficial during the conceptual phase as a strawman for trade-off studies, it is also useful in later design phases and even during the fabrication phase. For example, this process can be used to compare alternatives at the subsystem level, to evaluate the impact of potential descopes, or for the sizing of data handling and storage.

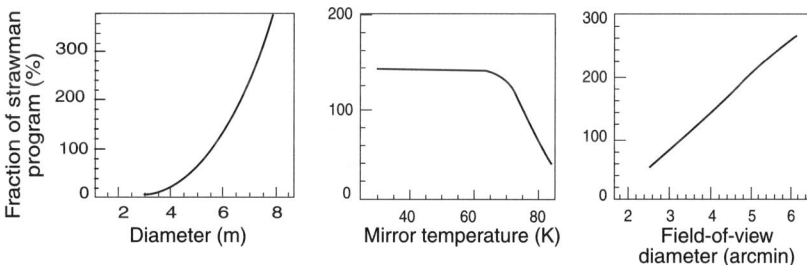

Fig. 3.6. Completion rate of the NGST core scientific program for 5-year mission duration, as a function of telescope diameter (left), mirror temperature (center), and field of view (right). These plots illustrate the rapid increase in program completion with increased telescope diameter and field of view, and the dramatic decrease when the temperature of the optics exceeds 70 K.

3.2.2 Requirements "flowdown"

The technical specifications of the observatory are developed during the initial design stage by combining the scientific requirements, operational considerations, and environmental factors into a comprehensive document defining all of the technical measures important to the observatory's performance.

This process is conveniently decomposed into "levels," going from the general to the specific (Fig. 3.7). Using terminology and definitions adapted from NASA, these levels can be as organized as follows.

- **Level 1** defines the fundamental requirements defining the science and programmatic goals of the observatory. This level may be considered the contract between the project office and that portion of the astronomical community that will use the observatory. An example of a set of Level 1 requirements is shown in Table 3.2.

70   3. Design Methods and Project Management

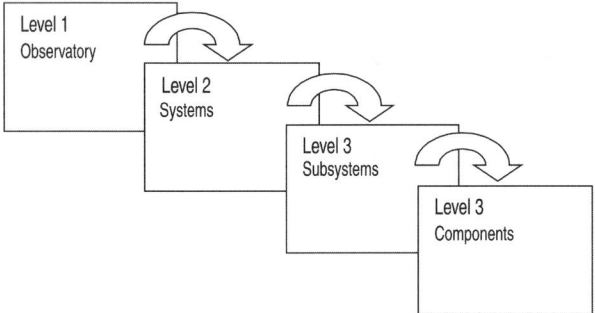

Fig. 3.7. Organization of observatory requirements in levels.

Table 3.2. Level 1 Requirements for NGST (adapted)

Ref.	Parameter	Requirement	Rationale
1	Scientific mission	The NGST Observatory shall be capable of implementing the scientific program defined by the Project DRP.	The DRP reflects the scientific program recommended by the HST & Beyond Report.
2	Science mission duration	The NGST science mission shall be a minimum of 5 years, with a duration goal of 10 years.	The DRP is designed to be executed in 2.5 years. Extending the mission duration to 10 years ensures a rich general observers program, and maximizes scientific return.
3	Cost	The total cost of Phase C/D shall be less than $850M (1996$). Operating costs over 10 years should be less than $400M (1996$).	Funding is provided by NASA "Origins" program and must allow for the timely development and operation of future "Origins" missions.
4	Schedule	NGST shall be launched in 2008.	The NGST science mission should overlap with and continue that of HST for a smooth transition between scientific and calibration programs.

– **Level 2** defines the observatory systems requirements derived from Level 1. These requirements are essentially science driven and result from trade-off studies made during the concept and Phase A periods. They may also reflect the recommendations of the project "science working group." Level 2 requirements typically include telescope diameter, optical quality, tracking accuracy, focus fields and f-ratios, wavelength coverage, suite of instruments, orbit or observatory site. Once the design has been firmed up, generally following the systems requirements review, the Level 2 requirements become part of the contracts between the project office and contractors. Making changes to Level 2 require-

ments during the telescope development stage is not unusual, due to maturing of the design, improvements in instrument technology, and advances in the science. But Level 1 requirements are fundamental conditions imposed on the design and manufacture which are modifiable only by the highest project authority or funding agency.
- **Level 3** defines the requirements for all observatory subsystems derived from Level 2. This is done through the use of preliminary science and engineering studies or by making judgments based on experience. The purpose is to provide sufficient detail to permit the design of each subsystem. These requirements may evolve as the design progresses and be the subject of trade-offs between subsystems. They are normally under the control of the project office's or the contractor's systems engineer.
- **Level 4** defines the requirements at the component level derived from Level 3 requirements. They are typically under the control of the engineer in charge of the subsystem.

In addition to the science-driven specifications which form the backbone of the design requirements, the requirements relating to the operation of the telescope after commissioning must not be overlooked. On the science side, operational requirements should cover issues such as telescope operation, remote observing, and science data processing.

Lifetime, safety, reliability, and maintainability are also important issues which are often ignored or considered only informally in telescope design. Failure to take them into account can have dramatic consequences later on, during the operation phase. The corresponding requirements should be established as part of Level 2.

Determining the required service life of the telescope is not easy. Ground-based telescopes are generally designed for long lifetimes, on the order of 30 years, and frequently continue to be in use considerably longer. A good example is the 5-meter telescope on Mt. Palomar, which was designed in the late 1930s and is still very much in operation. Designing for too short a lifetime is thus unrealistic and does not take into account the actual history of telescope use. On the other hand, requiring an unreasonably long service life can force designers to select materials and components that would otherwise not be required by technical performance objectives. In setting the service-life requirement for the telescope, therefore, these factors must be carefully balanced.

The safety requirements arise from the need to protect the personnel using and maintaining the telescope, the equipment itself, and the environment around the telescope. Hazards should be identified during the conceptual design stage and eliminated where possible by proper design. When safer alternatives are not available, guards and safety devices must be incorporated into the design. Warning devices, safety procedures, and personnel training are also used to reduce the risks from any remaining hazards.

Reliability and maintainability are two concepts that must be considered together in telescope design. Reliability refers to the telescope's ability to continue to meet performance specifications in the future. Maintainability refers to the frequency and complexity of maintenance tasks needed to ensure its continued performance. Telescopes are traditionally designed to be highly reliable but rarely have explicit reliability specifications and are not always formally analyzed for reliability. Similarly, maintenance plans are often devised only during or at the end of construction. Large telescopes have become quite complex, and these two issues should be addressed in the design in order to produce a reliable, easily maintained observatory.

3.2.3 Error budgets

Error budgets (also referred to as "performance budgets") are some of the most useful design tools in any engineering project. In the design of observatories, due to the multiplicity of error sources and the intricate nature of such systems, they are absolutely essential. Error budgeting consists of apportioning a given performance requirement among various subsystems and various components of a system. This then allows engineers in each discipline to design their own subsystems under the watchful eye of the systems engineer.

At the beginning of a project, allocations are made somewhat arbitrarily. They are then refined as the design progresses following trade-off analyses, detailed studies, and manufacturing tolerancing. Error budget allocation should be developed for all major observatory systems (e.g., optics, pointing, thermal, power). Two examples of such allocations, for optical quality and pointing stability, are shown in Fig. 3.8 and Fig. 3.9.

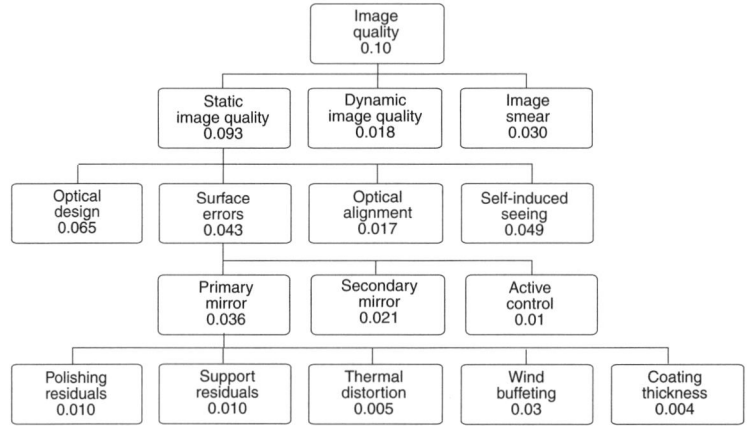

Fig. 3.8. Error budget for image quality of the Gemini telescopes. This budget is for the $f/16$ Cassegrain focus at 2.2 μm and for zenith pointing. Values are contributions to image diameter in arcseconds for 50% encircled energy. This example traces the errors contributions from the primary mirror only. (From Ref. [3].)

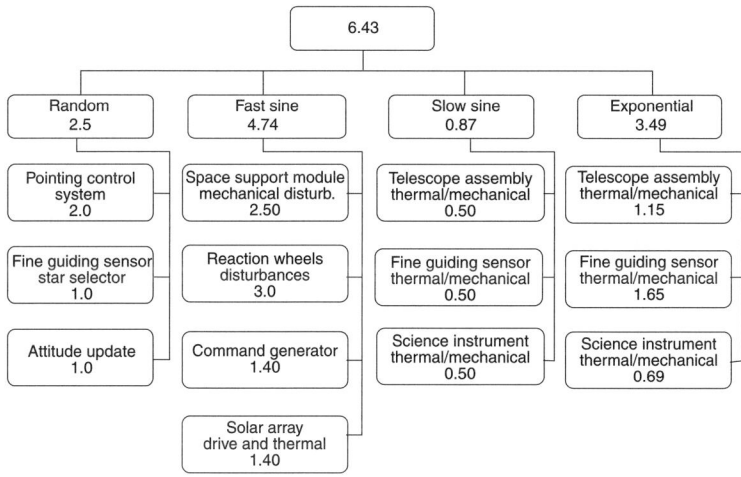

Fig. 3.9. Pointing error budget for HST for a 20-hour observation starting 20 minutes after worst-case overall slew (pre-launch estimates). "Random" refers to random contributors, "fast sine" to mechanical vibration contributions, "slow sine" to orbital time scale contributions, and "exponential" to slow contributions, such as repointing effects. All numeric values are in milliarcseconds rms.

Initially, error budgets are obtained in a "top-down" fashion, by allocating values for each potential source of error. These will be based on first-order analyses, experience with other projects, and on good engineering judgment. Where the magnitude of a contribution is unknown, the allowance will be set to an initial best guess. A summation of the error allowances is made and the budget is balanced to achieve the overall goal. Project goals must sometimes be adjusted to bring them into line with engineering feasibility.

Most components will contribute errors that are uncorrelated. These can be summed by the root sum of squares (rss) method. However, some sources of error are systematically linked and an rss summation is inappropriate. For example, the radius and conic constants of the primary and secondary mirrors are interrelated and should be balanced separately before entering the residual error into the overall wavefront error sum.

The initial error budget serves as a design guide for setting the performance goals for each component in the system. A component may sometimes perform better than the initial allowance or it may be unable to meet that goal. As the performance of each component becomes better understood, the budget is filled in "from the bottom up." It is good practice to document calculations or measurements and note the basis for each allowance in the budget spreadsheet. In this way, the error budget serves as an index to the collected engineering documentation. As initial guesses are replaced by measurements and calculations, the budget serves to predict the overall performance of the system. Periodically, the budget is rebalanced, with allowances being reallocated where they are most needed.

74   3. Design Methods and Project Management

If properly done, error budgeting results in an economy of design that concentrates engineering effort where it is most useful and helps control project costs.

3.2.4 End-to-end computer simulations

As observatories grow larger and more complex, the error budget method described above becomes insufficient because it is empirical and does not offer a deterministic way of finding an optimum. This is because the determination of budget components is done in an educated but "a priori" fashion instead of being directly traced to the physical characteristics and behavior of the subsystem in question. Also, by necessity, the error budget method fails to capture the complexity of environmental or operational conditions, typically addressing only nominal operational modes in average or worst-case conditions.

The availability of high-fidelity computer modeling offers an ideal solution. It is now possible to develop a detailed computer model of the entire system to simulate actual operating conditions and explore wide trade-off spaces. This provides a deterministic solution for the best configuration according to performance and cost criteria. Using traditional optical, mathematical, and finite element methods, individual models are produced for each of the relevant systems, which will typically include:

- the optical system with initial misalignment and figure errors,
- the wavefront control system (wavefront sensing, mirror positioning, figure control, control algorithms),
- the structure (geometry, dynamics, isolation devices),
- the thermal system (radiation, conduction and heaters),
- the pointing control system (motors and encoders for ground telescopes, reaction wheels and star trackers for space telescopes),
- the guiding system (guide star flux, centroiding), and
- the science instrument detectors (pixelization, dark current, readout noise, pixel cross-talk).

These models are then combined into an integrated, system-level model. One approach to creating integrated system-level models, employed during the design concept phase of the NGST program, is based on a set of tools developed at the Jet Propulsion Laboratory (Fig. 3.10). The first of these tools is IMOS, which stands for "integrated modeling of optical systems." IMOS is a collection of functions or subroutines allowing the analyst to combine the requisite subsystem models within the Matlab computing environment.[2] The

---

[2] Matlab is a popular commercial code for general-purpose, matrix-oriented numerical analysis. It also includes a powerful set of tools for control system analysis.

second JPL-developed tool is MACOS, standing for "modeling and control of optical systems." MACOS is an analysis code providing geometric and physical optics capabilities. Additional features making it uniquely useful in telescope design applications include support for segmented and deformable optics and a programming interface that allows other codes to access all of the capabilities.

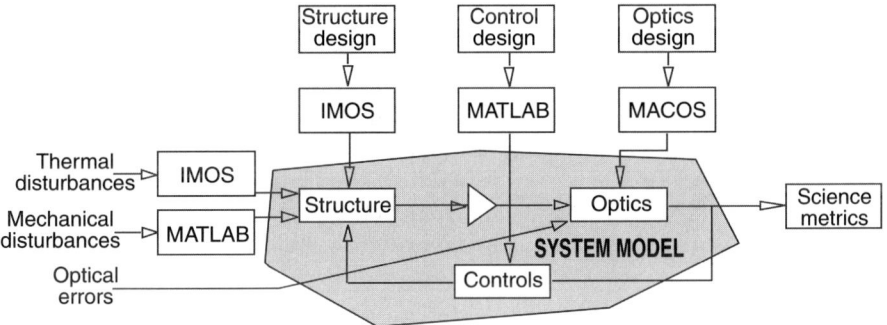

Fig. 3.10. End-to-end modeling for NGST.

Using these tools, analysts can rapidly determine system performance under a variety of operational conditions and carry out parametric design optimization. In addition, these integrated models are powerful tools for the simulation of jitter and wavefront control and for the analysis of specific phenomena such as thermal transients, microdynamic snaps, and stick-slip events.

3.2.5 Design testability and forgiveness

Design testability and forgiveness are two important characteristics which should be considered when comparing various proposed designs. Testability is the ability of a given design to be verified prior to final assembly (or deployment, for a space telescope). This factor is particularly critical for large space telescopes which cannot be thoroughly tested before launch or for very large ground telescopes to be located in remote sites. Given two design choices with the same performance, preference should be given to the one that can be best tested in the shop or before launch.

Design forgiveness is the ability of a given design to recover from design or fabrication errors, unforeseen changes in the environment, or damage or failure during operation. A robust, flexible approach that can work in a broad range of conditions to compensate for potential problems may ultimately be more cost-effective than a highly optimized design requiring extensive testing to ensure its validity.

3.2.6 Scaling laws

During the conceptual phase, it is useful to have at least an elementary understanding of the impact of size on the physical behavior, performance, and cost of a telescope. We will now take a look at some of the basic scaling laws.

Dependence of scientific performance on aperture diameter and wavefront errors

Since the aperture diameter of a telescope is the single parameter that most influences the cost of an observatory, it is important to understand its influence on scientific performance. The aperture diameter affects scientific performance in two ways: sensitivity and spatial resolution.

As we have seen in Chapter 1, sensitivity is best described by the integration time needed to reach a given signal-to-noise ratio. The dependence of integration time and flux limit on the aperture diameter is summarized in Table 3.3.

Table 3.3. Sensitivity dependence on aperture diameter (D)

	Seeing limited		Diffraction limited	
	Unresolved source	Extended source	Unresolved source	Extended source
Source-photon limited				
Exposure time	D^{-2}	D^{-2}	D^{-2}	D^{-2}
Flux limit	D^{-2}	D^{-2}	D^{-2}	D^{-2}
Background limited				
Exposure time	D^{-2}	D^{-2}	D^{-4}	D^{-2}
Flux limit	D^{-1}	D^{-1}	D^{-2}	D^{-1}

In space, or on the ground in the infrared where images are close to being diffraction limited, the gains in background-limited extended source observations due to a larger aperture are the most impressive, varying as D^4. This is because the increase in collected flux combines with a decreased image size that reduces the contribution of the background.

With respect to angular resolution, the theoretical diffraction limit of a telescope varies as λ/D, but, on the ground, image quality is affected by seeing which varies as $\lambda^{-1/5}$ (Chapter 1). Image quality is also affected by wavefront errors in the optics. Figure 3.11 shows how these two factors affect final image quality. As shown on the plot at left, median seeing and diffraction limits become comparable at a wavelength of about 10 μm. This indicates that at wavelengths less than 10 μm, the use of adaptive optics becomes mandatory if one is to benefit from a very large aperture. The Strehl ratio, a measure of image quality (Chapter 4), depends on wavefront error and wavelength, as shown in Fig. 3.11 (right).

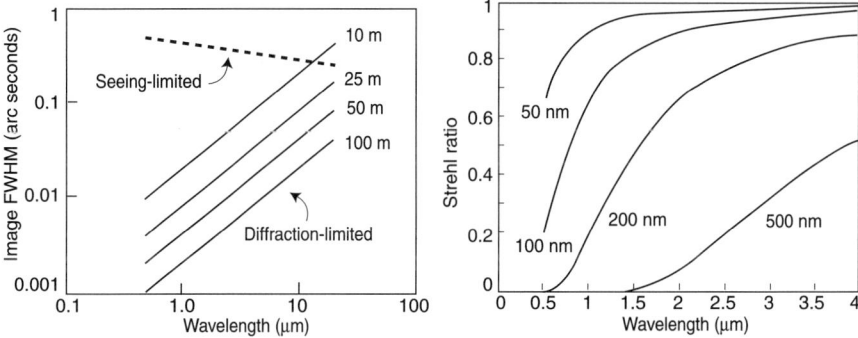

Fig. 3.11. At left, image size versus wavelength for seeing-limited and diffraction-limited conditions. The seeing-limited image assumes a seeing FWHM of 0.5″ at 0.5 μm. The diffraction-limited image is shown for aperture sizes of 10 to 100 meters. At right, the Strehl ratio versus wavelength as a function of wavefront error (assuming no atmospheric turbulence).

Engineering scaling laws

To get a feel for how gravity effects or natural frequency scale with telescope size, it is instructive to model the telescope structure as a simple beam of uniform cross section. Structural deflections have two origins: point loading due to the weight of the supported optics and self-deflection due to the mass of the structure itself.

In the first case, point loading, the maximum deflection is given by

$$\delta_{\text{pl}} = \frac{PL^3}{3EI}, \tag{3.1}$$

where P is the point load, L is the beam length, E is the Young modulus of the beam material, and I is the structural moment of inertia of the beam's cross section.

In the second case, self-weight, maximum deflection is given by

$$\delta_{\text{sw}} = \frac{wL^4}{8EI}, \tag{3.2}$$

where w is the mass of the beam per unit length. Assuming that all three dimensions of the beam grow in the same proportion, w will scale as s^2 and the moment of inertia as s^4, where s is the size scaling ratio. Generally, P will scale as s^2, as in the case of the mass of a meniscus primary mirror or a secondary mirror. As a result, δ_{pl} will scale as s and δ_{sw} as s^2.

Generalizing the above example, the scaling laws for the various factors affecting a telescope are given in Table 3.4 [4].

78   3. Design Methods and Project Management

Table 3.4. Scaling laws

Parameter	Law	Scaling
Length (L)		s
Gravity deflection (δ) due to self weight		s^2
Gravity deflection due to point load		s
Angular change due to self weight	δ/L	s
Angular change due to point load	δ/L	s^0
Mass (m)		s^3
Stiffness (k)	mg/δ	s
Resonant frequency	$\sqrt{\delta}$	s^{-1}
Stress	mg/A	s
Wind gusts – deflection	L^2/k	s^1
Wind gusts – angles	L^2/kl	s^0

A= area (L^2); l = lever arm

3.2.7 Cost models

Until the early 1980s, the cost of traditional telescopes followed a well established power law as a function of the aperture diameter (Fig. 3.12). If the technology used then were still in use today, 8 to 10-meter class telescopes would be unaffordable. But fortunately, a series of technological improvements has made it possible to build larger telescopes at lower cost: the mastering of the aspheric figuring process brought faster primaries (hence shorter telescopes and smaller domes); better testing methods and computer polishing transformed the optical figuring process from a black art to a deterministic science; the use of the stressed mirror or stressed lap techniques drastically reduced polishing time; the use of the alt-az mount instead of the more massive

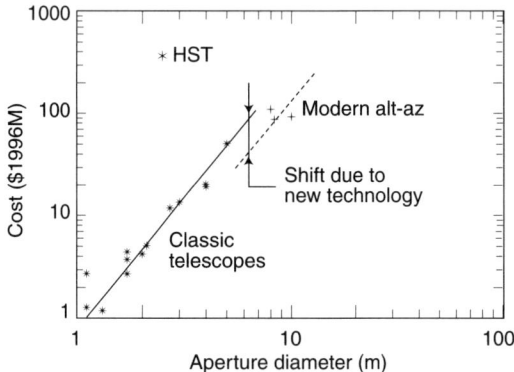

Fig. 3.12. Historical cost data for ground-based observatories. These data show that for observatories of the same type (equatorial mount) and technology, cost grows as the 2.6th power of the diameter. When new technology is used (e.g., faster primary mirror, alt-az mount, active optics), cost drops, but the growth with diameter should follow the same power law. The cost of HST is shown for reference. (From Ref. [5].)

equatorial mount reduced both cost and schedule; and the advent of computer control made active optics possible and led to more efficient pointing systems.

The effect of a new technology is not to lower the exponent of the cost versus diameter curve *power law* as is often believed, but to shift the power law *normalization*. The aperture diameter of a telescope is a fundamental scaling factor which is hard to break. Mirror area, hence cost, scales as $\sim D^2$; telescope tube and mount mass, hence cost, grow somewhere between a two-dimensional (D^2) and a three-dimensional (D^3) relationship to mirror size; enclosure mass and cost, similarly, scale somewhere between the square (area) and the cube (volume) of the enclosure diameter, which is a direct function of the tube length, hence of D. The exact cost/diameter power law coefficient depends on the mix of these cost elements, but does not change much with technology. What new technology does is reduce the cost of building a telescope of a given size. It lowers the constant in front of the power law but not the power law exponent. If one builds a larger telescope with this new technology, cost will still grow with more or less the same exponent.

Based on recent projects, the typical year-2000 costs of 2-, 4-, and 8-meter ground-based alt-az telescopes including enclosure and building are on the order of $5M, $18M, and $80M, respectively.

For space telescopes, one model developed by Technomics [6], based for the most part on military and surveillance missions, is of the form

$$\text{Cost} \propto \frac{D^{1.6} M_f D_f D'_f}{\lambda^{1.8} T^{0.2} e^{0.033(Y-1980)}}, \tag{3.3}$$

where M_f is a factor depending on the material of the optics and telescope structure equal to 1.0 for aluminum, 1.5 for glass and graphite epoxy, and 1.3 (optics) or 1.5 (structure) for beryllium, D_f is a factor depending on optical design (1.0 for on-axis, 1.33 for off-axis), D'_f is the mirror blank design factor (1.0 for solid or 1.3–1.4 for lightweight mirrors), λ is the operational wavelength, T is the operating temperature (in Kelvin), and Y is the completion year, 1980 being a reference year. This last factor takes into account the effect of technological advances on product costs.

One notes that, for space telescopes, the exponent of the scaling with diameter is smaller than for the ground. On the ground, significant costs such as the dome and telescope structure scale as the volume, whereas in space, the bulk of the fabrication costs scale as the aperture area. An additional reason is that the proportion of the cost which is weekly dependent on size (e.g., design, testing) is also larger for space telescopes.

3.2.8 Cost as a design variable

As emphasized at the beginning of this chapter, large observatories cannot be defined simply on the basis of technical and scientific requirements. Cost should also be part of the design optimization. The problem can be approached from three angles:

- maximizing performance within a given budget
- minimizing life cycle cost
- maximizing "cost-effectiveness"

The first case is the classic optimization performed by systems engineers: the science goals and the allocated financial budget are assumed to be given, and an attempt is made to meet those goals as closely as possible within the budget envelope. The second case is similar but gives additional weight to reliability, maintainability, and efficient operation.

In the third case, both the science capability and the total allocated budget are allowed to vary in order to determine the most cost-effective budget amount. For a scientific facility, cost-effectiveness is measured by the amount of science that can be accomplished for a given monetary amount (e.g., "science per dollar"). Cost-effectiveness is determined by assigning a scientific value to a particular set of science capabilities, evaluating the corresponding costs and taking the ratio.

As an example, the diagram at left in Fig. 3.13 shows the ratio of science return to cost for NGST as a function of the spectral coverage. The original requirement for this observatory was only to cover the near infrared (1–5 μm). Analysis showed that extending the coverage to the mid-infrared resulted in a higher science per dollar value. On the other hand, although extending spectral coverage into the blue would increase science productivity overall, the science per dollar value was lower because the science in the blue wavelengths was valued less than that in other bands and the cost was higher because of the more demanding image quality [7].

As a second example, the diagram at the right in Fig. 3.13 shows the notional science per dollar of a telescope as a function of diameter. Depending on the main thrust of the astronomical research and the capabilities of existing facilities, there will generally be a diameter, D_1, for which the value of the scientific discovery potential of a new facility starts to increase dramatically, thus leading to a rise in cost-effectiveness. Beyond a second diameter, D_2, however, the cost (or risk) may increase much faster than the expected value of the science return, leading to a slower gain and eventually to a decline in cost-effectiveness. Clearly, the location of this optimum changes with time as a function of technological progress and science research emphasis, but it is where one wants to be in a given set of circumstances. If there is some flexibility in the allowed budget, such a study helps to pinpoint the optimal budget.

This method of cost optimization is sometimes referred to as "cost as an independent variable" (CAIV). The expression is a bit of a misnomer since cost is, on the contrary, a function of other variables such as performance and schedule. The idea is still valid, however, in that one allows the *allocated budget* to vary in order to study its effects on the science return.

In doing such exercises, it is important not to base the project cost solely on the construction cost. The project's total cost should be considered as the

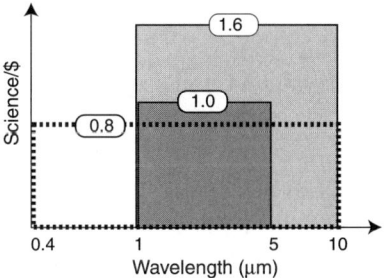 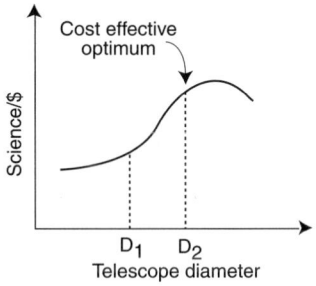

Fig. 3.13. At left, science per dollar of three possible total spectral coverage for NGST: 1 to 5, 1 to 10, and 0.4 to 10 μm, normalized to the 1–5 μm band (see text). At right, hypothetical cost-effectiveness of a telescope as a function of its size (see text).

value of all resources needed to design, carry out the research and development programs, build, and operate the observatory. This is what is referred to as "life cycle cost." Whether certain costs are actually charged to the project or not, a fair evaluation of the resources devoted to it, such as work by the internal staff and the use of testing facilities and computers within the agency, should be included in the optimization.

It may be justifiable for the funding agency to bear the costs of initial technological development on the grounds that the corresponding breakthroughs could benefit other projects. But discounting operation and maintenance is shortsighted, especially for space telescopes, because these costs can be comparable to the cost of constructing the observatory. It is also a misjudgment to discount Phase B design and technology development costs with the excuse, for example, that they are charged to different budget lines in the funding agency. With these costs accounting for up to 25% of the total cost, omitting these in the optimization is misleading. It can result in "overdesigned" systems where more money is spent on design and technology development than is justified, leading to less capability than if that misused money had been invested in construction.

3.2.9 Observatory performance metrics

It is sometimes necessary to evaluate the performance of a given observatory compared to that of other existing facilities. This may be to justify continued funding for the observatory or, on the contrary, to show that it may be more advantageous to replace it. Whereas the design reference program method discussed in Section 3.2.1 permits the comparison of various architectures or choices of main observatory parameters against a specific set of science goals, here one is addressing a higher-level question: "What is the value of the science made possible by this facility compared to that of other existing facilities?"

82   3. Design Methods and Project Management

Telescope performance can be characterized by two kinds of metrics: those reflecting *scientific productivity* and those which quantify *technical performance*, which itself eventually impinges on scientific productivity.

Scientific productivity

Scientific productivity is ultimately measured by the value of the scientific results obtained. Although it is impossible to fully predict the scientific return of a proposed facility, it is instructive to understand how the scientific productivity of *existing facilities* can be measured and the lessons one can draw from this. Scientific productivity can be measured in a number of ways, but the most common one consists of counting published papers or citations to published papers.

Tallies of published papers are easy to make, and several observatories maintain lists of those that are based on data obtained with their telescopes. But such counts give equal weight to papers of different scientific merit, and it is more useful to look selectively at the papers appearing in journals with the highest impact or at the papers with the highest citation rates.

Citation statistics provide a fair measure of the amount of interest generated by papers and are often used to assess the performance of individuals and organizations. Using this criterion, the result of one thorough study of telescope scientific efficiency is shown in Fig. 3.14. Although citation statistics suffer from biases stemming from geographical and language influences, this analysis clearly shows that the scientific productivity of ground-based optical telescopes is proportional to their collecting area.

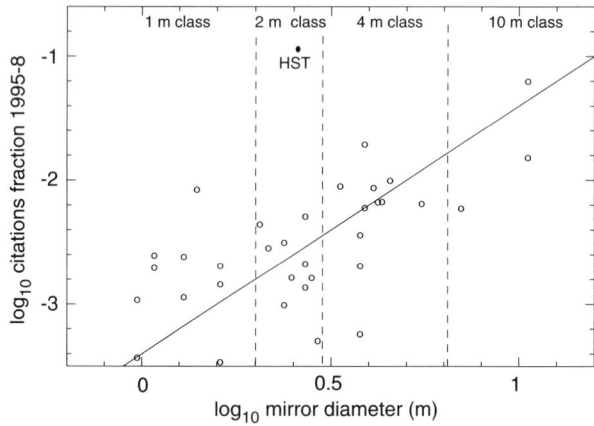

Fig. 3.14. Citations fraction for ground-based optical telescopes during the 1995 to 1998 period versus telescope diameter, D. The straight line corresponds to a D^2 law. HST is also shown for reference. (Data from the Benn and Sanchez study [8].)

Since the capital cost of telescopes is roughly proportional to their collecting areas, the cost-effectiveness of ground-based optical telescopes is largely independent of the diameter. This makes a good case for the small- to medium-sized telescope which is the university workhorse, but in no way implies that large telescopes should not built. Advances in fields especially hungry for photons, such as extragalactic astronomy, are critically dependent on ever larger apertures.

The Hubble Space Telescope generates about 50 times as many citations as does a ground-based telescope of the same aperture, but its cost was 400 times greater. Once again, however, a space telescope can tackle science that a ground-based telescope, however large, can only nibble at. The importance of space observatories is well demonstrated by the pie-plot in Fig. 3.15 which shows that HST and other space observatories supplied more than half of all citations in the 1995–1998 period.

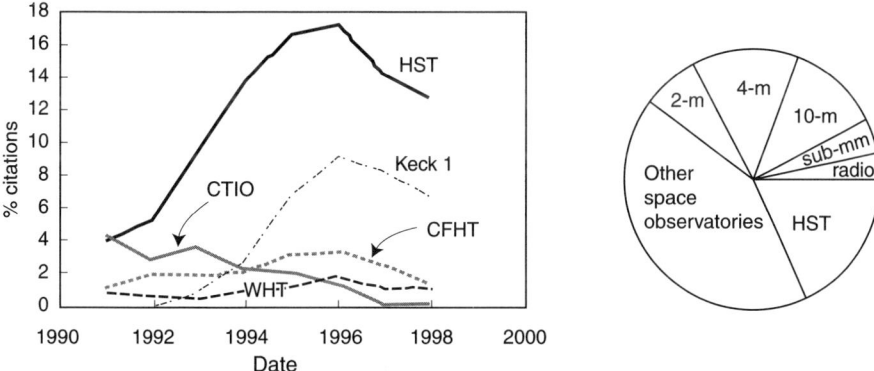

Fig. 3.15. At left, evolution of telescope scientific productivity as a function of time for selected 4-meter telescopes, Keck I and HST. At right, shares of citations for telescopes of different types for the 1995–1998 period. (Both plots are from the Benn and Sanchez study [8].)

Productivity analysis is also useful in assessing the continued scientific relevance of observatories. As an example, the scientific productivity of representative telescopes is shown in Fig. 3.15, left. The steady decline in citation fraction for most 4-meter telescopes during that period reflects the fact that pioneering work had shifted to other observatories, in particular HST, Keck I and II, and Hipparcos. The plot also suggests that HST had passed its peak productivity 6 years after launch, a value which may be typical for "frontier observatories."[3]

A similar study was performed by Leverington [9], this time for all telescopes of the same type combined, and is summarized in Fig. 3.16.

---

[3] The 5 m Hale telescope escaped this fate: it was still having the highest citation numbers thirty years after it saw first light.

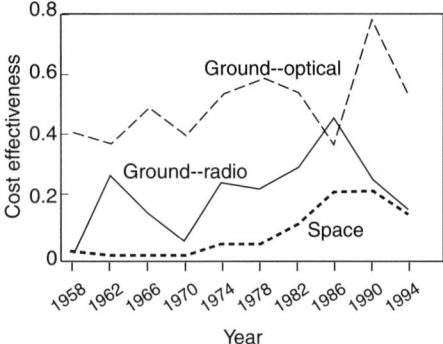

Fig. 3.16. Cost-effectiveness as a function of time for ground-based optical, radio, and space observatories. Cost-effectiveness is defined as the number of highly cited papers per year divided by the total annual costs in millions of 1992 US$ and where the total annual cost is the sum of annual operations costs plus amortized capital costs. The drop in space observatory cost-effectiveness in 1994 is due to the high cost of HST. (Courtesy of D. Leverington.)

Although cost-effectiveness is a useful tool, one must not rely on it exclusively. The starring roles that the Keck telescopes and HST have played in recent years attest to the importance of pushing back the frontiers of knowledge independently of cost.

Technical performance metrics

The scientific productivity of a telescope should, for a given site, instrumentation suite, and user community, be proportional to the amount of observing time available and be inversely proportional to the integration time needed to achieve the signal-to-noise ratio required for typical programs. Technical performance metrics can thus be considered of two kinds: those relating to observing time available (e.g., observing overheads, downtime due to technical problems), and those relating to integration time needed to reach the required signal-to-noise ratio (e.g., mirror reflectivity, seeing, readout noise). Examples of such metrics are given below, together with a brief discussion of how observatory resources can be invested to maximize scientific productivity.

Available observing time

Breakdowns for the use of available time at representative observatories are shown in Table 3.5. Available time is defined as the time between astronomical twilights[4] for ground-based telescopes and total time for space telescopes. In the latter case, it must be noted that space telescopes in low Earth orbit, such as HST, are at a disadvantage since their targets are obstructed by Earth for

---

[4]See glossary.

a large portion of the orbit. In spite of this, HST, which uses some of this dead time to perform calibrations, turns out to be remarkably efficient. If one considers the total time on target per year, HST is about twice as efficient as ground-based observatories, and high-Earth-orbit observatories such as NGST or SIRTF, which do not suffer from Earth occultation, would be about four times as efficient.

Table 3.5. Breakdown of available time in % for representative telescopes

Observatory	Bad weather	Technical problems	Engineering & calibrations	Overhead & occultation	Science
Keck I	16	4	11	26	43
VLT	12	2	5	15	63
Gemini	20	3	10	15	52
CFHT	16	5	11	15	53
WHT	22	4	10	19	45
HST	–	2	5	50	43

Technical downtime is small compared to downtime due to bad weather and observing overhead. This suggests that limited resources may be better invested in reducing readout overheads or improving other aspects of instrument performance, rather than in trying to further reduce technical downtime.

Time set aside for engineering and calibrations includes commissioning for new instruments, sky tests for telescope upgrades, calibration of instruments and telescope pointing system, realuminizing of mirrors, and so forth.

Overhead includes slewing, optical alignment, and phasing for active telescope systems, acquisition of guide stars, blind offsetting, waiting for instrument mechanisms to move, detector readout, planning for the next observation, and so forth. Pointing and blind-offsetting are generally so accurate that lost time is minimal. The largest contribution to overhead is detector readout time. For ground observatories, most of the remaining overhead is due to interactions between the observer and the system (e.g., verifying the field acquired on the slit-viewing TV and assessing data quality). But these interaction times are much reduced when observing is carried out in "service mode" by the observatory staff. Calibration exposures (e.g., of arc lamps, photometric standard stars) can take up to 10% of the observing night for some programs.

Factors affecting integration time

Recall from Chapter 1, Section 1.5, that, for background-limited point-source imaging, the integration time, t, to reach a given signal-to-noise ratio (S/N) is

$$t \sim \frac{(\text{S/N})^2 \sigma^2 \Phi_{\text{bkgd}}}{A\, n_s^2}, \qquad (3.4)$$

where n_s is the number of detected photons from the source per unit collecting area and unit time (the signal S being $S = n_s A t$), A is the collecting area of the telescope, σ is the equivalent angular diameter of the image of a point source, S/N is the desired signal-to-noise ratio, and Φ_{bkgd} is the background in photoelectrons per unit time, unit collecting area, and unit solid angle. This equation shows that an improvement of image quality (seeing, focus, etc.) brings large benefits. For example, improvement in image quality from $1.0''$ to $0.9''$ yields a 20% decrease in integration time (the comparable gain for spectroscopy is only 10%, however, because only one dimension is spatial, the other one being dispersion).

This formula also shows that the main performance factors affecting the integration time needed to reach the required signal-to-noise ratio for a given observation are overall throughput (telescope, instrument, and detector), background brightness (sky and scattered light), image quality and seeing, detector readout noise, and multiplexing gain (detector area, multiobject observing).

The reflectivities of the primary, secondary, and fold mirrors are critical and should be monitored. Inadequate cleaning of a mirror can lead to a loss of reflectivity of several percent per year. Instrument throughput and detector quantum efficiency are more stable, but should be monitored regularly via observations of standard stars.

Image quality is also critical and testifies to the importance of site selection and avoiding "dome seeing" (Chapter 9). Optical aberrations due to misalignment should be kept small (i.e., $< 0.2''$ for ground telescopes) through regular sky tests, so as not to contribute significantly to image size.

Sky brightness is generally low at the best observatory sites, but light pollution should be minimized by baffling scattered light from moonlight and by requesting nearby cities to control their street lighting. Diffuse scattered light within instruments can be due to moonlight leaking through spectrograph casings or past detector mounting rings, to sky light bouncing around inside poorly baffled cameras or scattering off dirty optical surfaces, and to light-emitting diodes. Accurate knowledge of sky brightness as a function of wavelength, ecliptic latitude, zenith distance, and phase of the sunspot cycle permits ready detection of extraneous scattered light in science exposures [10].

3.3 Project management

Over the last 40 years, the tools used in astronomy have multiplied in size and complexity. The optical, mechanical, and control engineering disciplines required in the past have expanded to include computers and software, cryogenics, solid state physics, active and adaptive optics, and atmospheric science. At the same time, facilities have grown far larger and the monetary cost of failure much greater.

The smaller, simpler telescopes and instruments of the past could be conceived and built by a few talented individuals with a mastery of all the disciplines involved. It is understandable, then, that for many past projects, the cost associated with formal project management procedures appeared to outweigh the advantages.

This is no longer true. Large telescope projects now cost in the $100 million to $1 billion range, and professional project management methods must be used to ensure technical success while minimizing the risk of costly overruns and delays.

3.3.1 General principles

Simply stated, the role of project management is to ensure that the project is completed to meet specifications while remaining within budget and on schedule. The techniques employed contain three main elements: *planning, controlling, and managing*. Planning is the process of identifying what needs to be done, in what order this should be executed, identifying the resources needed, and developing a budget. Controlling is the act of measuring performance and suggesting corrective actions. Managing is the communication of what has occurred, what may occur in the future, what will be done in the future, and what cannot be done.

It is important to approach project management with the proper philosophy. Too often, the initial attitude is, "We are innovators and we are building the best telescope ever built, regardless of what it may cost or how long it takes." In such a case, perfection is the real goal. But perfection is unachievable, and trying for it is expensive, time-consuming, and unwarranted. The better paradigm is, "We are building the best telescope possible with the budget and schedule resources we have." It is essential for project managers to convey this objective to the various members of the project and just as essential that it be embraced by all.

The temptation to perfection can also lead to the problem of *using performance as contingency*. This is reflected by setting requirements too high at the beginning of the project in the belief that it can be descoped if one runs into trouble. A variant of this belief is that the project can be "pushed" to achieve the best possible results if the goal is set just out of reach. In both cases, descopes are almost inevitable and result in greater loss of performance than if the design had been properly targeted from the start (Fig. 3.17).

In addition to applying the proper fundamental paradigm, the salient characteristics of a well-managed program are

- a strong management team, this being especially important for projects involving several institutions or countries,
- the early involvement of industry in order to capitalize on already existing technologies and to ensure that realistic costs and schedules are taken into account during the design phase,

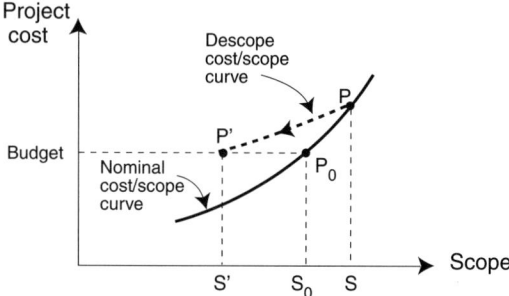

Fig. 3.17. Because of front-end expenses and wasted effort in design and technology development, a descope lowers the scope of the work (from S to S') more than it would have if the project had been properly scoped from the start (S_0).

- the use of fixed price contracts for the well-defined and standard technology elements of the project,
- strong systems engineering,
- timely decision making,
- financial discipline.

In the following subsections, we will examine the major tools and approaches of modern project management.

3.3.2 Project organization

The first step in project organization is to define and staff the project office. For many years, the standard approach was to use a "matrix" organization, with people and work elements being drawn from the functional line organization. Mechanical design work, for example, is assigned directly to a mechanical engineer in the mechanical department. Although matrix organization has been used successfully on numerous projects, it has several weaknesses. The primary one is that each worker on the project reports to both a functional supervisor and to the project manager. Unfortunately, the project manager and the functional supervisor often have different time horizons and conflicting goals, as well as different perspectives. A second weakness is potentially poor communication between project elements when the work is carried out several levels away from the project within the functional organization.

More recently, the concept of "teams," sometimes referred to as "integrated product teams" (IPTs), has been introduced to balance some of the weaknesses of matrix organization. Teams, usually colocated, take on the end-to-end responsibility for a function or product, which minimizes interfaces and improves communications. Such teams are composed of representatives from all of the disciplines required to develop or supervise the development of a given subsystem (product). This promotes communication between disciplines and also between the various disciplines and the project management.

3.3.3 Work breakdown structure

A key task in planning and control is breaking the project down into manageable units. The "work breakdown structure" (WBS) is a tool to assist in formalizing the decomposition of the project as a whole into smaller projects with well-defined tasks. For very large projects, each of these smaller projects may well have its own project manager and project office. At first sight, the WBS appears to be simply a hierarchical coding scheme, but its real value is threefold:

(1) it ensures that all elements of the project, including both products and services, have been identified down to work-package level and that each task has been clearly specified,

(2) it provides the structure to cross-check that all elements have been identified and assigned, and

(3) it provides the basis for cost estimating and detailed tracking of progress and spending.

The WBS should be the first step in the planning and control of any large project. It is the basis on which the project schedule and cost estimates are built. Figure 3.18 illustrates a typical WBS for a telescope project.

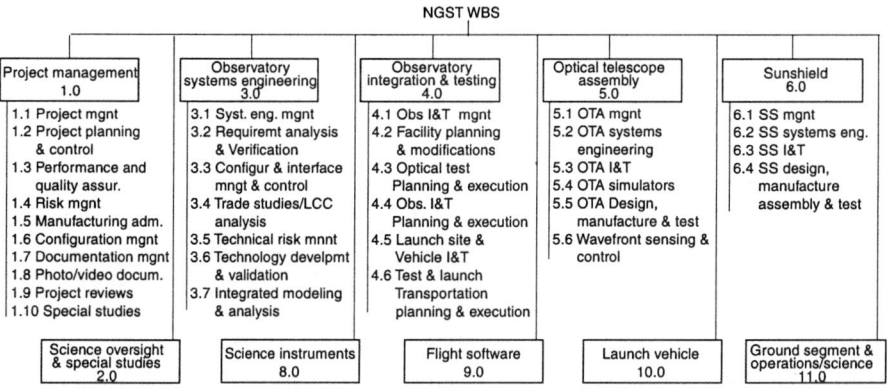

Fig. 3.18. NGST work breakdown structure. Details for the lower row are not shown.

A WBS should be simple and logical. Its value in identifying all of the project elements and tracking their cost and progress is quickly eroded if carried to the extreme. A good rule is that the level of detail of a work package should be understandable and manageable by a single individual, but that breaking it down further is unwarranted. During the study phase, WBSs are usually extended to Level 4, which is the level typically used for basis of estimates and budgets. During construction, it is wise to carry the WBSs down to Level 8 or 9 in order to adequately track budget and progress.

3.3.4 Project data base

During the design and construction phases, but more importantly during the operation phase, it is essential to maintain a widely accessible single repository of all the basic data concerning observatory specifications, critical dimensions, equipment characteristics, environmental conditions, command and control rules, and operational constraints. These data, conveniently stored as a computer data base accessible via traditional networks, are referred to as the "project data base" (PDB). A project data base should be "configuration managed," meaning that high-level approval is needed for changes and that all parties involved in the use of these data are made aware of approved changes. The use of such a controlled data base ensures that all designers work with the same data.

An excellent example is HST's PDB, which has been active for more than 20 years, has served countless users both within NASA and in industry, and is still very much used for flight operations and maintenance missions. It is a collection of simple, fixed-format ASCII files [11]. But not every project data base has been as successful. It is important that formats be well thought out from the start and that changes be implemented in a timely manner.

3.3.5 Procurement strategy

Procurement strategies cover a wide range of models schematically represented in Fig. 3.19. The two extremes of this range are:

- The **total systems authority model**, in which a prime contractor is in charge of the entire project and is entrusted with management and financial responsibility. This is the model sometimes used by the U.S. Department of Defense for major acquisitions. Little technical expertise is required within the project office, but the prime contractor must be given incentives to contain costs and guarantee performance.
- The **project office as prime contractor model**, in which the entire design work is performed by the project office and the work is then divided up into a relatively large number of work packages which are contracted out to separate contractors. This requires a strong project office with technical expertise and well experienced managers, and the availability of in-house integrating and testing facilities. The advantage of this approach is potentially lower cost because of more distributed competition.

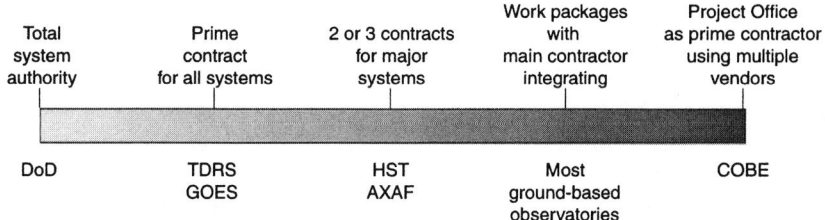

Fig. 3.19. The range of procurement strategies with representative examples of each type shown at the bottom.

Most ground- and space-based observatories are developed using an approach which falls somewhere between these two extremes. The choice depends on agency policies, strength of the observatory engineering staff, and existing competence in industry. The tendency is to break up contracts at the subsystem level for ground observatories, whereas space observatories are designed and integrated by a single (prime) contractor. The difference stems from commercial reasons and government policies. Ground observatories represent a relatively small market and facilities are widely dispersed geographically. Consequently, it does not pay traditional industries to become specialized in this field.[5] The aerospace industry, on the other hand, is concentrated in a handful of companies and is accustomed to "turnkey" productions under the impetus of the aircraft and defense programs. As a result, ground observatory technical expertise lies mostly within universities and government agencies, whereas the main aerospace companies have both the know-how and the equipment required for the design, construction, and testing of space observatories.

However, even in the case of space observatories, there is a crucial role for the observatory (government) team to play which cannot be fully transferred to the contractor. This is because space observatories are one-of-a-kind, and although similar to defense and civilian look-down systems, they contain an important scientific component which, for the most part, is foreign to the aerospace industry.

3.3.6 Technology development

Contrary to mid-sized, general-purpose observatories dedicated to solid but incremental research, very large observatories are often frontier instruments which require a major jump in capability. This typically requires new technology, either because the desired new capabilities are beyond the current state of the art or because they would not be affordable using current technology. But even for state-of-the-art observatories, a minimum of well-directed new

---

[5]This was not always the case. Historically, small to mid-sized professional telescopes were commercially produced by specialized companies such as Grubb-Parsons, Zeiss, and Boller & Chivens. EOST is one of the few companies continuing in this tradition today.

technology will generally pay for itself. This is because it is likely to result in improved performance for the same capital investment or in reduced construction schedule and overall cost. New technology can be classified in two categories: "enabling" and "enhancing."

Enabling technology is a technology that renders the new observatory possible. Without it, the observatory (or mission) would have to be descoped. For example, segmented mirror technology was required for building the Keck 10-meter telescopes, since 8 meters is the practical limit for producing and transporting monolithic mirrors. Similarly, lightweight deployable optics technology is necessary for a space telescope larger than about 4 meters because of current limitations on fairings and launcher mass-to-orbit capabilities.

Enhancing technology is, on the other hand, a new technology that is not strictly required for the realization of the observatory, but will increase its capability. If it is not successful, the capability of the observatory may be partially reduced, but the overall scientific goal of the observatory is not in jeopardy. Examples of enhanced technology may be improved detector quantum efficiency, or the use of cryocoolers instead of cryostats to increase the operational lifetime of space instruments.

Enhancing technologies must always have an "off-ramp," that is to say, a standard technology that can be used should the new technology development program not succeed. Otherwise, the enhancing technology will turn into enabling technology with all the associated performance, cost, and schedule risks. This **off-ramp technology** should not be left dormant, but actively developed in parallel with the enhancing technology so that it can be readily used at any time during the project should the enhancing technology program be abandoned for any reason.

A widely accepted definition of the various technology maturity levels is NASA's **technology readiness levels** or TRL which are defined in Table 3.6 and can readily be modified for ground-based applications.

Table 3.6. NASA technology readiness levels

TRL	Description
1	Basic principles observed and reported
2	Technology concept and/or application formulated
3	Analytical and experimental critical function and/or characteristic proof of concept demonstrated
4	Component and/or breadboard validation in laboratory environment
5	Component and/or breadboard validation in relevant environment
6	System/subsystem model or prototype demonstration in relevant environment
7	System prototype demonstration in space environment
8	Actual flight completed and "flight qualified" through test and demonstration
9	Actual system "flight proven" through successful mission operations

The "invention phase" of TRL 1 and 2 is normally the realm of governmental or university R&D programs, and it would be unwise to make serious proposals for a new observatory relying on technology at that level. The lowest technology readiness level that an observatory proposal can be soundly founded upon is TRL 3. The role of a technology program within the project will then be to bring available technology from TRL 3 to TRL 6. At TRL 6, technology is sufficiently validated to be safely handed over to mainstream project development (i.e., Phases C/D).

One common mistake is to allow Phases C/D to start while key technology development is only at TRL 4. A working laboratory model satisfying the desired performance goals is no guarantee that problems will not be encountered when going to the production mode. The tight meshing of project activities during Phases C/D can be gravely perturbed when an insufficiently validated key technology is found to require extra development time. The higher TRL levels, TRL 5 and TRL 6, which address issues of manufacturability rather than performance, must be worked through before one passes on to Phase C/D. They should not be shortchanged or, worse, bypassed.

Another mistake is throwing the project headlong into a broad technology program on the grounds that new technology cannot be bad. This distracts the engineering and management teams from the real problems and swallows funds that will be sorely missed later, in the legitimate execution of the project. It is important to focus technology on well-specified goals which are directly tracable to the science and systems requirements.

A similar mistake is to "set the bar too high." This unnecessarily increases risk because the chances of not reaching the goal are higher and funds may be spent with nothing to show for them in the end, whereas a less ambitious program would have succeeded at a lower cost and on a shorter schedule.

The underlying message of these last two points is that for an observatory project, technology development should not be an end in itself. The goal is to enable new capabilities or to reduce overall cost, not to be a stage for brilliance.

3.3.7 Reliability

Reliability is the probability that a given system will not fail during its scheduled lifetime. In most systems, the failure rate (i.e., the number of failures per unit time) is higher early on due to burn-in and debugging problems, then falls and stabilizes, and, finally, increases again. This is the classical "bathtub curve" shown in Fig. 3.20. During the useful life of the component, the failure rate, λ, is generally constant, and the mean time between failures (MTBF) is then $1/\lambda$.

Reliability analysis consists of using empirical data to obtain λ for each component, establishing, from the design, the "fault tree" of the system, then applying the rules of probability to determine the reliability of the overall system. If a component's failure rate is constant and failures are independent

94   3. Design Methods and Project Management

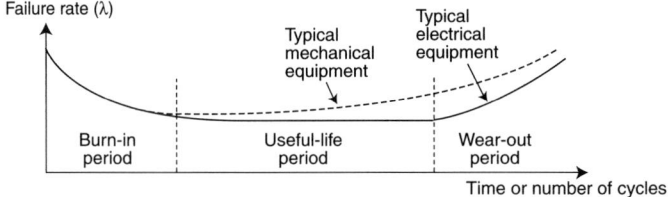

Fig. 3.20. Typical failure rate as a function of time. (From Ref. [12].)

events, the survival probability for a set of these components is equal to

$$\frac{N}{N_0} = e^{-\lambda t}, \tag{3.5}$$

where t is time, N_0 is the number of components one starts with, and N is the number of components that survive. The overall reliability of systems in series with each other is simply obtained by multiplying the individual reliabilities. But complicated systems must be calculated via reliability models [13].

3.3.8 Quality assurance, verification, and validation

Once a telescope component or subsystem has been designed to meet a given performance, fabrication specifications have been derived from it, and fabrication has occured according to these specifications, one would expect the product to meet its designed performance. This process is open loop, however, and each step is subject to the proper implementation of the previous step. Materials used may not meet the expected standards, fabrication processes may be subject to varying environmental conditions, workmanship may not be perfect, human errors may occur, and the design itself may be flawed. "Quality assurance" (QA) and "verification and validation" provide the necessary feedback in the chain.

Quality assurance consists of monitoring the fabrication process to make sure that standards and procedures are followed and to ensure that the delivered parts or subsystems will meet "form, fit, and function" as intended in their design. QA is oriented toward *prevention*. In order to provide objective assessment, quality assurance should be accomplished by project personnel acting independently of program management. That person, the QA engineer, must have access to all phases of the design and fabrication, attend the major reviews, and monitor all acceptance tests.

Verification consists of ensuring that components and subsytems accomplish their purpose, meaning not only that their fabrication was satisfactory but also that the design was adequate. Proof of compliance may be determined by analysis, inspection, or actual testing. Verification is oriented toward *detection* of problems at the part and subsystem level.

Validation consists of proof that the system accomplishes its purpose. In contrast to verification, which is accomplished at the part and subsystem level throughout the entire fabrication process, validation is performed at the system level after completion. Verification and validation are important project elements that will be the topic of Chapter 10.

3.3.9 Interface documents

Once the design has been refined to the point where the overall architecture is stabilized and the main technical "liens" have been resolved, it is efficient to abandon the collegial form of designing and let each engineering group or subcontractor work more independently. This is accomplished by formally dividing the overall system into subsystems, each of these *subsystems* then becoming a *system* from the viewpoint of the engineers involved.

To that end, the system requirements are recast in subsystems requirements and formally expressed in an "interface requirements document" (IRD). This document is normally presented for approval at PDR. As design progresses, requirements evolve into specifications, and the IRD into the "interface control document" (ICD) which is presented for approval at CDR.

The project partitioning should be performed so as to produce clean, simple interfaces. As a rule, overall project costs and difficulty in meeting deadlines increase sharply with the number and complexity of interfaces. Whenever possible, the division should be made for the benefit of the program, with the project's organizational structure and the distribution of contracts being derived from it, not the reverse, as is, unfortunately, often the case.

3.3.10 Configuration management

Once the project has reached maturity, generally some time during Phase B, and the various teams begin working somewhat independently, it becomes essential that the main documents defining the project baseline not be changed without formal approval. This is, *a fortiori*, the case for any portion of the work involving contractors and suppliers. This project document control function is the role of "configuration management." Configuration management consists of the following:

- Identifying which project documents need to be formally controlled. These documents typically include the project's high-level requirements (e.g., levels 1 and 2 and possibly level 3), the interface control documents, the verification procedures and requirements, the engineering standards, all engineering drawings, the optical designs, the software packages, and the project data base.
- Controlling these documents (i.e., approving or disapproving any change that is requested). For very large projects, this is performed by a configuration control board (CCB) composed of the project manager, the systems engineer, and the managers of each organization or contractor affected.

- Verifying, as part of the approval process, that the change will not alter the highest level intent of the project as it propagates through the various subsystems.
- Recording the change in an archival system.
- Disseminating the approved change to all parties concerned.

3.4 Project scheduling

Efficient management of the project requires that each project activity take place in a manner compatible with all other activities from both the time and resources points of view. The simplest scheduling tool is the familiar Gantt (or bar) chart. The activities are laid out against a calendar and are represented by a horizontal bar connecting the start and end dates. Progress is shown by a mark on each bar which, when compared to the current date, clearly show activities ahead or behind schedule.

Gantt charts are excellent representations for readily assessing project status, but they do not show the interdependence of activities. Attempts at doing this graphically quickly become intractable. The best approach consists of using a network logic diagram (Fig. 3.21), where the layout of tasks is based on *dependencies*, not *time*. Although the strong relation with time that the bar chart displays is lost, network diagrams have the advantage of allowing the determination of the minimum time necessary for completion of the project and the identification of critical activities. Network diagrams also lend themselves naturally to computer analysis. A network diagram is established by defining

- the complete set of activities required for the project, these activities typically being directly linked to the WBS structure,
- the most likely duration of each activity,
- the dependencies between activities, i.e., which activities must be completed before others can begin, and
- the project milestones passed upon completion of one or more specific activities.

Schedule networks are then customarily analyzed using relatively simple computer techniques, namely PERT and CPM.[6] Both programs lead to essentially the same results. Their main difference is in the way activity durations are treated. PERT uses a probabilistic estimate of time for completion,

---

[6]PERT stands for "program evaluation and review technique" and CPM for "critical path method." Both programs were developed in the 1950s for the management of large projects: PERT by the U.S. Navy for Polaris missiles and CPM by DuPont for chemical plants.

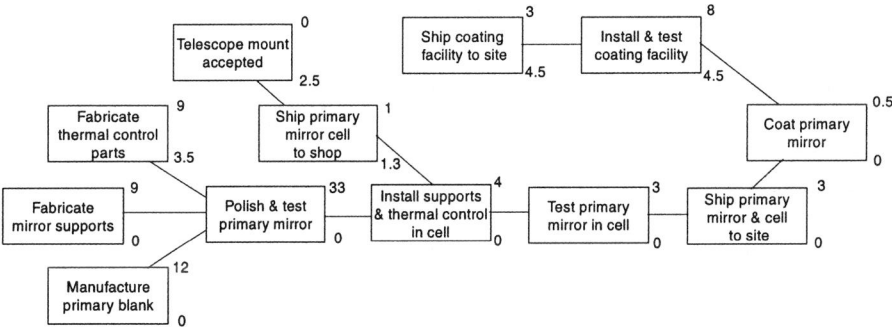

Fig. 3.21. An example of a schedule network (portion of an early Gemini schedule). The numbers at the upper and lower right of each task represent the duration and slack of the task in months, respectively. Tasks with zero slack are on the critical path.

whereas CPM uses the most likely estimate. Another difference is that PERT focuses exclusively on time, whereas CPM emphasizes trade-offs between the cost of the project and its overall completion time. Modern commercial software packages tend to blur the distinction between PERT and CPM and include options for uncertain activity completion times and project completion time/project cost trade-off analysis. There are many such software packages which are variously referred to as network analysis, PERT/CPM, critical path analysis, or project planning. In all cases, the computer analysis supplies

– the earliest time at which each activity can begin,
– the latest time at which each activity can begin without affecting the overall completion time (the difference between the earliest and latest start times being referred to as the *float* or *schedule slack*), and
– the *critical path*, which is the sequence of activities with zero float: any delay in the completion of these critical activities will delay project completion.

Tasks which fall on the critical path should clearly receive special attention, but one should be careful not to limit scrutiny to critical tasks. The critical path will often shift as the project progresses. This typically happens when tasks are completed either behind or ahead of schedule, causing other tasks which may still be on schedule to fall on the new critical path.

There should be enough margin between the critical path completion time found by network analysis and the official project schedule to cope with activity duration underestimates and unforeseen difficulties. In the case of observatory projects, it is good practice to allow a 1-month margin for each year in the schedule.

Network schedules should be updated regularly, on a monthly basis in the early phases of the project and then weekly during the peak construction

98   3. Design Methods and Project Management

period, in order to keep the project manager abreast of any difficulties and delays encountered.

Observatory project network schedules are composed of hundreds of activities and are not easily grasped. The data are usually summarized as Gantt charts (1) for the entire project, the so-called "master schedules" (Fig. 3.22), which are produced for various horizons (e.g., a near-term "90-day schedule," a mid-term "1-year schedule"), and (2) for each main subsystem. Another useful summary is a table of floats for key activities, together with a history of the schedule reserve.

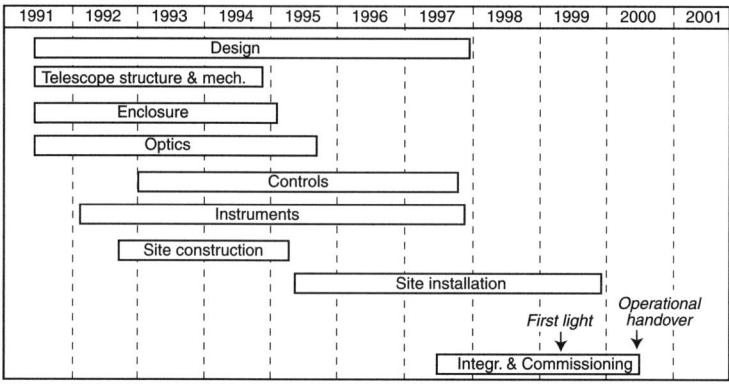

Fig. 3.22. An example of a Gantt chart: the master schedule for the Gemini North telescope.

Project completion time

Observatories are complex, difficult projects with optics which, by nature, take a long time to fabricate. To keep one's optimism within reasonable bounds and as a sanity check, it is useful to compare one's schedule for a new observatory project to past project completion times. As shown in Table 3.7, historical data indicate that a major observatory rarely takes less than 15 years from concept to commissioning.

Table 3.7. Observatory completion time (years)

Observatory	Inception	Commissioning	Duration
5 m Palomar	1927	1945	18
3.6 m CFHT	1968	1979	11
10 m Keck	1977	1993	16
8 m VLT	1983	2001	18
8 m Subaru	1984	2000	16
8 m Gemini N	1986	2000	14
HST	1972	1990	18
Chandra	1976	1998	22

3.5 Risk analysis

Risk represents the potential that something will go wrong in the execution of the program or in the actual performance of the observatory. Programmatic risks can involve funding shortcomings, contractual legal problems, a change in political landscape, failure to meet commitments in international cooperations, schedule delays of various origins, technical difficulties in meeting design requirements, and cost overruns. Performance-related risks include accidental breakage of critical parts, failure to meet performance expectations in actual use, and deployment or component failure in space missions.

A ubiquitous risk in the realization of state-of-art projects such as large observatories stems from the use of new and unproven technologies. Observatories are years in the making and are typically in use for several decades. In order to maximize the scientific benefits of a new observatory and avoid its early obsolescence, one is always tempted to incorporate promising new technologies and to avoid freezing the design until the last minute in the hope of new technological developments. This is a gamble with high potential payoffs but with the associated high risk that the new technology will not be ready in time. In the worst case, this can lead to massive cost overruns and important delays.

Risk management techniques are designed to identify and quantify risks and predict their impact on the success of the project. It is then up to the project manager to judge whether the risk is acceptable or not and, if not, to find a way of mitigating it.

The simplest method for analyzing risks is to model them as the product of the probability that an expectation will fail to materialize by a measure of the consequences of that failure:

$$\text{Risk factor} = (\text{Probability of occurrence}) \times (\text{Consequence})$$

Risks are identified and evaluated via interviews with engineers and managers, then quantified in terms of their probability and the severity of their consequences. It is customary to classify the consequence of risks according to the three major components of a project's success: performance, schedule, and cost. A score is assigned to the probability that each identified risk will occur and to the severity of the consequence in each impact category, using a scale such as 0.1 for low, 0.3 for minor, 0.5 for moderate, 0.7 for significant, and 0.9 for high [14].

The overall score for each identified risk is determined by multiplying the score assigned to the probability of occurrence by the score assigned to the severity of its consequence and then taking the weighted average of each risk category. Table 3.8 shows a simplified and hypothetical example of such a risk assessment for a space observatory.

Risk assessment highlights the areas requiring attention, and a plan should then be constructed to mitigate the identified risks. The highest risks demand

Table 3.8. Risk assessment scoring (hypothetical)

	Probability	Consequence factor			Score
		Performance	Schedule	Cost	
Weight		0.6	0.3	0.1	
Sunshield	0.3	0.9	0.1	0.1	0.17
Fine steering mirror	0.1	0.9	0.1	0.1	0.06
Mirror fabrication	0.5	0.1	0.5	0.7	0.14
Detector development	0.7	0.7	0.3	0.7	0.32

the most attention and are handled via technological development and testing, transfer, or complete avoidance. Risk transfer consists of shifting a risk area from one element to another where it can be controlled in a more effective manner. Avoidance consists of completely eliminating the risk by using a different design. Medium risks are handled by testing or by incorporating a margin sufficient to cope with the problem. Lowest risks are generally handled by providing a safety margin.

This exercise should be repeated on a regular basis during the design phase as the perception of risk gravity evolves and as development and testing in the high-risk areas hopefully demonstrate their acceptability. The goal is to have all major risks "retired" before construction begins.

3.6 Cost estimates and budgeting

3.6.1 Approaches to cost estimating

Cost estimates are of great importance in all phases of the project. They are tools for comparing alternate architectures and for optimizing particular concepts, they are the basis for the budget request at the time the project is sold, and they are one of the project control mechanisms during the construction phase.

With the extensive cost data now available for ground-based telescopes and space systems, there is no excuse for not developing proper estimates for new facilities. Unfortunately, many observatory projects are still undertaken with budgets determined by:

- ascertaining the maximum budget the funding agency will support, then developing a (usually optimistic) back-of-the-envelope estimate within this amount, or
- ascertaining what the funding agency or the community thinks the project should cost, and using that number as the estimate.

Both of these methods may result in "selling" the project, but the probable huge overrun is likely to result in long work-weeks and demoralization for the project team, dissatisfaction from the funding agency, and, in today's funding environment, cancelation of the project after years of hard work.

Making realistic cost estimates calls for a two-way approach: "bottom-up" and "top-down." In a bottom-up estimate, the facility is divided into relatively small elements (e.g., WBS) which are estimated individually. These estimates are then accumulated. In the top-down estimate, one relies on cost data for existing facilities and on scaling laws to obtain the final cost directly. The role of the estimator is to refine and compare these two estimates until a satisfactory match is obtained. During the early phases of the project, most of the weight is accorded to the top-down estimate, whereas more and more weight is given to the bottom-up estimate as the design progresses.

3.6.2 Budgets of main funding agencies

To place the cost of a proposed observatory in perspective, it is useful to compare it to the annual budget of the funding agency. Budgets for the main agencies are shown in Table 3.6.2.

Table 3.9. Annual budgets of selected institutions (circa 2000)

Agency	Total (M$)	Astronomy (M$)
NSF	3 900	120
NASA	13 600	2000
ESA	–	600
ESO	–	60
NOAO	–	27*

* part of the NSF budget.

3.6.3 Cost estimate accuracy

During the conceptual phase of the project (pre-Phase A or feasibility study), the scientific goals of the project and the corresponding requirements are not yet fully defined. Although major parameters such as aperture may be specified, details of the functional requirements are still too amorphous to generate an accurate estimate. At this point, estimates with an accuracy of ±50% are typical, and any budget proposal should include such a reserve.

The next project phase, preliminary design or Phase A, is concerned with developing a better understanding of the detailed requirements. At this point, an experienced estimator should be able to produce a cost estimate to an accuracy of ±25%. In government procurements, this is often done as part of the proposal process. One should be very wary of underestimating the cost of the project in order to "sell" it, hoping that some miraculous source of funds will later materialize to absorb the inevitable future overruns. Accurate estimates are crucial in today's cost-capped environment, as projects are usually forced to live within the original conceptual estimate.

102   3. Design Methods and Project Management

To further refine the cost estimate, the detailed design must be completed. In Phase B of the project, cost accuracy should improve to ±10%.

As a rough guide, representative apportionments of capital cost is shown in Table 3.10.

Table 3.10. Typical distribution of observatory costs (%)

Component	Ground	Space
Optics	23	15
Telescope structure	15	8
Instruments	15	18
Site development and auxiliary buildings	8	-
Enclosure	8	-
Space support system	-	8
Observatory software	3	8
Integration and testing	-	8
Design and R&D	20	25
Management and systems engineering	8	10

3.6.4 Construction of multiple units

Building several units of the same telescope or instrument can lead to substantial savings. For example, the cost of the second Keck 10-meter telescope was about 67% of the total cost of the first one. Similarly, it is estimated that the cost of the four VLT 8-meter telescopes was only about three times that of a single telescope.

Because research and development, design, and tests (i.e., nonrecurring costs) are important contributors to the overall cost of an observatory, most of the savings are in those areas which, for the most part, are unnecessary when building a second or third unit. Fabrication costs also decrease thanks to economy of scale and work familiarization, but less significantly than design costs unless more than a few of the *same* unit are produced. Greater savings in fabrication costs materialize when a large number of the same component, a mirror actuator for example, are produced.

The reduction in cost for multiple units is referred to as the "learning curve." Learning curves generally follow a power law, with the exponent being a function of the reduction in total cost when the number of units produced is doubled:

$$\text{Total cost} = \text{Cost}_{1\text{st unit}} \times N^\alpha, \quad \text{with} \quad \alpha = 1 - \frac{\log_e(1/s)}{\log_e 2}, \quad (3.6)$$

where N is the number of units and s is the learning curve "slope." For example, for $s = 0.95$, the individual cost of two units is 95% of the cost of a single unit, and the total cost will be 1.9 times the cost of building a single unit.

The value of s depends on the type of product being reproduced (electrical, electronics, mechanical, etc.) and on whether the parts are produced in parallel or sequentially. In the absence of previous experience, one can use a value of 0.95 for fewer than 10 units, 0.9 for between 10 and 50 units, and 0.85 for more than 50 units [15].

3.6.5 Budgeting and resource planning

Budgeting consists of predicting the funds required for the project as a function of time (monthly or yearly). This is done by determining the cost of each WBS activity and cumulating them on a monthly or yearly basis according to the network schedule. A notional funding profile for a large ground-based telescope is shown in Fig. 3.23.

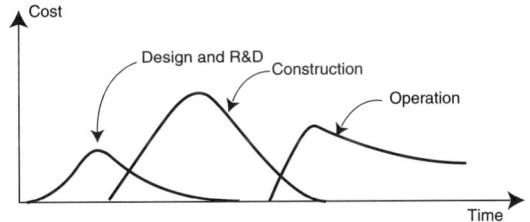

Fig. 3.23. Notional cost profile by program phase.

The same procedure can be applied to *resource planning* in order to determine the profile of resources, that is to say, staff, tools, and equipment. There are usually constraints on the funding and resource profiles which may require adjusting either float activities or the critical path.

An important component of an activity cost is its *fixed cost*, meaning the cost incurred once that activity has been triggered even if no work is done. This may be the cost of keeping a piece of equipment available (e.g., a polishing machine), storage costs for a system waiting to be worked on, or the cost of key personnel that cannot be reassigned to other tasks for fear of losing them.

If a schedule slippage occurs, this fixed cost is incompressible and will lead to an increase in the overall project cost. A sad example of this situation occured when the Hubble Space Telescope had to be placed in storage following the *Challenger* space shuttle accident. Storing the telescope in a prime-usage clean room and keeping a minimal team of technicians and engineers available cost about $1 million per month.

Schedule slippages are particularly critical during the construction phase, when a large number of interdependent activities are coalescing and the fund expenditure rate is at its peak. Any delay or stoppage due to technical difficulties may then be catastrophic (the "marching army" syndrome). As emphasized earlier, it is therefore important that the construction phase not be started until all required technological developments have been demonstrated.

To assess the effect of a slippage, one must divide the budget into fixed and variable costs, extend the fixed-cost profile according to the predicted slippage, then recompute the variable costs as per the new schedule (Fig. 3.24).

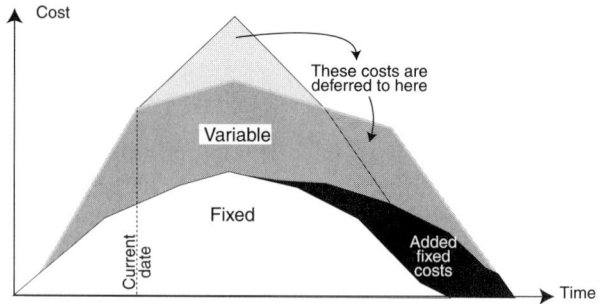

Fig. 3.24. Effect of a schedule slippage on the cost and budget profile. The additional fixed cost increases the overall cost, and the redistribution of variable costs will extend the high spending period. (From Ref. [12].)

References

[1] Presentation from the Office of the Comptroller, reproduced from *NASA Systems Engineering Handbook*, SP-610S, U.S. GPO, 1995, section 3.2.
[2] Stockman, H.S., ed., *The Next Generation Space Telescope — Visiting a Time When Galaxies Were Young*, AURA, p. 149, 1997.
[3] Gemini Project Newsletter, No. 6, p. 6, 1993.
[4] Nelson, J. and Mast, T., Giant optical devices, in Proc. of *Bäckaskog Workshop on Extremely Large Telescopes*, Lund Univ. & ESO, 1999.
[5] Schmidt, T.H. and Rucks, P., Telescope costs and cost reduction, SPIE Proc., Vol. 2871, p. 635, 1997.
[6] Horak, J., *Cost estimating relationship study*, Report TR-9114-01, Technomics, Inc., 1993.
[7] Bely, P., et al, *Implications of the mid-infrared capability for NGST*, NGST Monograph 3, NGST Project Office, Goddard Space Flight Center, 1998.
[8] Benn, C.R. and Sánchez, S.F., Scientific impact of large telescopes, PASP, Vol. 113, p. 385, 2001.
[9] Leverington, D., Optical telescopes – biggest is best?, Nature, Vol. 385, p. 196, 1997, Star-gazing funds should come down to Earth, Nature, Vol. 387, p. 12, 1997.
[10] Benn C. and Ellison, S., *La Palma night-sky brightness*, La Palma Technical Note No. 115, 1998.
[11] Hubble Space Telescope Project, ST ICD-26, NASA Goddard Space Flight Center.
[12] NASA, *Systems Engineering Handbook*, SP-610S, U.S. GPO, 1995, section 4.5.
[13] See for example: http://www.itl.nist.gov/div898/handbook/apr/section1

[14] Kerzner, H., *Project Management, A Systems Approach to Planning, Scheduling and Controlling*, Van Nostrand Reinhold, 1995, p. 877.

[15] Wong, R., Cost modeling, in *Space Mission Analysis and Design*, Larson W.J. and Wertz J.R., eds., Kluwer, 1992, p. 734.

Bibliography

Armstrong, J. E. and Sage, A. P., *Introduction to Systems Engineering*, J. Wiley & Sons, 1995.

Augustine, N.R., *Augustine's Laws*, American Institute of Aeronautics and Astronautics, 1983.

Badiru, A.B. and Pulat, P.S., *Comprehensive Project Management: Integrating Optimization Models, Management Principles, and Computers*, Prenctice-Hall, 1995.

Barter, N.J., ed., *TRW Space Data*, TRW Inc., 1999.

Blanchard, B.S., *Systems Engineering and Analysis*, Prentice-Hall, 1998.

Harwit, M., *Cosmic Discovery*, Basic Books Inc., 1981.

Kerzner, H., *Project Management, A Systems Approach to Planning, Scheduling and Controlling*, Van Nostrand Reinhold, 1995.

Larson W.J. and Wertz J.R., ed., *Space Mission Analysis and Design*, Kluwer Academic Publishers, 1992.

Modell, M.E., *A Professional's Guide to Systems Analysis*, McGraw-Hill, 1996.

NASA, *Systems Engineering Handbook*, SP-610S, U.S. GPO, 1995.

Sarafin, T.P. and Larson, W.J., eds., *Spacecraft Structures and Mechanisms — from Concept to Launch*, Kluwer Academic Publishers, 1998.

4
Telescope Optics

4.1 Optical design fundamentals

Telescope configurations are extraordinarily diverse. We limit ourselves here to the common types of reflecting system used in modern, general-purpose telescopes and to basic notions on image analysis and optical aberrations. An exhaustive coverage of telescope configurations, optical aberrations, image formation and optical design can be found in the pertinent books listed in the bibliography.

4.1.1 Fundamental principles

Most large reflecting telescopes are composed of conic surfaces (paraboloids, ellipsoids, hyperboloids) formed by rotating conic sections about their axes of symmetry. Working directly with reflection laws for conics is thus more informative than dealing with lens formulas. In the realm of optics, the fundamental property of conics is that the normal at a given point on any conic bisects the angle formed by the two radii joining that point to the two foci (Fig. 4.1). This means that all optical rays issuing from a source located at one of the foci will converge at the other focus and thus form a perfect image of the source. This perfect imaging of a point source is called "stigmatism."

The simplest case is that of the parabola, which is a degenerated ellipse with one of its foci at infinity. Rays issuing from this focus at infinity, that is to say, rays parallel to the parabola axis, will, after reflection, converge at the parabola focus.

4.1 Optical design fundamentals

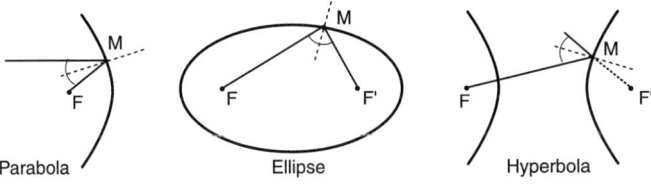

Fig. 4.1. The property of conics that is the foundation of reflecting telescopes: the normal at any point of a conic bisects the two radii issuing from that point (Apollonius theorem).

The next case uses a second conic surface with one of its foci coincident with the focus of the parabola. This second conic surface will reimage the original source at its second focus, again in a perfectly stigmatic way. When this second conic is an ellipsoid, the system is called a "Gregorian," and when it is a hyperboloid, it is a "Cassegrain" (Fig. 4.2). A system with three powered

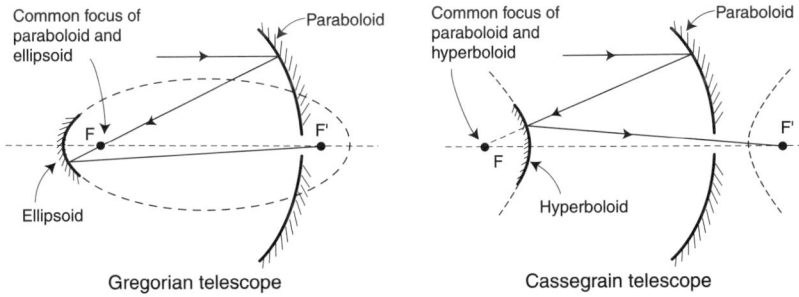

Fig. 4.2. A system of coaxial conic surfaces with coincident foci is "stigmatic."

mirrors follows the same principle. The first and second of a two-mirror system are called the "primary mirror" and the "secondary mirror," respectively. The third powered mirror of a three-mirror system is called the "tertiary," and so on. Fold-flat mirrors are not counted in the sequence.

It can be shown that, to first order, any system composed of several mirrors is equivalent to a single mirror the focal length of which is called "equivalent focal length" or "effective focal length." The focal ratio of the system, N, is defined as

$$f\text{-ratio} = N = \frac{f}{D}, \tag{4.1}$$

where f is the equivalent focal length and D is the diameter of the primary mirror.

The scale of the image at the focal plane is generally expressed in arcseconds on the sky per millimeter or microns on the image. It is proportional to $1/f$ and, using meters for f and micrometers as units on the image, is given by

$$\text{Plate scale} = \frac{0.206}{f} \text{ arcsecond per } \mu\text{m}, \tag{4.2}$$

Although these telescope configurations give (mathematically) perfect images of a source located on axis, the image quality deteriorates with increasing angular distance of the source from the axis. This phenomenon is referred to as "oblique aberration" or "off-axis aberration."

4.1.2 Equations of conic surfaces

The conic surfaces of revolution defining the mirror can be found using the conic vertex equation

$$\rho^2 - 2Rz + (1 - e^2)z^2 = 0, \qquad (4.3)$$

where R is the radius of curvature at the vertex, e is the eccentricity of the conic, and ρ and z are the running surface radius and sag, respectively. Extracting z from this second-degree equation gives

$$z = \frac{R - \sqrt{R^2 - (1+\kappa)\rho^2}}{1+\kappa}, \qquad (4.4)$$

where κ, called the conic constant, is equal to $-e^2$, and defines the conic family according to

$$\begin{array}{ll} \kappa < -1 & \text{hyperboloid,} \\ \kappa = -1 & \text{paraboloid,} \\ -1 < \kappa < 0 & \text{prolate ellipsoid,} \\ \kappa = 0 & \text{sphere,} \\ \kappa > 0 & \text{oblate ellipsoid.} \end{array}$$

Equation 4.4 can be expanded as a Taylor series to give

$$z = \frac{\rho^2}{2R} + (1+\kappa)\frac{\rho^4}{8R^3} + (1+\kappa)^2\frac{\rho^6}{16R^5} + \frac{5}{128}(1+\kappa)^3\frac{\rho^8}{R^7} + \cdots. \qquad (4.5)$$

For $\kappa = -1$, equation 4.5 reduces to the well-known paraboloid expression

$$z_{\text{paraboloid}} = \frac{\rho^2}{2R}. \qquad (4.6)$$

The above formulas apply to axisymmetric conic surfaces. It is sometimes necessary to describe conic surfaces in a coordinate system not aligned with the axis of symmetry (e.g., for an off-axis configuration or for a segmented primary mirror). Analytic expressions in this case are highly complex. The corresponding Taylor expansions have been derived by Nelson and Temple-Raston [1].

4.1.3 Stops and pupils

Two stops limit the ray bundle passing through a telescope: the "aperture" and the "field" stops. The aperture stop limits the rays that enter the telescope, whereas the field stop limits the extent of the image and, hence, the field of view. In general, the aperture stop will be the periphery of the primary or of the secondary mirror and the field stop will be the boundary of the detector (Fig. 4.3). There are cases where intermediate optical components (lenses, mirrors, baffles) in the telescope, or more likely in the instruments, are not large enough to accept all of the oblique rays entering the aperture. As a result, illumination of the image will gradually fade near the edges of the field, an effect referred to as "vignetting."

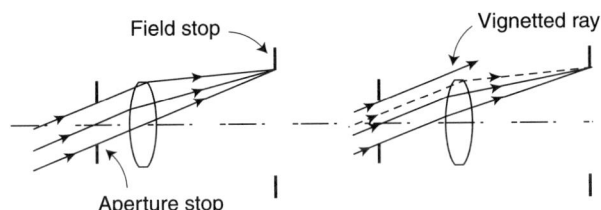

Fig. 4.3. Aperture and field stops (left) and vignetting (right).

The space in front of the first optical element of the telescope is called the "object space," and the space after the last optical element is called the "image space." The image of the aperture stop in object space is called the "entrance pupil," and its image in image space is called the "exit pupil." Intermediate images of the aperture stop are called simply "pupils." A defining characteristic of a pupil is that it contains all the rays that will reach the image, whatever the field angle (Fig. 4.4). The ray bundles corresponding to different field angles do not shift over the pupil as a function of the field angle, and the ray fan comprised of all field angles passes through a minimum diameter there. Pupils are good locations for deformable mirrors, fine steering mirrors, and filters. This minimizes the sizes of these elements and, in the case of filters, guarantees that their characteristics remain unchanged as a function of field angle.

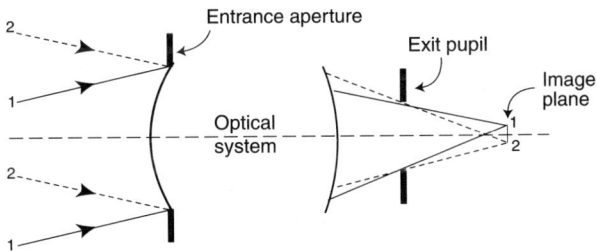

Fig. 4.4. Exit pupil.

4.1.4 Primary aberrations

Generally speaking, aberrations in an image are the result of an optical system failing to produce an exact point-to-point correspondence between the source and its image. There are five primary aberrations: spherical aberration, coma, astigmatism, field curvature, and distortion (Fig. 4.5). The last four are generally off-axis aberrations, whereas spherical aberration affects image quality even on axis. Another type of aberration, chromatic aberration, is not present in reflecting systems since the law of reflection, unlike that of refraction, is independent of wavelength.

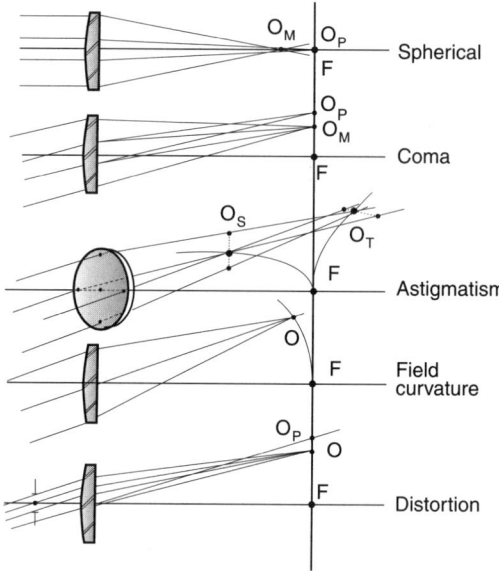

Fig. 4.5. Principal aberrations (the subscripts P, M, S, and T stand for paraxial, marginal, sagittal and tangential, respectively – see text).

Spherical aberration is due to the fact that rays issuing from a source at infinity on axis do not all converge at the same point. The focus of the marginal rays (those at the periphery) is different from the focus of the paraxial rays (those close to the axis). "Refocusing" (e.g., by displacing the detector in the image space) will not help: there is no single focus. This effect is independent of the field angle and inversely proportional to N^3. The term "spherical aberration" is employed because it is the aberration exhibited by a spherical mirror imaging a source at infinity. This aberration is eliminated by using a paraboloid. In two-mirror systems, spherical aberration is due to the fact that the mirrors do not have matching conic constants. This is the condition that affected the Hubble Space Telescope upon launch, and most existing large ground-based telescopes display the same problem to some degree [2]. This aberration is difficult to avoid completely in large telescopes because the primary and secondary mirrors are seldom tested as a system in the optical shop, as the collimators required to simulate a source at infinity are costly and difficult to make. The two mirrors are therefore generally tested individually against some reference, with the result that test errors can lead to an imperfect match.

Coma is due to the fact that rays issuing from an off-axis source do not converge at the same point in the focal plane. This creates a blur which resembles a comet, hence the name. It is the dominant aberration in classical Cassegrain systems used off-axis. In this case, the coma aberration is field dependent and increases linearly with the off axis angle. Fast (small f-ratio) mirrors are much more affected because the effect scales as the inverse of the square of the f-ratio. Coma can also appear when the secondary mirror axis is not exactly coaxial with the primary mirror axis. In such a case, this additional coma is "field independent": it has the same amplitude throughout the field.

Astigmatism originates from the fact that the focus of rays in the plane containing the axis of the system and an off-axis source (the tangential plane) is different from the focus of rays in the perpendicular plane (sagittal plane). Astigmatism scales as the square of the field angle and is inversely proportional to the f-ratio.

Field curvature occurs when the image does not form on a "plane," but on a curved surface. In the absence of astigmatism, the image would be formed on a curved surface called the "Petzval surface," with a sag that scales like astigmatism.

Distortion originates in the fact that the plate scale (scale in the image plane) is not perfectly constant but varies both with the field angle and the direction. In general, the effect is not of great importance because it can be calibrated out (i.e., measured and removed from the actual two-dimensional data). Distortion scales as the cube of the field angle.

The scaling laws of these aberrations are summarized in Table 4.1, where θ is the field angle and N is the focal ratio of the overall system.

Table 4.1. Primary aberrations scaling laws

Spherical	N^{-3}
Coma	θN^{-2}
Astigmatism	$\theta^2 N^{-1}$
Field curvature (sag)	$\theta^2 N^{-1}$
Distortion	θ^3

4.1.5 Wavefront errors

Another way of describing the aberration in an image is with the notion of "wavefront error." A light beam can be thought of as a wavefront propagating through the optical system. Before entering the system, the wavefront of a source at infinity is flat and perpendicular to the direction of propagation. When it exits the optical system to form the image, the ideal wavefront is spherical, with its center located at the image so that the optical path of all rays are of the same length (Fig. 4.6). If it is not exactly spherical, the image will not be perfect. The deviation of the wavefront from a plane or sphere is called "wavefront error."

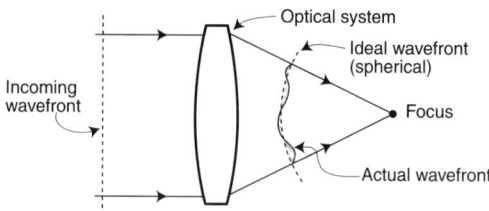

Fig. 4.6. A wavefront propagating through an optical system. The wavefront error is the deviation of the outcoming wavefront from a sphere.

The magnitude of the error is generally measured as the root mean square (rms) of the deviation over the entire surface of the wavefront and is expressed in nanometers or as a fraction of the wavelength or "wave." Each of the principal aberrations described above has its characteristic wavefront error (Fig. 4.7). Traditionally, a system is considered nearly perfect if the rms wavefront error of the exit beam is less than $\lambda/14$. This is discussed in more detail in Section 4.1.7.

A common analytic representation of the wavefront is by means of a set of orthogonal polynomials called the Zernike polynomials [3, 4]. Using polar coordinates where r is the normalized radius (the ratio of the running radius on the wavefront to the outer radius) and φ is the polar angle, any wavefront $W(r, \varphi)$ can be expressed as a linear combination of the Zernike polynomials:

$$W(r, \varphi) = \sum_n a_n Z_n(r, \varphi), \qquad (4.7)$$

where $Z_n(r, \varphi)$ is the nth Zernike polynomial and a_n is the coefficient of the wavefront error component corresponding to that term. The values of the Zernike polynomials depend on the shape of the aperture. They are shown up to the fifth degree in Table 4.2 for the case of a clear circular pupil. The Zernike terms for a circular aperture with a significant circular central obstruction have been derived by Mahajan [5].

The higher the Zernike term, the higher the spatial frequency and, usually, the lower the amplitude of the corresponding wavefront error. As a rule, 20 terms are amply sufficient to describe wavefront errors due to misalignment, mechanically or thermally induced deformations, and to figuring errors in the optics. The advantage of the Zernike polynomials is that they are easily related to the classical aberrations and can be fitted to any measured wavefront by

4.1 Optical design fundamentals 113

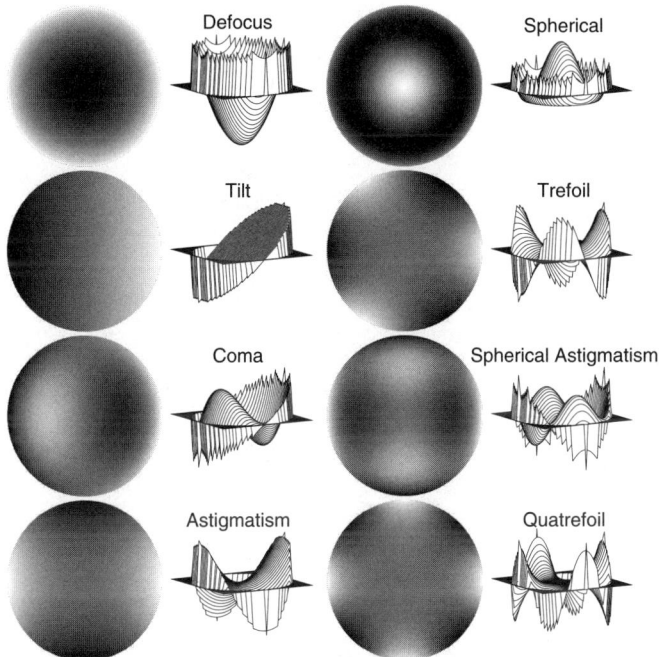

Fig. 4.7. Wavefront errors corresponding to the main aberrations.

best least square fitting. For an array of values of (r, φ), equation 4.7 may be written symbolically in matrix form as

$$\mathbf{W} = \mathbf{Z}\mathbf{a}, \qquad (4.8)$$

where $\mathbf{W}$ is the wavefront error for the array of (r, φ) values. Equation 4.8 may then be solved by least square fitting to obtain the Zernike polynomial coefficients according to

$$\mathbf{a} = (\mathbf{Z}^\mathbf{T}\mathbf{Z})^{-1}\mathbf{Z}^\mathbf{T}\mathbf{W}. \qquad (4.9)$$

The Zernike polynomial representation of wavefront errors is convenient for describing traditional aberrations and high spatial frequency defects in the mirror surfaces. It can also be used to describe atmospheric wavefront distortion [6], but is impractical for describing very high spatial frequency effects such as air turbulence or microroughness in mirror surfaces. The number of terms required would be too large. These effects are better treated in a statistical fashion, as indicated in Section 4.1.7.

During actual use of the telescope, the wavefront error will vary with time due to various effects. In general, higher terms are much more stable than those of low spatial frequency. The lowest terms such as tip-tilt (guiding) and focus vary quite easily because of overall temperature changes and gravity effects. The next term, coma, is introduced by optical misalignment or mirror

Table 4.2. First 22 Zernike polynomials for an unobscured circular pupil

Aberration	Term	Value
Piston	Z_1	1
x-Tilt	Z_2	$2r\cos\varphi$
y-Tilt	Z_3	$2r\sin\varphi$
Focus	Z_4	$\sqrt{3}(2r^2 - 1)$
0° Astigmatism	Z_5	$\sqrt{6}\, r^2 \cos 2\varphi$
45° Astigmatism	Z_6	$\sqrt{6}\, r^2 \sin 2\varphi$
x-Coma	Z_7	$\sqrt{8}(3r^3 - 2r)\cos\varphi$
y-Coma	Z_8	$\sqrt{8}(3r^3 - 2r)\sin\varphi$
x-Trefoil	Z_9	$\sqrt{8}\, r^3 \cos 3\varphi$
y-Trefoil	Z_{10}	$\sqrt{8}\, r^3 \sin 3\varphi$
Third order spherical	Z_{11}	$\sqrt{5}(6r^4 - 6r^2 + 1)$
Sphere astigmatism	Z_{12}	$\sqrt{10}(4r^4 - 3r^2)\cos 2\varphi$
Sphere astigmatism	Z_{13}	$\sqrt{10}(4r^4 - 3r^2)\sin 2\varphi$
Quatrefoil	Z_{14}	$\sqrt{10}\, r^4 \cos 4\varphi$
Quatrefoil	Z_{15}	$\sqrt{10}\, r^4 \sin 4\varphi$
	Z_{16}	$\sqrt{12}(10r^5 - 12r^3 + 3r)\cos\varphi$
	Z_{17}	$\sqrt{12}(10r^5 - 12r^3 + 3r)\sin\varphi$
	Z_{18}	$\sqrt{12}(5r^5 - 4r^3)\cos 3\varphi$
	Z_{19}	$\sqrt{12}(5r^5 - 4r^3)\sin 3\varphi$
	Z_{20}	$\sqrt{12}\, r^5 \cos 5\varphi$
	Z_{21}	$\sqrt{12}\, r^5 \sin 5\varphi$
Fifth order spherical	Z_{22}	$\sqrt{7}(20r^5 - 30r^4 + 12r^2 - 1)$

thermal deformation. Astigmatism is also usually caused by the deformation of mirrors resulting from improper support or thermal effects. Third-order spherical aberration can be introduced by the axial displacement of the secondary mirror. Higher terms are generally "built into" the optics during figuring and, short of adaptive optics devices, no action can be taken to correct them once the optics have been installed in the telescope.

4.1.6 Diffraction effects

Thus far, we have been dealing with geometrical optics. However, because of the wave nature of light, even a perfect optical system will not image a point source as a true point, but rather as a bright spot surrounded by faint concentric rings. This is caused by "diffraction," which is a spreading of light as it passes the edge of an opaque body. In a telescope, this happens at the edge of the aperture and at any obstacle within the aperture such as the secondary mirror and its supporting vanes. As illustrated in Fig. 4.8, the image of a point source at infinity produced by a perfect system is a bright disk surrounded by alternating dark and bright rings, the intensity of which falls as the inverse cube of the distance from the center of the image. A linear obstruction such as that of a secondary support vane will produce a spike in the image of a point source in a direction perpendicular to the vane and with a total intensity that is proportional to the area of the obstruction relative to the aperture area. A polygonal aperture will produce an image with a structured core and spikes.

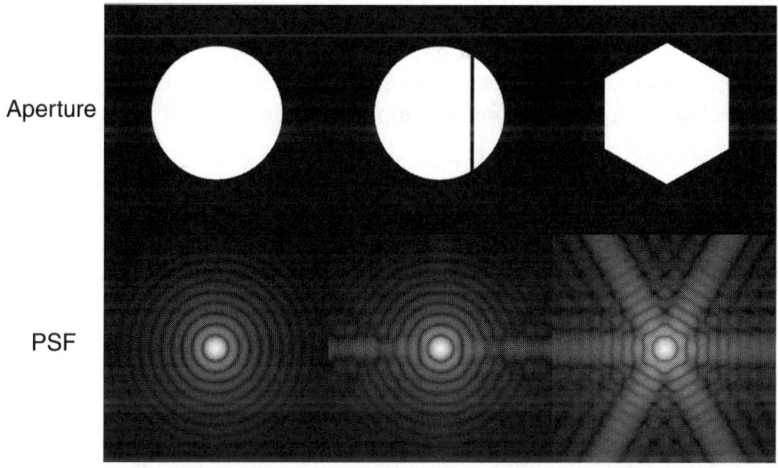

Fig. 4.8. Image of a point source at infinity formed by an optical system without aberration but with different aperture shapes: at left, circular, at the middle, circular with a bar (e.g., a secondary mirror support vane), and at right, a hexagonal shape. In the case of the vane (middle image), note that the diffraction pattern is independent of the position of the vane in the aperture (i.e., centered or not) and that the spike is perpendicular to the direction of the vane.

In addition to the diffraction effects created by the aperture, two other factors contribute to the diffractive degradation of the geometric image:

- **Dust** on the mirror surfaces scatters incoming light and creates a faint halo around the image. Dust particles are typically several tens of microns across and hence lead to very wide-angle diffractive scattering (on the order of degrees).
- **Mirror surface defects** of high spatial frequencies such as those due to roughness of the optical surface or a periodic pattern on the surface related to the internal structure of the mirrors. These defects also create a faint halo with an angular size determined by the spatial frequency of the defects.

4.1.7 Image formation

The best way to characterize the imaging capability of an optical system is by examining the image it forms of a point source. This is relevant for astronomical instruments since, for all practical purposes, stars are point sources. As for extended objects, they can be considered to the first order as a collection

116   4. Telescope Optics

of point sources. The distribution of light intensity in the image of a point source is called the "point spread function" (PSF).

Treating light as a wave allows for a complete mathematical description of the PSF. As discussed in the previous subsections, the distribution of light intensity in the image of a point source is a function of the shape of the aperture and obstructions, geometrical aberrations, and diffraction effects due to dust and defects on the optics surface. These factors can be incorporated into a single concept, the "complex pupil function" [7] defined as:

$$P(r,\varphi)\, e^{i\,k\,W(r,\varphi)}, \tag{4.10}$$

where $k = 2\pi/\lambda$ is the wave number for light with wavelength λ, $W(r,\varphi)$ is the wavefront error, and $P(r,\varphi)$ is the transmittance of the aperture, which is equal to 1 for the parts of the aperture that are completely unobscured and to 0 for obscured regions.

A theorem of capital importance shows that the PSF of the image of a point source at infinity is proportional to the two-dimensional Fourier transform of the complex pupil function.[1] This remarkably simple relationship is the foundation of modern image-formation analysis. The calculation can be done in closed form for the simplest cases only, but the advent of computers and of an extremely powerful algorithm to calculate Fourier transforms, called the "fast fourier transform," or FFT, now makes it trivial even for complex systems [8].

PSF of a system with perfect optics

In the simple case of a perfect optical system (no aberrations, perfect surfaces, no dust) with an unobstructed circular aperture and monochromatic light, the PSF can be determined analytically and is given by

$$I = C\left(\frac{J_1(x)}{x}\right)^2, \tag{4.11}$$

where I is the intensity of light in the image, C is a constant, J_1 is the first-order Bessel function, and $x = \pi D \theta/\lambda$, with D being the diameter of the aperture, θ the angular coordinate of the image spread, and λ the wavelength. This is the well-known "Airy function" named after George Airy who first derived this mathematical formulation. The shape and main characteristics of the Airy pattern are shown in Fig. 4.9, and the energy concentration in the image is plotted on the left in Fig. 4.10.

In a traditional Cassegrain telescope, the aperture is obstructed at the center by the secondary mirror. The point spread function of the resulting annulus is

$$I = C\left(\frac{J_1(x)}{x} - \epsilon^2 \frac{J_1(\epsilon x)}{\epsilon x}\right)^2, \tag{4.12}$$

---

[1] See bibliography, this chapter, under "Fourier optics."

4.1 Optical design fundamentals   117

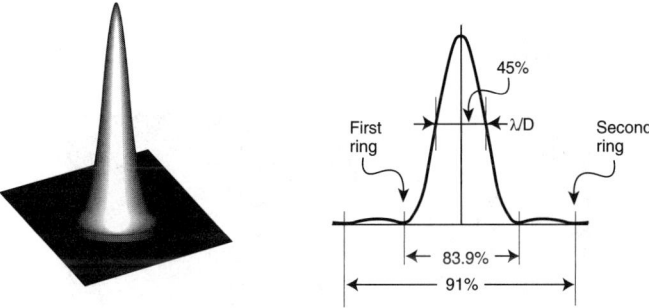

Fig. 4.9. The Airy pattern, the PSF given by a perfect telescope with a circular aperture, is shown at left. The cross section is shown on the right with the indication of the energy fraction contained within the first and second rings and within a circle with a diameter equal to λ/D, D being the diameter of the aperture and λ the wavelength.

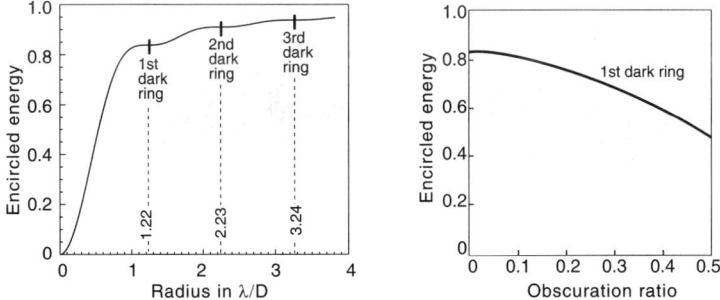

Fig. 4.10. Encircled energy for a perfect system with a circular aperture. At left, the fraction of energy contained within a given angular radius expressed in λ/D radians. At right, encircled energy within the first dark ring as a function of the linear obstruction ratio, ϵ. (Figure at right adapted from Ref. [9]).

where ϵ is the obscuration ratio of the telescope expressed as the ratio of the diameter of the central obstruction to that of the aperture. The effect can be significant when this ratio is much larger than 10% or 15%, as shown on the right in Fig. 4.10.

PSF of actual systems

For actual telescope systems with aberrated optics and where the shape of the aperture and other diffraction effects can be fairly complex, determination of the PSF is done via the Fourier transform method [8]. This is a simple numerical calculation once the aperture shape and wavefront errors map are known, but it is also available as part of commercial optical design programs as well as in software dedicated to the analysis of telescope imaging such as MACOS[2] or TIM and Tiny TIM.[3]

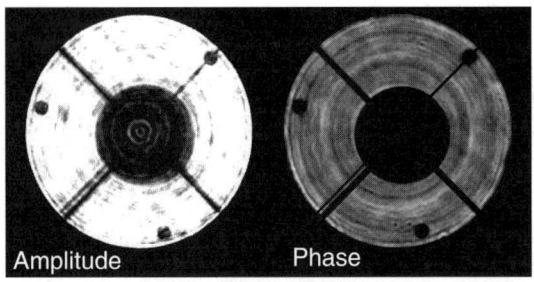

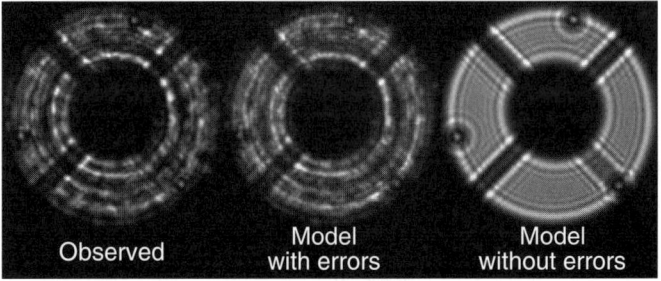

Fig. 4.11. HST pupil function and PSF. At the top, pupil amplitude and phase maps taken with the wide-field camera (WFPC2). Clearly visible are the shadows of the secondary mirror support vanes and three primary mirror support pads, as well as the WFPC2 secondary mirror obscuration and support vanes, which are offset from those of HST. At bottom left, defocused PSF in the ultraviolet ($\lambda = 170$ nm) taken with WFPC2. In the middle is a corresponding model PSF that included the HST pupil phase error map shown on top. At right is a model without the error map.

In addition to low-order aberrations, the scattering due to dust and surface microroughness and seeing (in the case of ground-based observations) can be modeled with excellent fidelity. Aberrations are obtained by hand calculation or a ray-trace program and are represented as Zernike polynomials. As for the effects of atmospheric seeing, mirror quilting, and microroughness and dust, they can be modeled in various statistical ways, which have been studied by Hasan and Burrows [11]. As an example of PSF modeling, Fig. 4.11 shows the pupil function and the modeled and actual PSF of HST. Figure 4.12

---

[2] Modeling and Analysis for Controlled Optical Systems, a software package available from JPL [10].

[3] TIM and Tiny TIM are telescope image modeling software developed by Burrows, Hasan, and Krist; Tiny TIM is available from the Space Telescope Science Institute at http://www.stsci.edu/software/tinytim.

4.1 Optical design fundamentals   119

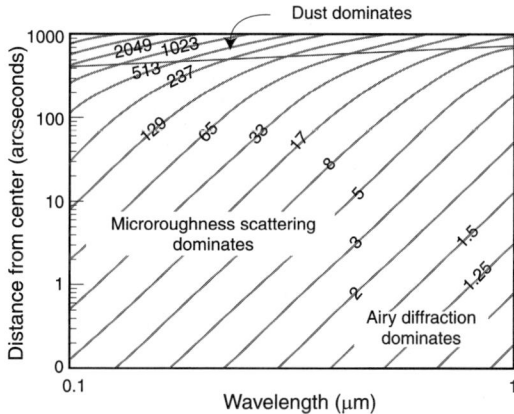

Fig. 4.12. Contour plots showing the variation of the ratio of the HST PSF to the diffraction-limited PSF as a function of wavelength and distance from the center of the PSF. Microroughness scattering dominates at shorter wavelengths for all angles, whereas, for longer wavelengths, it is important only in the wings of the PSF. Dust, which on the HST mirror is at the 1 to 2% level, increases the overall intensity in the extreme wings beyond about 400 arcseconds. (From Ref. [11].)

shows an example, also from HST, of the respective influences of aberration, microroughness, and dust.

Diffraction-limited system

No optical system is perfect, but as far as the effect of aberration is concerned, Rayleigh noted that image quality is not noticeably affected as long as the wavefront in the image space remains between two concentric spheres separated by 1/4 of the wavelength (refer to Fig. 4.6). This is called "the Rayleigh quarter wavelength rule." The problem with this rule is that the effect of the wavefront error on the PSF varies significantly according to the type of aberration involved. For example, the peak intensity of the PSF drops to 0.73 of that of the perfect PSF for third-order spherical aberration but to only 0.87 for third-order coma. Maréchal has thus proposed using a criterion directly linked to the PSF instead of to the wavefront peak to peak error. He considered as essentially perfect a system where the normalized peak intensity of the image is equal to 0.8 of that of the perfect image [12]. This is referred to as "the rule of Maréchal."

The ratio between the normalized peak intensity of the actual PSF to that of the perfect image is called the "Strehl ratio" [13]. An approximate value of the Strehl ratio is given by

$$\text{Strehl ratio} = 1 - \left(\frac{2\pi}{\lambda}\right)^2 (\Delta\Phi)^2, \qquad (4.13)$$

where $\Delta\Phi$ is the rms wavefront error expressed in wavelengths [3] (this approximation is valid only for Strehl ratios greater than 0.5). Requiring that the Strehl ratio be larger than 0.8 is then equivalent to the condition that the rms wavefront error be less than $\lambda/14$. An optical system which satisfies this widely accepted condition is regarded as quasi-perfect and is referred to as "diffraction limited."

Angular resolution

The ability of an optical system to distinguish details in the image it produces is called angular resolution. For a telescope, angular resolution can be quantified as the smallest angle between two point sources for which separate recognizable images are produced. For a diffraction-limited system, Rayleigh has proposed that this angle be defined as that for which the central peak of one image falls upon the first minimum of the other. With this condition, which is called "the Rayleigh criterion" (no relation to the 1/4 wavelength rule), the angular resolution $\Delta\theta$ in radians is defined as

$$\Delta\theta = 1.22 \frac{\lambda}{D}, \qquad (4.14)$$

where D is the diameter of the aperture and λ is the wavelength. For a perfect optical system and sources of equal brightness, this results in a dip of about 27% between the two peaks (Fig. 4.13).

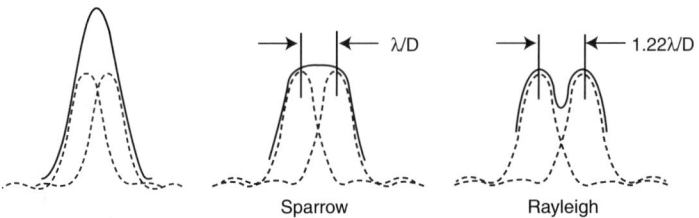

Fig. 4.13. Light distribution in the image of two point sources of equal magnitude. The dashed line represents the PSF of each source and the solid line represents the combined pattern. When the two sources are very close, as shown on the left, the resulting PSF resembles that of a single source. As the separation increases, the peak of the combined pattern first flattens (center), then exhibits a noticeable dip (right). The Sparrow and Rayleigh criteria correspond to a separation of λ/D and $1.22\lambda/D$, respectively.

Another widely used resolution criterion is the "Sparrow criterion" in which the resolution limit is defined as the angular separation when the combined pattern of the two sources has no minimum between the two centers. This occurs when

$$\Delta\theta = \frac{\lambda}{D}. \qquad (4.15)$$

In general, formulas 4.14 and 4.15 do not hold for ground-based telescopes because atmospheric turbulence degrades the actual resolving power below the theoretical value.

Depth of focus

The depth of focus is the tolerance on the axial position of the detector relative to the best optical focus. In the case of diffraction-limited optics, the light beam near the focus has a tunnel shape due to diffraction effects, as shown in Fig. 4.14, and the wavefront error at the distance Δz from the geometrical focus is given to the first order by $\Delta z/8N^2$, where N is the focal ratio at the corresponding focus. The depth of focus is generally defined using the Rayleigh $\lambda/4$ rule and is then given by [14]

$$\text{Depth of focus} = \pm 2\lambda N^2. \tag{4.16}$$

The above formula is only valid for diffraction-limited optics. For seeing-limited optics, the depth of focus is usually defined as the focus range within which the signal does not degrade by more than 2%.

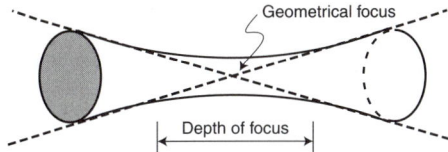

Fig. 4.14. Shape of the light beam near the focus of a diffraction-limited system. The depth of focus is the region where the rms wavefront error due to defocus is negligible compared to diffraction effects.

4.2 Telescope optical configurations

4.2.1 Single-mirror systems

A telescope with a single mirror is possible but has a very limited field. In such a system, the dominant aberration is coma and the angular length of the comatic image on the sky (in radians) is given by

$$\text{Coma} = \frac{3}{16}\frac{\theta}{N_1^2}, \tag{4.17}$$

where θ is the semifield angle (angle from the optical axis) in radians and N_1 is the mirror focal ratio ($N_1 = f_1/D$). On the ground, because of seeing, comatic images up to 0.5″ in length may be considered acceptable, so that the practical semifield angle expressed in arcminutes will be

$$\theta = 0.044 N_1^2. \tag{4.18}$$

Only very slow (long focal length) mirrors will thus have an acceptable field: a semifield of 1 arcminute would require a mirror slower than $f/5$; a more compact $f/2$ would have a semifield of only 10 arcseconds. The situation can be dramatically improved, however, by the use of relatively small refractive optics, called "correctors," located near the focal plane. A number of designs are available and produce excellent correction over a field as large as a degree in diameter [15]. These correctors can also be used to reduce or increase the focal ratio in order to match the plate scale to the detector's pixel size.

The advantage of a prime focus is its "purity": only one surface is used, maximizing throughput and eliminating image degradation due to optics misalignment. Its main disadvantage for ground-based telescopes is mass and size limitations for the focal instruments and difficulty of access. In order to service it, one has to use cranes or bring the tube almost horizontal. Also, the use of relatively large refractive optics to improve field generally precludes its use in the infrared.

4.2.2 Two-mirror systems

Focus access is greatly improved by folding the beam and bringing the focus behind the primary mirror. In one form or another, a two-mirror system is the most widely used telescope configuration. It benefits from minimal reflection and central obstruction losses, is compact, and offers an external and very accessible focus. In its classical form, the two-mirror system consists of a parabolic primary and a conical secondary relaying the common focus to the final focus. As explained in Section 4.1.1, if the secondary mirror is a convex hyperboloid located in front of the primary mirror focus, the combination is called "Cassegrain"[4]; if it is a concave ellipsoid located behind, it is called a "Gregorian." The two combinations are optically equivalent, but the Cassegrain version is more popular because its leads to a shorter tube. On the other hand, the Gregorian is easier to baffle since it has a real exit pupil. As noted earlier, this two-mirror combination is perfectly stigmatic but only on axis. It suffers from coma just as the single parabolic mirror does, but considerably less because the equivalent focal ratio is much slower than that of the primary due to the secondary mirror magnification. Typically, Cassegrain f-ratios are at $f/8$ or slower so that, with the assumption used in equation 4.18, the practical field can be 5' to 20' in diameter for a seeing-limited telescope.

With two surfaces, however, it is possible to depart slightly from the classical paraboloid/hyperboloid combination to correct for coma over a large field while retaining the main feature of the conic surface, freedom from spherical aberration. This configuration, first proposed by Schwarzschild in 1905 and

---

[4] "The advantage of this design are none, but the disadvantages so great and unavoidable, that I fear it will never be put in practice with good effect": judging by the success of this combination compared to the folded prime focus that he promoted, Isaac Newton's criticism of Sieur Cassegrain's invention was unfounded [16].

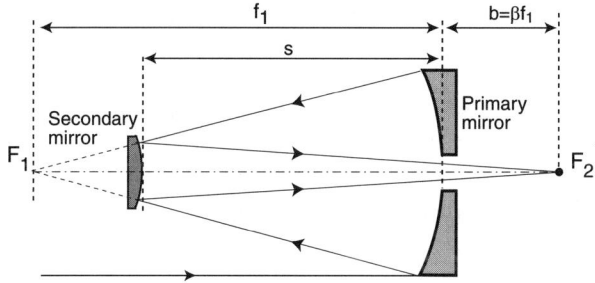

Fig. 4.15. Nomenclature for the Ritchey-Chrétien two-mirror system.

fully developed by Ritchey and Chrétien in 1910, is referred to as "aplanatic" (free of spherical aberration and coma) or Ritchey-Chrétien (R-C)[5]. In the R-C system, both the primary and secondary mirrors are hyperboloids.

Actually, this solution is part of a general principle proposed by Wilson as the "generalized Schwarzschild theorem"[17], which states that "for any geometry with reasonable separations between the optical elements, it is possible to correct n primary aberrations with n powered elements." Thus, with two powered mirrors, it is possible to correct spherical aberration and coma, but the correction of angular astigmatism and field curvature would require one or two additional mirrors.

The dominant remaining aberration of an R-C system is astigmatism. To the first order ($\beta \sim 0$ – see Fig. 4.15), it is equal to

$$\text{R-C astigmatism} \simeq \frac{\theta^2}{2N_1}, \tag{4.19}$$

so that the field of good quality will be proportional to the square root of the primary mirror focal ratio and, hence, becomes too small for fast primaries. When refractive elements are possible, correction of astigmatism is easily achieved over the entire coma-free field with a single lens near the focus. Alternatively, in the case of instruments used at a fixed off-axis position as in the Hubble Space Telescope, the correction can be built into the reflective relay optics of each instrument.

The main parameters of a Ritchey-Chrétien system are given in Table 4.3. The image quality of an R-C system is fairly sensitive to misalignment errors. Tilt and decenter of the secondary mirror introduce coma, and longitudinal (piston) error introduces spherical aberration and plate-scale change. A detailed investigation of the tolerances on the secondary mirror position can be found in Refs. [9] and [18]. To the first order, the tolerance on the decenter of the secondary mirror is proportional to the cube of the primary mirror focal ratio. Thus, telescopes with fast primaries are very sensitive to misalignments.

---

[5]The original papers by Schwarzschild, Chrétien and Ritchey are conveniently reproduced in the SPIE Milestone Series, Vol. MS 73, cited in the bibliography.

124   4. Telescope Optics

Both tilt and decenter introduce the same type of coma, and it is possible to cancel the coma due to tilt by an appropriate decentering of the secondary mirror. This has led Meinel to propose a particular structural telescope tube arrangement which satisfies this condition and make the optical system relatively insensitive to wind disturbances [19].

Table 4.3. Optical characteristics of the Ritchey-Chrétien configuration

Optical parameters

Primary mirror diameter	D_1
Primary mirror f-ratio	N_1
Primary mirror focal length	$f_1 = N_1 D_1$
Backfocal distance	$b = \beta f_1$
Normalized back focal distance	$\beta = b/f_1$
Magnification of secondary mirror	$m = f/f_1$
Primary–secondary separation	$s = (f - b)/(m + 1)$
Secondary mirror focal length	$f_2 = m(f_1 + b)/(m^2 - 1)$
Primary mirror conic constant	$\kappa_1 = -1 - \frac{2(1+\beta)}{m^2(m-\beta)}$
Secondary mirror conic constant	$\kappa_2 = -\left(\frac{m+1}{m-1}\right)^2 - \frac{2m(m+1)}{(m-\beta)/(m-1)^3}$
Secondary mirror dia. (zero field)	$D_2 = D_1(f_1 + b)/(f + f_1)$
Obscuration ratio (no baffling)	D_2/D_1
Final f-ratio	N
Final focal length	$f = ND_1 = \frac{f_1 f_2}{f_1 + f_2 - s}$
Field radius of curvature	$\frac{f_1 f^2 (f_1 - s)}{f f_1^2 + s(f^2 - f_1^2)}$

Aberrations

Angular astigmatism	$\frac{\theta^2}{2F} \frac{m(2m+1)+\beta}{2m(1+\beta)}$
Angular distortion	$\theta^3 \frac{(m-\beta)}{4m^2(1+\beta)^2}(m(m^2 - 2) + \beta(3m^2 - 2))$
Median field curvature	$\frac{2}{R_1} \frac{(m+1)}{m^2(1+\beta)}(m^2 - \beta(m - 1))$

For applications where cleanliness of the aperture is important (infrared telescopes, coronagraphy), it can be advantageous to avoid obstructing the primary mirror by using an off-axis design where the secondary mirror and its support system are outside of the incoming beam. Unfortunately, this design is inherently longer since, for a given final f-ratio, the diameter of the parent rotationally symmetric mirror is more than twice that of the off-axis mirror (Fig. 4.16).

4.2.3 Three- and four-mirror systems

The Ritchey-Chrétien is an excellent and widely used combination, but interest has recently shifted to three and more mirror combinations because of

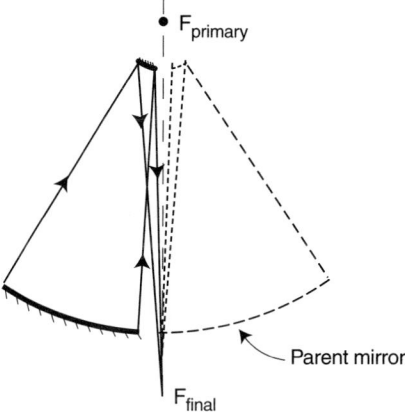

Fig. 4.16. An off-axis telescope design avoids obstruction of the primary mirror by the secondary mirror and its supporting vanes.

the need for beam steering or wavefront correction. Wavefront correction and line-of-sight jitter compensation can be done with the secondary mirror, but it is generally preferable to use small mirrors that can be oriented or deformed rapidly with minimal negative dynamic effects. If a deformable mirror or fine steering mirror is used, it should be placed at a pupil. The exit pupil of the R-C combination is located in front of the secondary mirror and is virtual, but one can reimage that pupil in order to create a small, real, accessible pupil. This can be accomplished by introducing a single powered mirror (tertiary), provided that the system is used slightly off axis to avoid beam blockage, or by the use of two judiciously placed extra powered mirrors (tertiary and quaternary) to remain on axis [20]. An example of a three-mirror combination is shown in Fig. 4.17.

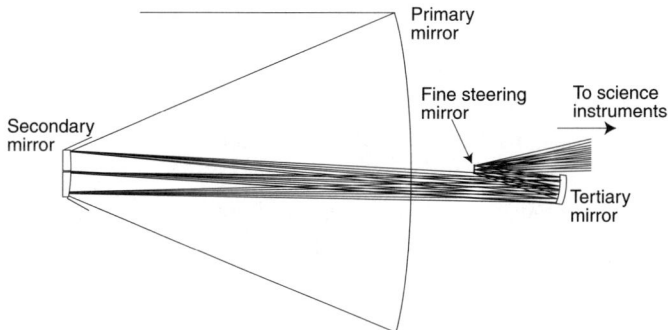

Fig. 4.17. Three-mirror system proposed for NGST. This is basically an R-C system with a third mirror added to create a real exit pupil where a fine steering mirror can be located. (From Ref. [21]).

4.2.4 Systems with spherical mirrors

Spherical primary mirrors have the advantage of low-cost fabrication. But correction of the massive spherical aberration over a reasonable field is not trivial and generally requires three additional mirrors. An example is shown in Fig. 4.18.

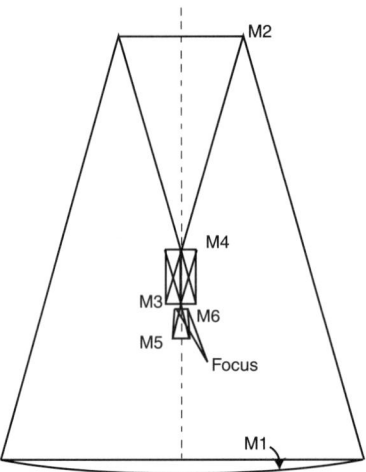

Fig. 4.18. Six-mirror system proposed for a 100-meter telescope with a spherical primary mirror. Two of the mirrors are simple flats. (From Ref. [22].)

4.2.5 Auxiliary optics

In addition to the main telescope mirrors, a number of auxiliary optical elements can be added for various applications such as:

- steering the output beam to correct for image motion,
- correcting wavefront errors due to atmospheric seeing or internally generated aberrations,
- improving imaging performance at the focus (e.g., via field flatteners and correctors),
- matching a given telescope design to specific detector conditions (e.g., focal reducers),
- relaying beam for packaging reasons (e.g., via an Offner relay),
- capturing part of the output beam to provide signals for guiding and wavefront correction (e.g., via a dichroic).

One auxiliary optical system usually needed on ground-based telescopes is a "field derotator." In a coudé configuration of an equatorial mount, or at any focus of an alt-az mount, the field rotates. When the field of the instrument is not too large, say less than 1 arcminute, it is possible to "derotate" the field. This can be accomplished by rotating a set of flat mirrors placed in the beam. These can be real mirrors or total reflection prisms (Fig. 4.19). Careful adjustment is required to keep the central image from wandering off-axis and to keep the beam properly collimated with respect to the instrument. The alignment of the mirrors is easier to control with prisms, but these will produce some lateral color. For fields wider than 1 arcminute, the derotator mirrors or prisms are impractically large and it is better to rotate the focal plane instrument instead.

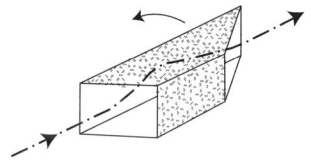

Fig. 4.19. Dove prism field derotator.

4.3 Optical error budget

As we have seen in Section 4.1.7, the final PSF of a system is a function of two factors:

- the aperture (outer shape of the primary mirror, gaps between primary mirror segments, central hole, support for secondary mirror, other obstructions in the beam), and
- wavefront errors due to imperfect optics or, in a ground telescope, to the atmosphere.

The first factor is intrinsic to a given telescope design, but the second factor, wavefront errors, is the result of mirror fabrication imperfections, misalignment of optical elements, and mechanical or thermal effects. These errors are generally divided into three categories according to their spatial frequency: low, mid-, and high frequencies. Low spatial frequencies are essentially due to the classical aberrations, and the spatial wavelength is in the range of $D/10$ to D, D being the aperture diameter. For mid-spatial frequencies, the spatial wavelength ranges between $D/10$ and $D/1000$. The high frequencies range from a spatial wavelength of $D/1000$ down to fractions of the wavelength of light. The various sources of these wavefront errors are described in more detail in Table 4.4.

The traditional approach to budgeting wavefront errors is to specify an upper limit to the mid- and high-frequency errors so that their impact is negligible. This means that the full error budget will be allocated to the

Table 4.4. Origin of the wavefront errors

Low frequency	Mid-frequency	High frequency
Optical design	Actuator print-through	Substrate material
Misalignment	Tool size and stroke	Polishing residuals
Polishing stresses	Tool conformity	Microroughness
Unblocking	Substrate structure	Cryoroughness
Test metrology errors	(e.g., quilting)	Dust
Dimensional instability	Sluice	
(e.g., desorption)	DM spline errors	
Mirror mounts		
Gravity release		
Bulk temp. effects		
Temp. gradients		
Coating stresses		
Segment position errors		

low spatial frequencies (i.e., low-order figure errors, misalignment, and mirror mount effects). This approach is generally justified by the fact that the low spatial frequencies are the most difficult to control, whereas the mid- and high frequencies are easier to be made negligible by proper fabrication.

For the high frequencies, a typical allocation is on the order of $\lambda/100$ rms. As for the mid-frequencies, a good starting point for a diffraction-limited system is to allocate a maximum wavefront rms error of $\lambda/20$ ($\lambda/40$ on the mirror surface) to the fabrication of each component of the optical train. The rationale for this condition is based on an estimate of how mid-frequency scatter affects the PSF. Surface error measurements made on typical large mirrors indicate that the surface defects power spectrum density (PSD) falls as $1/f^2$, where f is the spatial frequency of the defects (Fig. 4.20).

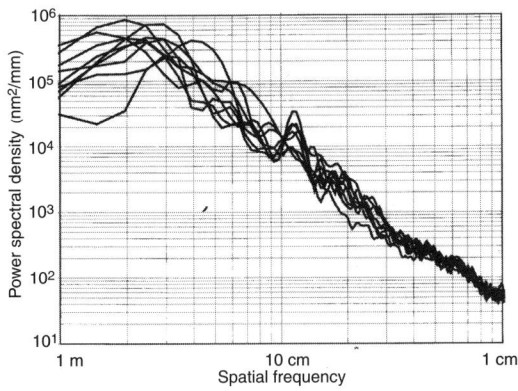

Fig. 4.20. Power spectrum density of wavefront errors for the Gemini mirrors. For mid- and high frequencies, wavefront errors follow a power law with an exponent close to 2. (Courtesy of REOSC and Gemini Project.)

Based on this empirical law, Fig. 4.21 shows the intensity of the scatter due to surface errors compared to that of a pure Airy pattern. To the first order, this graph shows that, in order to be negligible, the rms of mid-frequency figure errors must be less than $\lambda/40$ rms.

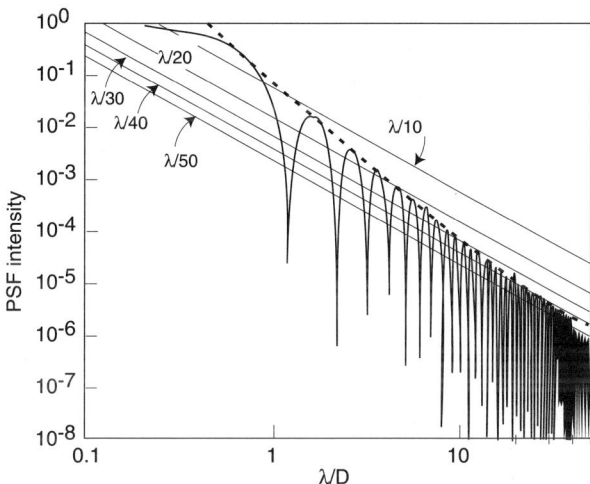

Fig. 4.21. Scatter due to the mid-spatial-frequency components of the surface errors from $\lambda/10$ to $\lambda/50$ rms on a mirror (straight lines) compared to the Airy pattern. The power law for surface errors has a starting frequency of 32 cycles (i.e., a wavelength of the mirror diameter divided by 32). This is based on the assumption that frequencies lower than that are either avoided by proper mirror figuring or corrected by actuators or a deformable mirror. To have a negligible impact on the PSF of a perfect image, the mirror surface errors should be less than $\lambda/40$ rms, i.e., $\lambda/20$ on the wavefront. (Courtesy of R. Lyon.)

4.4 Criteria for image quality

The imaging performance of an optical system is fully characterized by its PSF, but it is a two-dimensional function which can be very complex, depending on diffraction effects and instrumental effects. When comparing telescope concepts or defining specifications for design and fabrication purposes, one needs a practical metric. Many attempts have been made to characterize the PSF with simple functions and even a single number. Those most commonly used are described below.

- **Modulation transfer function (MTF)**. Using Fourier analysis, an optical object can be represented as the sum of an infinite series of sinusoidal components of increasing spatial frequency. As each of these components is transmitted through the optics to form the image, the spatial frequency is unchanged, but the amplitude is reduced. The MTF is a measure of this degradation as a function of frequency and can be viewed as a filter function applied to the object. The modulation, or contrast, of a sinusoidal component is defined as

$$M = \frac{I_{\max} - I_{\min}}{I_{\max} + I_{\min}}, \qquad (4.20)$$

where $I_{\max}$ and $I_{\min}$ are the maximum and minimum intensities as shown on the left in Fig. 4.22. The MTF is the ratio of the modulation in the image, M_i, to that in the object, M_o, versus spatial frequency:

$$\text{MTF} = \frac{M_i}{M_o}. \qquad (4.21)$$

The MTF is directly related to the PSF: mathematically, the inverse Fourier transform of the PSF is the optical transfer function (OTF), the amplitude of which is the MTF. As such, the MTF is an excellent measure of image quality since it contains the same information as the PSF itself. The other advantage of the MTF representation is that a system's MTF is simply the product of the MTF of its various components (optics, detector, atmosphere), at least when wavefront errors are spatially uncorrelated.

It is possible to define the MTF as a two-dimensional function, but, in general, it is used as a one-dimensional function averaged azimuthally and, as such, loses the description of the PSF azimuthal structure (due, in particular, to the diffraction of nonaxisymmetric obstructions in the aperture such as mirror support vanes). But these features can be analyzed separately.

The MTF of a perfect system (no aberrations) with a circular aperture is given by

$$\text{MTF}(\nu) = \frac{2}{\pi}(\Phi - \cos\Phi \sin\Phi) \quad \text{with} \quad \Phi = \arccos\frac{\lambda \nu}{D}, \qquad (4.22)$$

where ν is the spatial frequency (cycles per radian), λ is the wavelength of light, and D is the diameter of the aperture. The MTF becomes 0 at the "cutoff spatial frequency," which corresponds to the ultimate resolution of the system λ/D. It is then convenient to define the "normalized spatial frequency" as

$$\nu_n = \frac{\nu}{\nu_c}, \qquad (4.23)$$

where ν_c is the cutoff frequency λ/D in angular units. The normalized spatial frequency, ν_n, thus varies between 0 and 1. An example of a real system's MTF is shown on the right in Fig. 4.22.

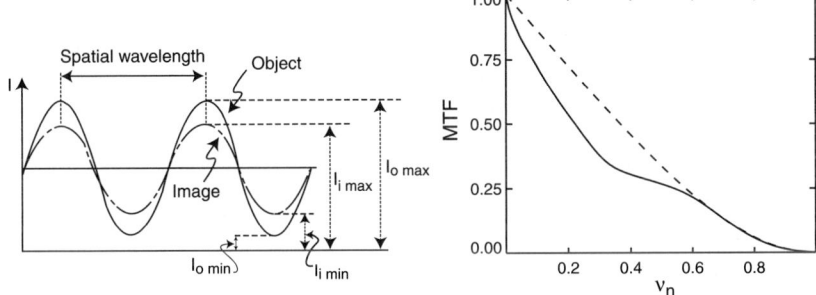

Fig. 4.22. At left, a sinusoidal component in an image compared to that in the object. At right, MTF of actual HST optics (solid line) compared to that of an ideal optical system with no central obstruction and no wavefront error (dashed line) as a function of the normalized spatial frequency.

- **"80% encircled energy" (EE).** This is the angular diameter containing 80% of the energy in the PSF. In the PSF given by a perfect optical system (no aberrations, no atmosphere) having a circular pupil with no central obscuration, 80% of the energy is contained in a diameter of $\sim 1.8\lambda/D$. This criterion, which is meant to represent the practical angular size of the image of a point source, is an excellent measure of the performance of a large telescope because it directly relates to the two main astronomically meaningful parameters: sensitivity and angular resolution. It is wavelength dependent, however, and must be set for the prime wavelength for which the observatory is intended, or else several values corresponding to various wavelengths must be supplied.
- **Full width at half-maximum (FWHM).** This is the width (average diameter) of the PSF at half the maximum intensity. The FWHM is a good measure of the image size, although not as telling as the 80% EE because it does not include the wings of the PSF.
- **Strehl ratio.** The Strehl ratio is the ratio of the peak intensity in the actual image compared to the peak theoretical diffraction intensity. The Strehl ratio is proportional to the area under the MTF curve. According to the Maréchal rule, an optical system is considered diffraction limited if the Strehl ratio is 0.8. The Strehl ratio is a good measure of image quality for a system which is close to being diffraction limited, but it does not capture features of PSF beyond the core. For example, strong mid-spatial frequencies in the wavefront error can seriously degrade sensitivity because they create a halo around the PSF core, while the height of the core, and thus the Strehl ratio, remain essentially unaffected.
- **Wavefront error rms.** Used as a single number, the root mean square of the wavefront error is directly related to the Strehl ratio as per equation 4.13. As such, it has the same drawback of not capturing the effect of mid- and high frequencies. But this can be alleviated by specifying the

rms wavefront error of, for example, three spatial frequency ranges: low, mid, and high, as indicated in Section 4.3. The rms wavefront error is convenient for optical error budgeting since the various components of the wavefront error can simply be broken down or recombined according to the rule of the sum of the squares.

- **Central intensity ratio (CIR)**. The CIR was introduced by Dierickx to quantify the image quality of ground-based telescopes, where degradation by atmospheric turbulence is a dominant factor [23]. It is defined as

$$\mathrm{CIR} = \frac{\mathcal{S}}{\mathcal{S}_0}, \qquad (4.24)$$

where $\mathcal{S}_0$ is the Strehl ratio of the telescope, assumed to be optically perfect and taking into account solely the effect of atmospheric turbulence, and $\mathcal{S}$ is the same quantity after telescope wavefront errors are taken into account. The CIR varies between 0 and 1, reaching 1 when the telescope is limited only by atmospheric turbulence. The CIR is wavelength and seeing dependent. For a given wavelength and seeing, it depends primarily on the rms of the wavefront slope error introduced by the telescope. To the first order, it is given by

$$\mathrm{CIR} = 1 - 2.9 \left(\frac{\sigma}{\theta_0}\right)^2, \qquad (4.25)$$

where σ is the rms wavefront slope error and θ_0 is the seeing angle at FWHM ($\theta_0 = 0.98\lambda/r_0$, where r_0 is the Fried parameter). As an example, the CIR for the VLT was specified to be greater than 0.82 at $\lambda = 500$ nm and for a seeing angle of 0.2″.

- **Sharpness.** "Sharpness" is an image-quality figure of merit for the detection of point sources in background-limited mode, which was introduced by Burrows (see Appendix C). Sharpness, Ψ, is defined as the sum of the squares of the pixelized intensity in the PSF (Fig. 4.23):

$$\Psi = \sum P_{ij}^2, \qquad (4.26)$$

where P_{ij} is the intensity in each pixel of the normalized point spread function ($\Sigma P_{ij} = 1$).

Sharpness is the best image-quality criterion for a near-diffraction-limited telescope primarily used for background-limited observations because it directly relates to the astronomical performance of the telescope in that mode. In the background-limited mode, the signal-to-noise ratio is given by

$$\mathrm{S/N} = I \frac{\sqrt{\Psi}}{\sqrt{B}}, \qquad (4.27)$$

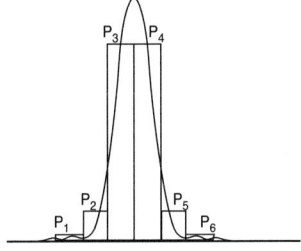

Fig. 4.23. Pixelized image used in determining sharpness (schematic).

where I is the total number of photons from the source and B is the mean background per pixel (including sky, telescope emission, detector readnoise, dark current, etc.). Sharpness is the second moment of the pixelized PSF. The idea behind this notion is that to extract maximum information from the image, one would weight the importance of each pixel in the image according to the square of its intensity. Although this is theoretically the ultimate metric for background-mode observations, sharpness is not systematically used, particularly for very faint extended objects. The reason is that it assumes fitting a model to the actual image which, although possible, is almost never done because of the uncertainties in the process.

Of all the criteria reviewed above, the only one offering a quasi-complete description of the image is the MTF. For convenience in summarizing imaging performance, however, the two most useful "global" or "single number" measures are the CIR for ground-based telescopes and the 80% EE for space (diffraction-limited) telescopes. For error-budgeting purposes, on the other hand, the rms wavefront error criterion is in common use.

For the end user, the astronomer, the tendency is, of course, to request the best imaging possible. But exquisite image quality comes at a cost and a compromise has to be found. As discussed in Chapter 3, performance specifications should be the result of a thorough study on how best to meet scientific goals within cost and schedule constraints.

Astronomical observational goals are highly diverse, with demands in sensitivity, spatial resolution, and spectral resolution which cannot commonly be cast in a simple requirement. The best approach for optimizing the image quality requirement is thus an *empirical* one in which one (1) establishes clear scientific goals, (2) models the proposed telescope with various choices for image quality, and (3) evaluates how each of these choices performs in the extraction of the scientific parameters of interest. Such an empirical study was conducted to define the optical quality requirement for NGST. A field of early galaxies typical of those expected to be observed was first modeled on purely scientific terms. A set of 32 PSFs exploring the range of low-, mid- and high-spatial-frequency wavefront errors that could reasonably be expected was then created and used to produce simulated images of the galaxy field

134   4. Telescope Optics

in several colors. The final step was to evaluate these images, not according to traditional optical criteria but in *scientific terms*. This was done by applying image processing software that observers would commonly use for real observations in order to extract the parameters relevant to the study, namely photometric redshift and galaxy size. This exercise allowed the pinpointing of the most relevant image-quality figure of merit (it turned to be the 80% EE criterion) and the wavelength at which it should be defined. Examples of simulated fields from that study are shown in Fig. 4.24.

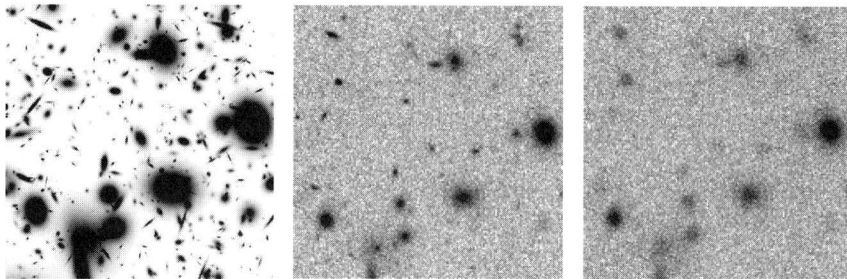

Fig. 4.24. Example of a simulation to evaluate NGST's scientific performance as a function of PSF quality. At left, an original noiseless simulation of a galaxy field typical of NGST's science goals. The center panel shows a 10-hour H-band exposure with a "good" PSF having low-frequency surface figure error and no mid-frequency surface error. The panel at right shows the same field imaged for the same exposure time, but with a poor quality PSF having three times the low-frequency wavefront error rms of the previous one and five times the HST rms mid-frequency wavefront error. (From Ref. [24].)

4.5 System issues

4.5.1 Focus selection

Space telescopes, for which operational simplicity is primordial, are equipped with only one focus, a Cassegrain focus, which offers good correction and proper plate scale. But for ground telescopes, a configuration with several foci increases flexibility. It offers a choice of several f-ratios and instrument mounting interface sizes and allows an instrument to remain mounted on the telescope while another is being used at a different focus. However, the larger the number of foci, the more complex the telescope structure and control system will be. A reasonable compromise has to be found. The choices, illustrated in Fig. 4.25, are as follows.

The **prime focus** offers the minimal number of surfaces, an important factor when scatter or thermal emissivity must be minimized and reflectivity maximized. The f-ratio is usually between $f/1.5$ and $f/3$. This gives an

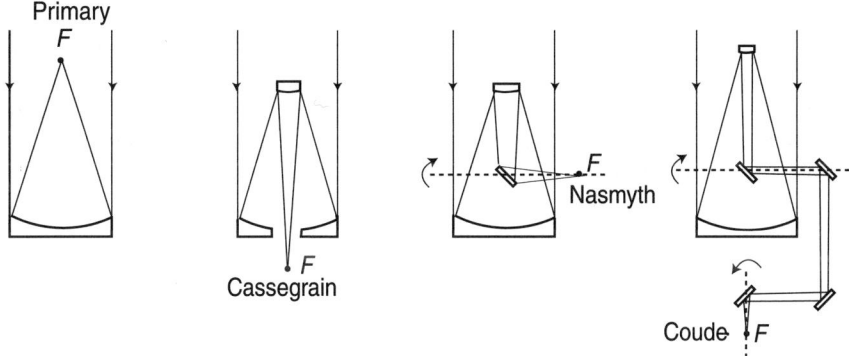

Fig. 4.25. Focal configurations.

appropriate plate scale for seeing-limited, wide-field observations but is too small for many instruments. The prime focus requires a corrector, and access is difficult.

The **Cassegrain focus** is, for the majority of observations, the preferred focus. It offers easy accessibility, good field without transmitting elements, and a small number of surfaces. The f-ratio is typically between $f/8$ and $f/15$, which supplies a plate scale well adapted to high-spatial-resolution imaging. It can accept fairly heavy, bulky instrumentation.

The **Nasmyth** focus, is simply a Cassegrain focus that remains fixed on the rotation (elevation) axis of the tube thanks a 45° fold-mirror. It offers the same advantages as a normal Cassegrain but with even easier accessibility. An additional advantage is that an instrument can remain fixed with respect to the gravity field provided that an optical field rotator is used (if not, the instrument has to rotate around its axis). The telescope tube does not need rebalancing when exchanging instruments. Nasmyth instruments can be left at the focus while the telescope is used at other foci.

The **coudé focus**[6] is a long focal-length Cassegrain focus that remains fixed in space thanks to fold mirrors located on the tube and mount rotation axes. This permits the use of heavy, bulky instruments such as high-resolution spectrographs. The field of view at the coudé focus is quite small, on the order of a few arcseconds. The field rotates, but this is not a problem when the target is a star or an extended object taken as a whole. If necessary, however, one can use an optical derotator as shown in Fig. 4.19. The usual range of f-ratios is from 30 to 100 in order to reduce the beam size, since the beam must be piped a considerable distance from the telescope. The corresponding large plate scale is well adapted to high spectral resolution spectroscopy but is too large for imaging.

---

[6]The word "coudé" comes from the French: "bent"

The number of mirrors required in a coudé arrangement varies with telescope configuration, but is usually between five and seven. Thus, the main disadvantage of the coudé focus is its low throughput, especially at lower wavelengths. Another shortcoming is the nonnormal incidence of light on the mirrors of the coudé train, which introduces phase changes complicating polarization measurements, especially if the fold mirror angles change when the mount rotates [25, 26]. An alternative to using a train of flat mirrors is to use fiber optics, fed from the prime focus, for example.

4.5.2 Selection of f-ratio

The choice of f-ratio for each telescope focus, and particularly for the primary mirror, is arguably the most difficult decision faced in telescope design. That is because this f-ratio selection has repercussions throughout the entire observatory system: the optical train, the telescope structure, the control system, the instruments, and the dome and building, all will be affected.

Since the cost of an observatory is strongly affected by the length of the telescope tube, whether on the ground or in space, the tendency is to have the primary mirror as fast as is technically possible. Fast optics used to be difficult to fabricate, but, over the last 30 years, enormous progress has been made in figuring methods, and the f-ratio of the primary mirror has been coming down dramatically (Fig. 4.26). As a result, limits on how fast primaries can be are now mainly driven by the tolerances on the position of the secondary mirror. Indeed, the misalignment tolerance of the secondary is a strong function of the f-ratio of the primary mirror (it varies as the cube). But this is tempered by two facts: (1) for a given final f-ratio, the mass and size of the secondary mirror will be smaller and (2) a faster primary leads to a shorter telescope tube with a corresponding gain in stiffness and reduction in wind-buffeting torques.

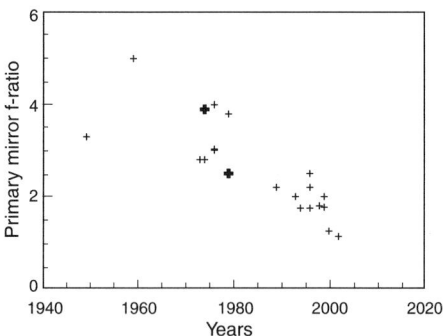

Fig. 4.26. Evolution of primary mirror f-ratios over the last 50 years.

Selection of the final Cassegrain or Nasmyth f-ratio also involves a number of trade-offs. To the first order, one would want an f-ratio which directly matches the plate scale to instrument needs (see Section 4.5.3); the choice will depend on whether the telescope is intended to be diffraction limited or seeing limited. Other factors also come into play. Generally, a slower final f-ratio will

- decrease the size and mass of the secondary,
- lengthen the tube, but only slightly,
- increase the size of the instrument entrance optics,
- increase the focal length of spectrograph collimators, making them bulkier overall,
- increase the size of beam steering mirrors for image-motion compensation.

4.5.3 Matching plate scale to the detector resolution

Whether obtained directly or via relay optics, the focal-plane plate scale must be adapted to the detector's spatial and noise characteristics in order to optimize sensitivity or spatial resolution. For the sake of argument, we assume that the pixel size of the detector is a given and that we want to find the optimal plate scale, that is to say, the optimal f-ratio of the focus where the detector is placed. This problem can be broken down according to the class of observation being performed.

In the **general case**, one wants to maximize sensitivity without losing spatial information in the object. Sampling too finely increases noise and has the drawback of accumulating redundant data, generally at the expense of field. It can be shown that virtually all of the spatial information can be recovered by sampling the image with an angular pixel size on the sky equal to half of the resolution element of the optics as defined by the Sparrow criterion (λ/D). This stems from Nyquist's sampling theorem, and the corresponding optimum is called "critical sampling" or "Nyquist sampling." Since the angular pixel size on the sky is p/f, where p is the linear pixel size and f is the final focal length, the plate scale to pixel size matching condition for critical sampling of diffraction-limited images is $2(p/f) = \lambda/D$, so that the optimal f-ratio is independent of the aperture diameter and simply

$$\text{optimal } f\text{-ratio} = \left(\frac{f}{D}\right)_{\text{opt}} = \frac{2p}{\lambda}. \quad (4.28)$$

Current detectors have pixel sizes ranging from 7 to 13 μm for CCDs and from 18 to 28 μm for near- and mid-infrared detectors. The resulting optimal f-ratios as a function of wavelength are shown in Table 4.5.

In the case of **high-spatial-resolution imaging** of bright objects, detector noise is no longer a factor and one gains by oversampling the image. Detector sizes are limited in practice, however, and this, in turn, sets a limit on oversampling because it comes at the expense of field.

138   4. Telescope Optics

Table 4.5. Typical f-ratios for critically sampled detectors

Wavelength (μm)	0.5	2	2	10
Pixel size (μm)	7	18	28	28
Optimal f-ratio	28	18	28	5.6

In cases where **sensitivity** is primordial, the optimal plate scale will be that which minimizes exposure time for a desired signal-to-noise ratio. This optimum depends on the solid angle of the source and the detector's dark current and readout noise, but it will generally correspond to some undersampling. An example of such a situation is shown in Fig. 4.27 (left). The problem with undersampling is a loss of spatial resolution and "aliasing," which is illustrated in Fig. 4.27 (right). Aliasing occurs when sampling is less than the Nyquist limit because the detector is unable to distinguish between several possible spatial frequencies. The result is increased noise in the signal at low spatial frequencies.

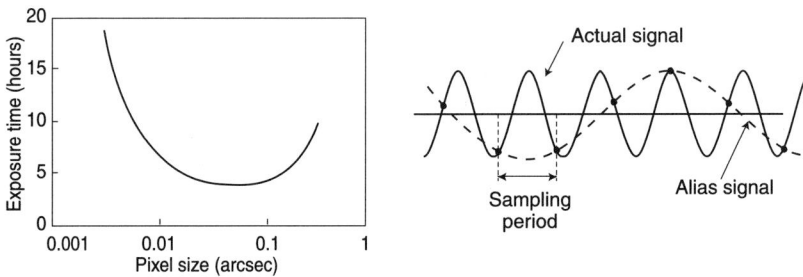

Fig. 4.27. At left, an example of how exposure time varies with pixel size in the case of a background-limited target. The calculation is made for the near-infrared camera of NGST. Near the minimum, exposure time is not greatly sensitive to pixel size over a range of a factor of 2. (From Ref. [27].) At right, an example of aliasing. When the sampling frequency of a sinusoidal signal is less than the Nyquist frequency, the sampled data can be satisfied by another sinusoid of equal amplitude but lower frequency (dashed line), the so-called "alias signal."

A technique which can be used to mitigate the loss of spatial resolution in undersampled images is "dithering." This consists of taking several exposures of the same field with the line of sight stepped by fractions of a pixel. The subexposures are then recentered and added, allowing the recovery of most of the spatial resolution afforded by the optics. This technique is particularly applicable to space observations, where exposures have to be broken up into short individual exposures in any case, so as to limit the effects of cosmic rays in each one. In such cases, there is no loss in signal-to-noise ratio.

On a historical note, when photographic plates were used, the optimal f-ratio was not set by grain size but by photographic speed. The problem was that photographic plates have poor sensitivity and using the ideal f-ratio would have led to prohibitive exposure times. The optimal f-ratio was then determined by the need to expose the sky background to the optimal density within a reasonable time (compared to the duration of the night), on the order of 4 to 6 hours. In the early 1900s, plate speed was such that this focal ratio was $f/4$ or $f/5$. When much faster plates became available, optimal f-ratios reached $f/8$ or $f/10$, which became the standard Cassegrain focal ratios for telescopes built in the 1960s and 1970s [28].

4.6 Mirror blank materials

4.6.1 Generalities

Mirror substrate materials have to satisfy a number of important conditions. They must:

- be dimensionally stable enough over time for the optical figure to be retained for decades,
- have low internal stress so as not to deform when material is removed during the figuring process, or over time due to stress relaxation,
- not deform when subjected to environmental temperature changes,
- be obtainable in large sizes,
- have sufficient mechanical rigidity and strength to permit both handling and mounting,
- take a fine surface polish and be coatable by vacuum-deposition methods,
- and finally, in the case of cryogenic applications, they must not undergo structural changes when cooled to very low temperatures.

Many materials have been used over the years for mirrors of various quality, but with current technology, the choices come down to those listed in Table 4.6.

Table 4.6. Properties of commonly used blank materials

Material	Density ρ kg/m^3	Young Mod. E GPa	Poisson ratio ν	Max stress σ_t MPa	CTE 273K α_{273} 10^{-6}/K	CTE 40K α_{40} 10^{-6}/K	Therm. cond. κ W/m K	Specific heat C_p J/kg K
Borosilicate	2200	63	0.20	78	3.3	-3.2	1.2	800
ULE	2200	68	0.18	50	0.03	-0.9	1.3	760
Zerodur	2500	91	0.24	57	0.05	-0.7	1.5	820
SiC (CVD)	3200	466	0.21	440	2.2	0.05	190	730
Beryllium	1850	300	0.08	240	11	0.05	210	1900
Aluminum	2700	70	0.33	310	23	2.5	170	890

Sources: Barnes [29], Paquin [30], manufacturer's literature.

Glasslike materials will generally take a better polish than metallic ones, the residual microroughness being in the 5–12 Å rms range for the former and 10–20 Å for the latter, except for bare aluminum. Bare aluminum residual microroughness is around 50 Å rms. This may be acceptable for infrared applications, but, if not, it can be overcoated with a nickel alloy for an excellent finish. All of these materials can thus make excellent mirrors, although they differ markedly in two important areas: stiffness and thermal behavior.

Stiffness

For large telescopes, whether on the ground or in space, mass is clearly an important issue and mirrors are thus usually designed to minimize mass in a given environment. On the ground, gravity is the dominant factor and deflection of the mirror must controlled. In space, a minimum natural frequency is usually the criterion in order to withstand acoustic loading during launch and satisfy operational constraints in orbit. Whatever the shape and internal structure of the blank and the characteristics of its supporting system, a general rule applies: deflection under self-load and fundamental frequency are a function of the ratio of the Young modulus,[7] E, to the density of the material, ρ (see Chapter 6). This ratio, which is a characteristic of a given material, is called "specific stiffness":

$$\text{Specific stiffness} = \frac{E}{\rho}. \qquad (4.29)$$

The higher this ratio, the less the deflection and the higher the fundamental frequency. It is an interesting fact of nature that denser materials are generally more rigid, so that, for most structural materials, this ratio does not vary greatly. Still, there are significant differences, especially in the case of composite materials.

Thermal behavior

The other important condition a mirror material must meet is extremely low sensitivity to thermal variations. On the ground, thermal changes occur early at night, when the mirror is trying to reach equilibrium with the night air temperature, and also throughout the night as the temperature drops. In space, thermal changes occur as a result of periodic eclipsing of the Sun by the Earth or changes in observatory orientation with respect to the Sun after repointing. As long as the mirror material is homogeneous and isotropic, bulk temperature changes will affect the focal length but not the figure. More troublesome, in general, are temperature gradients between the back and front surfaces of the mirror or gradients across the diameter, both of which can

---

[7]The Young modulus, also called the "elasticity modulus," is the ratio of stress to strain, a measure of the material's stiffness.

affect focal length and figure. These thermal effects will clearly be smaller as the coefficient of thermal expansion of the material, α, is lower. Figure 4.28 gives this coefficient as a function of temperature for major mirror substrate materials.

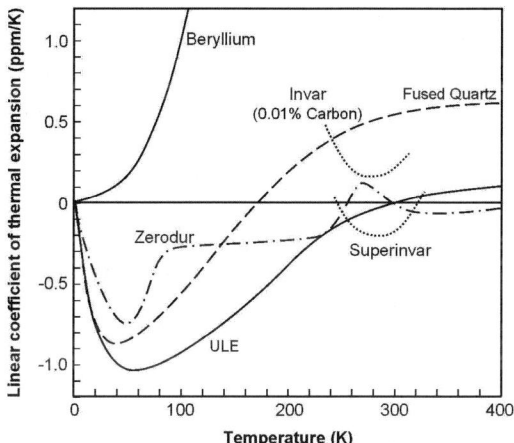

Fig. 4.28. Coefficient of thermal expansion as a function of temperature for the most common mirror blank materials.

An alternate to a low coefficient of thermal expansion (CTE) is to have lower *temperature gradients*. Gradients are reduced if the thermal conductivity, κ, is higher and if the specific heat, C_p, and density, ρ, are lower, that is to say, if the ratio $\kappa/C_p\rho$ is higher. This ratio, which characterizes how quickly a material will come into temperature equilibrium, is called "thermal diffusivity." These two conditions, low CTE and high thermal diffusivity, can be cast into a single figure of merit for thermal behavior, which one will want to maximize for passive optics:

$$\frac{\kappa}{\alpha\, C_p\, \rho} \quad \text{(passive optics)}. \qquad (4.30)$$

With active optics, the requirement that mirror figure not be affected by thermal change is much less stringent. This is because thermal effects are slow and well within the bandpass of active optics systems. In the case of ground-based telescopes, high thermal diffusivity is nevertheless useful for reducing "mirror seeing." This is the blurring of images due to air turbulence generated by a difference in temperature between the mirror's optical surface and the surrounding air (see Chapter 9). Mirror seeing is less for materials with high thermal diffusivity because they track ambient temperature variations well. For active optics systems on the ground, the figure of merit which one will want to maximize will then simply be the thermal diffusivity:[8]

$$\frac{\kappa}{C_p \rho} \quad \text{(active optics -- ground only)}. \tag{4.31}$$

The advantage of higher thermal diffusivity is significantly enhanced if the mirror is equipped with a thermal control system to keep the mirror's bulk temperature close to the predicted or actual night temperature (see Chapter 9).

Figure 4.29 compares the most common mirror materials according to the structural and thermal figures of merit developed above.

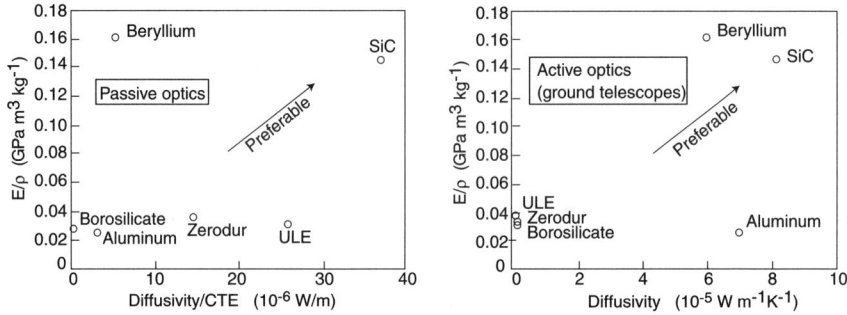

Fig. 4.29. Mirror blank materials plotted according to specific stiffness and thermal figures of merit for passive (left) and active (right) optics.

In what follows, we give a brief description of the most common materials used as mirror substrates.

4.6.2 Borosilicate glass

Low-expansion borosilicate glasses such as 7160 Pyrex (made by Corning) and Duran 50 (made by Schott) were the customary materials for large telescope mirrors until the 1960s. They have now largely been replaced by ultralow-expansion ceramic or fused silica, which have the same favorable polishing properties but coefficients of thermal expansion 100 times lower. If, however, temperature effects can be minimized by active thermal control, borosilicate glass offers the possibility of lightweighting by direct honeycomb casting and the added advantage of lower cost. This approach, proposed by Angel [31], has been successfully used for several large telescopes.

For the same mass as a zero-expansion ceramic meniscus, a honeycomb borosilicate mirror will be stiffer and, with air circulation inside the cells,

---

[8]This is less true for thinner mirrors, where the thermal coupling to the surrounding air is of greater importance. In such a case, the rate at which the mirror comes to equilibrium with the surrounding air is dictated more by convection than by conduction within the glass.

deformation due to temperature gradients can be reduced to an acceptable level.

4.6.3 ULE fused silica

Ultralow-expansion (ULE) fused silica is a product manufactured by Corning. It is fused silica doped with titanium, which results in an expansion coefficient at room temperature that is 20 times lower than that of pure fused silica. It is produced in "boules" of about 1.2 meters in diameter, and larger solid mirrors can be made by assembling boules. Lightweight mirrors can be made by building up the mirror blank from segmented top and bottom facesheets and cut plates for the internal ribs, and then fusing the pieces into a monolithic mirror by partial remelting at about 1500 °C (See Fig. 4.33 in Section 4.7.1). Lightweighting can also be achieved by milling a solid blank by the water jet process. When a large mirror is made by fusing segments, optical surface distortion may result from CTE differences between the various segments. This can be minimized by locating the individual segments optimally, according to their measured CTEs.

Ultralow-expansion fused silica has the useful property that its CTE is well correlated with the speed of sound in the material, so the CTE of a piece of ULE can be measured *in situ* to an accuracy of a few parts per billion.

4.6.4 Low-thermal-expansion glass ceramic

Low-thermal-expansion glass ceramics, known by the trade names Cer-Vit (by Owens-Illinois, now abandoned), Zerodur (Schott), and Astro-Sitall (Russia), are produced by including crystallization nucleating agents in the glass melt. These devitrified glasses are two-phase materials in which the balance between the crystalline phase (with negative CTE) and the amorphous phase (positive CTE) can be set to minimize the overall CTE in a given temperature range. The substrate is cast to a glassy state, cooled to ambient temperature, premachined, and reheated in a ceramization process to stimulate crystal growth. To minimize residual stresses, thermal gradients must be controlled to high accuracy throughout the whole process. With this precaution, residual stresses can be very low, with birefringence on the order of 3 nm/cm.

Glass ceramic is difficult to cast in complex shapes such as honeycombs. This is because next to the mold surface it develops a crystalline layer with a coefficient of expansion different from that of the rest of the ceramic, resulting in high stresses and breakage during cooldown. Any lightweighting must be done by conventional or water jet milling (an expensive process), or by building the mirror blank with facesheets and ribs fused together (a labor-intensive approach which also has the drawback of possibly introducing a foreign material, resulting in some temperature sensitivity).

4.6.5 Silicon Carbide

Silicon carbide (SiC), also known under the trade name Carborundum, is one of the hardest synthetic materials. It has excellent thermal diffusivity and its high specific stiffness makes it one of the best materials for dynamic applications such as chopping secondary mirrors. Bare silicon carbide also has good reflectance at extreme ultraviolet wavelengths, for which it is difficult to find suitable reflective coatings.

There are several methods of production, some that produce pure SiC, and others that produce a matrix of SiC with other materials, usually elemental carbon or silicon. Among the pure SiC forms, the production method most often used for mirror blanks is the "chemical vapor deposition" (CVD) process which consists of depositing gaseous chemicals on a graphite mandrel that is subsequently leached away. Mirrors with complex shapes and ribbed backing structures can be produced by a two-step process in which the facesheet is deposited in the first CVD operation, then the backing ribs are deposited in a second furnace run [32]. The deposition process is relatively slow but leads to an extremely pure SiC, which can be ground and polished with diamond grit to a surface roughness of less than 5 Å. Because of its extreme hardness, however, the polishing time is much longer than with traditional materials. Another drawback is that the CVD process tends to generate high internal stresses, and this is detrimental to deterministic figuring.

Among the production methods that produce a matrix of SiC with other materials, two deserve mention here. Reaction-bonded SiC is formed by casting a slurry of SiC grains in a sacrificial mold, baking the casting to burn off the mold material and fuse the grains together, then infiltrating the voids with molten silicon to form a solid structure that is 70–85% SiC, depending on the specific process [33]. This produces a solid material with very good material properties, although the specific stiffness is not as high as the CVD form. It is difficult to polish reaction-bonded SiC to a surface finish better than about 20 Å, so overcoats of pure silicon or CVD SiC are sometimes applied to provide a readily polishable surface. It is possible to cast reaction-bonded SiC into complex shapes, including honeycomb sandwich structures with continuous front and back sheets.

Another form of SiC matrix can be produced by infiltrating molten silicon into a shaped mass of chopped carbon fibers, which react to form SiC [34]. This type of material is called C/SiC (pronounced "seasic"). Before infiltration, complex shapes including honeycomb sandwich structures can be formed by machining and joining. By controlling the process conditions, the amount of carbon fiber remaining in the matrix can be tailored, which allows some control of the toughness of the material. The main drawback of infiltrated SiC is that it needs to be clad with a more polishable material if a fine polish is required.

In general, silicon carbide is brittle, and lightweight mirrors of this material are extremely fragile. Currently, the maximum size for SiC mirror blanks using any of these production methods is about 1 meter in diameter.

4.6.6 Beryllium

Beryllium, one the lightest and stiffest metals, has an expansion coefficient higher than that of vitreous mirror materials, but high specific stiffness and thermal diffusivity. This makes it an outstanding choice for mirrors when very low areal densities are desired, as for space applications or for chopping mirrors on ground telescopes.

Beryllium mirrors cannot be made by casting, as the metal loses its strength during the melting process. This is due to the fact that an uneven, dual (large and small) grain structure develops during solidification. To achieve highest strength, beryllium must have a fine-grained structure. Most beryllium is therefore produced by the powder metallurgy process, which consists of bonding particles solidly together by applying pressure and heat. This is accomplished by putting beryllium powder in a mold, then heating it to about 900 °C while compressing it by vacuum or pressure (1000 atmospheres). These methods are referred to as "vacuum hot pressing" (VHP) and "hot isostatically pressing" (HIP), respectively. With current tank size limitations, the largest piece of beryllium that can be produced is approximately 2 meters in diameter, with a maximal length of 2.5 meters.

Beryllium can be lightweighted by machining with conventional mills and lathes. But precautions must be taken to prevent very small beryllium particles, less than 10 μm in size, from becoming airborne and therefore potentially respirable. When breathed in, beryllium powder may cause a serious lung disease similar to silicosis. There is no such danger during polishing because it is a wet process.

Through variation in particle size, distribution, beryllium-oxide content, and temperature, it is possible to produce a variety of beryllium grades with different properties. For infrared applications, the best choice is the Brush-Wellman, O-30 grade, which takes a good polish with a residual microroughness of around 25 Å, so that such mirrors can be used bare (without optical coating). The O-30 grade also has the advantage of posessing very homogeneous thermal and mechanical characteristics thanks to the use of specially calibrated spherical powder grains. For improved polish quality, such as that needed for UV or visible applications, beryllium can be plated with electroless nickel, a nickel–phosphorus alloy.[9] Because the coefficients of expansion of nickel and beryllium are well matched, these mirrors are usable over a

---

[9]This alloy contains about 10% phosphorus to improve corrosion resistance, polishability, hardness, and coat adhesion. One of these coating processes is Kanigen, patented by Electro-Coatings of Iowa, Inc.

146   4. Telescope Optics

wide temperature range. For cryogenic applications, however, bare beryllium should be preferred in order to avoid bimetallic effects.

4.6.7 Aluminum

The main advantages of aluminum are low cost and high thermal conductivity. But aluminum's high coefficient of thermal expansion requires either temperature control in the case of space applications, or postcorrection with active or adaptive optics for ground telescopes. Bare aluminum is too soft to polish well and is generally overcoated with electroless nickel.

This makes aluminum difficult to use for cryogenic applications because the differences in coefficient of expansion between nickel and aluminum lead to large deformations during cooldown. Depending on annealing and the type of alloy used, blanks may exhibit dimensional instability in the long term, but this should be easily correctable if the mirror is used in an active optics system.

Although nickel-coated aluminum has thus far been shunned for ground-based optical telescope applications, it could become an outstanding choice for active optics systems. This is because its high CTE is no longer an issue, since thermally induced deformations are automatically corrected by the active optics system. The remaining characteristics are all favorable: aluminum is relatively inexpensive, easy to machine, and has a high thermal diffusivity conducive to low mirror seeing.

4.7   Mirror structural design

Once installed in the telescope, mirrors must not deform by more than a fraction of the optical surface tolerance when subjected to gravity and wind loading or, in the case of space mirrors, to excitation by spacecraft disturbances. Mirrors must also be strong enough for safe handling during manufacture and assembly, and withstand launch loads in the case of space telescopes.

Approximating a mirror to a thin, flat, circular plate, the maximum deflection, δ, of a mirror freely supported at the periphery and subjected to a perpendicular uniform load is given by

$$\delta = \frac{3}{16} \frac{q D^4}{E h^3} (1 - \nu^2) \frac{5 + \nu}{1 + \nu}, \qquad (4.32)$$

where D and h are the diameter and thickness of the plate, respectively, E and ν are the Young modulus and Poisson ratio, respectively, of the material, and q is the load per unit area. Since the total load applied is $q\pi D^2/4$, this formula shows that for a given material, the rigidity of the mirror, defined as the load to deflection ratio, is proportional to h^3/D^2:

$$\text{Rigidity} \propto \frac{h^3}{D^2}. \tag{4.33}$$

This result is general: the same dependence is found if support conditions are different. Now, if we are interested in deflection under self-weight, q for a horizontal mirror is equal to $g\rho h$, where g is the acceleration of gravity and ρ is the material density, so that the maximum deflection will be proportional to D^4/h^2:

$$\text{Deflection under self-weight} \propto \frac{D^4}{h^2}. \tag{4.34}$$

Opticians sometimes still use the "aspect ratio," the D/h ratio, as a measure of mirror flexibility. As pointed out by Couder as early as the 1930s [35], Equation 4.34 proves this to be wrong: when comparing mirrors, the true flexibility criterion is the D^4/h^2 ratio.

If we were interested in avoiding vibrations, the criterion would be the natural frequency, which is inversely proportional to the square root of the above ratio:

$$\text{Natural frequency} \propto \sqrt{\frac{\text{Stiffness}}{\text{Mass}}} \propto \sqrt{\frac{h^2}{D^4}} \propto \frac{h}{D^2}. \tag{4.35}$$

As will be seen in Chapter 6, ground-based telescope mirrors can be "floated" in a gravity compensation system to eliminate gravity effects. This is a "passive" open-loop solution which works well only if the mirror is rigid enough to minimize residual errors in the compensation system. As the diameter increases, it becomes more and more difficult to maintain high rigidity since this varies as h^3/D^2, and the mass of the mirror has to be decreased without affecting rigidity significantly. This is done by "lightweighting," that is to say, removing mass from inside the mirror where it least contributes to rigidity. This technique will be discussed in Section 4.7.1.

The alternative is to replace the gravity-compensation system support points by actuators which are commanded to maintain the mirror figure at all times. This technique, referred as "active optics," will be treated in Chapter 8. With active optics, the rigidity requirement is drastically relaxed, with a corresponding reduction in the total mass of the overall mirror assembly (Fig. 4.30).

Mirror blank structures can be classified into four categories according to their rigidity levels: "rigid," "semirigid," "low rigidity" and "very low rigidity." Rigid mirrors can maintain their shapes under gravity load or external disturbance without any support other than the defining support points. Low- and very low-rigidity mirrors, on the other hand, rely wholly on back supports and beds of actuators to maintain their shapes. Semirigid mirrors are in between these two extremes and are able to maintain their shapes with the help of gravity-compensation systems. But flexibility can be an advantage. Once

148   4. Telescope Optics

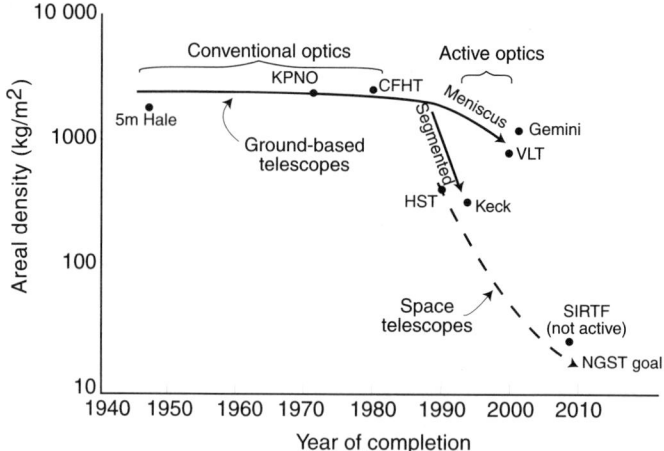

Fig. 4.30. Areal density of mirrors as a function of the year they went into service; mass includes the blank, mirror cell and actuators.

polished, a rigid mirror cannot be corrected for figure errors built in during the figuring process (e.g., because of inaccurate metrology) or due to stress induced by the mirror mounts or changing thermal environment. Low- and very-low-rigidity mirrors, on the other hand, can be corrected for their own figure errors and possibly even be used to correct for wavefront errors originating elsewhere in the optical system or in the atmosphere. These main types of mirror can then also be classified, as far as figure control is concerned, as *no* or *low authority, medium authority*, and *high authority*, as a function of how much shaping of the mirror surface is possible. Figure 4.31 summarizes their respective domains.

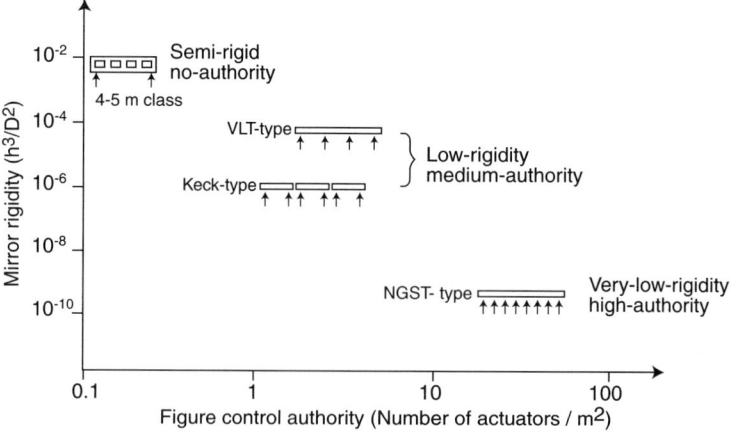

Fig. 4.31. Main types of structural mirror design plotted in terms of rigidity and figure control authority.

For space applications, the HST mirror (2.4 m in diameter, 0.3 m thick, 90% lightweighted) and the SIRTF mirror (0.9 m in diameter) can be considered of the rigid type in the sense that no or very little figure control is possible.

Under gravity loading, only small, thick mirrors are considered truly rigid.[10] Mirrors in the 1- to 8-meter range are either of the semirigid or low-rigidity type. The solid 4 m class traditional mirrors and the lightweighted 5 m Palomar are representative of the semirigid, no-authority type. These mirrors may be supported on gravity compensation systems, but only to "float" them so that they maintain their original figures, not to actively shape their surfaces. The VLT, Gemini, and Subaru 8-meter "meniscus" mirrors and the honeycomb LBT mirror are examples of the low-rigidity, medium-authority type. Although the individual segments of the Keck telescopes are relatively rigid, the mirror, taken as a whole, also belongs to that same low-rigidity, medium-authority class, since some figure control is possible at the scale length of the segments. In space, where disturbance levels are very low, this last (segmented) concept can be pushed to extremely low rigidity, as demonstrated by some of the mirror designs proposed for NGST.

On the ground, no-authority type mirror supports can be used up to a diameter of 4 or 5 meters. If larger than that, all mirror types require medium-authority mirror supports. Lightweight structured mirrors have the advantages of greater rigidity to resist wind loading and a thin-walled structure that can be ventilated to achieve a short thermal time constant.[11] The disadvantages are the risk of print-through due to the rib structure and challenging thermal control if borosilicate glass is used instead of ultralow-thermal-expansion material. The advantages of the lower-rigidity, zero-expansion meniscus mirrors are predictable figuring and a greater range of real-time control of the figure. This figure control can be used to correct primary mirror polishing imperfections and mirror mount and thermal effects, as well as errors in the secondary mirror or in the matching of conic constants. The disadvantages are a longer thermal time constant and a greater sensitivity to wind loading because of lower rigidity.

As for the segmented solution, its main advantages compared to the meniscus are a lower mass and shorter thermal time constant; its drawback is a possible image-quality degradation due to segmentation. But its true strength is in being extendable to unlimited aperture sizes, whereas monolithic mirrors

---

[10]This is deemed the case if $D^4/h^2 < 5$ m^2, where D and h are both expressed in meters [35]. Essentially, any mirror larger than 30 cm in diameter (4 cm thick) is not rigid.

[11]The thermal time constant of a slab of material is proportional to the square of the slab thickness. Large lightweighted mirrors have facesheet and rib thicknesses of around 3 cm, and meniscus mirrors are about 20 cm thick. Consequently, the time to reach thermal equilibrium is better than an order of magnitude shorter for a lightweighted mirror, assuming the use of forced convection to flow air into the cavities of the lightweighted mirror (Chapter 9).

are *de facto* limited in size by handling and transportation constraints and by blank inhomogeneities and residual stresses.

In the following sections, we look in more detail at two of the techniques mentioned above: lightweighting and segmentation.

4.7.1 Lightweighted mirrors

The basic idea behind lightweighting mirrors is to reduce mass without significantly affecting rigidity. Figure 4.32 shows a beam simply supported at the two extremities and subjected to a uniform load. Material is in compression on the upper surface and in tension on the underside. In the middle, along the so-called "neutral axis," normal stress falls to zero. Other than carrying shear loads, material near the neutral axis contributes little to bending stiffness, and most of it can be removed without much reducing rigidity. Since the total mass is reduced, deflection under self-weight will then actually be less than for a solid beam. This is the principle of the familiar "I-beam." The same can be done with mirror blanks, resulting in a "honeycomb" structure, as shown in the upper right in Fig. 4.32.

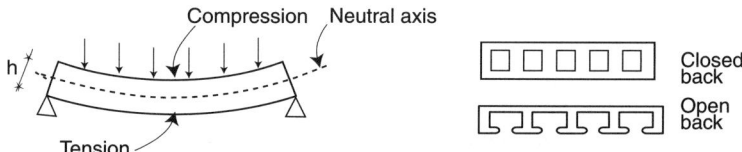

Fig. 4.32. At left, beam under load. At right, cross section through a lightweighted closed-back and open-back blank.

Some materials can be cast directly in a honeycomb, closed-back configuration by leaching away the mandrels via small holes on the bottom face.[12] This is the case for borosilicate glass and some types of beryllium. Other materials cannot be cast in complex shapes and the blank must then be hollowed out by drilling and milling. This is easier to do if the bottom facesheet is omitted as shown in the lower right in Fig. 4.32, a configuration referred to as "open-back." The reduction in stiffness due to lack of the bottom facesheet can be compensated by deepening the ribs. Another approach is to build the honeycomb structure from plate elements and fuse the pieces together. This is how the HST mirror was fabricated (Fig. 4.33).

First-order design and cell shape optimization of lightweighted mirrors are amenable to hand calculations [36, 37], but final design must be done by finite element analysis in order to take into account specific effects, such as those due to variable thickness and reinforcements for mounting points. In prelimi-

---

[12] Such a process, leading to a structure close to the desired final shape of the blank, is called "near net shape."

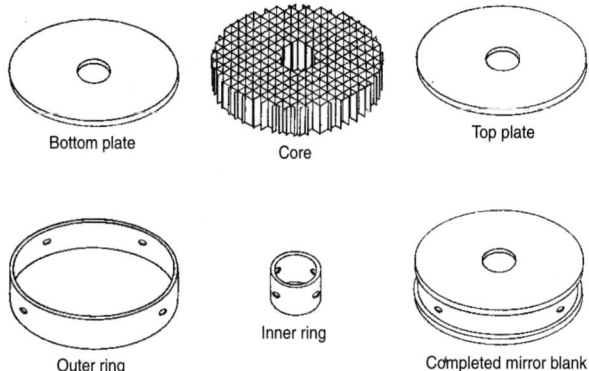

Fig. 4.33. The 2.4 m ULE primary mirror for HST. The mirror was built of components which were then fused together. A 90% lightweight ratio was achieved thanks to this method.

nary analyses, a useful notion is the "equivalent thickness" of a lightweighted mirror. This is the thickness of a solid mirror which would have the same bending stiffness [36].

The limit of how much lightweighting can be achieved is, in general, controlled by manufacturing and handling constraints rather than by structural requirements for actual use. Rib height is limited by shear during handling or launch loads, and minimum rib thickness is on the order of 1 mm. Thinner ribs are difficult to fuse in the built-up method and present a danger of breakthrough if machined out. Face-plate thickness is itself controlled by the grinding/figuring process. If too thin, the face plate will deflect under the weight of the figuring tool, then bounce back once the tool is removed, leading to a "print-through," or "quilting," effect.

This effect can be avoided to some extent by using a "zero-pressure" figuring method in which the mirror face plate is held up against the tool by vacuum instead of having the tool press down on it. However, it is generally better to size the face-plate thickness so that face-plate deflection under tool pressure is negligible.

4.7.2 Segmented mirror systems

The size of monolithic mirrors for ground telescopes is, in practice, limited to about 8 meters. This is principally due to limitations imposed by transportation by road, but also due to the current capacity of furnaces (for melting or annealing solid or assembled blanks) and of existing polishing machines. Although monolithic mirrors larger than 8 meters are not impossible, the cost of the required facilities makes this a breakeven point where segmented systems begin to be more economical.

152   4. Telescope Optics

In space applications, currently available launch vehicles have fairing diameters limited to about 4 meters (see Chapter 12). Launching a telescope with a monolithic elliptic mirror in the 4 x 8 m size range has been proposed [38], but if a larger collecting area is desired, a segmented, deployable primary mirror is the only option.

Modern segmented-mirror telescopes were born with the visionary work of Horn d'Arturo in the 1930s [39], which culminated in a 1-meter primary mirror composed of 61 hexagonal segments. Starting in the 1970s, a number of segmented-mirror systems were proposed and prototyped for military space application (LAMP, ALOT, Pamela), and three have been built for ground-based astronomy applications: the two 10 m Keck telescopes and the Hobby-Eberly 11 m spherical-mirror telescope.

Because segmented systems lose the advantage of "surface continuity" that a monolithic mirror affords, they must satisfy two important conditions. First, the segments must be figured so that they are all parts of the same overall parent shape and, second, once installed in the telescope, they must be positioned exactly on that parent shape and maintained there in spite of changing gravitation direction, thermal effects, and wind disturbance. The first condition requires nontraditional figuring to create the off-axis, or noncylindrically symmetric, surface of the segment. The second requires active control, which is described in Chapter 8.

The fabrication of off-axis optics is challenging and expensive. With the construction of the two Keck telescopes, however, fabrication of 2-meter class off-axis optics has been shown to be achievable at reasonable cost, and modern computer-controlled polishing techniques are likely to facilitate the task further.

Segmentation geometry

There are two categories of segmentation geometry: "petals" (also called "keystone"), which use segments that are slices in radial and azimuthal coordinates, and "hexagons," which use segments of nearly identical-sized hexagons (Fig. 4.34). Petals have the advantages of a circular periphery and fewer different surface shapes. Hexagons have the virtue of requiring only a single support type and lending themselves to more uniform distribution of active control elements. The petal geometry was used for LAMP and ALOT and hexagonal segmentation was used for the Keck and Hobby-Eberly telescopes. The remainder of this subsection emphasizes the hexagonal geometry.

The hexagonal geometry is built up as rings. To allow for the passage of a Cassegrain beam, one can either omit the central segment or use one with a hole in its center. The presence of a central segment improves the performance of an edge-sensing active control system. With two or more rings, this improvement is modest, but a central segment is essential if there is only one ring.

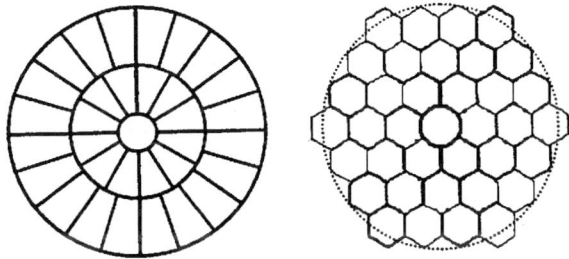

Fig. 4.34. The two main segmented mirror geometries: petal (left) and hexagonal (right).

With no central segment, the number of segments as a function of the number of rings, N_{rings}, is given by

$$N_{\text{segm}} = 3N_{\text{rings}}(N_{\text{rings}} + 1). \tag{4.36}$$

The side length, a, of the hexagon required for an equivalent area circular aperture of diameter, D, is given by

$$a = D\sqrt{\frac{\pi}{6\sqrt{3}\,N_{\text{segm}}}}. \tag{4.37}$$

Unless the primary is spherical, the number of different optical surfaces required is equal to $N_{\text{segm}}/6$ (but note that some segments differ only in the orientation of the hexagonal perimeter, rather than in the off-axis distance).

Table 4.7 gives the number of segments and different surface types as a function of the number of rings.[13] The table also gives the number of positioning actuators required, assuming only three-degree-of-freedom control (tip, tilt and piston — no lateral control), and the number of edge sensors required if such a sensing system is used (see Chapter 8).

Table 4.7. Segmented mirror parameters as a function of the number of rings

Rings	1	2	3	4	5	10	20
Segments	6	18	36	60	90	330	1260
Surface types	1	3	6	10	15	55	210
Position actuators	18	54	108	180	270	990	3780
Edge sensors	12	72	168	300	492	1848	7308

---

[13] Note that for a single ring of segments, the 12 edge sensors are not enough to define the desired lengths of the 18 actuators, or even the 15 degrees of freedom (ignoring the three rigid body motions that are not sensed by the edge sensors for any of the cases). Thus, six segments is not really a very good geometry and seven segments should be used. In such a case there are 24 edge sensors and 3×7 actuators (with 18 sensed degrees of freedom).

4. Telescope Optics

Segment size

Determining the optimal segment size is a complex matter involving many considerations. Smaller segments are

- easier to support against gravity, with deflection growing as d^4/h^2, where d is the segment diameter and h its thickness,
- easier to figure, with the asphericity growing as $d^2 D^2/R^3$, where D and R are the parent mirror's diameter and radius of curvature, respectively,
- easier to handle, requiring less crane capacity and a smaller coating facility,
- less sensitive to radial position error in the array, which grows as $d^2 D/R^3$,
- less sensitive to orientation error (rotation error about the segment center), which grows as $d^2 D^2/R^3$,
- and their large number per given collecting area decreases the impact of a segment active-system failure.

On the other hand, smaller segments require

- more types of optical surface to be figured,
- more spares,
- more actuators (three per segment), more edge sensors (if used),
- a more complex alignment and calibration procedure (with an increased edge sensor error propagation),
- a higher-spatial-resolution wavefront sensing,
- a control system which must compute and control more degrees of freedom,
- a backup structure with a higher density of support points,
- a more complex network of cables,
- and they increase the probability of active component failures.

With so many trade-offs to be made in the selection of segment size (and thus segment number), a detailed quantitative analysis is difficult. In practice, the segment size is selected using a mixture of some quantitative cost estimates and some experienced judgment. The current consensus is that the optimal segment size is in the 1- to 2-meter-diameter range.

From the blank manufacturer's point of view, sizes between 1 and 2 m are optimal. This is the typical size of "boules" and annealing furnace capacities. Larger pieces involve either joining and fusing or the use of larger melting and processing furnaces. From the optician's point of view, however, the optimal size is a function of the size and number of polishing stations, ion figuring tanks, and handling and testing facilities at his disposal. As a rule, cost per area will be lowest for the largest size that can be figured and tested at his facility, which is typically 4 meters or more. But handling risks and time increase with size.

Figure 4.35 shows the notional cost of manufacturing and figuring thin meniscus mirror blanks as a function of their diameter. The actual relationship depends on companies' manufacturing and figuring capabilities and may change with time due to technology advances, capitalization of equipment, and market forces.

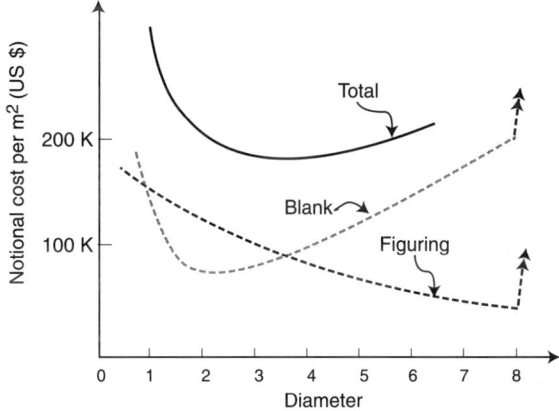

Fig. 4.35. Notional cost per m² of mirror blank manufacture and optical figuring as a function of mirror diameter. To that cost must be added the cost of the mirror's active control system.

4.7.3 Thermal effects

Unless the coefficient of thermal expansion is very near zero for the entire range of temperatures over which a mirror is used, it will deform when thermal conditions change. Clearly, deformation will occur if there is a change in the overall temperature of the mirror, referred to as the "bulk" or "soak" temperature change. This may result from a seasonal change or a difference between the temperature at which the mirror was figured and the temperature at which it is used. But deformation will also take place if a temperature gradient exists within the mirror (e.g., if the optical surface is radiatively cooled by the night sky while the back face remains at the mirror-cell temperature).

A bulk temperature change will, to the first order, affect only the radius of curvature. For a temperature difference of ΔT, the radius of curvature, R, will change by

$$\Delta R = R\alpha\Delta T, \qquad (4.38)$$

where α is the coefficient of thermal expansion. However, this is true only if the blank material is homogeneous. If not, the deformation will be a function of the geometrical distribution of the coefficient of thermal expansion within the blank. Typically, changes in CTE occur over large scales (e.g., from center to edge or across the thickness of the mirror).

In cases where the difference is between the front and back of the mirror and is linear, the change of radius of curvature is given by

$$\Delta R = \frac{R^2 \, \Delta\alpha \, \Delta T}{h}, \qquad (4.39)$$

where $\Delta\alpha$ is the difference in CTE across the mirror thickness h. As an indication, the maximum CTE variation in large Zerodur blanks is on the order of $0.01 \cdot 10^{-6}$ K^{-1} [40].

Nonaxisymmetric CTE distributions do more than just affect the radius of curvature and create wavefront errors, especially in the case of space mirrors used at cryogenic temperatures. The mirrors are figured at room temperature and will deform at the operating cryogenic temperature due to differences in the *integrated CTE* within the blank (integrated between room temperature and the operating temperature). Even with very homogeneous materials, this effect is usually so large that it must be compensated. This is achieved by figuring the mirror in the usual way (at room temperature), measuring the surface errors at cryogenic temperature, and then refiguring the mirror to create surface errors of opposite signs. Two or three such test/refiguring iterations are generally needed, a process referred to as "cryo-null-figuring."

As for the effect of temperature gradients within the blank, they can be complex and are best determined by finite element analysis coupled with thermal analysis.[14] For the simple case where there is a linear gradient between the front and back faces of the mirror, the change of radius of curvature is given by

$$\Delta R = \frac{R^2 \alpha \Delta T'}{h}, \qquad (4.40)$$

where $\Delta T'$ is the temperature difference between the front and back of the mirror.

We note from these equations that a *low* CTE is advantageous in the presence of temperature gradients, whereas, in the presence of soaks, it is low CTE *variability* that is important. We also recall that active optics greatly reduces the practical impact of the thermal issues just examined.

---

[14] A closed-form analytical solution for the distortion of the surface of any isotropic mirror with a parabolic surface caused by linear temperature gradients in all three directions has been derived by Pearson [41]. This solution provides the deformation in terms of piston, tilt, focus, coma, and spherical aberration. This hand calculation is very useful for checking thermal distortion finite element models.

4.8 Mirror production

With current technology, there are three approaches for obtaining high-precision surfaces such as those needed for optical mirrors: machining, lapping, and ion figuring:

- **Machining** consists of removing material with a cutting or grinding tool. The final surface is obtained directly, but the method relies on the intrinsic accuracy of the machine tool.
- **Lapping** consists of rubbing a tool over the workpiece while the latter is rotated. The tool stroke's velocity and pressure control the removal rate (see Preston's law in Section 4.8.1). The tool itself can be the wearing agent, but more uniform wear is obtained by using a loose abrasive slurry between the tool and the workpiece. If the tool is the same size as the workpiece, this automatically produces a spherical surface on the workpiece since a sphere is the only surface that can match itself after rotation and translation. Axi-symmetric aspheric surfaces can be produced by using a tool smaller than the workpiece and adjusting the stroke or pressure as a function of distance to the workpiece center. Since there is no direct measure of the material removed, this method requires periodic optical testing as the work progresses.
- **Ion figuring** consists of removing material by bombarding the workpiece with an ion beam. The amount of material removed can be closely predicted, but this method is capable of only very small corrections.

Conventional machine tools have an absolute accuracy of a few microns at best, which is not sufficient to produce optical quality mirrors. Ultraprecise machine tools have recently been developed but are still limited in size. In practice then, the production of large optical quality mirrors essentially depends on lapping, with possibly ion-figuring touch-ups. Typically, the mirror blank will be cut to rough dimensions at the mirror blank factory by conventional machining, then brought to the optical shop to be made into the final mirror by lapping, using abrasives of successively finer grit. This process is divided into three distinct phases: grinding, polishing,[15] and figuring (Fig. 4.36).

Grinding starts with the rough grinding phase, using a large tool and coarse grit to rapidly bring the surface of the blank to the "best-fit sphere." The best-fit sphere is the sphere from which the least amount of material has to be removed to obtain the desired aspheric surface. In general, this sphere is tangent to the desired surface at a radius of about 0.7 of the outer radius. In the second phase of grinding, a smaller tool is used ($\sim$ 1 meter diameter for

---

[15] The term "polishing" is commonly used to describe the entire process of making the optical surface of a mirror: grinding, polishing, and figuring, not just the polishing phase. To avoid confusion, we take that term in its strict sense: the act of making a mirror surface smooth enough to be specular.

158   4. Telescope Optics

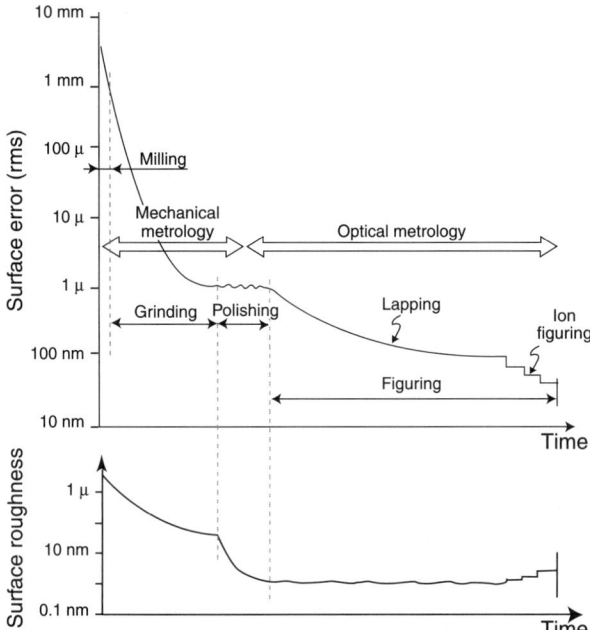

Fig. 4.36. Notional time history of surface errors (top) and surface roughness (bottom) during the milling, grinding, polishing, and figuring phases. Also shown is the method of surface testing used: mechanical until the blank surface is specular, and optical afterward.

an 8-meter mirror) to obtain a surface close to the desired aspheric surface. During the grinding phase, the surface error will drop from about 0.5 mm, as it comes from the blank supplier, to a few micrometers.

The mechanism of wet abrasive lapping is as follows. While the workpiece is rotated, the tool moves by translation and rotation across the optical surface coated with a sludge of abrasive grains and water. Various abrasive materials can be used, depending on the blank material and desired removal rate. Common choices are aluminum oxide, boron carbide, and silicon carbide. The abrasive grains roll between the tool and the surface and, because of their irregular shapes, wear away the surface by creating microfractures from which pieces of material are broken (Fig. 4.37, left). The result is a rough surface on top of a damaged layer which is about four to six times thicker than the peak-to-valley height of the relief surface (Fig. 4.37, right). Using successively finer grit, from 200 down to 5 μm, reduces both the roughness of the surface layer and the depth of the damaged layer. Surface roughness then drops from the 1 to 5 μm rms produced by machining to about 0.2 μm rms. Since this is still not smooth enough compared to the wavelength of visible light for the surface to be specular, mechanical means or infrared testing must be used during this phase to monitor the radius of curvature and surface error.

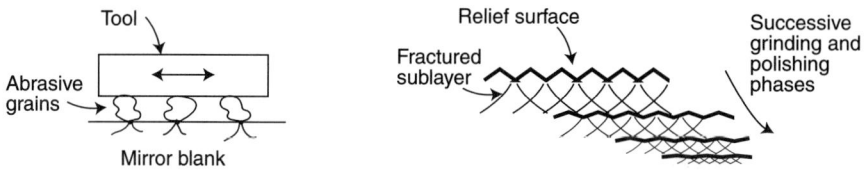

Fig. 4.37. During grinding, the glass is fractured by the abrasive grains under the high pressure occurring at contact points (left). The result is a relief surface on top of a layer with a large number of fractures (right). The roughness of the surface and the depth of the cracks decrease at each stage of the grinding process. At the end of the grinding phase, the thickness of the damaged layer is on the order of 5 to 15 µm.

The next step is to **polish** the mirror surface to render it specular and thus permit the use of more precise optical metrology. This is achieved by using still finer grit with a tool that is covered by blocks of a semirigid substance, usually pitch (a refined pine tar), that can slowly flow to adjust the shape of the tool to precisely fit the curved surface of the workpiece. In polishing, contact pressure and local temperature are such that a process different from grinding is taking place, involving hydrolysis and, possibly, fusion. The result is a surface where cavities and projections are reduced to 10–20 Å rms, allowing the use of optical means to monitor its figure.

The final phase is **figuring**, which consists of bringing the mirror surface to the desired aspheric shape or "figure." This can be achieved by one of several means:

– Using full-sized tools with a tailored contact distribution so that the combination of tool movement and contact area statistically produces the desired figure. This is the traditional technique used for more than a century. It works well as long as the asphericity is not too steep, because a large tool cannot comply to an optical surface with a strongly varying curvature. By nature, it produces surfaces with a low level of mid-frequency errors and cylindrically symmetric wavefront errors. The drawback is that it is difficult to produce off-axis mirrors as required for segmented primary mirrors. It is also a lengthy process requiring many test and contact shape adjustment iterations. These drawbacks have led to the development of new techniques described hereafter.

– Using mid- or small-sized tools having uniform contact with the surface, but using small strokes and adjusting tool pressure and velocity to remove material in a predictable fashion. This is a complex procedure that became possible with numerical control machines; it is called "computer-controlled lapping."

– Using a mid-sized tool which is actively deformed as a function of its position on the mirror blank so as to match the desired figure under it. This is called the "active lap" technique.

– Deforming the workpiece during the figuring process in such a way that, if figured as a sphere, it will bounce back to the desired shape when the deforming stresses are removed. This is called "stressed mirror" figuring.

Very fast optics used to be difficult or impossible to fabricate. This was because, with the old methods, the quality of the final optical figure of an aspherical surface was a function of the departure from the best-fit sphere, roughly inversely proportional to the cube of the f-ratio. Thanks to new figuring methods, optical quality is now independent of the mirror's f-ratio [42], as shown in Fig. 4.38.

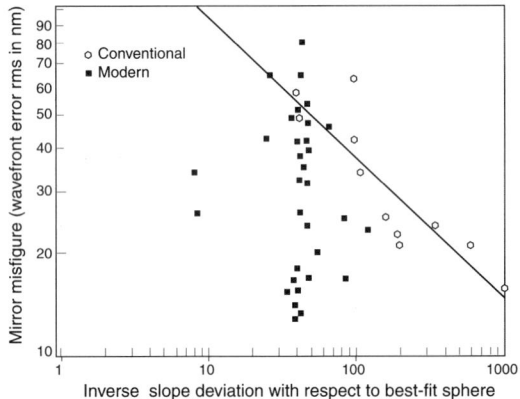

Fig. 4.38. Wavefront error for a large sample of conventional and modern mirrors as a function of the inverse slope of the departure from the sphere at the mirror edge ($dy = 8(f/D)^3/\kappa$, κ being the conic constant), the classical criterion for figuring difficulty. Traditional figuring and testing techniques were indeed limited by this criterion (open circle dots, line), but the lack of correlation for mirrors figured over the last 20 years (black squares) indicates that this difficulty has been fully mastered.

The main features of these last three new techniques are summarized next.

4.8.1 Computer-controlled lapping

Material removal by lapping is governed by Preston's law [43]

$$U = A p v, \qquad (4.41)$$

where U is the wear per unit time, p is the pressure of the tool on the workpiece, v is the relative velocity of the tool with respect to the workpiece, and A is the Preston constant, which depends on the blank and abrasive materials. In computer-controlled lapping (Fig. 4.39), pressure, velocity, and dwell time are automatically controlled as a function of the position of the tool on the workpiece to produce the desired figure in a deterministic way. The parameter to be varied is a matter of choice. Zeiss and Opteon typically vary the tool pressure via actuators on the back of the tool [44, 45, 46], whereas REOSC usually keeps pressure and velocity fairly constant and varies the dwell time.

Fig. 4.39. Computer-controlled lapping of an 8-meter mirror for the VLT (Courtesy of REOSC).

During the figuring process, the mirror should be supported so that it will not deform. Likewise, during optical testing, ground-based mirrors should normally be mounted on a support that matches the telescope zenith-pointing support. The similarity in the support requirements for figuring and testing has led to mirrors being tested and figured on their final active supports. When a different support system is used during figuring, it is important that the points of support be at the same locations as in the final telescope support system.

4.8.2 Stressed mirror figuring

The stressed mirror figuring method consists of deforming the mirror blank, using forces and moments around the edge to induce a surface shape with astigmatism and coma of the opposite sign to that desired for the final surface. While the blank is held in this deformed state, a spherical surface is ground and figured. After figuring, the forces and moments are removed and the mirror elastically deforms into the desired surface shape. The method is well suited to the production of off-axis mirrors, which are difficult to manufacture by conventional methods, and takes advantage of the low cost of figuring a spherical surface using a large tool.

This technique was pioneered by Schmidt to produce aspheric corrector plates [47] and was later systematically studied by Lemaitre for making axisymmetric aspheric mirrors [48]. It has since been used on a grand scale to produce the 36 off-axis 1.8 meter segments of each of the Keck telescopes [49, 50]. As shown in Fig. 4.40, forces and moments were applied around the edge of the blank to induce a deformation with coma and astigmatism components opposite to those desired for the final figure surfaces (up to 30 µm and 100 µm of coma and astigmatism, respectively). These forces and moments were adjusted during the figuring process to correct errors identified during periodic testing. The cutting of the hexagonal sides after completion of the figuring process caused a rebalancing of stresses inside the blank, thus introducing some small, low-frequency errors, which were corrected by ion beam figuring.

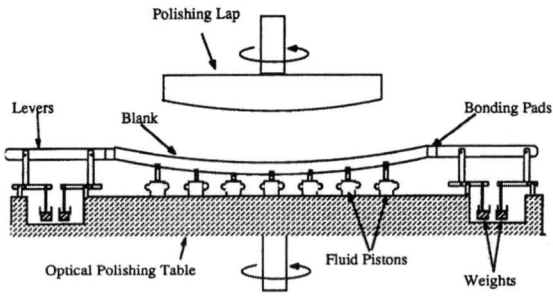

Fig. 4.40. Stressed mirror figuring.

A modification of this technique has been proposed for the figuring of future very large telescope segments [51]. Several segments, turned face down, riding on a large planetary polisher, could be figured simultaneously in order to increase the production rate.

4.8.3 Active lap figuring

An alternate method for figuring fast aspheric surfaces consists of stressing the lap instead of the workpiece. Here, the lap is deformed during its motion so that it matches the desired aspheric surface shape at all times [52] (Fig. 4.41). The advantage, compared to computer-controlled lapping, is that a large tool can be used, thus benefiting from the high removal rate and low level of mid-frequency surface errors inherent to the traditional method. This technique was successfully used at the University of Arizona Steward Observatory to fabricate a variety of secondary and primary mirrors up to 8 m in diameter [53].

4.8.4 Ultraprecision machining

As indicated in the introduction to this section, the accuracy of conventional machining is not sufficient to figure optical surfaces. But machining has clear advantages over lapping: the final surface is obtained directly without lengthy intermediate testing and, with numerical control, any surface shape can be produced. This has led to the development of "ultraprecise machining" to produce steep aspherical surfaces for military and space applications [54]. These

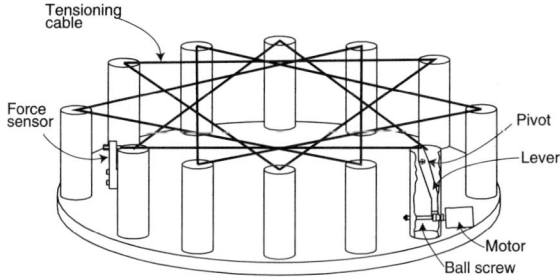

Fig. 4.41. Schematic view of the active lap, developed at the University of Arizona Steward Observatory. The lap plate is deformed by applying moments on its edge via a set of columns, actuators, and tensioning cables. The force applied to the cables is monitored by a sensor on a separate column. (Courtesy of the University of Arizona Mirror Lab.)

surfaces are difficult to fabricate by lapping techniques because the lapping tool needs to be very small to conform to the rapidly changing radius of curvature. The technique relies on an ultraprecise, laser-metrology-controlled, three-dimensional drive moving a diamond tool across the surface while the workpiece is being rotated. The tool is of the single-point type, which removes material in a planing action. The largest capacity and most accurate such machine is the "large optics diamond turning machine" (LODTM) at the Livermore National Laboratory [55]. It can produce mirrors up to 1.5 m in diameter with a surface accuracy of about 30 nm rms. This machine was used to produce secondary mirrors for the Keck telescopes.

4.8.5 Ion beam figuring

Ion beam figuring, which was accidentally discovered by Meinel and colleagues [56], involves the accurate removal of surface material by a beam of high-energy ions. The workpiece is placed in a vacuum chamber and bombarded by a beam of argon atoms which are ionized for acceleration to about 1.5 kV energy and then again made neutral for the beam. The desired removal is achieved by changing the beam's dwell time over different areas on the surface. Ion beam figuring requires that the optical surface be previously polished. The reason is that traditional polishing is a physical and chemical process that removes atoms and moves them around on the surface, smoothing it, whereas ion beam figuring simply removes atoms without any smoothing. Actually, the surface microroughness is slightly degraded by the process and there is even the risk that deep erosion could expose the subsurface damage. Hence, the erosion depth should not exceed a few microns and the number of ion beam figuring runs must also be limited.

Compared to lapping, the process exerts no force on the mirror surface, so that it can be used to figure very flexible optics [57].

Ion beam figuring works well with glass, ceramics and silicon carbide, but poor results are obtained with beryllium due to the close-packed hexagonal grain structure of that material. Current facilities are capable of figuring mirrors up to 2.5 m in diameter. The method can play a crucial role in the manufacture of segmented primary mirrors to correct for edge effects due to postfiguring cutting of the segments sides or to the lapping tool overhanging at the edge if the segments are already cut.[16] The segments of the Keck and Hobby-Eberly telescopes were finished this way.

4.8.6 Postfiguring mechanical deforming

The final figure of a mirror in the telescope may differ from that measured at the end of optical fabrication for a variety of reasons. These include optical testing errors, a mirror support system different from the one used during figuring, errors in the prediction of gravity effects or operating temperature, and installation position errors in a segmented primary mirror. An alternative to sending the mirror back to the optical shop is to deform it mechanically *in situ*.

This method has long been used in older telescopes with counterweight support systems to correct for localized small figure errors: some counterweights would be shifted away from their nominal balance to create a pulling or pushing force on the back of the mirror. Although the correction would only be valid near the zenith, the fix was still beneficial for most observations. A similar approach was used on the CFH telescope to correct the spherical aberration of the secondary mirror found after assembly at the site: the vacuum system used to support this mirror was modified to incorporate a pressure zone at the center so as to bend the mirror in its cell and eliminate the aberration [58].

As another example, the HST primary mirror was provided with a 24-actuator system acting on the back to correct for possible errors made in the prediction of gravity release once in orbit. Unfortunately, the magnitude of the compensation possible with that system was not enough to correct for the spherical aberration found in the mirror after launch, and the error had to be corrected by optical means in the second-generation instruments.

At the Keck telescopes, a "warping harness" is attached to the back of the primary mirror segments to correct most of the errors introduced by the cutting of the segment edges or due to material inhomogeneity (the remaining errors having been corrected by ion beam figuring). Adjustment of the forces and moment is done manually based on wavefront error measurements made with starlight. The current design of the Gran Telescopio Canarias calls for

---

[16]This last effect can be avoided by carefully framing the polygonal mirror with a cemented border of the same material. But it is essential that the cemented borders be stress free.

remote control of these compensating forces to speed up the process. For even larger segmented primary mirrors, such automated control will be essential.

With flexible, actively supported mirrors, the correction of postfiguring errors, even low-spatial-frequency figuring errors for that matter, becomes trivial. Indeed, the beauty of such active optics systems is that they automatically compensate for surface errors made in the figuring process or induced by mechanical or thermal effects occuring *in situ*.

4.9 Optical surface testing during manufacture

The vast subject of optical testing is extensively treated in numerous books and journal articles. The intent here is to summarize the basic optical tests commonly used for testing the telescope main optics during fabrication and, in particular, at acceptance. Exhaustive coverage of this important topic can be found in the standard texts listed in the bibliography at the end of the chapter. This section addresses only the testing of individual elements in the optical shop. Sky-testing of the assembled telescope is discussed in Chapter 8.

4.9.1 Testing philosophy

Compared to most optical testing, the testing of telescope mirrors is especially challenging because of their size, requiring very long or tall testing structures. The relative vibration between the mirror and the test equipment is hard to control and thermally induced variations in the index of refraction of air cause fluctuations in measurements or systematic errors due to air stratification. Another difficulty stems from the large number of iterations required to finish a large aspheric mirror and the associated handling and setup times. Space infrared mirrors are subject to their own set of problems: they must be tested in a vacuum and at cryogenic temperatures, which requires very large tanks found only at national facilities or in the defense industry.

All of this implies that the testing of large optics cannot be improvised and that the corresponding expense will be a major part of mirror manufacture costs. In general, such testing will require a dedicated building, well isolated from surrounding disturbances, and with thermal control to minimize air turbulence and temperature effects on the measuring equipment (Fig. 4.42).

Optical testing is delicate and prone to setup or interpretation errors. To protect against measurement errors which may be costly or even impossible to fix later, opticians have a golden rule:

*Always use **two** methods based on different physical principles.*

The Hubble Space Telescope primary mirror fiasco due to overconfidence in a single test was a painful reminder of the value of this rule.

166   4. Telescope Optics

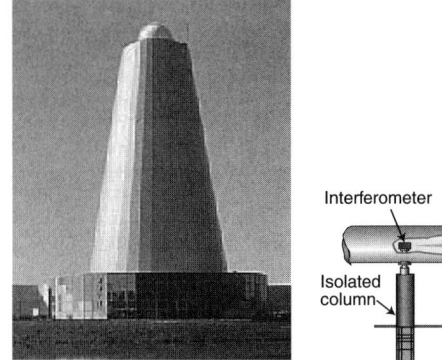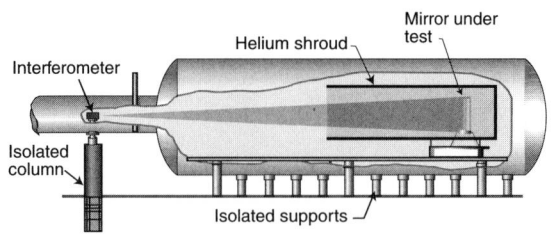

Fig. 4.42. At left, the 30 m test tower at REOSC used to test 8 m class mirrors. At right the 20 m long tank at NASA's Marshall Space Flight Center used to test cryogenic optics for NGST.

In addition to respecting this prime rule, it is prudent to abide by a few others as well:

- At least one method must provide continuous surface information to verify the smoothness and continuity of the test surface.
- When possible, auxiliary optics should be avoided, particularly if the test of such optics is difficult to perform. For any critical testing stage, at least one test method not requiring auxiliary optics should be used, even at the expense of reduced accuracy, in order to circumscribe potential errors.
- A mirror must be tested on a support that is identical in function to its final telescope support and, in order to separate the effects of the polished surface from those due to the support system, the mirror should be tested in several orientations (typically four or more).
- When different techniques are used to obtain complete information on the mirror surface (e.g., spatial frequencies), these techniques should overlap significantly so that intertest confirmations can be made.

Optical testing is the key to success in a telescope project. Optical components can only be made as good as they can be tested, which emphasizes the importance of measurement equipment and techniques. The corollary is that, in general, if the wavefront error can be measured, then the optics can be made better, although this may not always be warranted or economically justified. This is an important concept: limitations in optical quality are generally not due to the processes themselves, but to the surface quality measurement errors.

4.9.2 Main testing techniques

Optical testing techniques fall into three categories which are, in order of increasing accuracy: metrology by physical contact, ray-path testing, and interferometry, the difference between each being an order of magnitude in accuracy.

Physical contact metrology

Physical contact metrology permits the construction of a three-dimensional map of the surface using standard mechanical measurement means. The main tool used, called a "spherometer" or "profilometer," consists of three support points and several (e.g., five) electrical linear transducers attached to a rigid reference bench. Ultimate accuracy is obtained by using these tools in "zero mode," that is, by comparing the piece under investigation to a reference surface with quasi-identical curvature (Fig. 4.43). When measuring large surfaces, the tool is "walked" over the surface by steps, each one typically equal to half the width of the tool [59]. The accuracy of physical contact metrology is on the order of 0.1 μm over a base of about 1 m, leading to a determination of the surface map to about 5–10 μm rms and a determination of the radius of curvature of large optics to about 10 mm for a typical 30-meter radius of curvature. This technique is used during the grinding phase, when the mirror is not yet specular (infrared interferometry can also be used; see later in this section). After polishing, optical methods are preferred because of their higher precision, although mechanical metrology can still be used as a cross check.

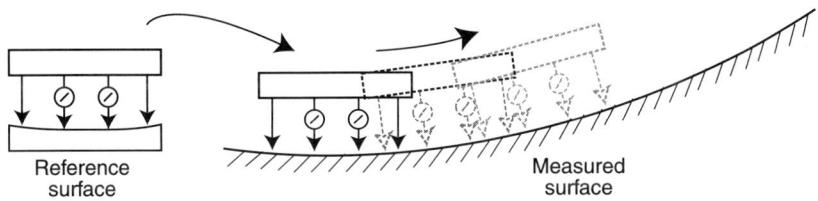

Fig. 4.43. Principle of the spherometer used for measuring surfaces by physical contact metrology.

Ray-path tests

Ray-path tests are the laboratory analog of ray-trace analysis. The most common is the Hartmann test. A mask with holes is placed over the surface to be measured and illuminated with a parallel beam (Fig. 4.44, upper left). Photographs are taken at two positions, ahead of and behind the focus (called intrafocal and extrafocal positions, respectively). By measuring the positions of the dots on these two photographs, one can reconstruct the ray paths near the focus and hence determine the slope of the mirror surface under the mask holes. From these measurements, a map of the mirror surface can be produced, provided that the surface is continuous (this method cannot be used to test a segmented primary mirror as a whole). This classical test requires a parallel beam feeding the entire aperture of the piece being tested, thus a collimator of the same size as that of the mirror. A variation of the test which does not require a collimator and hence can be used with very large mirrors consists of feeding a converging beam from the center of curvature as shown in Fig. 4.44 (lower left). The Hartmann test is a simple, robust technique

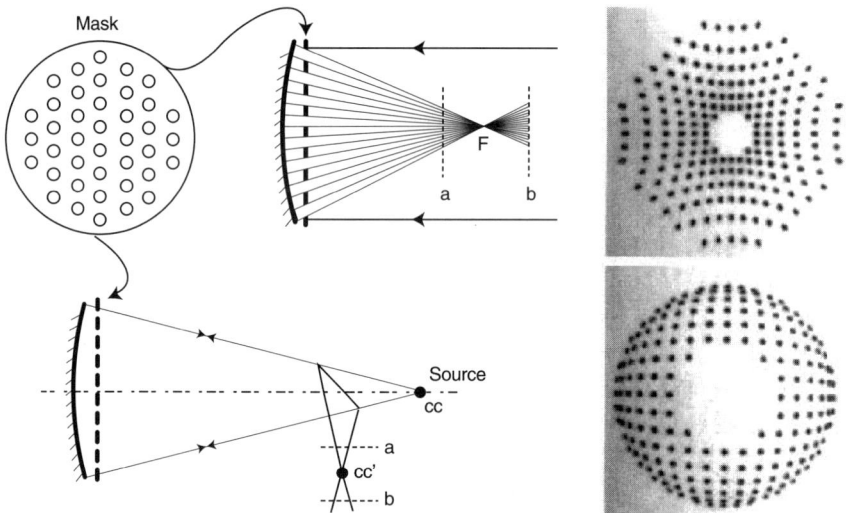

Fig. 4.44. Principle of the Hartmann test. The classical configuration using a parallel beam is shown in the upper left. A variation with a source at the center of curvature of the tested piece is shown in the lower left. As an example, extrafocal and intrafocal Hartmann pictures of a paraboloid mirror are shown to the right.

which is insensitive to vibrations in the setup and to atmospheric turbulence, since the exposures average out these effects. It can be used directly, without auxiliary correcting optics (e.g., null corrector — see below) and is therefore very trustworthy. The accuracy of the method is limited to a fraction of wavelength, mainly because the surface map is obtained by integration of the slope measurements: each successive calculation carries with it the accumulated errors from the previous points. Still, the accuracy is sufficient to serve as a confirmation for the more accurate but more involved interferometric tests.

Modern equivalents of the Hartman test, the "Roddier curvature sensing" and the "Shack-Hartmann" methods, do not require a mask at all. These are well suited for testing assembled telescopes on the sky and will be described in Chapter 8.

Interferometric testing

In contrast to ray-path tests, which provide the slope of the mirror surface at a discrete number of points, interferometric tests measure optical path length variations and provide the shape of the entire surface directly. The principle is to measure the difference between the surface under test and a reference surface by interfering wavefronts returned from the two surfaces (Fig. 4.45).

The technique was routine in optical laboratories but could not be used to test large optics until the advent of lasers. This is because, with incoherent light, the two arms of the interferometer have to be essentially equal, meaning that the reference surface would need to be as large as the mirror under test [60]. With the highly coherent light of lasers, the two paths may be unequal, thus allowing the reference surface to be small and the entire setup of to be of manageable proportions. The resulting device, called a "laser unequal path interferometer" (LUPI), is extremely sensitive, with an accuracy of 1/100 of the wavelength or better, and has become the standard in large optics testing.

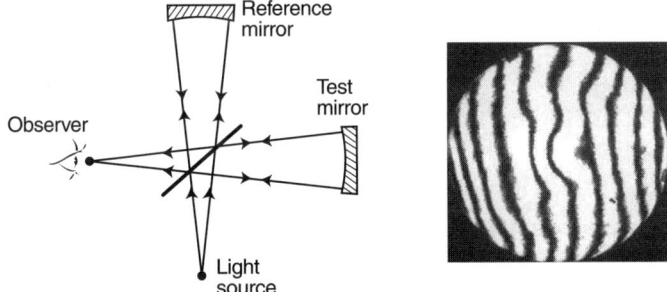

Fig. 4.45. Principle of interferometric testing (left). A source illuminates both the tested surface and a reference surface. The returned wavefronts are combined and the resulting interference pattern is a measure of the difference between the two surfaces (this particular interferometer configuration is called "Twyman-Green"). An example of the interference pattern is shown on the right.

The drawback of the high accuracy of interferometers is that they are highly sensitive to side effects such as air turbulence, temperature gradients, and vibrations. These effects can be reduced with sturdy setups, vibration isolation, air containment, and temperature control, but this is not enough. There are two main approaches to solving this problem.

The first one consists of taking very short exposures with four separate cameras with different phase shifts to explore the full amplitude of the fringe intensity (Fig. 4.46, left). The key is to take these exposures simultaneously, so that they are insensitive to vibration and air turbulence. The corresponding device is called a "simultaneous phase shift interferometer" (SPSI) [61]. The second approach consists of employing an air wedge to create a series of fringes

170   4. Telescope Optics

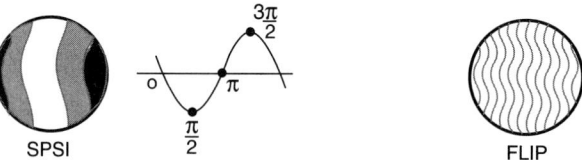

Fig. 4.46. In the standard LUPI interferometer, the fringe pattern reduces to shades of gray when, as shown on the left, the shape of the tested surface is close to that of the reference surface. But vibrations and air turbulence make it difficult to determine the exact intensity. One solution is to introduce deliberate phase shifts (e.g., 0, $\pi/2$, π, $3\pi/2$) to explore the full fringe intensity variation and permit interpolation (middle). Another approach is to use an air wedge to create a large number of fringes across the surface (right).

across the mirror surface to provide sufficient spatial sampling (typically every 10 cm), then taking a large number of very short exposure interferograms to freeze the fringe pattern and average out the effects of air instabilities and vibrations. This method is used by REOSC, for which they have developed an analysis software referred to as "flow interferogram processing" (FLIP) [62]. It typically produces a map of the wavefront error with data points every 3 cm or so and an accuracy of about 10 nm rms. The accuracy, though lower than that of the SPSI device, is sufficient, and the equipment is simpler to use and less costly.

Infrared interferometry

The ray-path and interferometry techniques described above are normally used with visible light and require a polished surface. This would not work for mirror surface testing during the grinding phase. At a wavelength of 10.6 μm (CO_2 laser), however, a mirror surface is optically smooth after fine grinding, allowing the use of interferometry techniques. Infrared interferometry is more accurate than spherometry and is used for checking the asphericity of the mirror surface before polishing. Its drawback is cost, since the required infrared interferometer and null correctors are very expensive.

Surface finish measurement devices

Microroughness of a mirror surface can be measured with a scanning white-light interferometric microscope. This commercially available device makes a three-dimensional map of the test surface by interferometric comparison to an internal reference surface. To reduce risk of damage to the mirror, a replica of the mirror surface is made using a resin compound, and the surface finish of the replica is measured in the laboratory. Typically, the size of the sample is on the order of a few centimeters, the map has a spatial resolution of 1.5 mm, and the vertical accuracy is in the Angström range.

4.9.3 Testing the figure of primary mirrors

It is simple to test a spherical concave mirror: one puts a source at the center of curvature and examines the returning image at that point. If the mirror is a perfect sphere, the image should be perfect, since incoming rays normal to the surface return along the same path. To do the same test with an aspheric primary mirror, one puts an optic called a "null corrector" near the center of curvature of the mirror.[17] The null corrector corrects for the difference between the aspheric mirror surface and a perfect sphere. In other words, seen through the null corrector, the mirror appears "spherical" (Fig. 4.47).

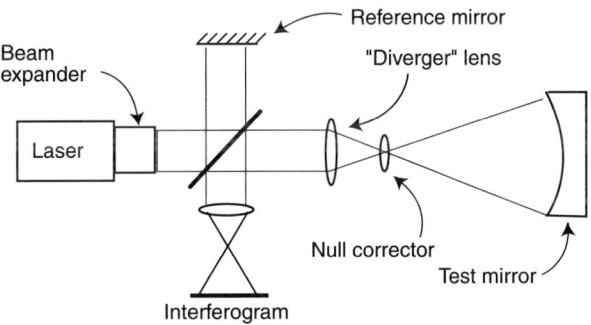

Fig. 4.47. Null test for aspheric concave surfaces.

This is a standard test, but it relies on the null corrector prescription and installation being exactly right. This is where the problem with the Hubble telescope primary mirror arose: the field lens in the null corrector was improperly positioned. As a result, the mirror was figured "perfectly" but to the wrong prescription. One would have caught such a problem with an "end-to-end test," that is to say, by using the telescope in its final configuration, with the primary and secondary, and looking at a source at infinity. The difficulty is in having a source at infinity. One can create such a source with a "collimator," but that collimator has to be as large as the mirror being tested.

An alternate to the end-to-end test consists of "double checking" the null corrector. This is now done quite simply by generating a "hologram" with a computer (Fig. 4.48). The hologram is used to create a wavefront that simulates a perfect primary mirror. If there is no error in the null corrector, the interferogram from the computer-generated hologram (CGH) will look perfect. In other words, this is like comparing the "physical optics" of the null corrector to a "mathematical" prescription [63, 64, 65].

---

[17] Null correctors are generally composed of two lenses or mirrors.

172   4. Telescope Optics

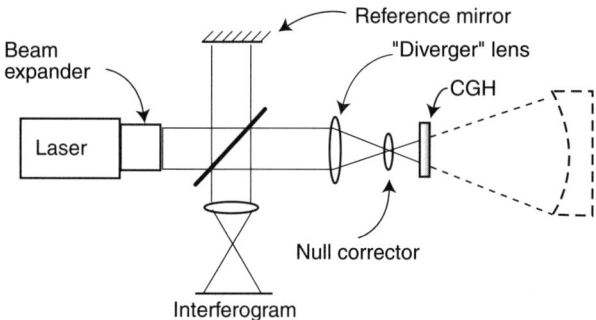

Fig. 4.48. Verification of a null corrector with a CGH. A computer-generated hologram is used to create a wavefront that simulates a perfect primary mirror. If there is no error in the null corrector, the interferogram from the CGH will look perfect.

4.9.4 Testing secondary mirrors

Contrary to concave mirrors, convex mirrors cannot form a real image of a real source. Auxiliary optics are always required. Secondary mirrors are hyperboloids, and the classic secondary mirror test, the "Hindle sphere test," makes use of the Apollonius theorem for conics shown in Fig. 4.1. By placing a spherical mirror with its center of curvature at the back focus of the secondary mirror, a point source located at the front focus will be imaged at that same point in a purely stigmatic way (Fig. 4.49, left).

The drawback here is that the Hindle sphere is somewhat larger than the secondary mirror, which creates a nontrivial manufacturing and supporting problem. Several variations of this method have been proposed [66], as well as new techniques using diffractive plates and computer-generated holograms [67]. An example of one such method, the Hindle-Simpson test, where the spherical mirror is replaced by a transmission meniscus located directly in front of the mirror to minimize its size, is shown in Fig. 4.49 (right).

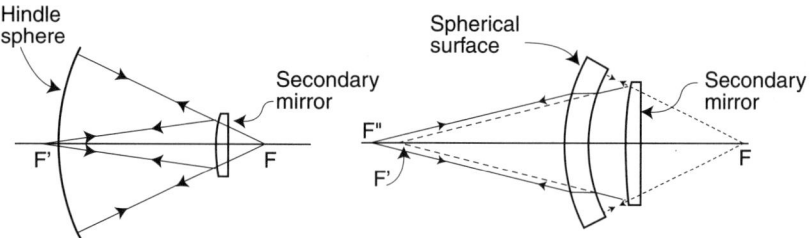

Fig. 4.49. At left, the classic Hindle sphere test. At right, the Hindle-Simpson test where the Hindle sphere is replaced by a transmission meniscus (with the source placed at F''').

4.9.5 Measuring the radius of curvature

The problem with interferometric testing of the primary mirror surface is that one cannot untangle the radius of curvature from the wavefront error measurement. To the first order, a radius of curvature which is slightly off can compensate for errors in the asphericity coefficients and lead to the same interferogram as one with correct values. The radius of curvature then has to be measured independently by mechanical means. Spherometers give an approximate value, but a more accurate method is to locate the center of curvature by optical means, then measure its distance to the mirror vertex by stacking reference Invar gauges. The accuracy of such a measure is on the order of ± 1 mm for a typical 30 m radius of curvature.

4.9.6 Eliminating the effect of gravity

Space telescope mirrors present a particular problem because they are designed to have the right figure in orbit in the absence of gravity, yet must be tested on the ground. There are two approaches.

If the mirror is rigid enough so that its deformation under gravity is not excessive, it is simplest to test it with the axis horizontal, to minimize sag, and then average the results of the measurements by rotating it around its optical axis (say six times with 60° rotations). An alternative, somewhat more involved method consists of measuring the surface first with the mirror facing up, then facing down, and averaging the results.

The second approach consists of compensating the effect of gravity by a series of counterweights pulling or pushing on the back and sides of the mirror, in a fashion similar to that of a ground-based telescope mirror.

4.9.7 Testing cryogenic mirrors

Testing mirrors at cryogenic temperatures is difficult because they must be placed in a vacuum tank, and this complicates the optical measurement setups and creates a noisy environment detrimental to interferometric measurements. The large temperature swing between ambient and cryogenic temperatures also plays havoc with mirror mounts, disrupting pretest alignments.

The solution used to test NGST mirrors at 40 K consists of placing them on a precision, remote-controlled, five-degree-of-freedom stage so that they can be realigned when the operating temperature is reached. The interferometric wavefront sensor is located outside the vacuum tank, viewing the mirror through a precision optical port (Fig. 4.42, right). To minimize uncorrelated vibrations, the interferometer is mounted on the same foundation as the tank itself. During optical tests, vacuum pumps and unnecessary equipment are turned off to reduce the vibration level.

4.10 Mirror coatings and washing

The best coatings for optical telescopes are aluminum, silver, and gold, which have good reflectance in the visible and infrared (Fig. 4.50).

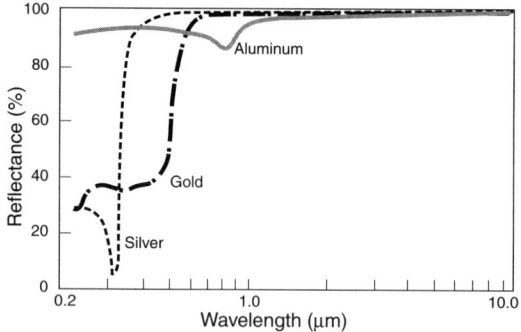

Fig. 4.50. Reflectance of evaporated metals in the visible and infrared.

Aluminum is the reflective coating of choice for general-purpose, ground-based telescope mirrors because of its very good reflectance from the ultraviolet to the infrared and because it can be easily deposited and removed. The reflectance of silver is greater than that of aluminum at all wavelengths longer than 400 nm, but it must be protected against oxidation [68]. This can lead to losses in reflectance at some wavelengths due to destructive interference in the overcoat and also requires special coating equipment.

Gold, on the other hand, is the coating of choice for mirrors dedicated to infrared use. This might be the case of a secondary mirror dedicated to infrared observations, on condition that the visible part of the spectrum is not required for auxiliary functions such as guiding. Bare gold has the disadvantage of being soft and easily damaged, and appropriate cleaning techniques must be used. Gold can be overcoated with a single layer for protection and with multilayer dielectrics for selective enhancement, but both options reduce reflectance over a portion of its useful range. The use of bare gold produces the most uniform high reflectance throughout the infrared spectrum.

Ideal coat thickness for bare aluminum is around 100 nm. If less than that, the coating is still somewhat transparent; if much more than that, uniformity suffers because this is generally a function of the amount deposited. Typically, variations in coating thickness across a large mirror are on the order of 5%.

4.10.1 Mirror cleaning

Mirror coatings degrade over time due to exposure to dust, pollen, and molecular contaminants (e.g., oil and water drops). On ground-based telescopes, reflectance can decrease by as much as 0.5% per month, with a corresponding increase in scatter and infrared emissivity [69], so that periodic cleaning is in order. Even in the very clean environment used for construction of space telescopes, accumulation of particulates over time may reach several percent of the mirror area, which warrants the cleaning of the optics before final assembly and launch. The principal cleaning techniques used in observatories are as follows.

Washing

Washing with water and a mild detergent is the traditional cleaning method for mid-sized telescopes. This is effective in removing dust and molecular contaminants. When the mirror cell design permits, the mirror can be left in the telescope for washing if inflatable seals are used to protect sensitive equipment. Washing is a simple operation, takes only a few hours, and can be performed every few months to maintain a high level of cleanliness between recoatings [70]. It works well for mirrors up to 4 m in diameter because the full surface can easily be reached from the edge and the center hole. For larger apertures, however, this is more of a problem.

Plastic film peeling

This method is routinely used in optical laboratories for cleaning small optics. It consists of brushing or spraying the mirror surface with a polymer liquid which dries to a rubbery film. The film is then peeled away, removing dirt and deposit down to the molecular level. The method works well with small optics [72], but is virtually impossible to use on large ones.

Blowing gas

Dust can be blown off using a jet of filtered air or nitrogen. This method was used to clean the HST mirror before launch and the residual dust coverage was less than 2%. Blowing gas only removes particles larger than 20 μm [71] and is not very effective in removing much larger particles unless high jet velocity is used, which risks scratching the surface. This technique has now been mostly abandoned in favor of CO_2 snow cleaning.

CO_2 snow cleaning

The method consists of spraying CO_2 snow across the surface of the mirror with an apparatus similar to a fire extinguisher [73, 74]. Liquid CO_2 is released through a fine nozzle and becomes a mixture of gaseous CO_2 and dry ice. Two effects occur: (1) snowflakes collide with dust particles and force them away and (2) dust particles freeze, contract and break away from the surface. Snow cleaning is efficient and safe, as it is a noncontact method. It can be applied over large areas and has become routine in major observatories, where it is applied on a more or less monthly basis.

CO_2 snow cleaning has also been proposed for space telescopes while in orbit. The aim is to guarantee perfectly clean optics for infrared observatories by removing the particulate contamination that inevitably occurs during integration and test and during the high-vibration launch phase [75].

4.10.2 Coating plant

Space telescope mirrors are by nature coated only once, usually in an industrial facility, and kept in a clean room until launch. In orbit, degradation is negligible. Ground telescope mirrors, on the other hand, are exposed to a much less clean environment. As shown in Fig. 4.51, their coatings degrade after a few years and need to be renewed. Because of the time and risk involved in transporting the mirror elsewhere for recoating, a dedicated coating facility is usuallly required at the observatory itself.

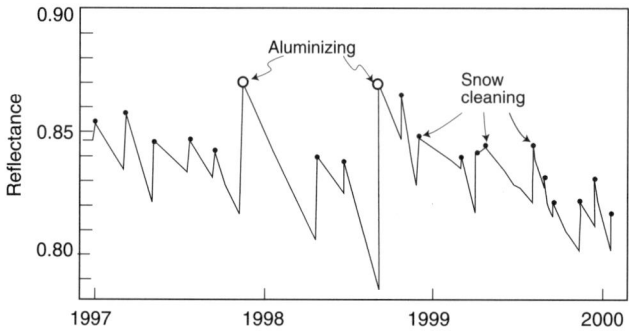

Fig. 4.51. Despite regular snow cleaning, mirror reflectance degrades with time, and realuminizing is required every 2 years or so. (Data from the William Hershell Telescope [69].)

Except for the old "mirror silvering" technique, which was used until the 1930s, all coatings are applied in a vacuum tank with a vacuum on the order of 10^{-6} torr.[18] There are three main methods for depositing metal coats.

The **thermal evaporation** method consists of heating the coating metal to sublimation a short distance from the mirror surface. The molecules travel unimpeded in the vacuum of the tank and condense on the mirror surface. Typically, small wire coils of the coating metal, aluminum for instance, are placed on an array of tungsten filaments positioned about 1 m away from the mirror surface. The filaments are heated electrically until all of the aluminum has evaporated [76, 77]. To eliminate the risk of molten drops falling on the mirror, it is best to place it vertically or upside down in the tank. The thickness and uniformity of the coat is controlled by weighing the coils of aluminum placed on each filament. Prior to coating, the mirror is given a final cleaning in the tank by glow discharge to improve adhesion. The advantage of this

---

[18]torr is a unit of pressure, named after E. Torricelli, equal to 1 mm of mercury (about 133 Pa).

technique is simplicity of the equipment and good coat uniformity. Typical reflectance of coatings obtained with this method under actual observatory conditions is shown in Fig. 4.52.

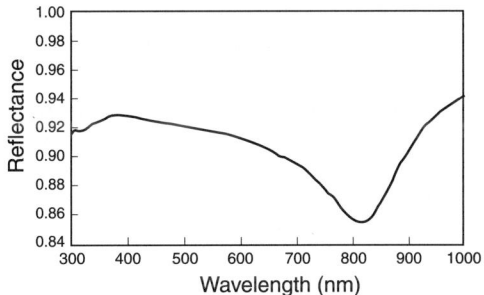

Fig. 4.52. Typical reflectance of a fresh aluminum coat obtained in an observatory coating facility. (From Ref. [77].)

The **electron beam method** consists of evaporating the coating metal by heating it with a beam of electrons instead of direct heat. The advantage is that the amount of evaporated metal can be well controlled.

The last method, **ion sputtering**, uses mechanical action to spray the coating metal rather than using heat. It consists of bombarding the coating metal with an ion beam of an inert gas (argon). The device used for that purpose is called a magnetron. Some of the particles of the coating metal which are detached during the process end up on the mirror and accumulate to form the desired coat. Ion sputtering has several advantages: (1) the mirror can be placed face up, since there is no risk of falling drops, (2) the height of the vacuum tank is reduced, as the coating metal source need only be a few centimeters above the mirror surface, (3) the adhesion of the coating to the substrate is better, and (4) multicoat deposits are possible. The magnetron can be a single source that spirals around the mirror, or a fixed radial slot under which the mirror is rotated.

Whatever the process employed, it is useful to have "coating witness slides" prepared whenever a large mirror is coated. These witness slides, which are simply microscope slides that are coated alongside the mirror, provide a means of measuring the quality of the coating. They can also be placed at strategic locations in the telescope to study how a number of variables affect the reflectance of mirrors *in situ*.

References

[1] Nelson, J. and Temple-Raston, M., *The Off-axis Expansion of Conic Surfaces*, University of California TMT Report No. 9, 1982.

4. Telescope Optics

[2] Roddier, C., Graves, J.E., Northcott, M.J., and Roddier, F., Testing optical telescopes from defocused stellar images, SPIE Proc., Vol. 2199, p. 1172, 1994.

[3] Born, M. and Wolf, E., *Principles of Optics*, Pergamon Press, 1989, p. 464.

[4] Malacara, D., *Optical Shop Testing*, Appendix 2, John Wiley & Sons, 1978, p. 489.

[5] Mahajan, V.N., Zernike annular polynomials for imaging systems with annular pupils, J. Opt. Soc. Am., Vol. 71, No. 1, p. 75, 1981.

[6] Roddier, N., Atmospheric wavefront simulation using Zernike polynomials, Opt. Eng., Vol. 29, No. 10, p. 1174, 1990.

[7] Wetherell, W.B., *The calculation of image quality*, Applied Optics and Optical Engineering, Vol. VIII, Academic Press, 1980, p. 199.

[8] Hayes, J., *Fast Fourier Transforms and Their Applications*, Applied Optics and Optical Engineering, Vol. XI, Academic Press, 1992, p. 55.

[9] Schroeder D. J., *Astronomical Optics*, Academic Press, 2000.

[10] Redding, D.C. and Breckenridge, W.G., Optical modeling for dynamics and control analysis, J. Guidance, Vol. 14, No. 5, p. 1021, 1991.

[11] Hasan, H. and Burrows, C.J., Telescope image modeling (TIM), PASP, Vol. 107, p. 289, 1995.

[12] Maréchal, A., Etude des effets combinés de la diffraction et des aberrations géometriques sur l'image d'un point lumineux, Rev. Opt. Théor. Instrum., No. 9, p. 257, 1947.

[13] Strehl, K., Ueber Luftschlieren un Zonenfehler, Zeitschr. Instrum., Vol. 22, p. 213, 1902.

[14] Wetherell, W.B., *The Calculation of Image Quality*, Applied Optics and Optical Engineering, Vol. VIII, Academic Press, 1980, p. 198.

[15] Wilson, R.N., *Reflecting Telescope Optics I*, Springer-Verlag, 1996.

[16] King, H.C., *The History of the Telescope*, Dover, 1979, p. 76.

[17] Wilson, R.N., *Reflecting Telescope Optics II*, Springer-Verlag, 1999, p. 232.

[18] Wetherell, W.B. and Rimmer, M.P., General analysis of aplanatic Cassegrain, Gregorian and Schwarzschild telescopes, Appl. Opt., Vol. 11, No. 12, p. 2817, 1972.

[19] Meinel, A.B. and Meinel, M.P., Wind deflection compensated zero-coma telescope truss geometry, SPIE Proc., Vol. 628, p. 403, 1986.

[20] Swanson, P.N., Meinel, A.B., et al., A system concept for a moderate cost Large Deployable Reflector (LDR), SPIE Proc., Vol. 571, p. 233, 1985.

[21] Bely, P.Y., The NGST 'Yardstick mission', Proc. 34th Liège International Astrophysics Colloquium, *The Next Generation Space Telescope*, ESA SP-429, p. 159, 1998

[22] Dierickx, P., Beletic, J., Delabre, B., Ferrari, M., Gilmozzi, R., and Hubin, N., The optics of OWL 100 m adaptive telescope, Proc. of *Bäckaskog Workshop on Extremely Large Telescopes*, Lund Univ. & ESO, 1999

[23] Dierickx, P., Optical performance of large ground-based telescopes, J. Mod. Opt., Vol. 39, No. 3, p. 569, 1992.

[24] Ferguson, H., et al., *Image quality guidelines*, NGST Monograph No. 7, Space Telescope Science Institute, 2001.

[25] Borra, E.F., Polarimetry at the coudé focus, PASP, Vol. 88, p. 548, 1976.

[26] Babcock, H.W., *Astronomical Techniques*, Vol. II of *Star and Stellar Systems*, Hiltner, W.A., ed., Univ. of Chicago Press, 1962, p. 107.

[27] Petro, L. and Stockman, H.S., *Optimal pixel scales for NGST*, poster presented at the AAS General Meeting, Atlanta, Jan. 2000.

[28] Bowen, I. S., in *The construction of Large Telescopes*, Crawford, D.L., ed., IAU Symposium 27, Academic Press, 1965, p. A7.

[29] Barnes, W. P., *Optical materials – reflective*, Applied Optics and Optical engineering, Vol. VII, Academic Press, 1979, p. 97.

[30] Paquin, R.A., Properties of Metals, in *Devices, Measurements and Properties, Handbook of Optics*, 2nd ed., Vol. II, McGraw-Hill, 1995, Chap. 35.

[31] Hill, J.M., Angel, J.R.P., Lutz, R.D., Olbert, B.H., and Strittmatter, P.A., Casting the first 8.4 meter borosilicate honeycomb mirror for the Large Binocular Telescope, SPIE Proc., Vol. 3352, p. 172, 1998.

[32] Knohl, E-D., Schoeppach, A., and Pickering, M.A., Status of the secondary mirrors (M2) for the Gemini 8 m telescopes, SPIE Proc., Vol. 3352, p. 258, 1998.

[33] Paquin, R.A., Magida, M.B., and Vernold, C.L., Large optics from silicon carbide, SPIE Proc., Vol. 1618, p. 53, 1991.

[34] Deyerler, M., Pailer, N., and Wagner, R., Ultra-lightweight mirrors: recent developments of C/SiC, SPIE Proc., Vol. 4003, p. 73, 2000.

[35] Danjon, A. and Couder, A., *Lunettes et Telescopes*, Blanchard, p. 572, 1935.

[36] Metha, P., Flexural rigidity characteristics of lightweighted mirrors, SPIE Proc., Vol. 748, p. 158, 1987.

[37] Valente, T. M. and Vukobratovich, D., A comparison of the merits of open back, symmetric sandwich and contoured back mirrors as lightweighted optics, SPIE Proc., Vol. 1167, p. 20, 1989.

[38] Angel, R., Martin, B., Sandler, D., Wolf, N., Bely, P., Benvenuti, P., Fosbury, R., Laurance, R., Crocker, J., and Giacconi, R., The Next Generation Space Telescope: a monolithic mirror candidate, SPIE Proc., Vol. 2807, p. 354, 1996.

[39] Horn d'Arturo, G., Altri esperimenti con lo specchio a tasselli, Pubb. Osserv. Univ. Bologna, Vol. V, 11, 1950.

[40] Müller, R., Höneß, H., Morian, H., and Loch, H., Manufacture of the first primary blank for the VLT, SPIE Proc., Vol. 2199, p. 164, 1994.

[41] Pearson, E. and Stepp, L., Response of large optical mirrors to thermal distributions, SPIE Proc., Vol. 748, p. 215, 1987.

[42] Dierickx, P., Optical fabrication in the large, Proc. of *Bäckaskog Workshop on Extremely Large Telescopes*, Lund Univ. & ESO, p. 224, 1999.

[43] Preston, F.W., The theory and design of plate glass polishing machines, J. Soc. Glass Techn., Vol. 11, No. 42, p. 214, 1927.

[44] Beckstette, K. and Heynacher, E., A new fabrication technology for large mirrors, Proc. ESO Conf. on *Very Large Telescopes and Their Instrumentation*, Vol. I, ESO, Garching, p. 341, 1988.

[45] Korhonen, T. and Lappalainen, T., Computer-controlled figuring and testing, SPIE Proc., Vol. 1236, p. 691, 1990.

[46] Korhonen, T. and Lappalainen, T., Computer-controlled figuring method for thin and flexible mirrors, SPIE Proc., Vol. 2199, p. 176, 1994.

[47] Schmidt, B., Ein lichtstarkes komafreies Spiegelsystem, Mitt. Hamburg Sternwart, Vol. 7, No. 36, p. 15, 1932.
[48] Lemaitre, G., New procedure for making Schmidt corrector plates, Appl. Optics, Vol. 11, No. 7, p. 1630, 1972.
[49] Lubliner, J. and Nelson, J.E., Stressed mirror polishing, Appl. Opt., Vol. 19, No. 14, p. 2332, 1980.
[50] Mast, T.S. and Nelson, J.E., The fabrication of large optical surfaces using a combination of polishing and mirror bending, SPIE Proc., Vol. 330, p. 139, 1990.
[51] Nelson, J.E. and Mast, T. S., Giant optical devices, in Proc. of *Bäckaskog Workshop on Extremely Large Telescopes*, Lund Univ. & ESO, 1999.
[52] Smith, B.K., Burge, J.H., and Martin, H.M., Fabrication of large secondary mirrors for astronomical telescopes, SPIE Proc., Vol. 3134, p. 5, 1997.
[53] Anderson, D., Martin, H., Burge, J., Ketelsen, D., and West, S., Rapid fabrication strategies for primary and seconday mirrors at Steward Observatory Mirror Laboratory, SPIE Proc., Vol. 2199, p. 199, 1994.
[54] Benjamin R.J., Diamond machining applications and capabilities, SPIE Proc., Vol. 433, p. 2, 1983.
[55] Donaldson, R.R. and Patterson, S.R., Design and construction of a large vertical axis diamond turning machine, SPIE Proc., Vol. 433., p. 62, 1983.
[56] Meinel, A.B., Bashkin, S., and Loomis, D.A., Controlled figuring of optical surfaces by energetic ionic beams, Appl. Opt., Vol. 4, p. 1674, 1965.
[57] Wilson, S.R., Reicher, D.W., and McNeil, J.R., Surface figuring using neutral ion beams, SPIE Proc., Vol. 966, p. 74, 1988.
[58] Bely, P.Y., Salmon, D.A., Wizinowhich, P.L., and Tournaire, A., Bending the CFHT Cassegrain secondary for optical figure improvement, SPIE Proc., Vol. 444, p. 253, 1984.
[59] Espiard, J., Controle par sphérometrie des grandes surfaces aspheriques, ESO/CERN Conference on *Large telescope design*, Geneva, 1971.
[60] Malacara, D., *Optical Shop Testing*, John Wiley & Sons, 1978, p. 48.
[61] Koliopoulos, C., Simultaneous phase shift interferometer, SPIE Proc., Vol. 531, p. 119, 1991.
[62] Esnard, D., Maréchal, A., and Espiard, J., Progress in ground-based optical telescopes, Rep. Prog. Phys. Vol. 59, No. 5, p. 601, 1996.
[63] Wyant, J.C. and Bennet, V.P., Using computer gernerated holograms to test aspheric wavefronts, Applied Optics, Vol. 11, No. 12, p. 2833, 1972
[64] McGovern, A.J. and Wyant, J.C., Computer generated holograms for testing optical elements, Appl. Opt., Vol. 10, p. 619, 1971.
[65] Wyant, J.C. and O'Neill, P.K., Computer generated hologram; null lens test of aspheric wavefronts, Appl. Opt. Vol. 13, p. 2762, 1974.
[66] Wilson, R.N., *Reflecting Telescope Optics II*, Springer-Verlag, 1988, p. 86 sqq.
[67] Burge, J.H., Measurement of large convex aspheres, SPIE Proc., Vol. 2871, p. 362, 1996.
[68] Jacobson, M.R. et al., Development of silver coatings optics for the Gemini 8-meter telescope projects, SPIE Proc., Vol. 3352, p. 477, 1998.
[69] Benn, C.R., Increasing the productivity of the WHT, SPIE Proc., Vol. 4010, p. 64, 2000.

[70] Magrath, B. and Nahrstedt, D., A cleaning process for the CFHT primary mirror, PASP Vol. 108, p. 620, 1996.

[71] Kozicki, M., Hoenig S., and Robinson, P., *Clean Rooms Facilities and Practice*, Van Nostrand Reinhold, 1991.

[72] Bennett, J.M. and Ronnow, D., Test of Opticlean strip coating material for removing surface contamination, Applied Optics, Vol. 39, No. 16, 2000.

[73] Zito, R., Cleaning large optics with CO_2 snow, SPIE Proc., Vol. 1236, p. 952, 1990

[74] Kimura, W.D., Kim, G.H., and Balick, B., Comparison of laser and CO2 snow for cleaning large astronomical mirrors, PASP, Vol. 107, p. 1, 1995.

[75] J. Clark, J., *Active Cleaning Experiment for SBIRS*, Presentation by Raytheon Systems Co. to Goddard Space Flight Center, March 18, 1999.

[76] Sabol, B.A. et al., Evaporative coating systems for very large astronomical mirrors, SPIE Proc., Vol. 1236, p. 940, 1990.

[77] Atwood, V. and Sabol, B.A., Studies of some aspects of aluminizing large astronomical mirrors, ESO Conference on *Progress in Telescope and Instrumentation Technologies*, 1992.

Bibliography

Optical Design

Born, M. and Wolf, E., *Principles of Optics*, Pergamon Press, 1989.

Driscoll, W.G. and Vaughan, W., *Handbook of Optics*, McGraw-Hill, 1978.

King, H.C., *The History of the Telescope*, C. Griffin & Co., 1995.

Korsch D., *Reflective Optics*, Academic Press, 1991.

Maréchal, A., *Traité d'Optique Instrumentale*, published by Revue d'optique theorique et instrumentale, 1952.

Riedl, M.J., *Optical Design Fundamentals for Infrared Systems*, SPIE Tutorial Texts in Optical Engineering Vol. TT20, SPIE Press, 1995.

Schroeder, D.J., ed., *Selected Papers on Astronomical Optics*, SPIE Milestone Series, Vol. MS 73, SPIE Press, 1993.

Schroeder D. J., *Astronomical Optics*, Academic Press, 1999.

Wilson, R.N., *Reflecting Telescope Optics I*, Springer-Verlag, 1996.

Wyant, J.C. and Creath, K., *Basic Wavefront Aberration Theory for Optical Metrology*, Applied Optics and Optical Engineering, Vol. XI, Academic Press, 1992.

Mahajan, V.N., ed., *Selected Papers on Effects of Aberrations in Optical Imaging*, SPIE Milestone Series, Vol. MS74, SPIE Press, 1994.

Fourier optics

Goodman J.W., *Introduction to Fourier Optics*, McGraw-Hill, 1996.

Hayes, J., *Fast Fourier transforms and their applications*, Applied Optics and Optical Engineering, Vol. XI, Academic Press 1992.

Mariotti, J.-M., Introduction to Fourier optics and coherence, in *Diffraction-limited Imaging with Very Large Telescopes*, Alloin, M. and Mariotti, J.-M., eds., Kluwer Academic Publishers, 1989.

Reynolds, G.O., DeVelis, J.B., Parrent, G.B., and Thompson, B.J., *Tutorials in Fourier Optics*, SPIE, 1989.

Image quality criteria

Wetherell W.B., *The calculation of image quality*, Applied Optics and Optical Engineering, Vol. VIII, Academic Press, 1980.

Mirror blank materials

Barnes, W.P., *Optical Materials – Reflective*, Applied Optics and Optical Engineering, Vol. VII, Academic Press, 1979.

Musikant, S., *Optical Materials*, Marcel Dekker, 1985.

Paquin, R., *Optical Materials — Advanced Materials for Optics and Precision Structures*, SPIE CR 67, SPIE, 1996.

Mirror blank manufacture

Kumanin, K.G., *Generation of Optical Surfaces*, Focal Library, 1962.

Marioge, J.-P., *Surfaces optiques: Méthodes de Fabrication et de Controle, Recherches*, EDP Sciences, 2000.

Wilson, R.N., *Reflecting Telescope Optics II*, Springer-Verlgag, 1998.

Yoder, P R., ed., *Optomechanical Design*, SPIE CR 43, SPIE, 1992.

Optical testing

Geary, J.M., *Introduction to Optical Testing*, SPIE VTT15, SPIE, 1993.

Malacara, D., *Optical Shop Testing*, John Wiley & Sons, 1978.

Wilson, R.N., *Reflecting Telescope Optics II*, Springer-Verlag, 1998.

5
Stray Light Control

5.1 Causes of stray light

Stray light is any light (visible or infrared) that does not come from the celestial source of interest and yet illuminates the detector, creating an unwanted "background" and lowering sensitivity. Stray light affecting optical telescopes and instruments has two origins:

(1) light from celestial sources outside the field of view, referred to as "off-axis" sources, that is scattered or diffracted from various observatory surfaces into the detectors, and

(2) thermal emission of the telescope and surrounding surfaces that hits the detectors, either directly or through scattering.

Stray light from off-axis sources is curtailed by installing baffles and stops to prevent direct illumination of the detector. The impact of self-emission is reduced by cooling the surfaces seen by the detector where possible, and minimizing the view of those which cannot be cooled.

This chapter deals with the methods used to reduce the effects from these sources to an acceptable level. For stray light originating in off-axis sources, "acceptable level" is usually taken to mean a level lower than that of the other, natural sources of background that one cannot control. For a space telescope, this would be the zodiacal light, and for a ground-based telescope, the sky background. Ideally, the same criterion should also apply to the second source of stray light, thermal emission, but as we will see later, this is only possible in cryogenically cooled space telescopes.

5.2 Finding and fixing stray light problems

The basic principle for finding and eliminating or reducing stray light in telescopes and instruments is not to analyze the system from the object space downward, as one might first assume; this would be intractable. Rather, one should visualize what would be seen if one were looking out of the system from the detector's surface. It is thus important to do this analysis with the full system, telescope and instruments together, not separately for each subsystem.

One therefore imagines oneself positioned at the detector, gazing outward [1]. The first step is to determine the sources of stray light that can be seen directly. These sources can be celestial objects (e.g., Moon, stars, planets), the bright Earth in the case of a space telescope, and any radiating source in the nearby environment (the inside of the dome, the telescope structure, etc.). It is imperative that the *direct paths* between these sources and the detector be eliminated by placing adequate stops and baffles. An obvious example is the case of Cassegrain systems, which have their focal planes facing up and must be baffled to avoid a direct view of the sky.

The next step is to make a list of every object, optical or structural, visible from the detector, either directly or by reflection in the mirrors of the telescope (Fig. 5.1). These objects, called "critical objects," are important because all of the stray light comes from them; if an object is not seen from the detector, then it does not contribute to stray light.

The last step is to make a list of the "illuminated" objects, which are all those objects that receive power from potential stray light sources. Because different objects are illuminated by sources at different locations, there will be a separate list for each such location.

To find the stray light paths through a telescope, one then compares the lists of critical and illuminated objects. Any object that is common to both lists is on a *first-order stray light path*, since stray light propagates from the source to the detector after a single scatter from the object. The first-order paths are generally the largest sources of stray light, and if these are eliminated, stray light can be reduced by factors of 100 or more.

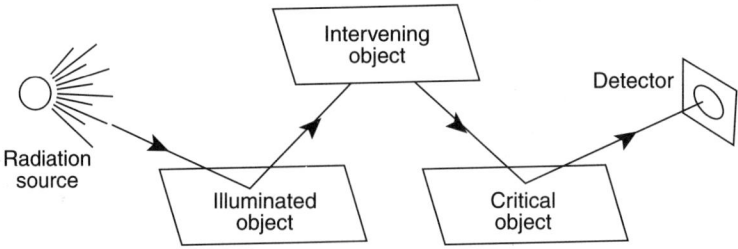

Fig. 5.1. Critical and illuminated objects as seen from the detector.

To eliminate the first-order paths, the rule for illuminated critical objects is "move them or block them" [1]. One can sometimes eliminate a first-order path simply by moving the object (i.e., changing its location or shape) so that it is no longer visible from the detector or illuminated by the source. When this is not possible or desirable, the alternate is to place a baffle or stop between the detector and the critical object or between the source and the illuminated object.

Once the first-order paths are blocked, additional improvement is obtained by working on second-order stray light paths. To find these paths, one goes back to the lists of critical and illuminated objects and determines all of the possible connections between the objects on these two lists. Each connection identifies a stray light path in which light from a source illuminates an object, then scatters from the illuminated object to a critical object, and then scatters from the critical object to the detector. The paths with the most power are again removed by placing baffles or by moving objects, as was done for first-order paths. Second-order stray light paths are much more numerous than first-order paths, which makes it difficult or impractical to block them all. This is where a stray light computer analysis can help. Such analysis provides a quantitative estimate of the power contributed by each path and identifies which paths must be blocked and which can be safely ignored.

Later in this chapter, we will take a brief look at how these computer programs operate. But first we examine the basic types of baffle and stop, as well as the scattering properties of surfaces that serve as input to these programs.

5.3 Baffles and stops

The most useful tools for controlling stray light are baffles and stops. These are mechanical walls and apertures that block the propagation of unwanted light from a source to a detector. To be effective, these components must be properly placed and sized. Stops in optical systems were defined in Chapter 4, but we now examine them from the point of view of stray light elimination.

5.3.1 Aperture stop

The aperture stop, or entrance pupil, is the stop which limits the size of the incoming beam that eventually converges at the focal plane. Objects in the space preceding the aperture stop outside of the desired beam are not seen by the detector. On the other hand, objects downstream of this stop may be seen (Fig. 5.2). The aperture stop is usually formed by the periphery of the primary mirror. In chopping infrared systems, however, this aperture is usually located at the secondary mirror.

186   5. Stray Light Control

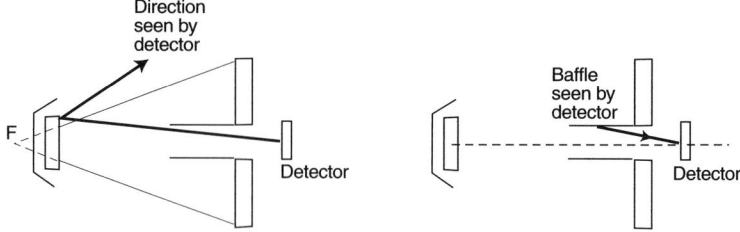

Fig. 5.2. Objects downstream from the aperture stop, here the primary, can be seen by the detector. This is true of the telescope's close surroundings when imaged through a secondary mirror oversized for field coverage (left), or the inside of the lower baffle (right).

5.3.2 Field stop

A field stop prevents off-axis sources at infinity from reaching the detector, but it will not block closer sources of energy (Fig. 5.3, left). When possible, a field stop should be placed at the first unused focus to block diffracted light produced by the front light-baffle. By slightly oversizing the field stop, light diffracted at the stop itself will fall off outside the area covered by the detector.

5.3.3 Lyot stop

A Lyot stop is a stop limiting the beam at the exit pupil. It prevents the detector from seeing any surface preceding the stop other than the optics itself. This is critical for all infrared instruments. In infrared cameras, for example, cooling the stop and the relatively small environment immediately around the detector will minimize the thermal energy affecting the detector (Fig. 5.3, right). In demanding, high-contrast observations, a Lyot stop is placed at an intermediate pupil to block the light diffracted by the edge of the entrance pupil (coronagraph).

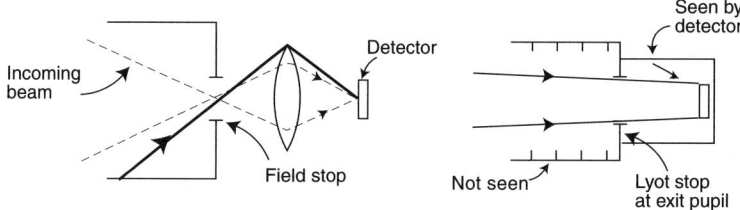

Fig. 5.3. A field stop prevents distant sources outside of the field of view from reaching the detector, but will not block close sources of radiation (left). This is accomplished by placing a stop located at the exit pupil; such a stop (Lyot stop) prevents the detector from seeing any nonoptical surface preceding it (right).

5.3.4 Baffles

Baffles are conical or cylindrical objects designed to block unwanted radiation paths. To further suppress scattered light, the baffle sides facing the region of unwanted stray radiation are generally provided with a series of concentric rings called "vanes" (Fig. 5.4). The geometry of these vanes, height, angle, spacing, and tip bevel, are variables that must be optimized for every design. This is best done by computer analysis, but a first-order analysis can be performed by estimating the apparent reflectivity of the fictitious surface formed by the tip of the vanes and assuming the tip scatter is Lambertian [2, 3, 4]. As a rule, however, tilted vanes are of little benefit over perpendicular ones, and the cavity between vanes only needs to be deep enough to block single scatter paths.

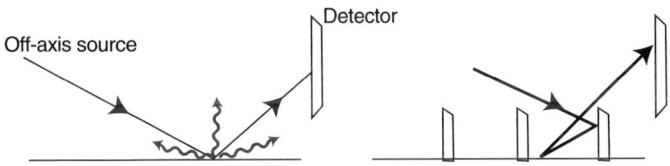

Fig. 5.4. Off-axis sources can scatter directly off baffle surfaces and be seen by the detector (left). With vanes, light from such sources can reach the detector only via multiple scattering and is greatly attenuated (right).

5.3.5 Baffles for Cassegrain systems

One troublesome feature of Cassegrain focus is that, unlike the case of a primary or coudé focus, the focal plane faces the sky and, hence, the sky or celestial sources outside the field of view can directly illuminate the detector.

This direct illumination is prevented by installing two sets of baffles, as shown in Fig. 5.5, one around the secondary mirror and the other above the primary, and sizing them so as to align the edges of these baffles (points A and B on the figure) with the field stop at the focal plane (point C). Of all the solutions possible, only one will result in the minimum obstruction of the telescope aperture. For this optimal solution, the longitudinal positions x_u and x_l and radii r_u and r_l of the upper and lower baffle edges, respectively, are given by the following formulas:

$$x_u = \frac{-b - \sqrt{b^2 - 4ac}}{2a}, \tag{5.1}$$

$$r_u = x_u(\theta - \theta_0) + \theta_0 f_1, \tag{5.2}$$

$$x_l = \frac{-c_1 b_2 + b_1 c_2}{a_1 b_2 - a_2 b_1}, \tag{5.3}$$

$$r_l = \frac{-c_1 a_2 + c_2 a_1}{b_1 a_2 - b_2 a_1}, \tag{5.4}$$

188 5. Stray Light Control

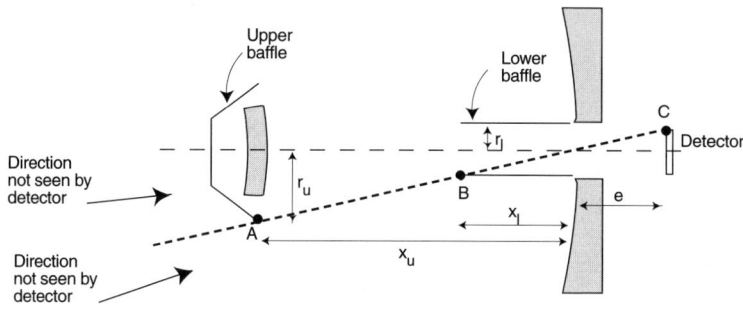

Fig. 5.5. Baffles for Cassegrain systems.

where f_1 is the primary mirror focal length, θ is the semifield angle one wishes to protect (in radians), and the remaining constants are defined by

$$\theta_0 = D/2f_1,$$
$$a = \theta_0^2(f_1+e)^2(m+1) + \theta_0\theta(f_1+e)(mf_1(m-1) - e(m+1)),$$
$$b = -(f_1+e)^2\theta_0^2((2m+1)f_1 - e) - \theta_0\theta(f_1+e)((mf_1)^2 + e^2)$$
$$\quad + f_1\theta^2(f_1^2(m^3-3m^2) - 2f_1e(m^2-m) + e^2(m+1)),$$
$$c = \theta_0^2(f_1+e)^2 f_1(mf_1 - e)$$
$$\quad - f_1\theta^2((mf_1)^3 + 2f_1^2 e(m^2-m) - f_1 e^2(m^2-m) - e^3),$$
$$a_1 = \theta_0(f_1+e) - \theta(m^2 f_1 + e),$$
$$b_1 = -(f_1+e)m,$$
$$c_1 = \theta_0(f_1+e)e + \theta(m^2 f_1^2 - e^2),$$
$$a_2 = \theta_0 x_u - \theta_0 f_1 - f_1\theta,$$
$$b_2 = -f_1,$$
$$c_2 = -\theta_0 f_1 x_u + \theta_0 f_1^2,$$

where D is the primary mirror diameter, m is the magnification of the secondary mirror, and e is the backfocal distance (i.e., the distance between the primary mirror vertex and the focal plane).

5.4 Scattering processes

With proper use of the baffles and stops defined above, no off-axis source will reach the detector directly. But the detector obviously sees the optical surfaces and possibly also surrounding surfaces, such as the inside of the baffles. Strong off-axis sources can scatter from these surfaces and reach the detector via a single or multiple bounce. The dominant sources of scatter are the optical elements themselves. The purely specular part of the optics does not contribute much because its microroughness is generally very small, except in the ultraviolet. However, dust, which is unavoidable on large optics, will scatter in a diffuse fashion and a small portion of the scattered rays will be propagated through the rest of the optical system like bona fide rays coming from the observed field (Fig. 5.6).

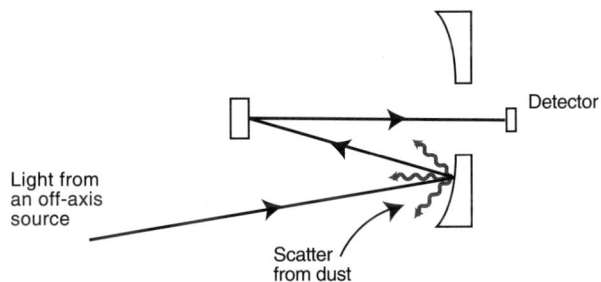

Fig. 5.6. Dust on mirrors is the dominant source of scatter.

Even if the off-axis source does not illuminate the optics directly, it can do so indirectly by scattering off baffles and other surfaces in the telescope. Using HST as an example, Fig. 5.7 shows the main paths of light from strong off-axis sources, such as the bright Earth and Moon, scattering off the inside of the light shield.

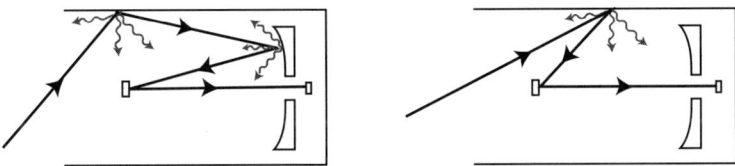

Fig. 5.7. Scattering processes in the HST baffle system: scattering at large angles from the baffle and then from dust on the primary mirror (left) or via the secondary mirror for smaller angles (right).

5.5 Stray light analysis

After an initial baffle design is complete, an analysis has to be performed to determine the power of stray light propagated to the detector and ascertain which of the paths contribute the most. The purpose of this analysis is to find out if the telescope meets its stray light requirements and, if not, to identify and correct the areas of the design that are causing the problem.

The scattering properties of surfaces are at the heart of any such evaluation. If an incident beam falls upon an ideal flat surface, the reflected beam is concentrated in specular fashion (Fig. 5.8, left). If, on the other hand, the surface is a perfect diffuse reflector, light is scattered uniformly, but with the power of the scattered beam varying as the cosine of the angle when the apparent cross section of the scattering surface is taken into account (Fig. 5.8, center). Such a surface, whose radiance is independent of angle and obeys Lambert's cosine law, is called "Lambertian." Real surfaces fall between these two extremes. If the surface is not too rough, the scattered light is concentrated in the specular direction, but a significant portion of it is distributed around it (Fig. 5.8, right).

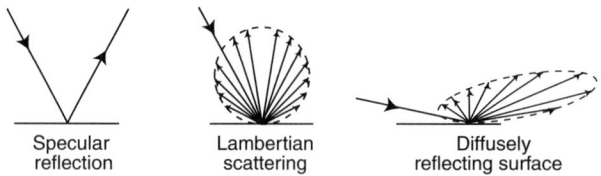

Specular reflection   Lambertian scattering   Diffusely reflecting surface

Fig. 5.8. Reflection and scattering.

Surface scattering is fully described in two dimensions by a function which takes into account the angle of incidence of the incoming light. This function is called the bidirectional reflectance distribution function (BRDF), and the flux transferred from a small scattering surface of area dA into an elementary solid angle $d\omega$ can then be expressed as

$$d\Phi = \text{BRDF}\ E\ dA\ \cos\theta_i\ \cos\theta_o\ d\omega, \qquad (5.5)$$

where E is the impinging flux density and $d\omega$ is the elementary projected solid angle subtended by the collector as seen from the scattering surface (Fig. 5.9, left). The inclusion of the cosines of the input angle (θ_i) and of the output angle (θ_o) in the definition removes the geometrical dependence of the impinging and scattered power caused by the apparent shrinkage of the area when viewed obliquely. The BRDF depends on $\theta_i, \Phi_i, \theta_o, \Phi_o$, polarization, and on wavelength. This description of surface scattering is very general and applicable to both specular (mirrors) and nonreflective surfaces. This function must not be confused with the "hemispherical reflectance," ρ, also called "albedo," which is the dimensionless ratio of the total reflected flux to the incident flux. Whereas ρ is always less than or equal to 1, the BRDF may take values much greater than 1 in some circumstances, over a small solid angle. A perfect specular surface, for example, has a BRDF of infinity in the specular direction and zero elsewhere. When the scattering surface obeys Lambert's law, the BRDF is constant and its value is obtained as a function of the total reflectance by integrating equation 5.5 for $d\omega$ over half a sphere. This gives simply

$$\text{BRDF} = \frac{\rho}{\pi}. \qquad (5.6)$$

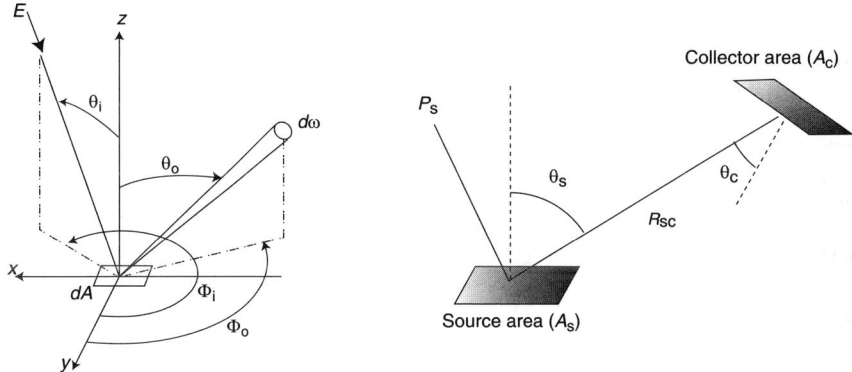

Fig. 5.9. Geometry for the definition of the BRDF (left) and GCF (right).

The scatter characteristics of lenses, windows, and other transmissive surfaces are usually described in terms of their bidirectional transmission distribution function, or BTDF. This quantity is defined in the same way as the BRDF, except that it refers to scatter on the transmitted, rather than the reflected, side of a surface.

When the BRDF of a surface is known, one can calculate the amount of power that is scattered from one surface to another. Let the power incident on a small area A_s be P_s (Fig. 5.9, right). Then, the power, P_c, scattered to a small collecting area, A_c, is given by

$$P_c = \pi P_s \, (\text{BRDF}) \, (\text{GCF}), \tag{5.7}$$

where GCF is the geometrical configuration factor defined as

$$\text{GCF} = \frac{A_c \cos\theta_s \cos\theta_c}{\pi R_{sc}^2}, \tag{5.8}$$

with R_{sc}, θ_s and θ_c as defined in Fig. 5.9.

The GCF has a physical interpretation: if the source area were turned into a Lambertian emitter, then the fraction of its power that was incident on the collecting surface would be equal to the GCF.

Equation 5.7 shows that there are only three ways to reduce the power on the collecting surface: reduce the incident power P_s, reduce the BRDF, or reduce the GCF. Of these three, reducing the GCF is usually the most effective, for by blocking the propagation path from the source area to the collector area, one can make this term vanish completely. This is the foundation for the "move it or block it" approach to stray light reduction introduced in Section 5.2.

In principle, stray light estimates for first-order paths are straightforward. First, the power from the source to the illuminate and critical object is evaluated. Then, the geometry of the telescope is used to evaluate the GCF of the detector. If there are intervening optical components, then it is the image of the detector that is used to calculate the GCF. Equations 5.7 and 5.8 are then used to evaluate the power on the detector. Second-order paths are evaluated in two steps, with the same equations used to calculate the power scattered from an illuminated object to a critical object, then used again to calculate the power from the critical object to the detector. Because equations 5.7 and 5.8 assume small areas, it may be necessary to break a large surface into smaller sections and calculate the power from each section separately.

Approximate stray light analysis can be obtained by hand calculations as shown by Greynolds [3], but for a thorough analysis, a ray-tracing program is used. Two such programs, APART and ASAP,[1] are the industry standards in this domain.

APART performs stray light calculations in a deterministic manner using the approach summarized above [5]. The program breaks down the surface of each optical and structural component of the telescope into small areas and calculates the geometrical configuration factors and BRDFs between the areas of each object. Equations 5.7 and 5.8 are applied to all possible connections between these areas to obtain the stray light power on the detector.

ASAP is a ray-trace program that uses a statistical, Monte Carlo approach. Rays are launched from a source through a three-dimensional model of the optical and structural components of the telescope. Each time a ray intersects an object, additional scattered rays are generated. The power of the scattered rays is weighted in proportion to the power of the incident ray and the BRDF of the scattering surface. Scattered rays are collected on the detector to calculate the irradiance distribution.

A common measure of stray light transmission is known as the normalized detector irradiance (NDI). If the entrance to the telescope is illuminated by a distant point source (collimated light), then the NDI is defined as the ratio of stray light irradiance (power per unit area) on the detector to the source irradiance at the entrance to the telescope:

$$\text{NDI} = \frac{E_{\text{det}}}{E_{\text{source}}}. \tag{5.9}$$

The source irradiance is measured on a plane that is normal to a line connecting the telescope with the distant point source. The value of the NDI is a function of both the location of the source and the position on the detector. For a given point source, the actual detector irradiance is obtained by simply multiplying the source irradiance at the entrance to the telescope by the NDI. For an extended source, the detector irradiance is obtained by multiplying the NDI by the radiance (power per unit area and steradian) of the source and integrating it over the solid angle subtended by the source.

---

[1] APART and ASAP are maintained and distributed by Breault Research Organization, Tucson, Arizona.

5.6 Surface scattering properties

5.6.1 Scatter from mirrors

No mirror is perfectly smooth, and small amounts of roughness on a mirror surface will scatter light around the specular reflection. The roughness of a mirror is usually specified in terms of the root-mean-square (rms) variation in the surface height. However, this single number is not adequate to characterize scatter distribution because the spatial frequencies of the microroughness is left undetermined. In practice, one must resort to direct measurement of the BRDF.

Dust particles on mirrors add to the light scatter. In fact, at infrared wavelengths, scatter from dust particles, or particulates, is often the dominant source of stray light from optical components. Particulates are of random sizes, however, and scatter is again best determined by direct measurement.

It has been experimentally shown by Shack and Harvey [6] that the angular dependence of the BRDF is solely a function of the difference of the sines of the incident and output angles and that a good approximation of the BRDF is given by

$$\text{BRDF} = b(100\,|\sin\theta_s - \sin\theta_o|)^m,\qquad(5.10)$$

where θ_s is the angle of the specular ray, θ_o is the output angle, and b and m are empirical constants. The difference between the sines of the two angles is commonly denoted as β. This formula, called the "Harvey-Shack" law, is a straight line when plotted in log-log, and b is the intercept for β=0.01 radian. As a rule, m does not vary much with microroughness or wavelength; a fairly typical value is -1.8 (Fig. 5.10). The intercept b, on the other hand, is a strong function of microroughness, dust coverage, and wavelength.

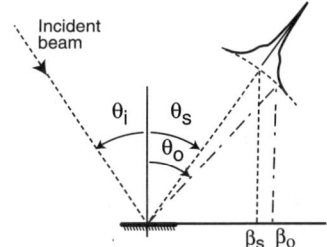 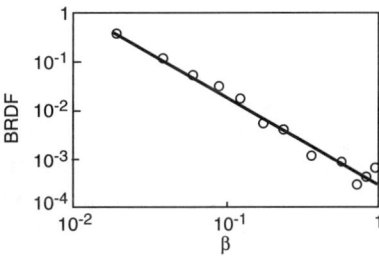

Fig. 5.10. At left, mirror scatter geometry and notations. At right, Harvey-Shack approximation with b =1.1 and $m = -1.8$ (straight line), compared to the theoretical estimates for the cleanliness level 500 calculated by Spyak and Wolfe at a wavelength of 3.39 μm (data points), where β is the cosine of the angle of incidence. (From Ref. [7].)

In the ultraviolet, mirror scatter is dominated by microroughness. In general, the intercept b associated with surface roughness decreases as the wave-

194   5. Stray Light Control

length increases. If the slope m does not change much with wavelength, then b can be approximated by the formula

$$b(\lambda) = b(\lambda_0) \left(\frac{\lambda_0}{\lambda}\right)^{4-m}, \qquad (5.11)$$

which is plotted for various wavelengths in Fig. 5.11.

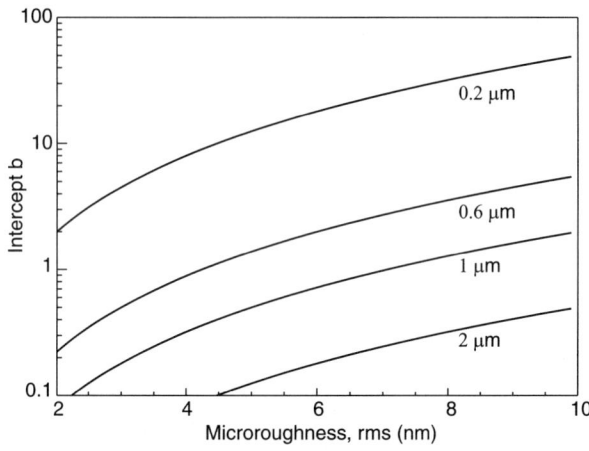

Fig. 5.11. Shack-Harvey intercept as a function of mirror microroughness for various wavelengths.

At more than 1 μm, mirror scatter is generally dominated by dust, not surface roughness. Dust coverage can be estimated from the cleanliness level of the environment where the optics is kept. The cleanliness level for optical systems is best defined by the U.S. military specification MIL-STD 1246C [8]. In this specification, the distribution of particle sizes is the same for all cleanliness levels, even though the number of particles changes. Cleanliness depends on clean room class, duration of exposure, whether the optics are vertical or horizontal, and whether they are fully exposed or bagged and purged.

It is convenient to describe the cleanliness level in terms of dust coverage areal fraction, since the BRDF scales linearly with this value. This relationship is plotted in Fig. 5.12.

It is very difficult to maintain a dust coverage lower than a few percent in large optics. HST's primary mirror, which was cleaned after integration and maintained in a clean room environment until launch, has a dust coverage approaching 2%. Even if more stringent precautions are taken, it is not realistic to expect a cleanliness level of under 500 ($\sim$ 1% dust coverage) for large optics. Based on Mie scatter theory and experimental confirmation [7], the intercept and slope of the Harvey-Shack BRDF representation are given in Table 5.6.1 for a selected set of infrared wavelengths.

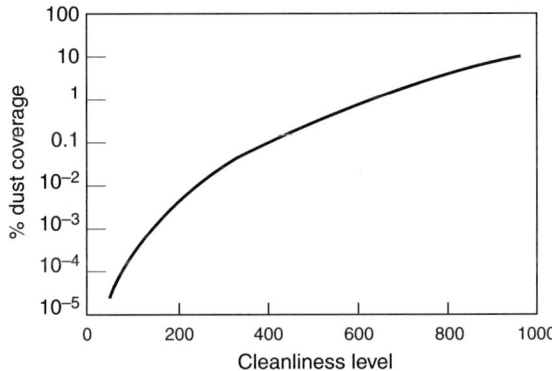

Fig. 5.12. Areal dust coverage percentage versus cleanliness level.

Table 5.1. Level 500 Harvey-Shack parameters and total integrated scatter

$\lambda(\mu m)$	b	m	TIS
3.39	1.1	-1.8	$9 \cdot 10^{-3}$
5.0	0.5	-1.5	$6 \cdot 10^{-3}$
10.6	0.15	-1.2	$5 \cdot 10^{-3}$

Table 5.6.1 also gives the total integrated scatter (TIS), which is the total amount of light scattered by the mirror in all direction. It is obtained by integrating the BRDF over the hemisphere:

$$\text{TIS} = \int_0^{2\pi} \int_0^{\frac{\pi}{2}} \text{BRDF} \cos\theta \sin\theta d\theta d\phi, \qquad (5.12)$$

where θ and ϕ are the polar and azimuthal angles with respect to the surface normal. The integral of the BRDF Harvey-Shack approximation can be evaluated in closed form if the slope m is greater than -2 and is given by

$$\text{TIS} = 2\pi b \frac{100^m}{m+2}. \qquad (5.13)$$

Evaluating the TIS is useful as a check of the BRDF value, since the TIS must be roughly equal to the fraction of the mirror aperture covered by particulates (neglecting absorption, the total light scattered is equal to the portion of the incoming beam falling onto the particulate).

5.6.2 Scatter from diffuse black surfaces

Baffle surfaces in telescopes usually have a diffuse black coating to absorb as much incident light as possible. It is important to remember that none of these coatings will absorb all of the incident light. Figure 5.13 shows the BRDF of Aeroglaze Z306, a diffuse black paint [9]. At normal incidence, the BRDF

196   5. Stray Light Control

of this paint is nearly constant as a function of scatter angle; the surface is nearly Lambertian. The behavior is very different at high angles of incidence, however. In directions around and forward of the specular direction, the BRDF rises to values larger than 1, orders of magnitude above the normal incidence value. This large increase in the BRDF for the forward scatter direction is characteristic of many diffuse black surfaces.

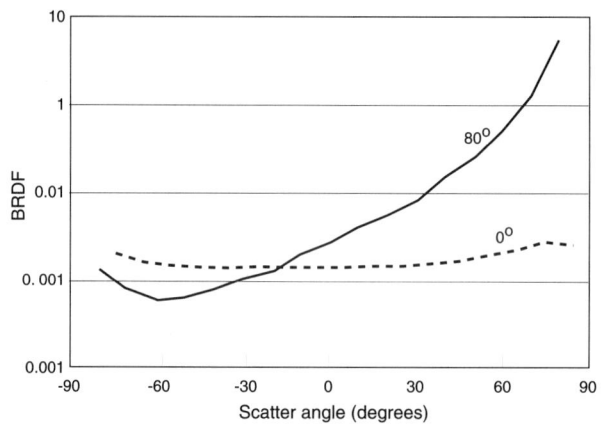

Fig. 5.13. BRDF of Aeroglaze Z306 for 0° and 80° angle of incidence at 633 nm.

Such diffuse black surfaces should therefore not be used to control stray light in the forward scatter direction. The large BRDF in this direction causes them to appear very bright. Instead, black surfaces should be positioned so as to be illuminated at or near normal incidence. This is accomplished by placing vanes on the surface, as discussed in Section 5.3.4. Incident light now strikes the back sides of the vanes at angles close to the normal. In addition, scatter from an adjacent vane is needed before light can propagate to the detector, which further reduces the stray light. Where vanes are not practical, a less effective but still useful measure is to machine small grooves into the surface, as shown in Fig. 5.14. This is referred to as "threading" the surface. The intent is to capture most of the light at and around normal incidence and force an additional bounce off the opposing side of the grooves before the light propagates to the detector or other surfaces.

Fig. 5.14. Threading a surface to reduce the effective BRDF.

5.7 An example of protection against off-axis sources: HST

The stray light environment of the Hubble Space Telescope is particularly severe because observations must be made in the presence of the Sun and Moon and while flying only about 500 km above the bright Earth. The telescope itself is baffled in the unusual manner, with an upper baffle around the secondary mirror and a lower one around the Cassegrain return beam. This prevents any off-axis stray light from hitting the focal plane directly. The inner sides of the tube and lower baffle have vanes to prevent direct bounces, and all surfaces are coated black to reduce scatter.

In addition, the telescope is preceded by an extensive light shield about 4 m in length, coated black and provided with vanes (Fig. 5.15). The aperture door, always kept on the Sun side, is dimensioned such that the Sun cannot shine on the light-shield entrance as long as it is more than 50° away from the viewing axis.

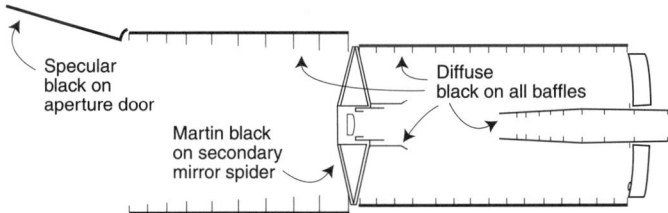

Fig. 5.15. HST's baffling arrangement. The aperture door is tilted by about 15° to avoid being viewed by the primary mirror.

The effectiveness of the baffling system and stray light suppression for off-axis surfaces can be described by a quantity referred to as the "attenuation factor." This factor is defined as the ratio of the flux density reaching the focal plane to the flux density impinging on the telescope aperture.[2] The flux Φ impinging on the focal plane due to stray light from an off-axis point source over a unit area of the detector (one pixel, for example) per given bandpass $\Delta\lambda$ is then given by

$$\Phi = E\, A(\alpha)\, f_e^2\, d\omega, \qquad (5.14)$$

where E is incoming flux density per $\Delta\lambda$ from the source, A is the attenuation factor, α is the angle of the source with respect to the axis of the telescope, f_e is the effective focal length of the telescope, and $d\omega$ is the solid angle subtended by one pixel on the sky.

HST's baffle attenuation factor is shown in Fig. 5.16, left. The attenuation is extremely high except for low angles (less than 15°), where the off-axis source can penetrate the tube and illuminate the primary mirror.

---

[2] Also used is the "rejection ratio," which is the inverse of the attenuation factor.

198   5. Stray Light Control

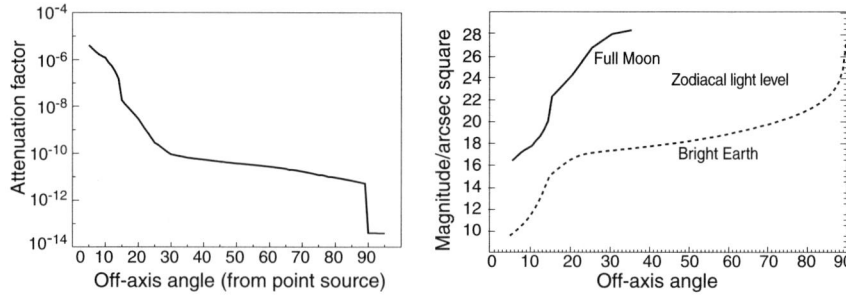

Fig. 5.16. HST baffle attenuation for off-axis sources (left) and stray light in the focal plane of HST as a function of off-axis angle from the Moon or the bright limb of the Earth (right).

Equation 5.14 must be integrated over the corresponding α range when the illuminating source is extended, as, for example, in the case of the bright Earth for a low-Earth-orbit space telescope. Figure 5.16 (right) shows the intensity of stray light from the Moon and the bright Earth in the case of HST. It is seen that the stray light level is less than the zodiacal light ($\sim$ 23rd magnitude per arcsecond square) as long as the Moon is more than 20° away and the bright Earth limb more than 80° away.

5.8 An example of minimizing stray light from self-emission: NGST

In the infrared, thermal emission of a room-temperature telescope and instruments is several orders of magnitude larger than the flux of sources to be observed (see Chapter 1). Reducing this "instrumental background" requires cooling. This is usually feasible for instruments which can be kept in a controlled environment, but cooling the telescope itself is another matter. Not much can be done on the ground, aside from going to a naturally cold site (e.g., Antarctica). Actively cooling the optics would inevitably create frost on the optical surfaces.

In space, small telescopes can be immersed in a cavity refrigerated by cryogens or active coolers. But this is totally impractical for large telescopes. Passive means must be used. It is fairly easy for a space system to reach very cold temperatures by natural means. The IRAS observatory, which was in low Earth orbit, stabilized at about 80 K once its cryogen was exhausted. Far from Earth, such as at the Sun-Earth L2 point, a system placed in the shade of a simple screen will reach a temperature of under 50 K, provided it is allowed to radiate fully to space.

Although a simple sunshield can keep the telescope optics cold enough to prevent significant background in the near or mid-infrared, a secondary process which is at play is the limiting factor. The back of the sunshield irradiates the unbaffled optical surfaces of the primary and secondary mirrors, the radiation scatters off the small amount of dust covering them, and the scattered radiation makes its way to the detector just like bona fide rays from an astronomical source in the field of view (Fig. 5.17, left). In the case of a single-layer sunshield, the back of the shield would be at about 250 K and the scattering process would create an instrumental background larger than the zodiacal light beyond 4 µm. To limit the instrumental background to less than the zodiacal light below 10 µm requires a multiple-layer shield, with about six layers.

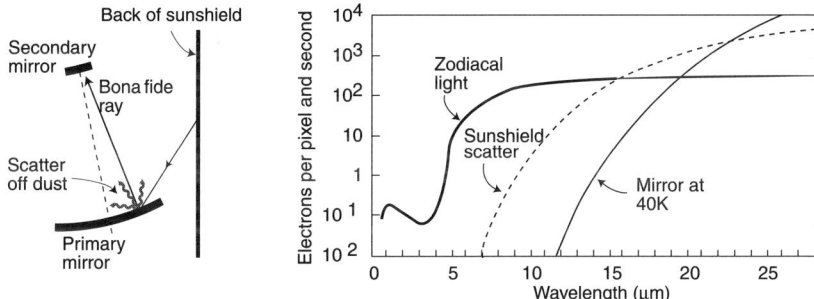

Fig. 5.17. On the left, scatter from the sunshield is the main source of instrumental background in "open-tube" space infrared observatories. On the right, predicted thermal backgrounds for NGST due to scatter from the back of the sunshield and thermal emission from the optics. The calculation is based on a 20% bandpass and thermal properties at end of life (10 years).

Cleanliness of the mirrors is also essential. Dust coverage should be less than 1%, a level which will probably require cleaning immediately before launch. On the other hand, molecular contamination, which does affect ultraviolet observations, is not a factor here.

With these assumptions, the thermal backgrounds predicted for NGST are shown on the right in Fig. 5.17 and indicate that NGST should be zodiacal-light limited up to a wavelength of about 16 µm [10]. For wavelengths beyond that, the background is first dominated by the scatter from the sunshield, then by the optics emission after 23 µm.

5.9 Minimizing thermal background in ground-based telescopes

On the ground, the telescope itself is a major source, sometimes the dominant source, of thermal background emission received by a detector. Telescopes and instruments for use in the thermal infrared must therefore be carefully designed to reduce the emissivity of the parts visible to the detector.

The main sources of thermal emission are the optical surfaces of the mirrors, vanes supporting them (unless the telescope is of an unobstructed design), warm lenses or filters in the instrument, and the window of the cryostat containing the instrument.

Reducing the emissivity of the optics is simply a matter of making the reflective surfaces as shiny as possible and the warm lenses and mirrors as transparent as possible, at the working wavelength. This needs to be done in any case to optimize the throughput of the instrument and telescope; further optimization to the infrared just intensifies that requirement. When possible, surface coatings should be selected to maximize reflectivity (i.e., minimize emissivity) at the working wavelength, and all optics should be maintained in the best possible condition and dust-free.

Supporting vanes should be as narrow as possible. They should not be shiny, since they will then reflect into the instrument radiation from sources well outside the telescope beam (such as the inside of the dome, passing observers, etc.).

Because the backgrounds encountered in the thermal infrared are so large relative to the signals from astronomical sources, a very small instability or modulation of the average infrared background level will suffice to prevent the signal's detection (see Chapter 1, Section 1.4.4). *Stabilizing* the background is as important as its overall reduction.

In conventional Cassegrain telescope designs, the entrance pupil image contains that of the central obstruction, including the hole in the primary (or equivalently, the Nasmyth fold mirror). The central hole is effectively black, and thus highly emissive, and can subtend a large fraction of the solid angle of warm world seen by the detector. It is therefore customary to eliminate the central obstruction as a background source, either by covering the center of the secondary (where the central hole would be seen imaged) with a conical or slightly tilted plane mirror, which, instead, reflects into the instrument a patch of sky seen via the primary. As discussed in Chapter 1, if the sky is transparent, its overall emissivity (and emissions) will be low. In highly optimized infrared telescopes, the center of the secondary mirror and its support system may be bored out so that the view of the sky is direct, rather than via two reflections from shiny, but nevertheless emissive, surfaces.

References

[1] Breault, R.P., Control of stray light, in *Handbook of Optics*, McGraw-Hill, 1995, Vol. 1, Chap. 38.

[2] Breault, R.P., Problems and techniques in stray radiation suppression, SPIE Proc., Vol. 107, p. 11, 1977.

[3] Greynolds A.W., Formulas for estimating stray-radiation levels in well-baffled optical systems, SPIE Proc., Vol. 257, p. 39, 1980.

[4] Breault, R.P., *Suppression of scattered light*, Ph.D. thesis, Univ. of Arizona, 1979, p. 57.

[5] Lange, S.R., Breault, R.P., and Greynolds, A.W., APART, a first order deterministic stray radiation analysis program, SPIE Proc., Vol. 107, p. 89, 1977.

[6] Harvey, J.E., *Light scattering characteristics of optical surfaces*, Ph.D. Thesis, Univ. of Arizona, 1976.

[7] Spyak, P.R. and Wolfe, W.L., Scatter from particulate-contaminated mirrors, Opt. Eng., Vol. 31, No. 8, p. 1746, 1991.

[8] U.S. Military Standard 1246C, *Product Cleanliness Levels and Contamination Control Program*, 1994.

[9] Ames, A.J., Z306 black paint measurements, SPIE Proc., Vol. 1331, p. 299, 1990.

[10] Bely, P.Y., Lallo, M., Petro, L., Parrish, K., Mehalick, K., Perrygo, C., Peterson, G., Breault, R., and Burg, R., *Stray light analysis of the yardstick mission*, NGST Monograph No. 2, NGST Project Study Office, Goddard Space Flight Center, 1999.

Bibliography

Breault, R.P., Problems and techniques in stray radiation suppression, SPIE Proc., Vol. 107, p. 2, 1977.

Breault, R.P., *Suppression of scattered light*, Ph.D. Thesis, Univ of Arizona., 1979.

Breault, R.P., Control of Stray Light, in *Handbook of Optics*, McGraw-Hill, Vol. 1, Chap. 38, 1995.

Greynolds, A.W., A consistent theory of scatter from optical surfaces, SPIE Proc., Vol. 967, p. 10, 1989.

Stover, J.C., *Optical Scattering*, McGraw-Hill, 1990.

6
Telescope Structure and Mechanisms

The role of the telescope structure and mechanisms is to maintain the optics figure and alignment during observations. Traditionally, this has been accomplished by "passive" means, that is, by a combination of design measures and the selection of favorable materials. But these passive techniques have their limits. As telescopes become larger and the need for mass reduction increases, it becomes advantageous, even compulsory, to adjust the position or figure of the optical elements in real time to compensate for the effects of changing gravity or temperature. This approach is referred to as "active optics."

Active optics relaxes the tolerances for telescope structures and mechanical systems enormously, engendering a significant reduction in mass and cost. But active optics has its own limits. The bandwidth of the correction is constrained by the wavefront sensor's sensitivity, the amplitude of the correction by the optical and mechanical limits of the active elements, and the response of the control systems by the mechanical nonlinearities. It is thus best to view the traditional passive structure and the modern active optics systems as complementary and distribute the tasks according to what each system does best: passive structures can assure the first line of defense against gravity and large thermal loads, whereas active optics can be left to handle the rapid, small-amplitude corrections.

In this chapter we review the general principles conducive to proper optomechanical design, look at the conditions imposed on telescope structures and mechanisms, and examine typical implementations. Active optics will be the subject of Chapter 8.

6.1 General principles

6.1.1 Kinematic mounting

The structures supporting the telescope optics deform under the weight of the optics, their own weight, and thermal effects by amounts which are orders of magnitude larger than the optics can tolerate. It is thus indispensable to isolate the optics to avoid subjecting them to undue stress. This is accomplished by mounting the optics in a *statically determinate* manner, using what is called a "kinematic mount."

A rigid body in space has six degrees of freedom: translation and rotation along each of three orthogonal axes. The body is fully constrained when each of these possible movements is singly prevented. If any one movement is constrained in more than one way, then the body will be deformed by external forces. A kinematic mount is a mounting system which does not constrain more than six rigid-body degrees of freedom. When an optical element is mounted kinematically, the structure supporting it can deform in response to a thermal load or changing gravity vector without affecting the optical figure: the optical element can move as a rigid body, but will not deform. Kinematic mountings are not limited to the support of optics. They should also be used for the mounting of all sensitive equipment, such as the science instruments.

The simplest form of kinematic mount supports all six rigid-body motions at a single point. In practice, however, this solution is seldom used for optical elements because it creates localized stresses. Kinematic mountings are generally designed to constrain at least three separate points in the body. One common example is the "point/V-groove/plane" support shown in Fig. 6.1.

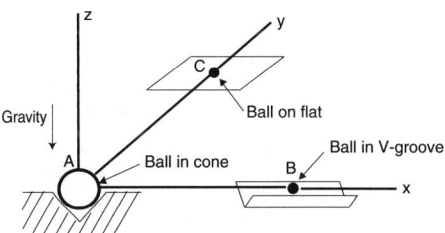

Fig. 6.1. Point/V-groove/plane kinematic mount. For clarity, the implementation shown here assumes that points A, B, and C remain pressed against their supports thanks to gravity, but the same principle can be applied by using loading springs.

The ball joint at A fixes the body in translation at that point in the x, y, and z directions. Point B is a V-groove constraining the body in translation in the y and z directions. Acting together, the constraints at points A and B thus prevent rotation about the y and z axes while allowing expansion in the x direction. Point C then constrains motion normal to the x, y plane, preventing rotation around the x axis. This arrangement is referred to as "3-2-1" in reference to the number of translational degrees of freedom constrained at each of the three points. An example of this arrangement is shown in Fig. 6.2.

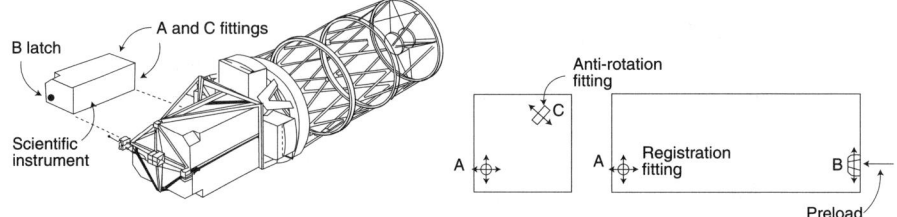

Fig. 6.2. The "3-2-1" kinematic mounting of the science instruments on HST. The system was designed for simple guided insertion and latching by astronauts during instrument change-out.

Another common form of kinematic mount for optical elements is the tangent bipod, or "2-2-2" configuration. This configuration has the advantage of minimizing decentering errors even in the presence of large differential contractions between optical elements and their support structures, as found in cryogenic systems. The tangent bipod is most commonly implemented using flexures that allow radial motion at the mount points, thus avoiding the need for mechanical joints (Fig. 6.3).

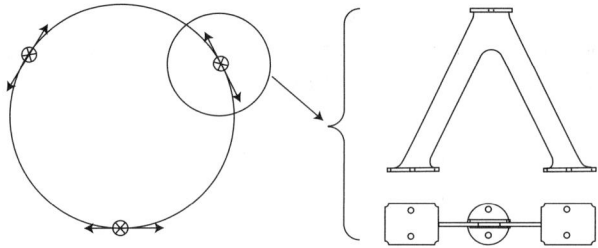

Fig. 6.3. Tangent bipod kinematic mount commonly used for optical elements not requiring adjustment.

Another approach to kinematic mounting is the "hexapod" mount (Fig. 6.4), which is gaining popularity with the advent of computer control. This type of support can be used for mounting mirrors that must be adjustable in all directions, such as secondary mirrors. It has also been proposed as a telescope tube mount for the DGT [1]. This type of mount, sometimes referred to as the "Stewart platform," was first developed for flight simulators.[1] The length of each leg of the mount is adjustable with a linear actuator. To avoid constraining the supported system when adjusting the legs, it is essential that each of

---

[1] This design was first described in a paper by Stewart in 1965[2]. Unpatented, it is in the public domain.

the 12 attachment points (2 at the end of each leg) be either ball joints or flexures. The adjustment of the six degrees of freedom allowed by the hexapod is not mathematically "orthogonal": adjusting for one direction will affect the other degrees of freedom. This is of no consequence in automated systems, however, where desired movements can be obtained by simultaneous actuator control.

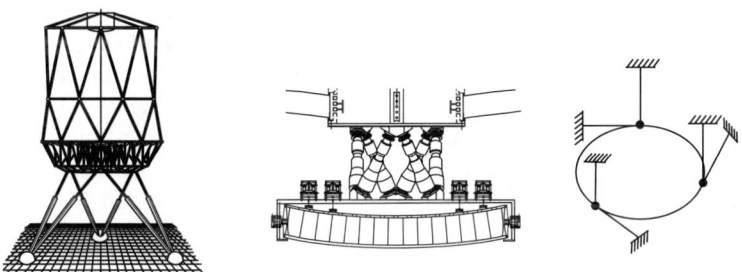

Fig. 6.4. Hexapod mount proposed for the German telescope "DGT" (left) and used in the MMT secondary mirror (center). In these implementations, the displacements are not independent of each other, but in the modified configuration shown at right, they are.

6.1.2 Minimizing decollimation

The optical train must remain "collimated" (i.e., aligned in all directions) in spite of changing thermal gradients and gravity direction as the tube rotates to follow a target. The optics must also remain collimated when one element is displaced, for example, for refocusing. Solutions to this problem fall into three categories:

- *Intrinsic rigidity.* Deflection under gravity is minimized by the use of a very rigid structure. This becomes increasingly difficult as telescopes grow larger.
- *Compensation.* Here, the structure is allowed to deflect, but is configured or dummy masses are moved in such a way that the optical elements do not move significantly with respect to each other,
- *Active optics.* Here, actuators are used to move the key optical elements so that they remain aligned with respect to each other at all times. This can be done in an open-loop fashion (after calibration of the effect) or in closed loop, using some sort of sensing system.

Two examples of the "compensation" approach often used on large telescopes are presented below. As indicated earlier, active optics techniques will be examined in detail in a later chapter.

The "Serrurier" truss

Small ground-based telescopes often use an intrinsically stiff, cylindrical shell to mount the optics. But deflection at the two ends of such a tube increases rapidly with mass and size and cannot be kept small enough for large telescopes. One can size the two extremities of the tube so as to have the same deflection at both ends, but the extremities are still tilted with respect to each other and, consequently, the optics are misaligned (Fig. 6.5, left). The solution is to abandon the use of a shell and replace it with a truss of simple geometry consisting of linear members arranged in isosceles triangles on a square base. When the triangles in the vertical plane deflect, the parallelogram formed by the two horizontal triangles constrains the two "tube" ends to move in parallel plane (Fig. 6.5, right). This design, called the "Serrurier truss" after Marc Serrurier [3], who invented it for the 5-meter Hale telescope at Mt. Palomar, has been used in many mid-sized telescopes and will be studied in more detail in Section 6.4.1. However, its use is now being abandoned for very large telescopes because of its mass inefficiency and the advent of active optics.

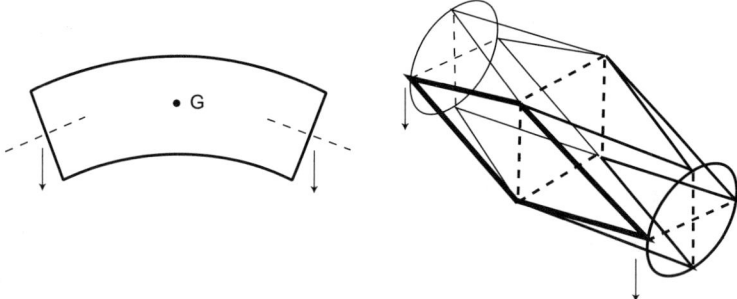

Fig. 6.5. Under the effect of gravity, the deflection of a cylindrical shell tube induces a tilt in the extremities which will decollimate optical elements (left). This is avoided in the Serrurier design, where the two end rings deflect parallel to their own plane (right).

Balanced Cassegrain "top unit"

The secondary mirror of a Cassegrain system must be supported on thin structural members, forming what is called a "spider," so as to minimize obstruction of the incoming beam. Using simple radial members in the plane of the mirror itself is not a satisfactory solution, as such members have little rigidity in the direction of the optical axis; this results in a focus change whenever the tube's zenith angle varies. One common solution consists of using the triangular configuration shown in section in Fig. 6.6 (left). When the tube is vertical the piston motion is much reduced, but the secondary mirror will droop when the tube is horizontal. This droop can be compensated by placing a counterweight on the other side so as to locate the system's center of gravity in the spider's plane of symmetry (Fig. 6.6, right).

If the secondary mirror is heavy and must be displaced by large amounts to accommodate different instruments at the Cassegrain focus, the counterweight must also be moved to keep the system's center of gravity within the spider's plane of symmetry.

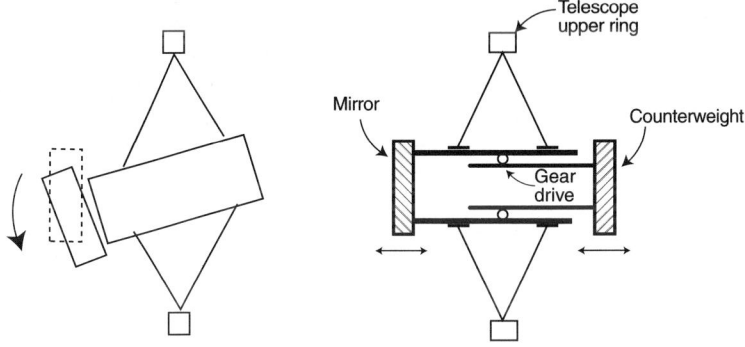

Fig. 6.6. The droop of a Cassegrain mirror spider (left) can be compensated by a counterweight (right).

6.1.3 Use of preload

Preloading, also called "prestressing," consists of applying a force or moment to a structural or mechanical system which is reacted "internally." Preloading has several applications in telescopes:

(1) It can be used to **increase the natural frequency** of a thin structural component by tensioning it, as in the case of a stringed instrument. Increasing the natural frequency of thin members may be useful in avoiding resonance when they are excited by disturbances such as wind.

(2) It can be applied to **prevent buckling** of thin structural members. For example, in a secondary mirror spider, the lower member is in compression when the tube is horizontal. Introducing a preload equal to at least half the weight of the secondary mirror unit will keep that lower member always in tension and thus prevent buckling.

(3) It is often used to **avoid free play, dead zones, and backlash** in mechanisms and joints. For example, deployment latches and hinges have inherent free play, such as is found in a nut on a lead-screw mechanism or between a ball-ended link and its socket. Preloading can eliminate this free play in cases where the self-weight effect of gravity is either too variable (ground-based telescopes) or nonexistent (space-based systems). Preloading joints also reduces the low-stiffness complications

associated with small contact areas, thus making structures behave in a more linear and well-behaved manner. As another example, gear drives are often split into two counteracting systems (Fig. 6.7, center), forcing the teeth of the spur pinion to remain in contact with the driven gear, thus eliminating backlash (see Chapter 7 for additional details).

(4) Finally, preloading can be used as a **load limiter**. For example, the support points used to define the position of a mirror are often designed with a preloaded spring, as shown on the right in Fig. 6.7. The defining point offers a precise reference for the mirror when it is "floated" on its support system, but will collapse when the floating support system is deactivated, thus limiting local stresses on the mirror to a safe level.

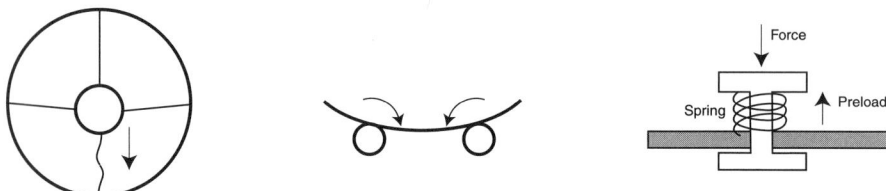

Fig. 6.7. Three applications of preload: preventing buckling of thin members (left), avoiding mechanical backlash in drive systems (center), and limiting load (right).

6.1.4 Load paths

Optical elements and science instruments are delicate, sensitive systems which may deform or become misaligned when subjected to even small loads. Great care should be taken to remove these critical elements from the main load paths in the overall telescope. On the ground, this means mounting the optics and instruments in such a way that they never support loads other than their own weight. In the case of space telescopes, this means making sure that the optics and science instruments will not be used to support other parts of the telescope during launch (Fig. 6.8), nor to transfer loads for attitude control of the observatory during normal operations.

Beyond these obvious cases, both ground and space telescopes will benefit in terms of performance by having direct, continuous, and deterministic load paths throughout. This becomes even more important for large telescopes because any detrimental effects of local load paths are amplified as telescope dimensions, mass, and flexibility increase.

6.1.5 Designing out "stick-slip" and "microlurches"

Telescopes must have exquisite pointing performance and are consequently ultrasusceptible to mechanical disturbances. This sensitivity is exacerbated

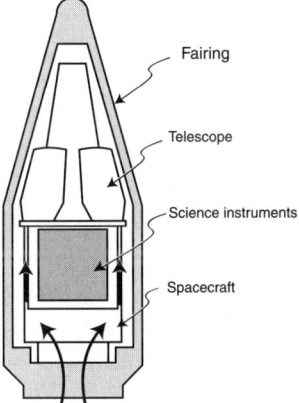

Fig. 6.8. During launch of a space telescope, the main telescope structure must be supported directly by the payload interface, not through the science instruments.

by the optical magnification engendered by powered optics. One troublesome disturbance is the sudden release of stored energy when two contacting surfaces move with respect to one another. This can occur when the two surfaces are in motion, as in a telescope drive system, or when they are not supposed to move, but do, as in a friction joint.

At the microscopic scale, elements such as rollers or gears in a telescope drive or balls in a ball bearing do not roll in the "mathematical" sense: the load on the rolling element deforms the two contacting surfaces, resulting in contact over a small area rather than at a point or on a line. Velocity is not completely uniform over that area and this results in microscopic "sliding" instead of pure rolling. Solid-to-solid sliding is jittery due to the sudden relaxation of local stresses (Fig. 6.9) so that, at slow speed, rolling elements experience a stop-and-go behavior referred to as "stick-slip." As we will see in Chapter 7, stick-slip is a nonlinear disturbance that pointing control systems do not cope with very well.

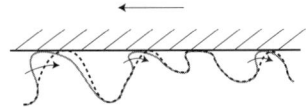

Fig. 6.9. Microscopic-level stress relaxation during sliding action is at the origin of the "stick-slip" effect.

The release of stored energy can also occur when two contacting surfaces are stationary but subjected to variable loads. This is the case with articulation joints: instead of "giving" smoothly when gravity changes direction or because of differential thermal expansion, stresses build up in the joint, then suddenly release, resulting in an abrupt movement of the supported component. The same phenomenon can occur in bolted structural elements or prestressed joints when the preload is insufficient. This effect is sometimes referred to as "microlurch."

It is prudent to invest considerable effort in eliminating stick-slip and microlurching by adequate design, rather than attempt to accommodate them after the telescope has been built. Some useful guidelines are as follows:

- When components move at slow speed with respect to one another, as in the case of a telescope tube and mount rotation axes, hydrostatic pads should be preferred over rollers or ball bearings.
- When mounting critical components in ways that must accommodate relative displacement, one should use flexures rather than ball joints or other mechanical devices (Fig. 6.10).
- When joining structural elements or mounting components using bolted or preloaded joints, one should allow a large safety margin (say about 3) over the level at which microlurching is determined to occur; the joint should also be oriented so that a microlurch will not move the supported component.

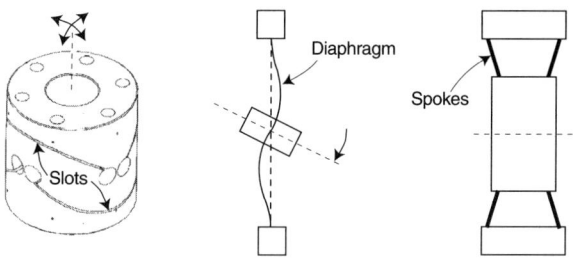

Fig. 6.10. At left, a "universal joint" which relies on structural flexure rather than hysteresis-prone mechanisms. Similarly, a diaphragm (center) or a spoke wheel (right) can be used to allow relatively free rotation over small angles. Compared to a diaphragm, the spoke wheel has greater axial stiffness. Spoke wheels were used on the Hale and CFH telescopes to avoid transmitting bending moments from the mount to the telescope tube (a detailed view of the CFH spoke wheel is shown in Fig. 6.40).

6.1.6 Choice of materials

Unlike most structures, telescopes are generally not driven by *stress* conditions, but by *deflection* under gravity, wind, or dynamic effects.

The deflection under gravity of a truss supporting optical components of a mass that is negligible compared to the mass of the truss itself (roughly the case for the tube supporting the secondary mirror in a Cassegrain telescope) is given by

$$\delta = k\frac{\rho}{E}L^4, \tag{6.1}$$

where k is a function of the truss geometry and support, L is the length of the beam, ρ is the material density, and E is the Young modulus (modulus of elasticity) of the material.

As in the case of mirror deformation (Chapter 4), we note again that the material properties enter as the ratio of Young's modulus to density (E/ρ), and the larger this ratio, the smaller the deflection. This ratio, referred to as "specific stiffness," also appears in the laws governing dynamic effects due to wind and internal disturbances.

For large telescopes, whether on the ground or in space, weight is a dominant issue and the most advantageous materials will be those with high specific stiffness. As shown in Table 6.1, aluminum and steel have nearly the same E/ρ. Since steel is less expensive, it is the material of choice for ground-based telescope structures. The other two materials, carbon fiber reinforced plastic (CFRP — a generic family that includes graphite epoxy, or GrEp) and beryllium, stand out for their high E/ρ. For the same stiffness, they offer a gain of 2 to 10 times in mass over steel and aluminum. They are more expensive to produce, however, and their use is limited to applications where mass is at a premium, such as space telescopes or active secondary mirrors on ground telescopes.

Table 6.1. Properties of major materials used in telescope structures

Material	ρ (kg/m^3)	E (GPa)	E/ρ (10^6m^2/s^2)	Poisson ratio	α 300K (10^{-6}/K)	$\int \alpha$ (10^{-6})
Beryllium	1850	300	160	0.08	12	1300
Aluminum	2700	70	26	0.33	25	4000
Steel	7800	210	27	0.30	12	2100
CFRP	1740	90–520	50–300	0.3–0.4	0.09	–

ρ is the density, E is Young's modulus, and α is the CTE
$\int \alpha$ is the integrated contraction from 40 to 300 K
Source: Paquin [4]

Unlike the other materials in Table 6.1, CFRP can be tailored in terms of its directionally dependent properties and fiber-resin selection. The laminate plies can be oriented to maximize stiffness in a given direction (e.g the axial direction for a secondary mirror tower), resulting in the most mass-efficient structure possible.

Carbon fiber reinforced plastic can also be tailored to provide a near-zero or even negative coefficient of thermal expansion (CTE) in one of the laminate's principal directions. In such a case, all plies of the laminate have identical CTEs, but these differ for directions running parallel to and perpendicular to the fiber axis. By varying the orientation of the fiber axis in each ply, one can obtain a laminate with a near-zero CTE in one principal direction (Fig. 6.11).

One issue with the use of CFRP in space telescopes is that the material absorbs moisture during the fabrication process and then degasses and changes dimension once in vacuum. This effect lasted about a year on HST

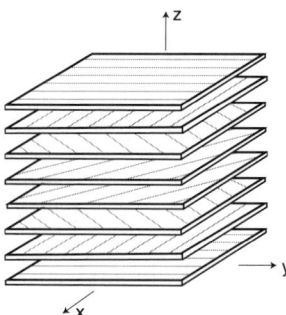

Fig. 6.11. The thermal expansion of CFRP can be controlled by laying down plies of linear graphite fibers at various angles to produce a zero CTE in one principal axis within the plane of the layers (e.g., y). The CTE in the second in-plane principal axis (x) and in the direction perpendicular to the plane of the sheets (z) remains unchanged, however.

and required readjusting the position of the secondary every 2 months or so to maintain focus. Absorption can be reduced by baking out each critical part or assembly at about 100 °C for several hours and then storing it in dry nitrogen until launch. Another solution is to replace the epoxy with cyanate ester, which substantially reduces the coefficient of moisture expansion.

6.1.7 Athermalization

Athermalization consists of designing an optomechanical system such that changes in bulk temperature do not affect optical performance. By bulk temperature change we mean a uniform temperature change throughout the object, without any temperature gradient. Bulk temperature changes can result from diurnal or seasonal temperature changes, or from a difference between the temperature at which a system has been aligned (room temperature in the shop, for example) and that at which it is operated (mountaintop or space). In the case of cryogenic space applications, the amplitude of temperature change can be very large, 200 K or more.

Bulk temperature change can modify the separation or orientation between optical elements or change the figure of the elements themselves. It causes degradation in image quality primarily through defocus. The means used to counteract this effect fall into three categories:

– **Passive.** The materials of the optics components and their support systems are chosen so that the overall optical system is insensitive to temperature. This can be achieved either by (1) selecting materials with a near-zero CTE or nearly identical total thermal cooldown strain (e.g., CFRP for structures and ULE for optics) or (2) by selecting the same material for both optics and structure (e.g., beryllium optics on a beryllium structure). In the latter case, focus changes due to the expansion

of the mirror blank are automatically compensated by the expansion of the structure supporting the mirror and the focal plane.
- **Passively compensated.** The alignment of the optics is maintained by passive systems, using thermal expansion itself to compensate for temperature effects.
- **Actively compensated.** Sensors and electromechanical actuators maintain the alignment via feedback loops.

An example of passive compensation is the position-defining pad used to hold a mirror centered in its cell regardless of changes in ambient temperature (Fig. 6.12). The pad is made of a material with a large coefficient of expansion (e.g., aluminum), and its length is determined so as to expand by the same amount as does the radius of the mirror cell.

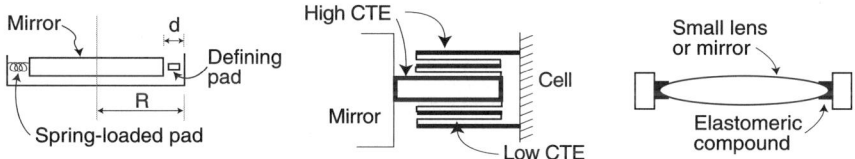

Fig. 6.12. Principle of passive compensation for a position-defining mirror pad (left). If the mirror's CTE is negligible and the CTE of the mirror cell material and of the defining pad are α and α', respectively, a pad with a length $d = R\alpha/\alpha'$, where R is the radius of the mirror, will keep the mirror centered in the cell. To minimize the overall length of the pad, it can be constructed of a series of concentric tubes of alternating low- and high-CTE materials (center). Another approach, shown at right, is the common "potted optics mount," which can be made athermal by selecting the correct thickness of the elastomeric compound [5].

6.1.8 Structural design

Two basic approaches are used in designing large structures and both have been used for large telescopes: "monocoque" and truss. In the monocoque structure, loads are carried by thin, continuous panels. This is the most common approach for mid-sized ground-based telescope mounts because it is a relatively mass-efficient structure and lends itself well to carrying loads from the elevation bearings to the azimuth bearings. It often also leads to larger, simpler pieces, which can be prefabricated in the shop and are easier to install on-site. One drawback of the monocoque design is that any large holes through the structure, such as those required for maintenance access or for passing a large-diameter light beam, can significantly decrease stiffness if not adequately reinforced. For a similar reason, areas where heavy equipment (e.g., drive motors, instruments) is mounted must be carefully designed to avoid local deformation. This increases the need for detailed analysis during the design phase.

The other design approach uses a "truss" (also called a "space frame") where all structural loads are carried through beams, columns, and rod elements. A truss can be simple and cost-effective and is an obvious choice when loads enter the structure at several discrete points, as in a mirror cell, where each support point introduces a load. Trusses also tend to be preferable from wind excitation and thermal viewpoints: they offer less wind attack area, do not impede ventilation airflow, and have lower thermal inertia because of their larger surface-area-to-mass ratios.

The choice between a monocoque and a truss depends on many factors, including the overall shape and size of the structure, worker skills in the countries of manufacture and final installation, thermal inertia, maximum size of structural elements that can be shipped and handled, and so forth. As a rule, however, the larger the structure, the more advantageous the space frame structure becomes. For very large telescopes, where wind effects, shipping and assembly constraints, and cost are critical, space frames are the solution of choice.

6.2 Design requirements

Design requirements for the telescope structure and mechanisms can be classified under "operational" and "survival." Operational requirements are those that must be met during observations, whereas survival requirements relate to exceptional conditions, such as earthquakes and launch.

6.2.1 Operational requirements

The structure design requirements are set by tolerances on the figure and alignment of the optical train elements. Typically, mirrors should not deform by more than $\lambda/20$, where λ is the shortest wavelength observed. Alignment tolerances are somewhat looser, generally by one to three orders of magnitude. A typical set of alignment tolerances for a large Cassegrain telescope is given in Table 6.2.1, with angles and deflections as defined in Fig. 6.13.

The deformation and alignment tolerances described above must be satisfied while the telescope is subjected to the following:

- static loads (gravity, static wind force, preloads),
- dynamic loads (wind gusting), and
- thermal changes (diurnal, seasonal, orbital or operational).

These loadings are treated no differently than in traditional engineering practice, but because wind loading constitutes a major disturbance for ground-based telescopes, we will look at it in some detail in Chapter 7.

Table 6.2. Typical tolerances for linear and angular displacements of the optical components of a large Cassegrain telescope

Optical element	Displacement	Tolerance
Secondary mirror	ΔX	10 µm
	ΔY	10 µm
	ΔZ	150 µm
	$\Delta \theta_x$	0.8″
	$\Delta \theta_y$	0.8″
Cassegrain or Nasmyth instrument	ΔX	100 µm
	ΔY	100 µm
	ΔZ	150 µm
	$\Delta \theta_x$	1′
	$\Delta \theta_y$	1′
Nasmyth folding flat	ΔX	20 µm
	ΔY	None
	ΔZ	150 µm
	$\Delta \theta_x$	20″
	$\Delta \theta_y$	20″

Data for the Subaru 8 m telescope from Ref. [6].

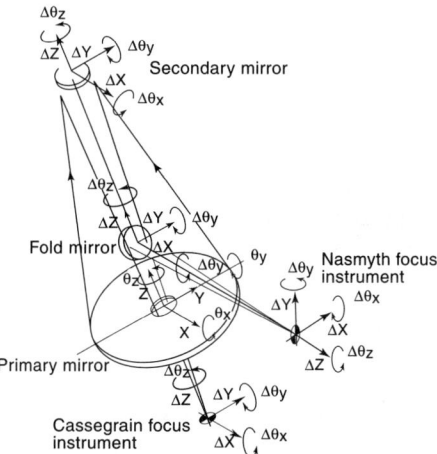

Fig. 6.13. Definition of the various deformations relevant to a Cassegrain/Nasmyth optical system.

6.2.2 Survival conditions

Telescopes must be able to resist exceptional conditions such as emergency braking and earthquakes for ground telescopes and launch and gravity loading during assembly for space telescopes. Whereas operational conditions primarily put constraints on maximum *strain*, that is to say, on how stiff the telescope structure and mechanisms must be, survival conditions have to do with the maximum *stress* that can be permitted.

Seismic load

As it happens, many first-class astronomical sites are located in zones of high seismicity (California, Chile, Hawaii) [7]. Seismic activity manifests itself by an acceleration of the ground, and the load on the telescope will depend on the mechanical properties of the soil and on the stiffness and damping of the pier and telescope. Design accelerations applicable to a given site are usually specified by the local construction code or a national geophysical institute. From the designer's point of view, it is convenient to define two seismic load levels: the "operational base earthquake" (OBE) and the "maximum likely earthquake" (MLE). These two levels are determined by a judgment call, taking into account the probability of occurrence and the consequences of extended loss of observation time or even loss of the facility.

The OBE is the earthquake level a telescope should be able to sustain without losing functionality: only alignment checks or replacement of inexpensive items are needed to recover the system. The MLE, on the other hand, is the maximum earthquake level that the facility can be expected to sustain. Major damage is acceptable, but not to the point where it would be uneconomical to recover the facility (e.g., breakage of the main mirror or permanent deformation of the mechanical structure).

As an example, the horizontal ground acceleration of the OBE for the VLT located at Paranal, in Chile, has been defined as 0.24 g with a 50% probability of occurrence within 25 years, and the MLE ground acceleration is 0.34 g with a 10% probability of occurrence in 100 years. Vertical accelerations are usually taken as two-thirds of the horizontal acceleration and are combined simultaneously with the horizontal acceleration.

Earthquake analysis proceeds as follows. The first step is to determine the frequency spectrum of the design earthquake in order to take into account possible amplification by the structure. Earthquake frequency content is a function of the geographical zone and soil conditions at the site and must be developed either theoretically or experimentally for each particular site. The response to an earthquake is defined as the acceleration that a single-degree-of-freedom system of a particular natural frequency would exhibit if subjected to that earthquake. A typical acceleration response spectrum is shown in Fig. 6.14. Generally, there is little frequency content beyond 20 Hz. Hence, objects will not suffer amplification of the earthquake if they are very rigid and firmly anchored to the ground. The object and its attachment to the ground must then simply be designed to withstand the horizontal force corresponding to the nominal earthquake acceleration.

Large telescopes have natural structural frequencies in the 1–10 Hz range, however, and they will suffer a larger acceleration than the ground does. For a first approximation, one can calculate the force acting on the telescope by multiplying its mass by the amplified acceleration derived from the above response spectrum for the lowest modal frequency. Instruments and auxiliary devices mounted on the telescope can experience much larger amplification

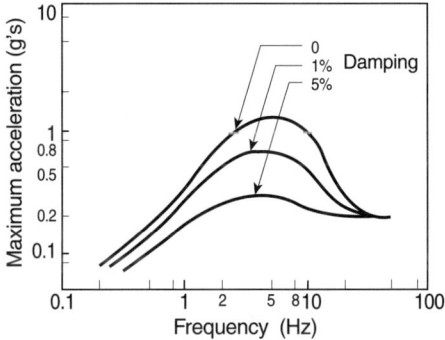

Fig. 6.14. Typical earthquake acceleration response spectrum for three values of damping of the structure. (After Ref. [7].)

factors if their natural frequencies are in the same range as those of the telescope. This can be avoided by specifying that all equipment mounted on the telescope must have a natural frequency significantly higher than that of the predominant mode of the telescope — 30 Hz or more, for example.

The detailed earthquake calculations should include not only the tube and mount, but also the bearings, drives, concrete pier, and soil stiffness. It is important that all of these elements in the chain be included because they will generally reduce overall stiffness significantly compared to the stiffness of the tube and mount alone. This is particularly true of the azimuth drive, where tooth-to-tooth contact and motor mounting can be the weakest link, a condition which can be mitigated somewhat by using a direct drive or by placing the drive at a larger radius [8]. The characteristics of the soil and the geometry of the foundation are also important factors. Figure 6.15 (right), shows how the soil's Young modulus influences the telescope's natural frequency in the case of the VLT.

Emergency braking loads

Every time the telescope starts and stops moving, the structure and optical components are subjected to dynamic loads. This type of load can be significant in large telescopes because of the potentially large distances from the rotation axes. As an indication, for the VLT, the maximum acceleration during repointing is about $0.4°/s^2$ for both altitude and azimuth, which leads to negligible loads. During emergency braking deceleration, however, this value is as high as $10°/s^2$, which creates accelerations of about $0.2\,g$ at the top of the tube.

Launch loads

Launch, which begins with engine ignition and ends with spacecraft separation from the launch vehicle, imposes a short-lived but highly stressful environment on a payload. During this period of a few minutes, it is subjected to severe

218   6. Telescope Structure and Mechanisms

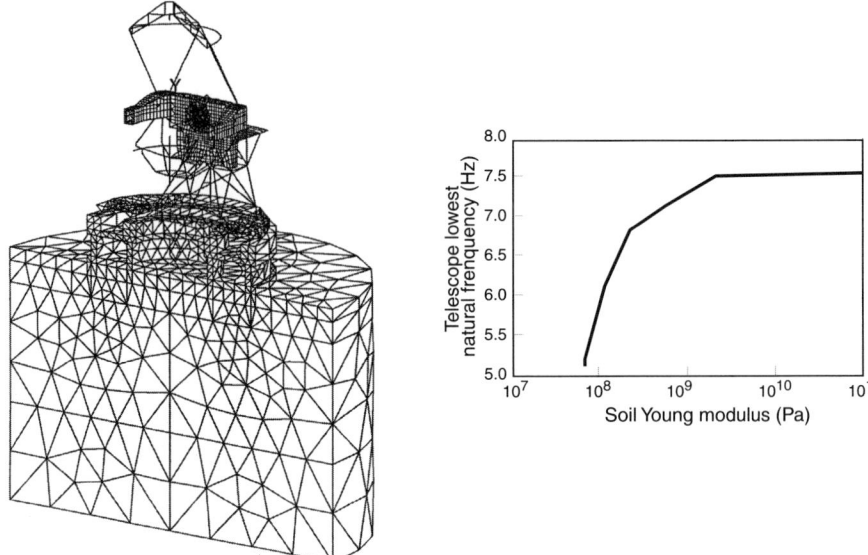

Fig. 6.15. At left, structural model of the VLT telescope, pier, and foundation used for complete dynamic analysis under wind action and seisms. At right, an example of the results of these studies showing the influence of soil stiffness on the overall frequency of the telescope.

structural loads, thermal transients due to heating of the payload fairing, and rapid loss of pressure. Of all these, the structural load environment is the most severe. This environment is a combination of quasi-static loads, low- and high-frequency dynamic loads, and shock loads. The quasi-static and low-frequency dynamic loads are due to the acceleration of the launch vehicle, wind gusts, steering, and engine transients. The high-frequency dynamic loads are predominantly of acoustic origin. Acoustic loads are most severe at lift-off, when the sound energy of the rocket engine exhaust is reflected by the launch pad, and again during the transonic portion of the flight, due to aerodynamic shock loads. The duration of peak acoustic loads is typically not more than 10 seconds. Low-areal-density components (i.e., those items with a relatively large surface area and low mass) are the most sensitive to acoustics loads. Typically, the only launch vehicle shock load of concern to the payload is that due to payload separation from the launch vehicle. Separation shock loads are highest at the payload mounting interface, then decay rapidly as they travel through the payload structure. Shuttle-launched payloads must also be designed to withstand emergency or normal landing loads. Launch-vehicle user guides and payload planners' guides provide data on the above environments for preliminary design purposes.

6.3 Mirror mounts

6.3.1 Mounts for single mirrors

Up to ~30 cm in diameter, a ground-based telescope mirror can be made stiff enough to be considered a rigid body and mounted kinematically (Fig. 6.16).

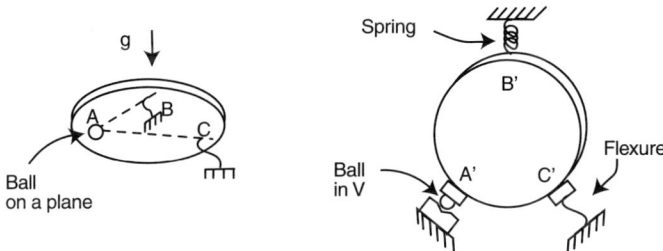

Fig. 6.16. Typical radial kinematic mirror mount for a small mirror. The mirror back is supported on three points, A being a ball on a plane and B and C being flexures acting along the directions AB and AC (left). The radial support is shown at right. Such a system allows for differential expansion between the mirror and its mount and prevents the mirror from being affected by mount deformation.

For a solid mirror of uniform thickness supported horizontally on three points, the maximum deflection is given by

$$\delta = \frac{\beta q a^4}{E h^3}, \tag{6.2}$$

where h is the thickness, q is the weight per unit area (proportional to h), and β is given in Fig. 6.17 as a function of the ratio of the three-point-support circle radius, r_s, to the radius of the mirror, a. Minimum sag occurs for supports located at about two-thirds of the mirror radius.

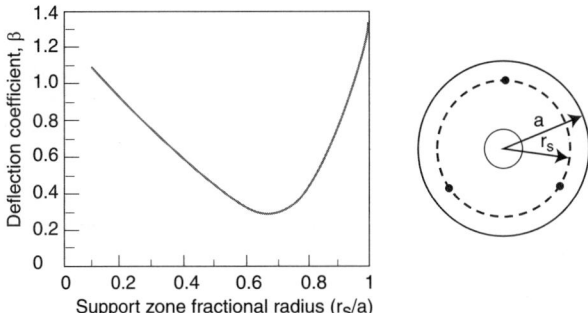

Fig. 6.17. Normalized maximum deflection of a mirror supported on three points with the axis vertical, as a function of the ratio of the support radius to the outer radius.

220   6. Telescope Structure and Mechanisms

Mirrors much larger than 30 cm would bend beyond tolerable limits under gravity if supported on only three points; additional supports are necessary to compensate for gravity, but without interfering with the stress-free condition afforded by the kinematic mount. Ideally, such a system would support the mirror as if it were floating in a liquid of its own density. Mirror support systems attempt to mimic this condition by applying a combination of forces at discrete points on the mirror's back and side. The number and spacing of the support points are chosen so that the mirror will not sag between points by more than a given allowance.

The simplest solution is to maintain the basic principle of the kinematic three-point support system, but spread the load supported by each of the three points over a larger number of points on the mirror. To preserve the kinematic nature of the system, support points on the back of the mirror are grouped by twos or threes and mounted on pivots. They can be arranged on multiple tiers to form what is called a "whiffletree." This principle has been used successfully to support mirrors up to 2 m in diameter (Fig. 6.18).

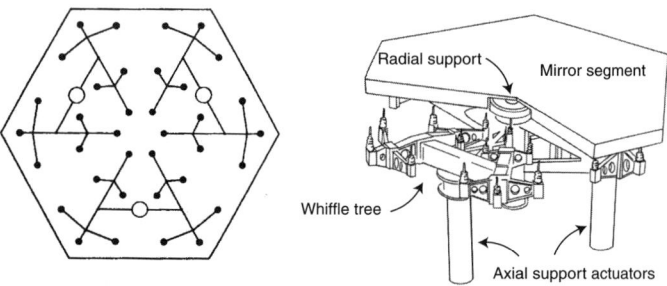

Fig. 6.18. The whiffletree support system used for the Keck primary mirror segments. The 75 mm thick, 1.8 m in diameter hexagonal mirrors are supported on the back at 36 points (left). These support points are grouped by three's and then connected to the three overall support points via levers forming a whiffletree. One of the three whiffletree assemblies, with its 12 support points, is seen in the cutaway view at right.

An alternate solution consists of "floating" the mirror so as to balance its weight while it is still kinematically mounted in traditional fashion (Fig. 6.19). This can be accomplished by applying to the back of the mirror either pressure (vacuum if it is facing down) or a set of discrete forces. It is important that this pressure or these forces be unaffected by deflection of the mirror cell. Figure 6.20 shows an example of the use of vacuum to support a secondary mirror. Force-generating devices include counterweight lever systems, air or oil jacks, or electro-mechanical actuators with feedback control (Fig. 6.21).

The pressure/vacuum or discrete forces must be continuously adjusted to compensate for the weight component in the vertical direction as the telescope tracks its target. In the past, this compensation was achieved "passively," using counterweights on a lever system (also called "astatic lever") or

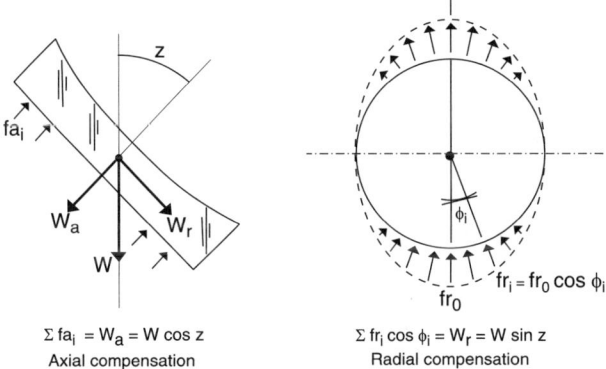

$\Sigma fa_i = W_a = W \cos z$
Axial compensation

$\Sigma fr_i \cos \phi_i = W_r = W \sin z$
Radial compensation

Fig. 6.19. Forces are applied to the back and edge of the mirror to compensate for gravity and minimize deflection.

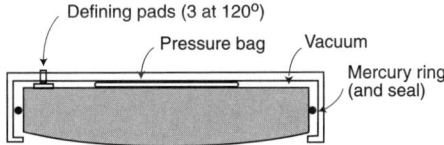

Fig. 6.20. The secondary mirror mount used on the CFHT. The back of the mirror is supported by vacuum, with a seal formed by a mercury tube along the mirror edge. The mercury exerts hydrostatic pressure to support the mirror radially. A pressure bag, acting on the back of the mirror, corrects for the spherical aberration found in the secondary mirror after its installation. Both vacuum and pressure are adjusted as a function of the zenith angle with a simple piston-driven regulator. (From Ref. [9].)

a pneumatic system with open-loop regulation, but such systems are limited in accuracy to about 0.1% by friction and the build-up of fabrication tolerances. Consequently, traditional passive support systems required careful tuning and, even so, their performance placed a limit on the optical performance of many older telescopes.

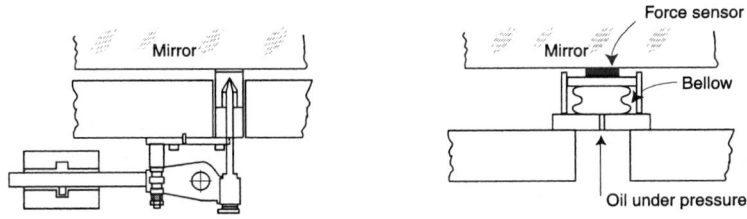

Fig. 6.21. Counterweight lever system at left, hydraulic support at right.

222   6. Telescope Structure and Mechanisms

Today, it is more convenient to use a feedback system based on load cells. For example, simple *local* feedback systems have been used on the MMT, Magellan, and LBT mirrors [10]. Each kinematic support point is motorized (for collimation adjustments) and equipped with a force sensor and a servo system. This servo system adjusts the total flotation force vector provided by the pneumatic force actuators, so that the forces on the kinematic support points are essentially zero at all telescope elevation angles.

For very large monolithic mirrors, which require greater compensation accuracy, the loop must be closed on the image itself. A wavefront sensor is used to measure the optical performance of the telescope; then, a computer calculates the necessary changes and issues commands to the support actuators to obtain the best possible mirror figure. With this, one enters the realm of fully active optics, which is the subject of Chapter 8.

Mirrors are supported on their periphery using the same principles as for back supports: counterweight levers, hydrostatic pressure (as in Fig. 6.20), or actuators with feedback. Ideally, the edge forces should push up in the lower half of the mirror and pull up on the upper half (Fig. 6.22a). The forces should also act in the plane containing the center of gravity of the mirror in order to avoid creating an overall moment. This condition cannot be satisfied in the case of strongly concave mirrors, however, with the result that local bending moments distort the optical surface (Fig. 6.22b). The effect can be greatly reduced by introducing shear forces to counteract the bending moment (Fig. 6.22c) [11, 12]. This is possible with an alt-az mount in which, unlike the equatorial mount, the primary mirror tilts in one direction only.

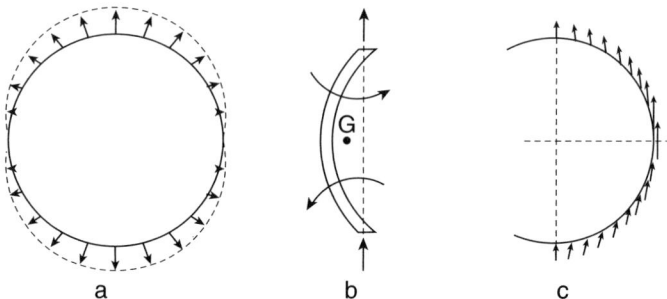

Fig. 6.22. Force distribution for the edge support of a mirror (see text).

When the mirror is small and light enough, a single radial support located at the center of gravity can be used, provided that it only acts radially in order to form a kinematic mount in combination with the back supports (Fig. 6.23).

Space telescopes do not suffer from gravity deflection. But since the effect of gravity release is difficult to measure on the ground, force actuators may be needed on the back of the mirror to make minute corrections to its figure once in orbit. The HST mirror has 24 such actuators which were intended

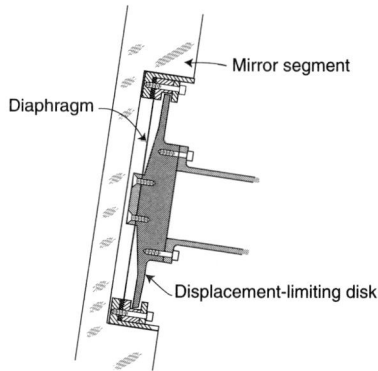

Fig. 6.23. Single radial support used on the Keck telescope mirror segments. A diaphragm gives the mirror freedom to move in the axial direction to avoid interfering with the back-support system.

to correct potential figure errors (principally astigmatism) due to imperfect a priori compensation of gravity release. The gravity release correction was found to be unnecessary. It is unfortunate that the location and capacity of these actuators were not adequate to correct for the spherical aberration discovered in the mirror once in orbit.

Mirror support theory and practice have been extensively studied and are well reported in the literature. A detailed account of mirror mount techniques and first-order design calculations can be found in Wilson [13] and Yoder [14].

6.3.2 Mounts for segmented-mirror systems

All methods described above for mounting single mirrors also apply to the support of individual segments of segmented mirrors. However, an additional requirement has to be satisfied here: each segment must be positioned so that its optical surface exactly matches that of the parent mirror surface.

This imposes two conditions: (1) the back actuators must be able to position the segments with an accuracy of a fraction of the operating wavelength, typically about 10 to 50 nm, and (2) the supporting structure of the segmented mirror must either be stable to that same accuracy in a passive fashion (a condition only possible in space), or an active system must be used.

Several actuators that meet the resolution requirement have now been developed. Those used on the Keck telescope primary segments have a resolution of 4 nm and a full range of 1 mm to correct for deflection of the mirror cell when the telescope moves from zenith to horizon (Fig. 6.24). Each has a mass of 11 kg and a power dissipation of 0.5 W [15].

The Hatheway actuator is one of several types developed for NGST. It has a resolution of 7 nm and a range of 25 mm and can operate at temperatures down to 40 K. The Hatheway actuator consists of a pair of stepper-motor-driven lead-screw assemblies attached in series to each other through a differential spring coupling. The two motor/lead-screw assemblies are actuated individually, with one being the fine stage and the other the coarse stage.

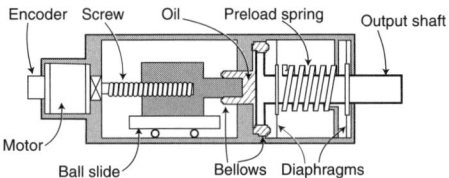

Fig. 6.24. A schematic view of the actuators used to position the primary mirror segments of the Keck telescopes. The hydraulic system demagnifies the stroke of the driving piston by a factor of 24.

In the weightless, benign environment of the best space orbits (e.g., L_2, which does not suffer from eclipsing or gravity gradient), thermal changes and dynamic disturbances are so small that a purely passive solution for the segmented-mirror supporting system is possible. This is the solution proposed for NGST. Thermal changes when the telescope is repointed are less than 1 K, and with the use of a material with a very low coefficient of expansion for the mirror's backup structure, this leads to an expected wavefront error of less than 20 nm. Once the mirror segments are adjusted using a bright star and phase retrieval techniques, the segmented mirror is expected to be stable for weeks.

On the ground, however, there is no hope for a mirror cell to remain stable to the required accuracy. Thermal changes, gravity, wind, and potential structural instabilities create disturbances with a complex frequency content and amplitudes up to the millimeter level.

Several methods have been proposed and tested for the active control of segmented mirrors, all of which rely on internal metrology. These include interferometry, laser metrology, and edge sensors and will be described in Chapter 8.

6.4 Telescope "tube"

Originally, the "tube" of a telescope was a genuine tube surrounding the incoming optical beam, holding the primary mirror at the bottom and the secondary mirror at the top. The term is now loosely used to refer to any structure supporting the primary and secondary mirrors, even if the structure surrounding the beam is an open truss or if some other design is used, such as the tripod employed for radiotelescopes.

6.4.1 Tube truss

Serrurier truss

As discussed in section 6.1, cylindrical or conical shell tubes are not used on large ground-based telescopes because they cannot be made rigid enough to keep the optics collimated. Moreover, an enclosed tube is highly susceptible to wind shake due to its large cross section, and it also prevents air from flowing across the primary mirror, which is essential to minimize "mirror seeing." An open truss is therefore decidedly preferable. Most ground-based 4 to 6 m class telescopes built before 1975 use the "Serrurier truss" (Fig. 6.25), which is designed to maintain the collimation of the primary and secondary mirrors in spite of relatively large deflections due to gravity.

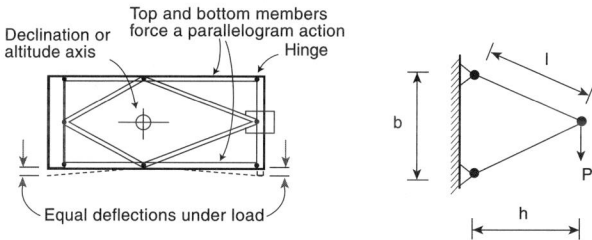

Fig. 6.25. With a "Serrurier" truss, the primary and secondary mirrors undergo parallel translation with no rotation (left). Geometry of the V-truss triangle (right).

Disregarding the weight of truss members, which is generally small compared to the load at the front and back of the truss, the deflection of the vertex of each V-truss is of the form

$$\delta = \frac{2Pl^3}{EAb^2} = \frac{P}{4EA}\left(\frac{4h^2}{b^2}+1\right)^{3/2}, \qquad (6.3)$$

where δ is the lateral deflection of the vertex, P is the load supported at the vertex, l, h, and b are the truss dimensions defined in Fig. 6.25, E is the Young modulus of the material, and A is the area of the beam section. The cross-section areas of the front and rear members are designed such that the deflections of the two ends are identical. An additional condition is that the natural frequency of the tube must be high enough to minimize dynamic wind effects.

Important conditions must be met for the Serrurier concept to work as intended. The center of gravity of both ends must be exactly in the plane of the V-truss vertices; if not, moments induced in the truss will result in decollimation (Fig. 6.26). The effect is more pronounced when the truss angle is very large, as is generally the case on the primary mirror side. It is then important to avoid using excessively heavy Cassegrain instruments. Another condition is that the truss attachments must act as "hinges" in order to avoid creating moments at each end. This can be achieved by using trusses with small enough diameters.

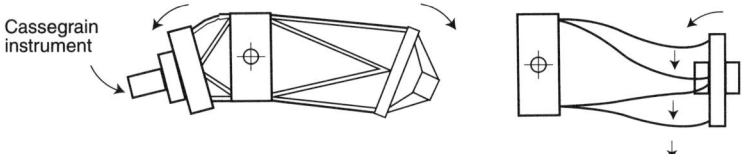

Fig. 6.26. Two situations where the Serrurier conditions are not satisfied: at left, moments created at the top end, due to an offset secondary, or at the bottom end, due to Cassegrain instruments that are heavier or longer than originally designed for, will decollimate the optics; at right, the deflection of the truss under its own weight also creates a moment on the upper end.

As the size of the telescope increases, the primary mirror mass generally increases proportionally more than that of the secondary mirror, resulting in the tube's center of gravity moving closer and closer to the primary mirror. For this reason, some designers now forego the lower Serrurier truss completely and attach the primary mirror firmly to the center box section of the tube while continuing to put the Cassegrain secondary on a "Serrurier like" V-truss. When this is done unintentionally, it is a misapplication of the Serrurier truss principle. With such a configuration, the secondary mirror will indeed remain parallel with the primary mirror, but the law of equal deflections will not be met, potentially resulting in a comatic image.

The Serrurier V-truss structure is an elegant solution to optical collimation, but it is inherently structurally inefficient. This is because the tube end weight is carried by only two compression and two tension members, whereas the other four members do not participate. Also, the rings supporting the primary and secondary mirrors are large and heavy and suffer from being supported at only four points, as imposed by the Serrurier system. These rings must then have high in-plane stiffness, resulting in even heavier structures. Another problem that arises as telescopes become larger is that the natural frequency of the individual Serrurier truss members approaches the vortex shedding frequency in the presence of wind (see Chapter 7). This results in disturbances in the telescope structure which exceed the frequencies that the pointing system can correct.

Multibay truss

With active optics, the ability to maintain collimation by passive means has lost much of its attraction. Consequently, the main design goal for the tube structure will be to increase the resonant frequencies and reduce wind attack. From this point of view, the most efficient structures are composed of straight members acting essentially in compression and tension and arranged in multiple triangles as shown in Fig. 6.27.

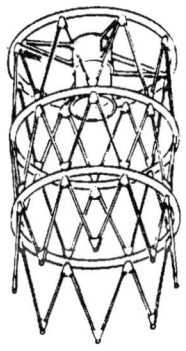 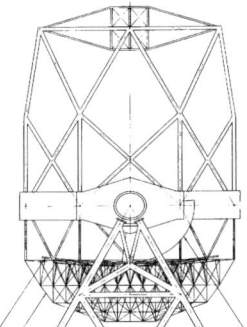 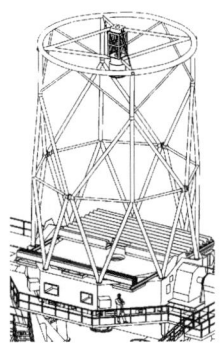

Fig. 6.27. Three-bay truss used on HST (left), and two-bay truss used on the Keck (center) and Gemini (right) telescopes.

Multibay structures are best optimized in an iterative fashion using finite element analysis [16]. As a first-order approximation, the maximum stiffness-to-mass ratio is generally achieved when the number of bays, n, satisfies the relationship

$$\sqrt{n(n-1)} < \frac{L}{D} < \sqrt{n(n+1)}, \qquad (6.4)$$

where L is the overall truss length and D is the diameter [17, 18] (Fig. 6.28).

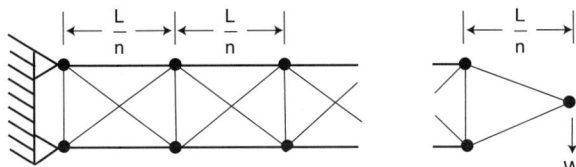

Fig. 6.28. Truss bay geometry.

The individual members should be thin-walled, hollow tubes with diameters sufficient to minimize bending under their own weight, but still thin enough to minimize wind effects. It is also structurally efficient to abandon the heavy rings used in the Serrurier design to support the primary and secondary mirrors, replacing them with straight members that act only in tension and compression, not in bending.

Tube natural frequency

A coarse estimate of the natural frequency of a telescope tube can be obtained by treating it as a cantilevered beam [19]. The natural frequency of a cantilevered beam with a constant distributed mass m and a concentrated mass m_c (secondary mirror) at the free end is

228   6. Telescope Structure and Mechanisms

$$f = \frac{1}{L^2}\sqrt{\frac{E}{\rho}\frac{I}{A}\frac{1}{(1+0.23\,m_c/m)}}, \qquad (6.5)$$

where L is the length of the beam, I is its moment of inertia, A is its cross section, E is the Young modulus of the beam material, and ρ is its density. To the first order, the ratios m_c/m and I/A are independent of overall tube size, so that the tube's natural frequency is essentially inversely proportional to L^2. This basic relationship is plotted in Fig. 6.29.

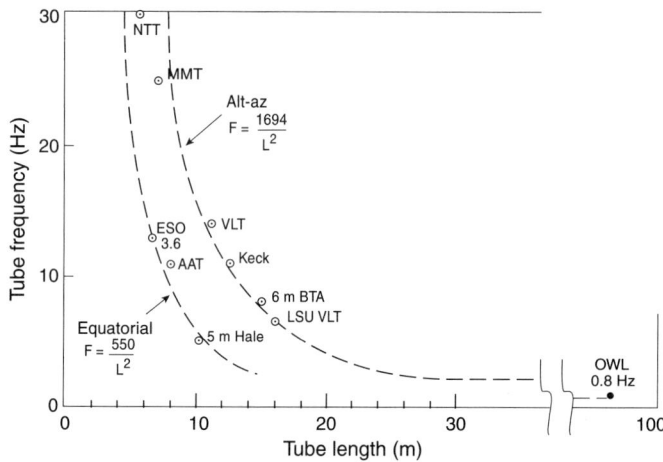

Fig. 6.29. General trend of natural tube frequency as a function of the length of the tube. The natural frequency of telescope tubes is inversely proportional to the square of the tube length, the proportionality constant depending mostly on the type of tube structure employed. One notes that the natural frequency decreases very little for tube lengths greater than 20 m.

One must not forget that the overall frequency of the tube around the altitude axis (the so-called "locked-motor" frequency) is not simply the tube's structural frequency; the stiffness of the altitude drive is also a factor. This is actually often the limiting factor unless a direct drive or, for a friction drive, a large journal radius is used.

6.4.2 Tripod and tower-type supports for secondary mirrors

Instead of the conventional tube, a tripod or central tower can be used to support the secondary mirror, as shown in Fig. 6.30. The major drawback of such solutions is their inherently large obstruction of the optical beam. However, they are structurally advantageous when the distance between the primary and secondary mirrors approaches the diameter of the primary mirror (e.g., for a fast primary).[2] The tripod scheme also lends itself to relatively easy deployment for space telescopes.

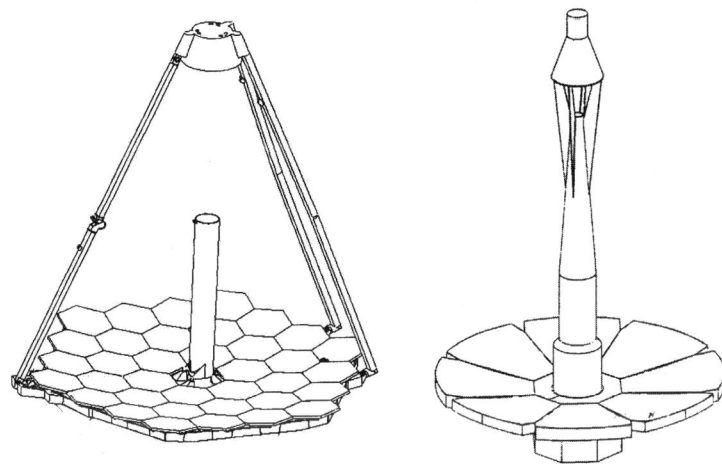

Fig. 6.30. Two secondary mirror support systems, alternatives to the traditional truss: tripod (left) and tower (right).

The tower support has been used on small telescopes (e.g., SIRTF) and has also been proposed for NGST. The support tower can be combined with the central stray light baffle, resulting in structural economy. However, its bending stiffness is low and the blades supporting the secondary mirror, which are in the "concentrated" beam going toward the Cassegrain mirror, create a significant obstruction.

6.4.3 Thermal effects

The effects of temperature change on optics alignment can be reduced by applying the principles discussed in Section 6.1.7. One solution consists of using near-zero CTE materials for the tube structure (e.g., CFRP), as was done for HST. Another approach is to use a combination of materials of different CTEs that cancels out overall thermal expansion effects.

With active optics, thermal expansion problems have now essentially disappeared, as both alignment and focus can be maintained in real time thanks to direct sensing of wavefront errors (Chapters 8 and 9).

---

[2]Note, however, that a tripod constrains only three degree of freedom at its apex, thus leading to low stiffness for the secondary mirror in tip and tilt. See also the discussion on pyramidal top ends in Section 6.4.4.

6.4.4 Cassegrain mirror "spider"

A "spider" is a structure which, in classical tubes, holds the secondary mirror to the ring at the top of the tube truss. Since the spider is in the incoming beam, its members must be made as thin as possible in order to limit light obstruction and diffraction, thus giving rise to the terms "vanes" or "knife edges" for these members.

Although composed of thin members, the spider must nonetheless provide a stiff support for the secondary mirror unit, either to minimize collimation in passive systems or to permit a high bandwidth for tip-tilt adjustment in active systems.

As already explained in Section 6.1.2, simply supporting the secondary mirror with an in-plane spider does not provide enough rigidity along the optical axis (Fig. 6.31, left). It is preferable to use a triangular geometry with a large base as shown in Fig. 6.31 (center). To avoid buckling, every member should be prestressed so as always to be in tension regardless of the inclination of the tube.

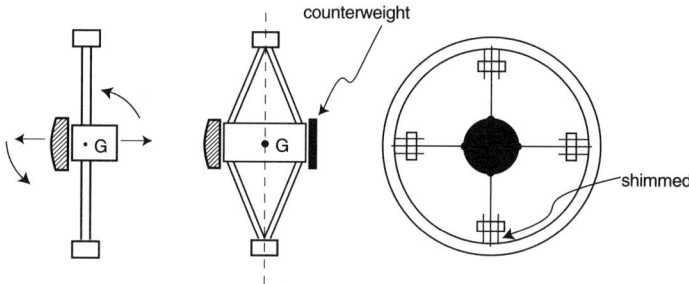

Fig. 6.31. A spider in the plane of its supporting ring has low rigidity in the axial direction (piston mode) and in tip-tilt (left). The situation is improved by extending the body of the secondary mirror support in the axial direction and supporting it with vanes arranged in a triangle (center). Prestressing is usually accomplished by having the vanes slightly shorter than the inner diameter of the ring and shimming the gap appropriately to obtain the desired tension (right).

In theory, prestressing the spider vanes also increases their natural frequency. In practice, however, the gain will not be significant. This is because the natural frequency of the secondary mirror unit radially in the plane of the top ring depends only on the lateral stiffness of the vanes. To the first order, this natural frequency, ω_0, is given by

$$\omega_0 = \sqrt{\frac{k}{m}} \simeq \sqrt{\frac{D^3}{48\,E\,I\,m}} \;, \tag{6.6}$$

where m is the mass of the secondary mirror unit, E is the Young modulus of the material, I is the area moment of inertia of the beam, and D is the diameter of the ring. The prestress in the vane increases the natural frequency, ω, according to the relation

$$\omega = \omega_0 \sqrt{\frac{1 + P_{\text{prestress}}}{P_{\text{Euler}}}}, \text{ with } P_{\text{Euler}} = \frac{\pi^2 EI}{L^2}, \qquad (6.7)$$

where $P_{\text{prestress}}$ is the preload applied at the end of the beam, P_{Euler} is the first critical load of the vane, and L is the length of the beam. Significantly increasing ω requires fairly high prestress, which is impractical due to the relative weakness of the top ring.

Finally, we note that prestress does not affect the natural frequencies of the secondary mirror unit in the optical axis direction because the axial stiffness of the spider depends only on the sectional area of the vane and the Young modulus of the material.

The system is also weak in torsion around the optical axis. The rotation of the secondary mirror around its optical axis is a priori of no consequence in on-axis optical systems, but low torsional frequency can be a problem when the secondary mirror is used for chopping. Torsional stiffness can be improved by positioning the vanes so that they do not converge at the center of gravity or by adding a fifth vane (Fig. 6.32).

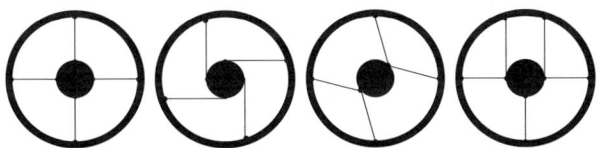

Fig. 6.32. Vanes arranged in a cross are weak in torsion (left). Offsetting the vane intersection points from the center of gravity (second and third from left) or adding a fifth vane (right) will significantly improve torsional stiffness.

The vanes should be arranged by groups and in parallel so as to minimize the number of diffraction spikes. Each straight obstruction in the beam produces two diffraction spikes 180° apart in the direction perpendicular to the direction of the vane, and the total amount of energy in the spike is proportional to the amount of light intercepted (for small obscurations). Since the spikes always issue from the center of the PSF, vanes do not need to be in line, simply parallel to each other. A three-vane spider will create six spikes. On the other hand, a six-vane geometry such as that used on the Keck telescopes will not create additional spikes since the vanes are parallel, two by two (Fig. 6.33, left).

In the case of segmented primary mirrors, it is best to have the vanes overlap the gaps between segments in order to minimize overall loss of light. This was the rationale for the six-vane spider used on the Keck telescopes, but it also fitted in well with the desire to increase the structural efficiency of the overall tube by abandoning the circular top ring and replacing it with an articulated hexagon.

Although the spider "tension" design described so far leads to the smallest beam obstruction, it requires that the tube extend at least to the secondary mirror. Since the tube's top ring and the truss offer a nonnegligible area to the wind, it can be advantageous to use a "pyramidlike" secondary mirror support system in order reduce the length of the main tube and, consequently, reduce the wind moment with respect to the altitude axis. The VLT, which is in a windy site, follows this strategy (Fig. 6.33, right). But this comes at the price of increased light obstruction. The pyramidlike configuration cannot be prestressed to place all elements in tension and thus prevent buckling; hence, the spider members must have larger cross sections.

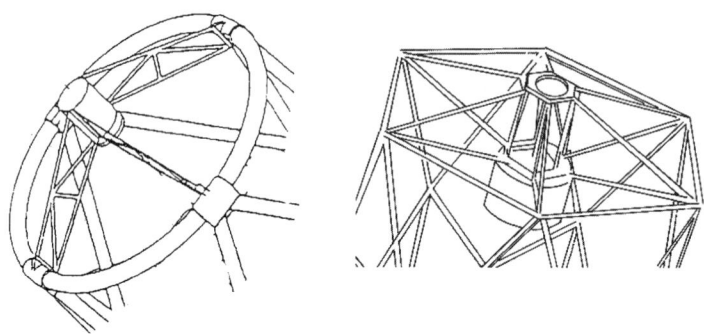

Fig. 6.33. At left, the pyramidlike structure used to support the secondary mirror of the VLT reduces the length of the tube and thus overall wind buffetting. At right, the six-vane geometry of the Keck telescopes' secondary mirror spiders.

Cables feeding the secondary mirror may be conveniently placed in a channel on top of the support vanes, as shown in Fig. 6.34, left. When the telescope field is large and one wants the obstruction to be independent of the field angle, T-sections should be used in place of flat vanes, as shown on the right in Fig. 6.34.

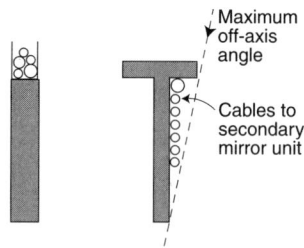

Fig. 6.34. Cable routing to the secondary mirror unit over the flat support vanes (left). T-cross-section vanes may be used to keep the obstruction constant over the entire field (right).

6.4.5 Primary mirror cell

In older telescopes, the primary mirror cell was a cylindrical box-type structure topped with a flat plate to which the mirror support system was mounted. The cell had to be fairly shallow to provide good access to the Cassegrain. An example of such a design is shown in Fig. 6.35 (left).

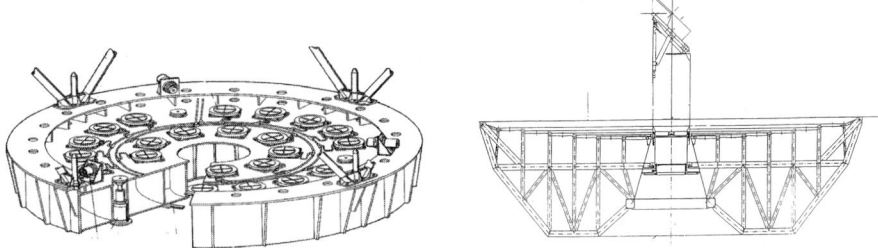

Fig. 6.35. Mirror cells. At left, the conventional box-type mirror cell used on the CFH telescope and at right the space frame mirror support structure for the VLT.

Since this design is intrinsically heavy, it is being replaced in large telescopes by the more structurally efficient space frame structure, as shown on the right in Fig. 6.35. The drawback of this configuration is poor access to the Cassegrain focus.

The main design criterion for the primary mirror cell is that its deflection under gravity must be within the acceptance range of the mirror actuators, which is typically on the order of 1 mm. Another condition is that the natural frequency of the mode along the optical axis must be high enough so that wind gusts do not excite that mode.

6.5 Mounts for ground-based telescopes

The purpose of a "telescope mount" is to support the telescope tube and allow for its rotation during pointing and tracking. This can be accomplished in one of four basic ways, which we will now examine.

6.5.1 Equatorial mount

Nearly all telescopes built before 1980 used the equatorial mount, the principle of which is to neutralize the rotation of the Earth by moving the telescope tube around an axis parallel to the Earth's, but in the opposite direction (Fig. 6.36, left). An equatorial mount provides for motion about the "polar axis," which is parallel to Earth's rotation axis, and the "declination axis," which is perpendicular to it. Once the tube has been pointed toward the target by rotating it around these two axes, the telescope can be kept pointed at the

234   6. Telescope Structure and Mechanisms

target by simply rotating it around the polar axis at the Earth's diurnal rate, but in the opposite direction.

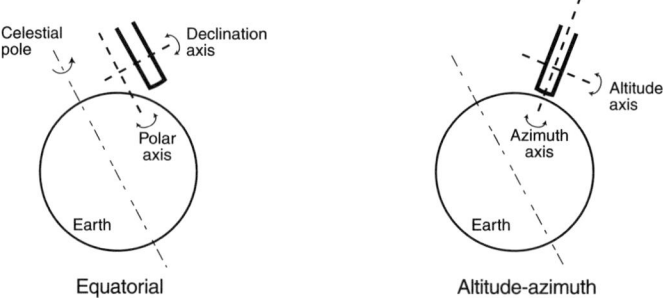

Fig. 6.36. Principle of the equatorial mount (left) and the alt-az mount (right).

This extremely simple means of tracking a celestial target requires rotation of only a single axis, and that at constant speed. This simplicity together with the absence of field rotation are the main advantages of the equatorial mount.

As shown in Fig. 6.37, equatorial mounts can be implemented in a variety of ways, depending on the type of support used for the declination axis.

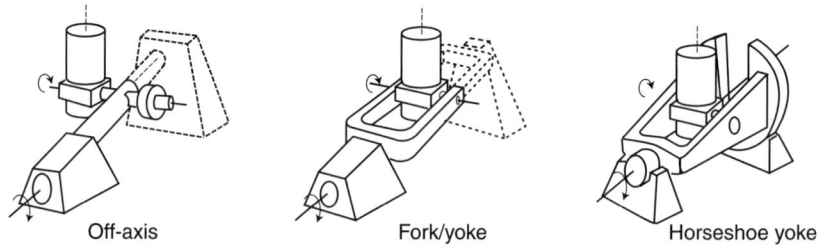

Fig. 6.37. Types of equatorial mount. The off-axis mount is shown on the left. A counterweight balances the weight of the tube. This configuration is called a German mount if supported as a cantilever and an English mount if supported at the north and south ends (dashed line). The symmetric mount shown at center is called a "fork mount" if supported from one side and an "English yoke" if supported on both sides. The "horseshoe mount," a variation of the yoke mount that allows access to the pole, is shown on the right.

Equatorial mounts are intrinsically heavy because, due to the inclined polar axis, they must be structurally either cantilevers or beams. At 5 meters, the Hale telescope is regarded as the practical upper limit for this type of mount. Larger telescopes must be supported more efficiently, by placing the center of gravity of the tube above the base of the mount. This is the principle behind the altitude-azimuth mount.

6.5.2 Altitude-azimuth mount

In the altitude-azimuth mount, "alt-az" for short, the tube is oriented by rotation around a vertical axis (the azimuth axis) and a horizontal axis (the altitude axis) as shown on the right in Fig. 6.36. As opposed to equatorial mounts, neither of the two axes supporting an alt-az mount changes direction with respect to gravity. Structurally, it is the sturdiest and simplest mount. The reduction in mass (and cost) is so significant (Fig. 6.38) that it has now become the standard mount even for mid-class telescopes.

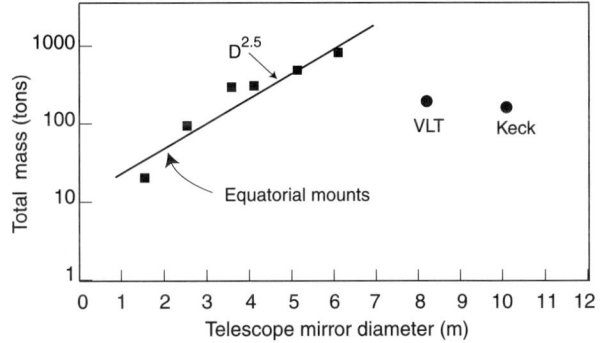

Fig. 6.38. Mass of telescope versus aperture size.

Three axes of rotation are needed, however, not just two as with equatorial mounts. Two are required to orient the tube (alt and az) and a third to compensate for field rotation at the focus. Also, during tracking, each of these three axes must be rotated at *variable* speeds. With the advent of computer control, however, this complication can be easily managed.

The transformation of equatorial coordinates to altitude, h, azimuth, A, and parallactic angle, q, is given in Chapter 1, Section 1.7.[3] The drive rates of each of these axes are obtained by differentiation with respect to the hour angle, HA, and the observatory latitude, φ by

$$\frac{dh}{d\mathrm{HA}} = \sin A \cos \varphi, \tag{6.8}$$

$$\frac{dA}{d\mathrm{HA}} = \sin \varphi - \tan h \cos A \cos \varphi, \tag{6.9}$$

$$\frac{dq}{d\mathrm{HA}} = -\frac{\cos \varphi \cos A}{\cos h}. \tag{6.10}$$

From the above equations, it is seen that the azimuth and parallactic angle drive rates become infinite at the zenith. The maximum allowable velocity

---

[3] The field rotation at the Cassegrain focus is given by the parallactic angle. At a Nasmyth focus, the field rotation is equal to the parallactic angle ± the altitude angle of the target, the sign depending on whether the east or west Nasmyth port is used.

depends on the inertia of the mount and tube and on the torque capabilities of the motors. For 8 to 10-meter telescopes, the drive rates are typically limited to 2° per second. This results in a blind spot at the zenith with a semiangle of about 0.5°.

6.5.3 Altitude-altitude mount

The altitude-altitude (alt-alt) mount can be viewed as an equatorial yoke mount where the yoke axis is horizontal, rather than parallel, to the Earth's axis [20].

Compared to the equatorial mount, it suffers from both image and pupil rotation, although to a much lesser extent and rate than the alt-az configuration. The alt-alt configuration avoids the blind spot at the zenith, but does not benefit from the structural advantages of the alt-az mount. As a result, the alt-alt configuration is not advantageous for large telescopes.

6.5.4 Fixed-altitude and fixed-primary-mirror mounts

For very large telescopes, it may be advantageous to further simplify the mount at the expense of limited observational capability. This can be done by fixing the primary in altitude and either rotating it in azimuth between observations or fixing it permanently in azimuth. An example of the former solution is the Hobby-Eberly telescope (HET) [21] shown on the left in Fig. 6.39. An example of the second solution is the famous Arecibo radio telescope, which has yet to be implemented in the optical (Fig. 6.39, right).

The main advantage of these designs is that the primary mirror does not change direction with respect to gravity, thus greatly simplifying the primary mirror support structure and control system. A second advantage is that there is no need for a formal tube or altitude axis. Another advantage is that the primary mirror is spherical. This is inevitable because, as the secondary mirror sweeps across the primary, it must always face a primary with a "vertex" on its axis and with the same "figure." Only a sphere fulfills this condition. The advantage here is that the spherical primary mirror segments are all identical and easier to make than conical surface mirrors. Thanks to all of these simplifications, this concept is considerably less expensive than the alt-az mount, but it suffers from the several limitations listed below:

- *Small field of view.* The spherical primary mirror results in a large spherical aberration at the focus which can be corrected with additional optics, but only over a relatively small field of view, typically on the order of a few arcminutes.
- *Reduced sky coverage.* With a fixed primary, the amount of sky coverage is a function of the angle between the main optical axis and the polar axis. For an angle of 35°, the sky coverage is about 70%.

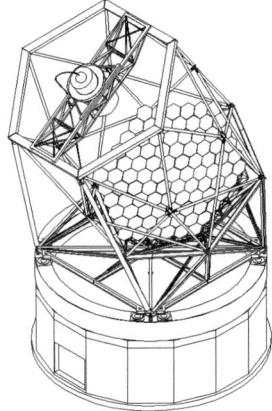 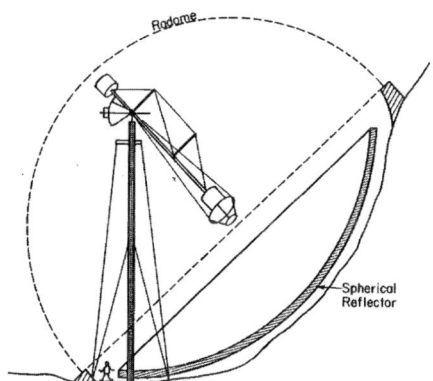

Fig. 6.39. At left, schematic view of the Hobby-Eberly telescope with a fixed altitude mount. Tracking is on the focal surface at the top end. The azimuth is changed between observations to select a different zone of declination for the field of view. At right, a concept for a fixed primary optical mirror of the Arecibo type (from Ref. [22]).

- *Limited exposure time.* This is on the order of 40 minutes to 2.5 hours, depending on declination.
- *Numerous reflecting surfaces.* Two to four additional mirrors are required for the correction of spherical aberration.
- *Complex instrument feed.* Instruments must be fed from the primary focus (no Cassegrain focus is possible, since the axis is always moving across the primary mirror). The instruments must either be at the prime focus, which limits their size, allowable mass, access, and optical scale, or they must be fed with fiber optics or a mirror system similar to that of a coudé focus, with the inherent throughput losses and field rotation.

6.6 Bearings for ground telescopes

Bearings are used to orient the tube and mount and to derotate the focal plane in alt-az configurations. Three primary technical considerations must be kept in mind when selecting bearings: stiffness, accuracy, and low friction. Two bearing technologies are used on large telescopes: rolling element bearings and hydrostatic bearings. Air bearings and magnetic levitation are not practical solutions because of the large masses requiring support.

6.6.1 Rolling bearings

Rolling element bearings, roller or ball type, are widely available commercially and have adequate stiffness and accuracy. They have been successfully used in 4 m class telescopes (Fig. 6.40), but their inherent friction makes them undesirable for larger telescopes. Friction in very slowly rotating rolling bearings is a particular problem because the friction torque is nonlinear, making control difficult (see Section 6.1.5).

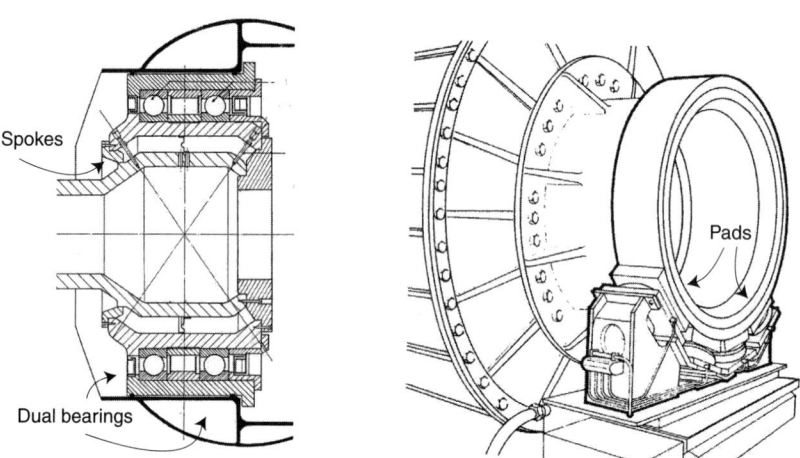

Fig. 6.40. Examples of rolling elements and hydrostatic bearings. At left, a dual-ball-bearing arrangement for a tube declination axis. The two bearings are mounted on spokes which flex to avoid inducing stresses in the tube as the mount deforms during rotation. At right, a typical hydrostatic pad system for supporting a tube in an alt-az mount.

6.6.2 Hydrostatic bearings

Most very large telescopes rely on hydrostatic bearings. They are practically friction free and do not exhibit nonlinearity at low speeds. They have high load capacities, are compact and extremely stiff, and can be at least as accurate as rolling element bearings.

Although hydrostatic bearings are intrinsically more expensive than rolling element bearings, the cost of the bearings, pumping units, and piping is only a small portion of the overall cost, and their low friction and high stiffness make them preferable.

Hydrostatic pads were first used on the 5 m telescope at Mt. Palomar [23]. The principle of this system is shown in Fig. 6.41. Oil is supplied under pressure to a pad with a central recess. This lifts the load and the oil flows out at the periphery. The sliding surface is then completely separated from the

6.6 Bearings for ground telescopes

supporting pad by a film of oil. The thickness of the film is not affected by the speed of rotation. The oil flow is laminar over the gap and the oil film thickness, h, is given by the Poiseuille equation

$$h = \sqrt[3]{12\frac{Q\eta l}{b\Delta p}}, \qquad (6.11)$$

where Q is the oil flow, η is the dynamic viscosity of the oil, Δp is the drop in pressure over the gap, l is the width of the gap, and b is the length of the gap $[b = 2(a_1 + a_2)]$. Supply pressure is generally in the range of 10 to 40 bars.

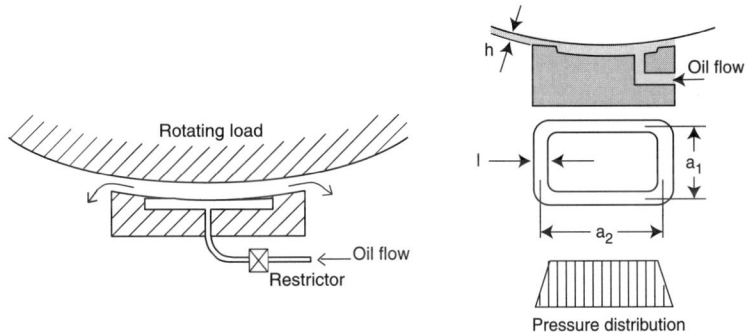

Fig. 6.41. Principle of hydrostatic bearing operation (left) and pad geometry and pressure distribution (right).

If a single pump is to supply oil to more than one bearing, one must limit the flow feeding each bearing; otherwise, the entire flow would go to lift the first bearing and the others would not lift. Equalization can be accomplished by throttling the flow via capillary restrictors or flow-control valves. The advantage of capillary restrictors is that the flow in them is laminar, as it is in the pads, so that changes in film thickness due to the viscosity changing with temperature is automatically compensated.

Film stiffness can be approximated from the formula

$$k_f = 3\frac{W}{h}(1 - \beta), \qquad (6.12)$$

where W is the load, h is the film thickness and β is the pad pressure ratio, which is the ratio of the pressure in the recess with the load lifted (i.e., when oil is flowing) to the pressure required to lift the load [24]. Typically, β is around 0.7 and the film thickness on the order of 50 µm.

Pressurized oil films are very stiff, approaching the modulus of elasticity of metals. A typical pad stiffness is about 5 kN/µm ($5 \cdot 10^9$ N/m).

The friction from a pad is given by

$$T = \frac{A\mu v}{h}, \qquad (6.13)$$

where A is the support area, μ is the viscosity (kg s m^{-2}), v is the linear velocity, and h is the film thickness. On 8-meter class telescopes, the total friction torque is on the order of 100–200 Nm per axis.

In case of a power failure, the pressurized oil supply should be sufficient to maintain the oil film during the time required for the braking system to stop the telescope (typically about 10 s). This is accomplished by installing accumulators on the high pressure side of the pump.

Unless cooled, oil at the pad could reach temperatures of 20 °C or more. It should be brought down to nighttime ambient temperature to minimize seeing effects (see Chapter 9).

One oil commonly used in hydrostatic bearings is ISO VG21, a mineral oil with low water absorption and a viscosity that does not vary unacceptably over the common temperature range, so that no seasonal oil change is required. Its viscosity is given in Table 6.6.2.

Table 6.3. Typical oil viscosity (ISO VG 24)

Temperature (°C)	Viscosity (kg s m^{-2})
−10	0.030
0	0.014
15	0.006
25	0.004

The pads must be able to accommodate misalignments between the telescope journals and the pad support due to fabrication tolerances and settling. This can be accomplished by the use of a secondary oil film incorporated within the pad, as shown in Fig. 6.42.

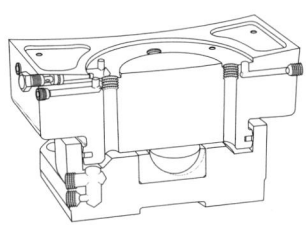

Fig. 6.42. Typical hydrostatic bearing design (cross section). The bearing is composed of two layers: the top pad supports the load and the lower one serves as a self-leveling device to cope with misalignments. (Courtesy of SKF.)

Pads are usually overlayed with bronze to avoid marring the journal in case of accidental contact. Pads can be replaced fairly easily, whereas remachining the track or a journal attached to the rest of the telescope would be a costly affair.

Pads are usually arranged to support the mount or the tube in a kinematic fashion. However, it is sometime advantageous to overconstrain the supported load in order to increase the overall stiffness of the system. This was done on the altitude axis of the VLT, where pads at each end of the axis holding the tube's center section in the horizontal direction counteract each other. The two sides of the yoke, rather than just one, collaborate in holding the tube, thus stiffening the yoke/tube in the transversal direction. A slight drawback is that differential expansion between tube and mount tends to stress the tube, but the higher stiffness is an important benefit because it raises the corresponding frequency above the range of wind disturbances. The VLT azimuth bearings were also designed to overconstrain the mount for this reason. Figure 6.43 shows two examples of azimuth hydrostatic bearing arrangements.

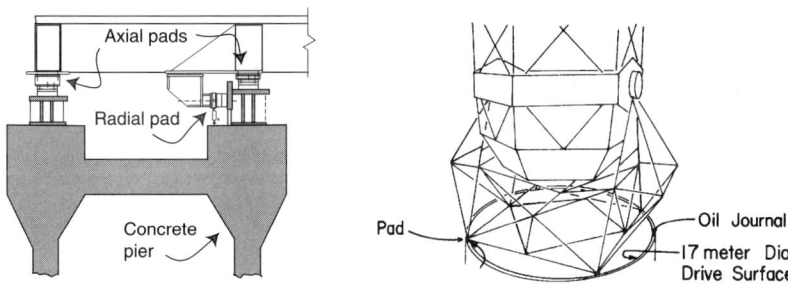

Fig. 6.43. Azimuth bearings. At left, the double track used for the VLT azimuth mount is mounted directly on the concrete pier, then shimmed and ground flat *in situ*. The double track overconstrains the mount in order to increase its lateral stiffness. The track and four hydrostatic pads of a Keck telescope azimuth axis are shown on the right.

Bearings should be located so as to minimize flexure in the supported element. As an example, Fig. 6.44 shows how placing the bearings inside the central section of an alt-az tube prevents its bending.

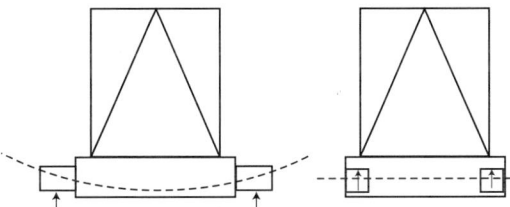

Fig. 6.44. Location of bearings supporting an alt-az tube. At left, the bearings are outside the central section; this provides better access, but the central section needs to be fairly rigid to avoid flexing under the self-weight of the tube. At right, the bearings are located inside the central section and take the truss load directly.

6.7 Miscellaneous mechanisms

6.7.1 Overall telescope alignment

Most equatorial telescope mounts are supported on a "base" which, itself, is supported on a flat concrete pier. Precise alignment of the telescope's polar axis can be achieved simply by adjusting the base support, without requiring separate, complex adjustments to the mount bearings (this also allows for in-shop preassembly of the telescope). The base is made of steel, so that differential expansion between the mount and the base is minimal. Differential expansion between the base and the concrete pier is dealt with by mounting the base kinematically on the pier.

Alt-az mounts do not require precise adjustment of the azimuth mount support plane, as small errors can be corrected by rotation of the tube in altitude. The only condition is that the azimuth track be sufficiently planar to fall within the tolerances of the mount's hydrostatic bearings. This can be accomplished by shimming the azimuth track on the concrete pier to make it as true as possible, then grinding it flat by using the mount itself as a machine tool.

6.7.2 Optics alignment and focusing devices

The optical train must be aligned with respect to the science instruments. The primary mirror is generally adjusted first, so that its optical axis is coaligned with the rotation axis of the Cassegrain focus turntable.[4] This is done by mechanical means, activated manually or remotely. The secondary mirror is then adjusted in tip, tilt, decenter, and piston to satisfy the optical collimation and focus conditions.

Focus can be adjusted by moving either the detector or the secondary mirror. When several instruments are mounted simultaneously at the same focus, it is convenient to have an individual focus adjustment for each instrument, to avoid having to refocus the secondary mirror when switching between instruments.

Alignment of the secondary mirror is usually accomplished with the use of actuators. One solution consists of using a hexapod mount, which provides all of the required five degrees of freedom in one set of mechanisms (plus rotation around the optical axis for off-axis systems). Care must be taken to avoid play in the hexapod actuators, by preloading joints and actuators. One solution, used on the Subaru telescope, consists of installing tensioning springs between the Cassegrain mirror cell and the top inner ring, so as to keep the mirror cell pulled toward the base (Fig. 6.45).

---

[4] If there is no Cassegrain focus, the primary mirror optical axis should be coaligned with the folded axis of the Nasmyth turntable.

6.7 Miscellaneous mechanisms   243

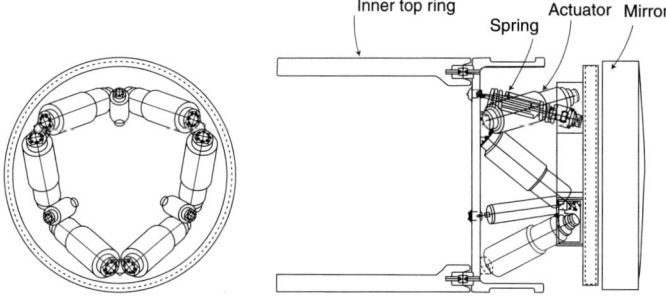

Fig. 6.45. Hexapod mount used to align and focus the secondary mirror of the Subaru telescope. Play in the actuators and joints are avoided by the use of three tensioning springs. (From ref. [6].)

An alternate solution, used on the Hubble Space Telescope, is to employ three pairs of eccentric motors to provide essentially the same type of adjustment (Fig. 6.46)

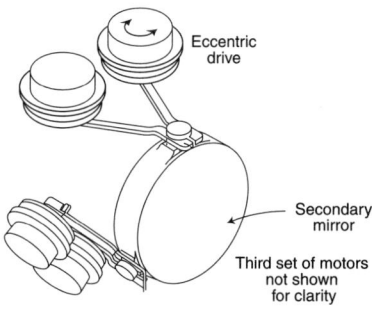

Fig. 6.46. Eccentric drives used to control tip, tilt, decenter, and focus on the secondary mirror of the Hubble Space Telescope.

6.7.3 Active secondary mirror for infrared chopping and field stabilization

Infrared observations generally require that the secondary mirror be used for "chopping" (see Chapter 1). The mirror must be chopped in a square-wave manner with a frequency of a few Hertz and an angular amplitude up to about 1 arcminute.

To avoid exciting the secondary mirror support system and the rest of the telescope, the secondary mirror unit must be "reactionless"; that is, generating no dynamic load at its mounting interface. This is accomplished by using a compensation mass moving in synchrony with it, but in the opposite direction, so as to keep the overall system dynamically balanced. The requirement for a reactionless chopping mirror is threefold. The first condition is that the

244   6. Telescope Structure and Mechanisms

rotations of both the mirror and reaction mass be about their respective centers of mass. The second condition is that the inertia/spring-rate ratios should be equal:

$$\frac{I_1}{I_2} = \frac{k_1}{k_2}, \tag{6.14}$$

where I_1, I_2 and k_1, k_2 are the inertias and spring stiffnesses of the mirror and reaction mass, respectively (Fig. 6.47). The third condition is that the same driving force be applied to the mirror and reaction mass. This condition is always satisfied when the two masses are driven by the same actuators.

Since the angular displacements are small, the mirror and reaction mass can be mounted on flex pivots, thus avoiding the play and friction effects inherent in regular bearings.

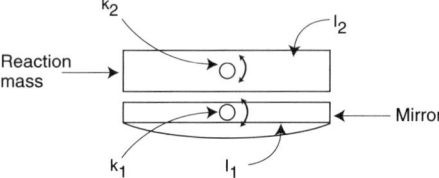

Fig. 6.47. Principle of force compensation on a secondary mirror chopping unit.

First-order correction of atmospheric turbulence and wind shake can also be made with a fast-steering secondary mirror. Examples of such implementations are shown in Fig. 6.48. Field stabilization requirements are typically as follows: 10 Hz bandwidth, 10″ maximum amplitude, and 0.01″ accuracy.

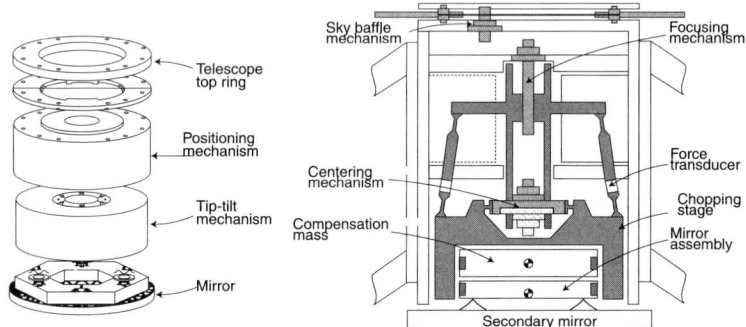

Fig. 6.48. At left, secondary mirror assembly of the Gemini telescopes with tip-tilt capability up to 40 Hz and rapid focusing at 3 Hz. In addition, this module can be chopped up to 10 Hz for infrared observations. At right, the secondary mirror unit for the VLT. Mirror actuation is provided by six linear motors, three for the mirror and three for the compensating mass located at 120°.

6.7.4 Balancing systems

Ground-based telescopes need to be extremely well balanced around their axes of rotation in order to minimize driving torques and differential flexure. This is accomplished by adding or removing fixed weights or by using motor-driven balancing weights. The procedure must be carried out at each optical configuration change (e.g., exchange of the telescope top unit when going from Cassegrain to coudé) and each time a focal instrument is replaced or installed. An example of a motorized balancing system is shown in Fig. 6.49.

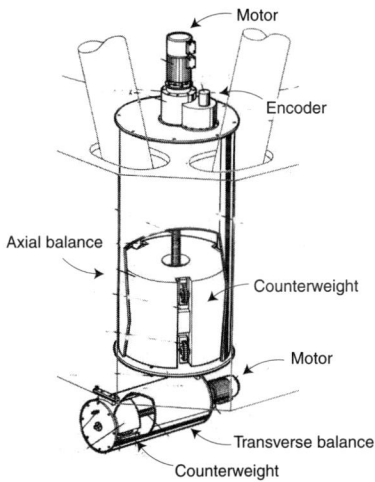

Fig. 6.49. Motorized axial and transversal counterweights used to balance the CFH telescope tube.

6.7.5 Cable wrap and cable twist

A large number of power cables, signal cables, helium, air and vacuum hoses, and hydraulic pipes must be connected from a fixed station to the telescope. These lines must pass through two or more moving interfaces, between the telescope tube and the mount, between the mount and the pier, and between the telescope and the instruments.

Slip joints are generally not used because they create electrical transients. The most common way of getting lines across these interfaces is to form a large-radius loop with a slack take-up. This system is referred to as "cable wrap." Alternatively, when space permits, lines can be arranged along the axis of rotation and twisted around it, a system referred to as "cable twist." In both cases, the ends of the cables, hoses, and pipes should be provided with connectors for easy replacement of defective elements.

During telescope tracking, cable wraps (or twists) can be a major source of disturbance because of friction occuring between lines and also in the articulated conduits. These disturbances can be reduced by motorizing the cable wraps separately instead of pulling them with the telescope or instrument rotator drives. Figure 6.50 shows an example of a cable-wrap system with an independent drive.

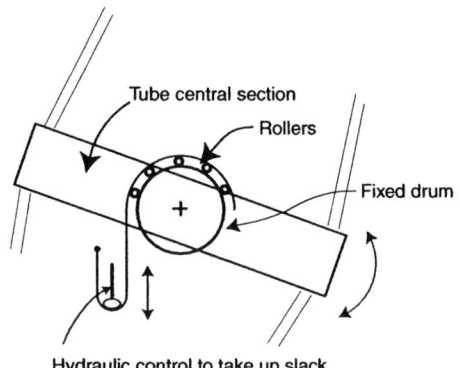

Fig. 6.50. Cable wrap with motorized hydraulic drive used on the altitude axis of the VLT. A potentiometric sensor determines when to activate the cable wrap.

6.7.6 Mirror cover

When not in use, primary mirrors are generally kept covered for protection against dust and for local thermal control. During maintenance, covers also prevent dropped objects from damaging the mirror. It is common practice to design the cover so as to withstand the impact of an average tool or part with a mass of 10 kg falling from the top of the tube or dome. Covers also prevent snow, ice, and accumulated dust from falling from the top of the dome when the shutter is opened.

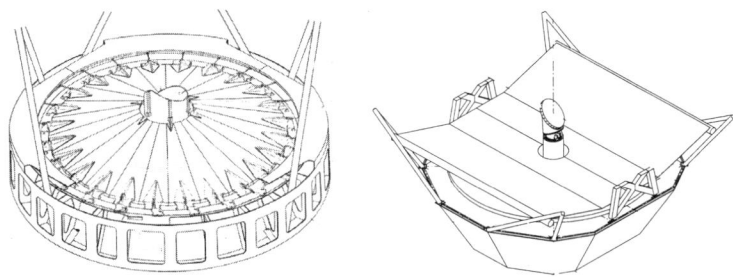

Fig. 6.51. Mirror cover with hard petals used on the CFH telescope (left) and with retractable cloth panels (right).

Finally, a cover is useful in keeping the Sun from shining on the mirror when the dome is open for maintenance. Although the Sun would not harm the mirror itself, the power generated at the focus can lead to high temperatures and damage nearby equipment.

Figure 6.51 shows a mirror cover composed of hard petals. The petals are opened or closed simultaneously by means of a chain drive. This solution ensures the best protection against damage to the mirror, but it is mechanically complex and the petals in the open position obstruct airflow across the mirror surface. The same figure depicts the principle of a cloth-type mirror cover as used on the VLT. It is mechanically simpler and retracts completely away from the mirror environment.

The Hubble Space Telescope does not have a mirror cover per se, but uses, instead, an "aperture door" located at the top of the tube. This door was kept closed during assembly and testing on the ground to minimize contamination of the optics (Fig. 6.52). In orbit, it is closed whenever the telescope is in deep "safe mode" to prevent the possibility of the Sun shining on the primary mirror and the tube interior.

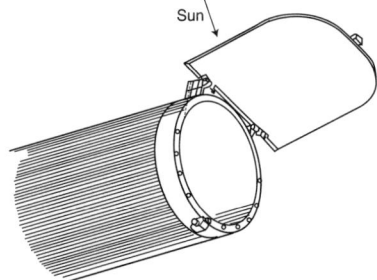

Fig. 6.52. Aperture door of the Hubble Space Telescope.

Secondary mirrors, which face downward, are less susceptible to dust and damage and are usually left uncovered when not in use.

6.8 Safety devices

6.8.1 Brakes

Brakes should be applied whenever the altitude, azimuth, or instrument rotators are not active. This is a safety measure in case of accidental activation of the drives or temporary imbalance in the telescope (e.g., during instrument change-out). Inhibiting accidental telescope motion is especially important when direct and friction drives are used, as the moving body is essentially free whenever the drive motor is deenergized.

Disk brakes are generally preferred over shoe types because their braking action is a pure torque and does not induce an overall force on the body being stopped. They also have the advantage of being self-aligning. Electrical, hydraulic, or pneumatic systems can be used for actuation, but, for safety, the brakes should be fail-safe.

Brakes should be sized so as to be able to stop the telescope tube or mount rotating at maximum speed without causing undue damage. They should also be able to overcome a reasonable imbalance in the rotating body, as well as counteract the drive motor at full power. This is to cover the eventuality of the drive motor control system failing in the full-power mode.

In addition to the primary braking system, it is advisable to provide an emergency braking system using different actuation and interlocking devices. The emergency brake should be designed for a much higher braking power than the primary system in order to cover human emergencies or accidental imbalances in the moving body. Indeed, a very high imbalance may result from even small counterweights or instrument parts pulling free from the telescope, due to the long lever arms. The emergency brake should be designed to stop the telescope almost instantly, even at the expense of some damage. One possible solution is simply to relieve the oil pressure at the hydrostatic pads (it will be remembered that the pads should be covered with bronze to avoid damage to the steel journal). This solution risks damaging the pads, but is a very effective braking method in situations of dire emergency. Several "red button" emergency stops should be located strategically around the telescope chamber and control room.

6.8.2 End stops

The tube should have mechanical stops to prevent it from being driven below the design altitude limit in case the software and electrical limits fail. End stops, also called "overtravel stops," may also be placed on the mount to prevent cables and pipes from being torn away should the normal limits fail. Stops should have an absorbing system (compliant material or spring) so that the deceleration does not damage the optics or instruments.

6.8.3 Locking devices

The altitude and azimuth axis must be positively locked during maintenance and when exchanging a Cassegrain instrument or removing the primary mirror for realuminizing (at which time the tube is top heavy). This can be done somewhat automatically with "locking pins" (Fig. 6.53) or simply with mechanical connectors installed by hand.

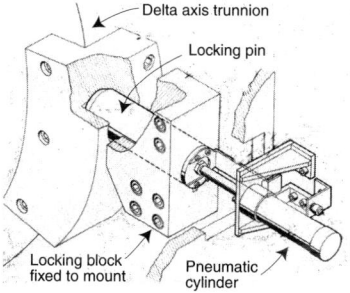

Fig. 6.53. Pneumatically actuated locking pin used on the declination axis of the CFH telescope.

6.8.4 Earthquake restraints

The altitude and azimuth bearings should be located such that the resultant of gravity and the maximum design earthquake force will stay within the supporting base. If this is the case, the bearing force will not become negative, and the tube and mount will not separate from the supporting base. Still, it is advisable to include some way of limiting the motion of the tube and mount should the earthquake exceed design limits. This is to prevent catastrophic instability, but it is also a good way to ensure that critical parts with small clearances such as rotors/stators in motors or tapes/reading-heads in encoders are not damaged by accidental contact. This can be accomplished by using mechanical clamps with clearances designed such that the seismic movement will be smaller than the air gap in the motor and encoders (Fig. 6.54).

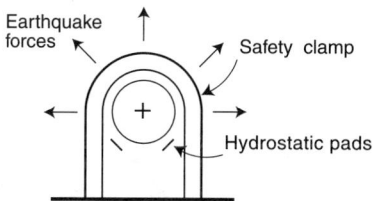

Fig. 6.54. Safety clamp for altitude axis of the VLT.

When direct control of earthquake forces is impractical, the energy induced in the telescope during an earthquake event can be dissipated by the breakage of calibrated pins and by letting massive parts slide against one another. Such a solution was proposed for the VLT pier, but it was abandoned because it lowered the telescope pier stiffness below acceptable limits.

References

[1] Schnur, G.F.O. and Stenvers, K.-H., The hexa-pod telescope (HPT), a contemporary concept for large telescopes, ESO Conference on *Very Large Telescopes and Their Instrumentation*, Garching, 1988, p. 151.

[2] Stewart, D., A platform with six degrees of freedom, UK Institution of Mech. Engineers Proc., Vol. 180, Pt. 1, No. 15, 1965-66.

[3] Serrurier, M., Structural features of the 200-inch telescope for Mt. Palomar Observatory, Civil Eng., Vol. 8, No 8, p. 524, 1938.

[4] Paquin, R.A., Properties of metals, in *Handbook of Optics*, 2nd ed., Vol. II, Devices, Measurements and Properties, Bass, M., ed., McGraw-Hill, 1995, Chap. 35.

[5] Mast, T., Elastomeric lens mount, SPIE Proc., Vol. 3355, p. 144, 1998.

[6] Miyawaki, K., Itoh, N., Sugiyama, R., Sawa, M., Ando, H., Noguchi, T., and Okita, K., Mechanical structure for the Subaru telescope, SPIE Proc., Vol. 2199, p. 754, 1994.

[7] Haskell, R.C., Protecting large telescopes from earthquakes, SPIE Proc., Vol. 628, p. 334, 1986.

[8] Davison, W., Design strategies for very large telescopes, SPIE Proc., Vol. 1236, p. 878, 1990.

[9] Bely, P.Y., Salmon D.A., and Wizinowich, P.L., Bending the CFHT Cassegrain secondary for optical figure improvement, SPIE Proc., Vol. 444, p. 253, 1983.

[10] Gray, P.M., Hill, J.M., Davison, W.B., Callahan, S.P., and Williams, J.T., Support of large borosilicate honeycomb mirrors, SPIE Proc., Vol. 2199, p. 691, 1994.

[11] Mack, B., Deflection and stress analysis of a 4.2 m diam. primary mirror on an altazimuth-mounted telescope, Appl. Opt., Vol. 19, No. 6, p. 1001, 1980.

[12] Schwesinger, G., Lateral support of very large telescope mirrors by edge forces only, J. Mod. Opt., Vol. 38, No. 8, p. 1507, 1991.

[13] Wilson, R.N., *Reflecting Telescope Optics II*, Springer-Verlag, 1999.

[14] Yoder, P.R., *Opto-mechanical Systems Design*, Marcel Dekker Inc., 1992.

[15] Meng, J.D., et al., Position actuators for the primary mirror of the W.M. Keck telescope, SPIE Proc., Vol. 1236, p. 1018, 1990.

[16] Del Vecchio, C., Optimization of the elevation structure of the Large Binocular Telescope, SPIE Proc., Vol. 2871, p. 159, 1996.

[17] Vukobratovich, D., Rugged yet lightweight: how can we achieve both in optical instruments, in *Optomechanical Design*, Yoder, P.R., ed., SPIE Critical Reviews, Vol. CR43, SPIE 1992, p. 32.

[18] Lubliner, J., *Preliminary design of tube for TMT*, TMT (Keck) Report No. 50, 1981.

[19] Schneermann, M.W., Structural design concepts for the 8-meter unit telescopes of the ESO-VLT, SPIE Proc., Vol. 628, p. 412, 1986.

[20] Vassileski, in *The Construction of Large Telescopes*, Crawford, D.L., ed., IAU Symposium No. 27, Academic Press, 1966, p. C25.

[21] Ramsey, L.W., et al., The early performance and present status of the Hobby-Eberly Telescope, SPIE Proc., Vol. 3352, 1998.

[22] Drake, F.D., Large telescopes utilizing fixed primaries, Proc. KPNO conference on *Optical and Infrared Telescopes for the 1990s*, Hewitt, A., ed., KPNO, 1980, p. 649.

[23] Karelitz, M.B., Oil-pad bearings and driving gears of 200-in telescope, Mech. Eng., Vol. 60, p. 541, 1938.

[24] Rippel, H.C., Design of hydrostatic bearings, Parts 1, 2 and 3, Mach. Design, Aug. 1, 1963, p.108, and the following two issues: Aug. 15, p.122 and Aug. 29, p 132.

Bibliography

Barlow B.V., *The Astronomical Telescope*, Wykeham Publications Ltd, 1975.

Buchanan, G.R., *Schaum's Outline of Finite Element Analysis*, McGraw-Hill, 1995.

Clough, R.W. and Penzian, J., *Dynamics of Structures*, McGraw-Hill, 1993.

Crawford, D.L., ed., *The construction of large telescopes*, IAU Symposium No. 27, Academic Press, 1966.

Krim, M.H., Design of highly stable optical support structure, Opt. Eng., Vol. 14, No. 6, p. 552, 1975.

Nelson, J.E., Mast T.S., and Faber, S.M., eds., *The design of the Keck Observatory and telescope*, Keck Observatory Report, No. 90, 1985.

O'Shea, D.C., ed., *Selected Papers on Optomechanical Design*, SPIE Milestone Series, Vol. 770, SPIE, 1988.

Thomson, W., *Vibration Theory and Applications*, Prentice-Hall, 1965

Wilson, R.N., *Reflecting Telescope Optics II*, Springer-Verlag, 1999.

Yoder, P.R., *Opto-mechanical Systems Design*, Marcel Dekker Inc., 1992.

Yoder, P.R., ed., *Optomechanical Design*, SPIE, CR43, 1992.

7
Pointing and Control

The role of the pointing system is to point the telescope toward the desired target and track that target in spite of external and internal disturbances and, in the case of ground telescopes, of the rotation of the Earth.

Early ground telescopes were all of the equatorial type, which, once pointed at the desired target, merely requires rotation at a constant rate ($15''$/s) around the polar axis in order to track the target. This hour-angle motion was produced in an "open-loop" fashion by a clock mechanism and, more recently, by a synchronous motor.

With the need for better accuracy and regulation, all modern telescopes now use "feedback control," where the actual motion of the rotating axis is constantly compared to the desired value and the error is fed back to the drive system. Feedback control is intrinsically more accurate and eliminates the need for high-precision gearing and drive systems. It also permits correction of disturbances that fall within its bandwidth. On the ground, the main sources of disturbance are gravity, thermally induced distortions, and wind. In space, disturbances include gravity gradient, solar pressure, and excitations internal to the spacecraft.

In this chapter, we will examine the various approaches used in modeling telescope structures and optics for the purpose of designing control systems, review the basics of servo systems, and look at the ways in which telescope control systems are usually implemented. We will also describe the various types of actuator and sensor and examine the sources of disturbance and methods for mitigating them by isolation.

7.1 Pointing requirements

The requirements for the pointing system are set by the scientific program of the observatory, the desired observing efficiency, and the ultimate angular resolution to be achieved. Table 7.1 lists typical values for ground-based and space telescopes based on the Keck [1] and Hubble telescopes (HST) [2].

Table 7.1. Typical pointing system requirements for ground-based (Keck) and space telescopes (HST)

	Keck	HST
Blind absolute pointing accuracy ($''$)	1	12
Blind offset absolute pointing accuracy ($''$)	0.1	0.01
Tracking accuracy ($''$)	0.020	0.007
Maximum tracking velocity ($''$/s)	15 (alt) 1600 (az)	0.21*
Small maneuver (10$''$) time (s)	1	15
Maximum slew velocity (deg/s)	1	0.22
Maximum slew acceleration (deg/s^2)	0.1	0.0008

* For tracking planets.

The absolute pointing-accuracy requirement for optical telescopes is typically about 1$''$ rms. This is easily achieved after calibration and suffices for the majority of observations. When better accuracy is needed, it is best to take reference directly on the sky by using "blind offsets" from calibration stars, as explained in Chapter 1. These are accurate to better than 0.1$''$ and 0.01$''$ for ground and space telescopes, respectively.

Tracking stability of ground telescopes is typically better than 0.1$''$ over periods of several minutes without recourse to a guide star. By closing the loop on a guide star, drifts are eliminated and the line-of-sight jitter is typically under 20 milliarcseconds rms for ground telescopes and a few milliarcseconds for HST.

7.2 System modeling

The design of a telescope control system consists of creating a mathematical model of the physical aspects and characteristics of the telescope including sensors and actuators, modeling also the disturbances, and then applying traditional control-system analysis techniques to determine the optimal control-law parameters that will maximize pointing and tracking performance.

For smaller telescopes, a simple model limited to the structure is sufficient, and one can even get by without modeling disturbances. As the size and complexity of the telescope increase, however, it becomes necessary to model the optomechanics in more detail and analyze the effects of disturbances. This is absolutely necessary for very large ground telescopes and space telescopes, which need careful optimization to reduce mass and maximize performance.

For structural analysis, a telescope is best analyzed with a finite element model (FEM). For the purpose of designing the pointing control system, however, it is preferable to use a "lumped-mass model," that is, a representation by masses mounted on springs and dampers simulating the flexibility of the structure and drive mechanisms and the friction in the supporting axes. Why create a lumped-mass spring model when a finite element model has already been developed for structural analysis? A first reason is that lumped-mass models lend themselves well to traditional dynamic and control analysis. But there are several other practical reasons:

- In practice, the design of a telescope control system involves no more than the lowest three to four dominant modal frequencies, so the use of a lumped-mass spring model replicating the lowest mode shapes is entirely appropriate and adequate.
- Mode shapes and frequencies of a structure are global properties, whereas stresses and deflections are local properties. This means that fairly significant localized variations in structure have little effect on the global properties, other than slight changes in frequency. This allows control system and structural refinement efforts to be independent and progress along separate but parallel paths.
- Once the lumped mass model is created, it can be quickly tuned to a range of natural frequencies simply by changing the value of the springs. Thus, the controls designer can bound his design problem by exploring a wide range of potential structural frequencies. This helps him create a robust design insensitive to potential errors in the finite element analysis prediction of the real-system frequencies.
- In the lumped-mass model, the placement of nonlinearities such as backlash, command and sensor quantization, time delays, nonlinear springs, friction models, and so on can be readily accomplished because there is a direct, physical, one-to-one correspondence between the masses and springs in the lumped-mass model and the real features of the system where these components can be attached. Simulating these in a finite element model can be more challenging. Such nonlinearities often lead to nonconverging numerical integration problems if the number of degrees of freedom is large, as is the case in a typical FEM model.

On the other hand, once the control system design has been firmed up, it is beneficial to return to a representation based on the finite element model and combine the structural and control models with the optical and thermal models to form an "integrated model" of the telescope. This creates a more faithful representation of the telescope system, allowing verification of the findings of the lumped-mass model and exploration of second-order effects involving the optics and the thermal environment.

7.2 System modeling

In what follows, we will first look at how lumped-mass models are created and then examine the more exhaustive integrated models.

7.2.1 First-order lumped-mass models

The first iteration in lumped-mass models is to assume that the optical elements are infinitely stiff and rigidly mounted on the structure. In other words, although the fundamental flexible modes of the structure are taken into account, no attempt is made to determine what is really happening to the respective position of the optical elements or to the line of sight. Still, such a basic model is useful in the preliminary design of the servo system and drives. Figure 7.1 shows such a first-order model for a ground-based alt-az telescope. A separate model must be created for each axis of rotation.

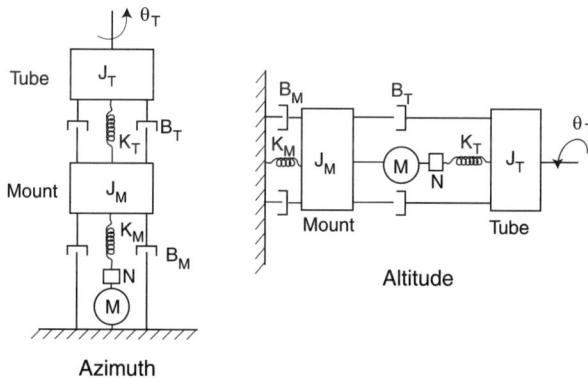

Fig. 7.1. First-order lumped-mass model of an alt-az telescope composed of two elements: tube and mount. Notation is defined in the text. For simplicity, the inertias, spring constants, and damping terms are shown with the same notation for the two axes (altitude and azimuth), but the corresponding values will generally be different.

The tube and mount are represented as lumped masses with moments of inertia, J_T and J_M, where the indices T and M refer to the tube and mount, respectively. These masses are mounted on torsional springs, K_T and K_M, simulating the flexibility of the structures and drives, with dampers B_T and B_M simulating the static (Coulomb) or dynamic (viscous) friction in the axes. Each axis is driven by a motor which applies a torque τ, via a gear or roller system of reduction ratio N, to control the angular position of the tube and mount.

These two spring constants, K_T and K_M, can be derived from the "locked rotor" resonances f_T and f_M (i.e., the fundamental frequency of the vibration of the tube and mount when the drive motors can be considered as locked) by

$$f_T = \frac{1}{2\pi}\sqrt{\frac{K_T}{J_T}}, \qquad (7.1)$$

256   7. Pointing and Control

$$f_M = \frac{1}{2\pi}\sqrt{\frac{K_T}{J_T} + \frac{K_M}{J_M}}. \tag{7.2}$$

The damping terms, B_T and B_M, are determined from the structural damping in the structure, friction in any rolling device (ball bearings, roller drives, cable wraps), and viscosity in hydrostatic bearings.

As an example, the values for the moments of inertia, torsional stiffness, and torsional damping terms for a large alt-az telescope are given in Table 7.2.

Table 7.2. First-order model parameter values for the Keck telescopes [3]

	Drive ratio	Moment of inertia 10^6 kg m^2		Tors. stiffness 10^9 N m/rad		Tors. damping 10^3 N m/(rad s)	
	N	J_T	J_M	K_T	K_M	B_T	B_M
Azimuth	111	2.4	10.2	6.5	4.0	25	90
Altitude	53	3.9	–	–	-	25	49

The system has three degrees of freedom: the angular rotation of each of the two masses and that of the motor. If solid friction is neglected and assuming small motions, the system can be considered linear. For each degree of freedom, the dynamic behavior is then governed by an equation of the form

$$J\ddot{\theta} + B\dot{\theta} + K\theta = \tau, \tag{7.3}$$

where θ, $\dot{\theta}$, and $\ddot{\theta}$ are the rotation angle, angular velocity, and angular acceleration, respectively, J is the moment of inertia of the body considered, B is the damping factor, K is the restoring torque coefficient due to flexibility in the system, and τ is the applied external torque. This equation simply states that the external torque is used to accelerate the mass in rotation, fight viscous friction, and tauten any flexibility in the system.

There are several ways to establish these mathematical equations, but arguably the most methodical and practical is by the use of Lagrange equations [4]. These equations are based on the equilibrium of kinetic energy, potential energy, and, if viscous friction is involved, dissipation energy. The method is widely used to develop the equations of motion, not just for telescopes but also for high-performance tracking devices for communications, radar, and military applications. The equations of motion for the altitude axis of the first-order model shown in Fig. 7.1 are derived in Appendix E, and are as follows:

$$J_m\ddot{\theta}_m + B_m(\dot{\theta}_m - \dot{\theta}_M) + \frac{K_T}{N}\left(\frac{\theta_m}{N} + \frac{N-1}{N}\theta_M - \theta_T\right) = \tau_m. \tag{7.4}$$

$$J_T\ddot{\theta}_T + B_L(\dot{\theta}_T - \dot{\theta}_M) - K_T\left(\frac{\theta_m}{N} + \frac{N-1}{N}\theta_M - \theta_T\right) = \tau_d, \tag{7.5}$$

$$J_M\ddot{\theta}_M + B_M\dot{\theta}_M + K_M\theta_M - B_m(\dot{\theta}_m - \dot{\theta}_M) - B_T(\dot{\theta}_T - \dot{\theta}_M)$$
$$+ K_T\frac{N-1}{N}\left(\frac{\theta_m}{N} + \frac{N-1}{N}\theta_M - \theta_T\right) = -\tau_m, \qquad (7.6)$$

where J, B, τ, and θ are as previously defined, the subscripts m and M refer to the motor and mount, respectively, τ_d is the disturbance torque, and N is the reduction ratio of the drive. Note that the external forcing function τ_m is an independent parameter because its value varies with the electric current applied to the motor. Similar equations are found for the azimuth axis.

These differential equations which govern the dynamic behavior of the system can then be incorporated into a control-system simulation model, as will be described in Section 7.3.2.

7.2.2 Medium-size lumped-mass optomechanical models

As the design progresses, it becomes necessary to model the telescope structure and mechanical devices in more detail and to monitor the optical line of sight instead of simply the tube boresight. A lumped-mass model is still used because of all the advantages mentioned in the previous section, but it is expanded to cover the structure in more detail and include the optical elements. The optical elements (mirrors or mirror segments) are still considered infinitely rigid, however. When this is clearly not the case, as for meniscus primary mirrors, for example, it is assumed that a separate control system is in charge of maintaining the mirror shape independently of the deformation of the structure, so that it appears as rigid.

Medium-size lumped-mass models are generally all that is needed for the design of servo systems. As an example, the model used for the design of the control system of the VLT is an eight-degrees-of-freedom model which includes the linear displacement of the primary mirror and its cell, the secondary mirror, the tube's center piece and mount, and the angular rotation of the tube's center piece, mount, and altitude drive [5, 6]. The model supplies the position of the primary and secondary mirror, from which the direction of the line of sight can be rather simply determined [7].

7.2.3 Integrated models

If the simple mathematical models described so far are sufficient for the design of the control system, they do not have enough fidelity to provide an end-to-end analysis of the telescope's performance. For this, one must represent the telescope as a complete system.

The foundation of this refined design approach resides in the well-meshed combination of structural, control, optical, and thermal models. Command

258   7. Pointing and Control

and disturbance torques serve as input to a high-fidelity finite element model of the structure and mechanical devices which, in turn, supplies the motion of the optics support points to an optical ray-trace program from which the instantaneous line-of-sight and wavefront error can be derived. These models are then combined with those of the thermal and control systems and of the actuators and sensors, so as to form a complete mathematical representation of the entire telescope system (Fig. 7.2). This symbiosis of models is referred to as "integrated modeling". The integrated model then allows one to find the response of the overall system when simulated thermal or mechanical disturbances are applied.

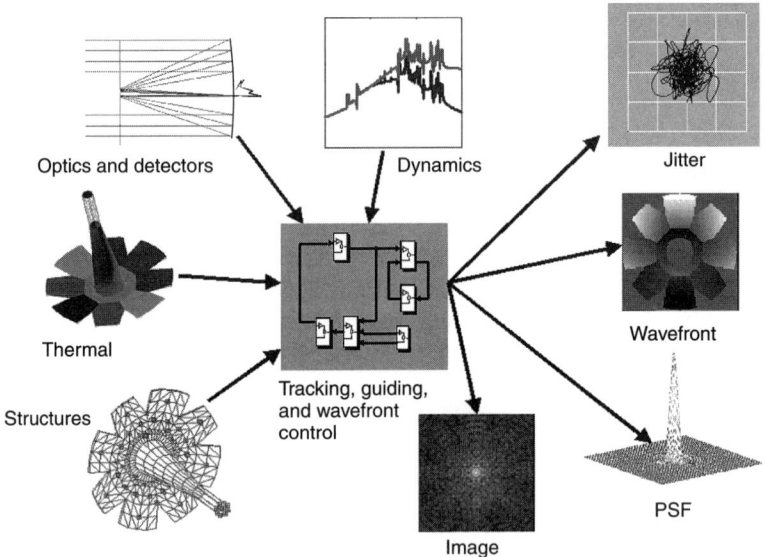

Fig. 7.2. Integrated modeling assembles optical, structural, dynamic, and thermal models so that a complete representation of the physical effects on line of sight and image quality can be simulated.

Integrated modeling is not a design tool in itself: traditional optical, structural, thermal, and control system design methods must still be used in each of these fields. However, integrated modeling provides a means of investigating the behavior of the system to disturbances of various origins acting concurrently, play "what-if games," and tune up the design in a semiempirical fashion. These various models are constructed as follows.

Structural model

Conceptually, the structure can still be represented as a series of springs, masses, and dampers, but now with all of the degrees of freedom needed to determine the position and deformation of all of the optical elements. The three equations which govern the motion of the three-degree-of-freedom (two masses + motor) model of Section 7.2.1 (i.e., Eq. 7.4, 7.5, and 7.6) can be generalized to a multiple-degree-of-freedom system. It can be shown that the motion of such a linear n-degree-of-freedom system is described by a set of n linear second-order differential equations written in matrix form as [8]

$$\mathbf{m\ddot{x} + b\dot{x} + kx = f}, \tag{7.7}$$

where $\mathbf{m}$ is the mass matrix, $\mathbf{b}$ is the damping matrix, $\mathbf{k}$ is the stiffness matrix, $\mathbf{f}$ is the disturbance (force/torque) vector, and $\mathbf{\ddot{x}}$, $\mathbf{\dot{x}}$, and $\mathbf{x}$ are the acceleration, velocity, and displacement vectors, respectively. Vector $\mathbf{x}$ contains the displacements for all six degrees of freedom (three translations, three rotations) for every mass element in the system. The $\mathbf{m}$ and $\mathbf{k}$ matrices are obtained from a finite element model of the system.

The eigenvalues (modes, or undamped natural frequencies), ω_i, and eigenvectors (mode shapes), Φ_i, of this system are found as solutions to the equation

$$(\mathbf{k} - \omega_i^2 \mathbf{m}) \Phi_i = 0. \tag{7.8}$$

For convenience, the eigenvectors can be mass-normalized using the condition

$$\mathbf{\Phi}^T \mathbf{m} \mathbf{\Phi} = \mathbf{I}, \tag{7.9}$$

where $\mathbf{I}$ is a unity matrix and where the modal matrix $\mathbf{\Phi}$ is given by

$$\mathbf{\Phi} = [\Phi_1 \; \Phi_2 \; \cdots \; \Phi_n]. \tag{7.10}$$

Applying the coordinate transformation $\mathbf{q} = \mathbf{\Phi} \, \mathbf{x}$, the equation of motion in generalized coordinates is then

$$\mathbf{\ddot{q}} + 2\mathbf{\Omega}\zeta\mathbf{\dot{q}} + \mathbf{\Omega}^2 \mathbf{q} = \mathbf{\Phi}^T \mathbf{f}, \tag{7.11}$$

where ζ is the modal damping and $\mathbf{\Omega}$ is the diagonal matrix formed by the eigenvalues.

It is important that the FEM faithfully represent, not only the main structural element, but also the mechanical devices such as drives, attitude sensor mounts, mirror supports, hydrostatic bearings, and so forth. These devices are often weak links in the system and should be modeled carefully. When possible, it is best to use actual test data.

A typical FEM dynamic computer run will happily crunch out hundreds of modes, but these should not be used blindly as a basis for dynamic analysis. In the first place, model errors easily creep in, and it is important to check the physical reality of the modes and the validity of the assumptions and simplifications made. One helpful tool in this respect is the "modal density plot," an example of which is shown in Fig. 7.3. As a rule, the density of modes is fairly uniform, so that any anomalous jump or large discontinuity in the scattered plot should be investigated to determine if it is due to errors in

260   7. Pointing and Control

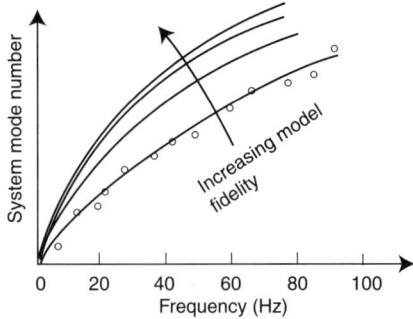

Fig. 7.3. Example of a structural modal density plot.

the model. This plot is also useful in determining the optimal fidelity of the FEM model when human analysis and computer processing times are taken into account. As the model's fidelity increases, modal frequencies generally go down. Past a certain point, however, they tend to stabilize, signaling that improving the model further will lead to diminishing returns.

Second, it is often necessary to "weed" the dynamics model before using it in the control system analysis. Because low-amplitude, high-frequency *structural* modes can be greatly amplified *optically* by a powered mirror, it is necessary to push the dynamic analysis to fairly high frequencies, say 100 Hz or more. A very large number of modes would lead to excessive complexity in the control system analysis, and it is better to cover a large frequency range by eliminating insignificant modes than to keep all of the modes and limit the frequency range. This weeding is done by consulting the table of mode characteristics and eliminating modes with low optical sensitivity and low kinetic energy.

Damping

The other component of the dynamic analysis is damping. Telescopes have stringent requirements concerning pointing repeatability. As discussed in Chapter 6, mechanical play and solid friction, which can cause sudden jumps during tracking, must be minimized. This requires that construction or deployment joints be avoided or highly preloaded and that all sources of internal friction be eliminated. When these precautions are taken, telescope structures are close to being monolithic. This, in turn, implies that telescopes have low damping. In mechanical or electrical jargon, they "ring" and have a high "Q." For ground telescopes, the damping coefficient is around 0.1 (this is ζ in formula 7.11). The damping of space telescopes is even lower, around 0.01, due to the use of high-modulus-of-elasticity materials and the absence of bearing friction. As a matter of fact, the damping of space telescopes approaches that of pure material damping.

Cryogenic temperature space telescopes face an aggravated situation because the material damping coefficient decreases with temperature and approaches zero when the temperature nears absolute zero.

Material damping models have been formulated by Zener for crystalline and metallic materials under low-strain conditions [10] and have been shown to be a function, among other parameters, of temperature, vibration frequency, material stiffness, and coefficient of thermal expansion. Experiments have demonstrated that the models are good predictors for room-temperature conditions, but diverge for cryogenic environments [9]. Although material damping is a linear function of temperature and is expected to approach zero at cryogenic temperatures, the quadratic effect of the coefficient of thermal expansion and the change of thermal conductivity and thermal relaxation at cryogenic temperatures counterbalance that trend. Typical variation of material damping for aluminum as a function of temperature is shown in Fig. 7.4, left. It decreases from about 0.1% at room temperature to about 0.001% at 30 K. For beryllium, the drop in damping at cryogenic temperatures is more pronounced since, for this material, the coefficient of thermal expansion also decreases sharply at cryogenic temperatures. Damping is also strongly dependent on vibration frequency, as shown in Fig. 7.4, right.

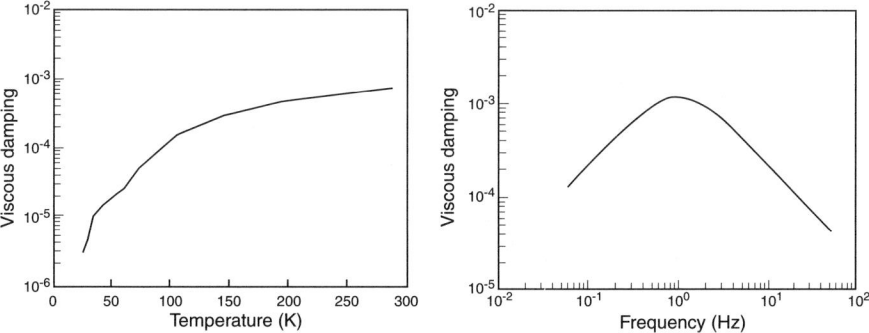

Fig. 7.4. Variation of material damping of aluminum with temperature (left) and frequency (right). (From Ref. [9].)

Low damping results in very pronounced resonant modes, which makes it difficult to avoid instabilities in the servo design. In some cases, it may be necessary to raise damping by the use of a hydraulic damper or higher-viscosity oil in the hydrostatic bearings.

Thermal model

For thermal modeling, one usually relies on two models: a system-level model of the entire observatory based on a relatively coarse grid and a telescope model covering the structure in more detail and including conductive and radiative paths within the structure and optics. The boundary nodes for the telescope model are derived from the system-level model.

These models are then exercised to determine the temperature field under various operating conditions (slews, transients, so-called "hot and cold conditions," etc.), and the output serves as input to the structural finite element model to determine the thermally induced displacements of the telescope structure and optics.

These displacements are then taken into account in equation 7.11 by adding the appropriate terms.

Optical sensitivities

Displacements of the structure supporting the optics and deformation of the optical elements result in a shifting of the line of sight, with potential degradation of the wavefront. The transformation of physical displacement into line-of-sight and wavefront error is referred to as "optical sensitivity."

Optical sensitivity is defined as the partial derivative of line-of-sight motion and wavefront error with respect to structural displacements due to static, dynamic, and thermal loads. In simple instances such as a Cassegrain configuration with rigid mirrors, it is possible to derive these sensitivities analytically. But this is impractical when the number of optical elements is large (e.g., segmented optics) or when the mirrors are not rigid bodies and are supported at multiple points. Optical sensitivities are then obtained numerically using an optical ray-trace model. Each degree of freedom in the structure is perturbed by a small amount (e.g., less than one wave), and the optical ray-trace program is exercised to derive the optical path difference and image centroid. This calculation is repeated for all degrees of freedom.

State-space representation

It can be shown that the second-order linear differential equation 7.11 can be recast into a set of two first-order linear differential equations of the type

$$\dot{\mathbf{X}} = \mathbf{A}\mathbf{X} + \mathbf{B}\mathbf{U}, \tag{7.12}$$
$$\mathbf{Y} = \mathbf{C}\mathbf{X} + \mathbf{D}\mathbf{U}, \tag{7.13}$$

where $\mathbf{A}, \mathbf{B}, \mathbf{C}$, and $\mathbf{D}$ are matrices which are functions of the system's characteristics, $\mathbf{X}$ is a vector representing the state of the system (position and velocity of nodes), $\mathbf{U}$ is a vector representing the input and disturbance forces and torques, and $\mathbf{Y}$ is a vector representing the output, that is to say, the linear and angular displacements of the optical elements or subelements in the system. This form is called "state-space representation" [8, 11].

The advantage of this formulation is that a large number of methods and tools are available to obtain solutions in both time and frequency domains. Once the solutions are found, the optical sensitivities are used to obtain the corresponding effects on image motion and wavefront error.

7.3 Pointing servo system

7.3.1 Fundamentals of servo systems

The basis of a telescope's closed-loop servo system consists of continuously comparing the actual pointing direction to the desired pointing and "feeding back" any difference to the drive system for correction. This process is represented graphically in Fig. 7.5.

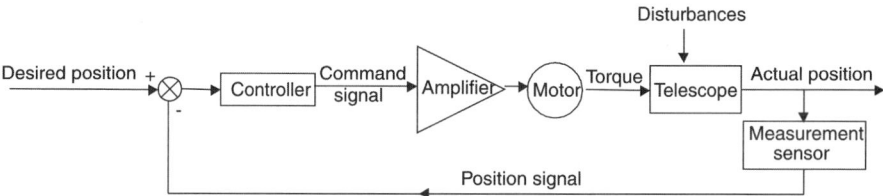

Fig. 7.5. Basic functional block diagram of the pointing servo system. The actual position is compared to the desired position; then, the corresponding error is fed to the drive system to correct the position.

The main components of the control system are the position sensor (or encoder), the controller, and the drive system chain (amplifier and motor). With the advent of computers, the desired position signal is now produced digitally, and the measurement of the actual position is generally also digital. The low-level-signal part of the control system is thus digital, and a digital-to-analog converter is used to feed the command signals into an amplifier, which, in turn, drives the motors.

The purpose of the "controller" is to supply a command signal based on the error detected. It should be designed to minimize that error as quickly as possible without creating instabilities. In its simplest form, the controller could supply a command signal proportional to the error signal: the greater the error, the greater the command torque, and as long as the error remains, the controller will generate a corrective command. Increasing the controller gain (the proportionality coefficient) will boost the correction signal and reduce the time needed to reach correction. This can be done up to the point where the overshoot becomes too large and the system is unstable. This type of controller is called a "proportional controller." The problem with proportional controllers is that they suffer from steady-state error. Because of friction in the drive chain (in gearboxes, rollers, motor shafts, etc.), the motors cease to act when the error signal is just below the point required to break friction. The telescope will sit there with that error and the corresponding command torque, but will not move. The solution is to introduce an *integrator* term in the control law. With integral action, the controller's output is proportional to the time during which the error is present and will eventually force the telescope to move. This type of controller is called "proportional-integral" (PI).

Integral action has a destabilizing effect due to the increased phase shift, however, so that a lower gain must be used, resulting in longer response time. The proportional-integral controller is thus more accurate and will correct low-frequency errors well, but will be less responsive at higher frequencies. The remedy is to introduce a *differential* term in the control law so that controller output is proportional to the rate of change in the error. The differential term will permit swift correction of rapid changes in the error signal and thus inhibit overshoots and allow for higher gain. This results in a faster settling time and higher correction bandwidth. Most controllers are of this type, called "proportional-integral-derivative" (PID). The price one pays is higher sensitivity to sensor noise, since the derivative of a fluctuating signal also fluctuates. Most PID controllers are thus equipped with noise filters to minimize extraneous fluctuations.

The general behavior of these three types of controller is well illustrated in Fig. 7.6, which shows their typical response to a step input.

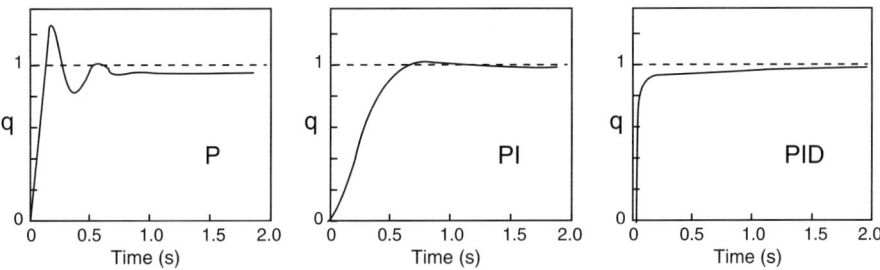

Fig. 7.6. Typical responses of a servo system to a step input. The response ($q =$ outut/input) of a proportional controller is shown on the left. The response time is good but a steady-state error is present and there is some overshoot. Adding an integral term eliminates the steady-state error and reduces the overshoot, but at the expense of a slower settling time (center). Adding a derivative to form a PID controller leads to a system with no steady-state error, no overshoot, and rapid response (right).

The control law of a PID controller is of the form

$$u = K_p e + K_i \int e \, dt + K_d \frac{de}{dt}, \tag{7.14}$$

where u is the corrective command signal, e is the error signal, and K_p, K_i, and K_d are the proportional, integral, and derivative gains, respectively.

The optimal values for these control-law coefficients are obtained by combining a mathematical model of the system being controlled with the control law, then performing an analysis of the control system. This mathematical representation leads to a set of simultaneous differential equations which are linear if the system is linear. The most popular approach for solving these differential equations is by the use of the Laplace transform, which transforms derivatives and integrals into algebraic expressions. Various techniques are then used to determine the stability domains, the response to command and disturbances, and the optimal control law. A description of these techniques can be found in standard textbooks such as those listed at the end of this chapter.

7.3.2 Telescope control system implementation

Control law for space telescopes

The block diagram of a typical space telescope control system is shown in Fig. 7.7 [12]. The three-axis torque commands are developed via a digital PID controller in series with low-pass filters for attenuation of the structure's flexible modes. Gyroscopes and star trackers serve as sensors for attitude determination and their data are optimally combined with the guiding system information via a Kalman filter to minimize the effects of sensing noise and drift (see discussion later in this subsection).

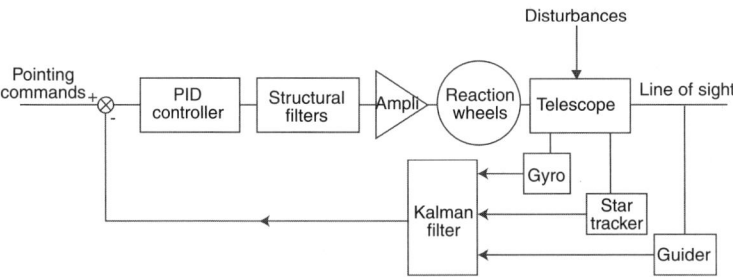

Fig. 7.7. Block diagram of a space telescope control system.

The tuning of the controller is fairly straightforward. The telescope is first modeled as a rigid body and the PID gains are set to produce adequate gain and phase margins[1] (e.g., 12 dB and 30°, respectively). The flexible body dynamics are then added to the model, and the cutoff frequencies and orders of the structural filter are determined such that the closed-loop system is stable (e.g., with a 10 dB margin).

Ground-based telescopes

The control systems of ground-based telescopes differ somewhat from those of space telescopes because of the different nature of disturbances encountered. The space environment is benign and friction effects are almost nonexistent

---

[1] See glossary.

due to the absence of gravity. On the ground, control systems must be more aggressive to combat larger torques and react better to nonlinearities (see Section 7.3.3). The solution consists of emphasizing the control of velocity and is generally accomplished by using two separate feedback loops as shown in Fig. 7.8: a velocity loop and a position loop.

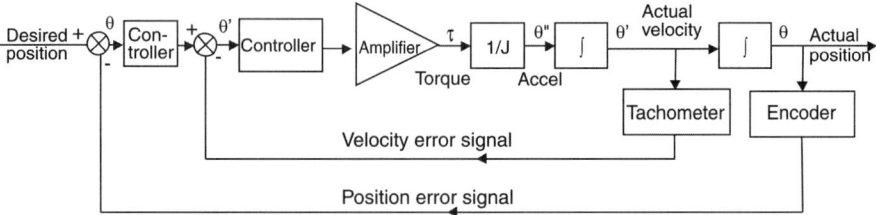

Fig. 7.8. Position loop with nested velocity loop. The velocity signal is either obtained with a dedicated tachometer or derived from the position encoder.

The velocity loop tightly controls the dynamics of the telescope structure and rejects disturbances such as wind and friction. The loop is usually implemented with a PI controller to maximize responsiveness, combined with a filter to avoid exciting resonance frequencies in the telescope structure. In practice, it is only possible to extend the bandwidth up to about 60% of the lowest locked rotor frequency of each telescope axis. A typical velocity-loop frequency response is shown in Fig. 7.9.

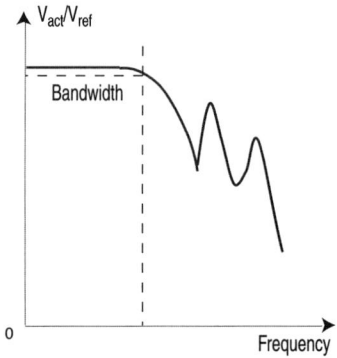

Fig. 7.9. Notional frequency response of the velocity loop for a ground telescope. The loop acts as a low-pass filter with a bandwidth of about 60% of the first locked rotor frequency.

With velocity control solved, the position loop's task is reduced to maintaining zero tracking error and to handling large changes in the desired position. During tracking (small excursions), a PI controller is used to maximize responsiveness. Large steps in position commands (large offset requests, repointing) are handled with a special algorithm.

The respective roles of the velocity and position loops can then be summarized as follows: the velocity loop takes care of the *dynamics* of the telescope structure, whereas the position loop takes care of pointing *accuracy*.

Guiding loop

The control systems discussed above are inherently limited in pointing accuracy by the *absolute* accuracy of the encoders or attitude sensors. Not only may their angular resolution not be high enough to track celestial objects, but a number of effects of optical (distortion), mechanical (optical/boresight/encoder alignment), thermal (alignment changes), and astronomical origin (atmospheric refraction) introduce additional errors. Although some of these effects can be calibrated out, residual errors are generally still significant. Fortunately, the sky itself can provide a final check on the pointing position by the observation of a star inside the telescope's field but outside the science field. Until the 1960s, this process, called "guiding," was carried out by the observers themselves, who made corrections manually. Guiding is now done automatically using an image centroiding device called a "guider." Details on how the corrective signal is obtained will be given in Section 7.5.5. This error signal can be used in two ways. It can either be fed to the position loop as additional position-error information or form an additional correction layer totally independent of the main control servo system (Fig. 7.10).

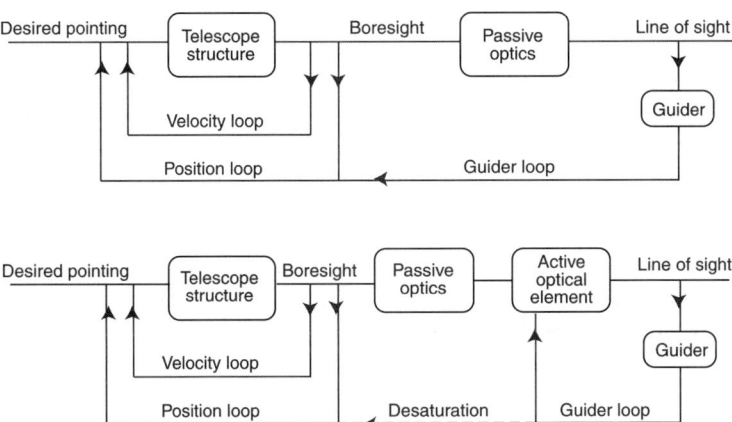

Fig. 7.10. The two schemes for using the guiding system error signal. On top, the guiding error signal is blended into the position error signal of the main control loop. At bottom, the guiding error signal is used to drive an active optical element which corrects the line of sight while leaving the telescope structure pointing unchanged. When the active optics correction becomes too large and the optical angular correction reaches physical or optical quality limits, the active optics loop is "desaturated" by feeding an error signal to the main pointing system.

268   7. Pointing and Control

The first scheme has been employed in most large ground telescopes so far, and also for the Hubble Space Telescope. The guiding-error signal is blended into the position-error signal coming from an encoder or attitude sensor. Since noise from the two types of sensors exhibits different power-spectrum density characteristics, the combined noise error can be reduced by estimating the attitude error from both sensors on the basis of frequency. In the case of HST, the attitude is determined by gyroscopes having fairly low noise levels on a time scale of tenths of seconds, whereas their drift is inherently significant over several seconds. On the other hand, the photon noise of the guiding system is large on a short time scale due to the faintness of the guide stars used,[2] but decreases as the square root of the integration time and is much reduced on a time scale of seconds. The power-spectrum density (PSD) of these two sensor noises is shown in Fig. 7.11. The crossover frequency for the blending of the two sensors was selected near the crossing point of the two PSDs. The pointing system uses the gyro data at a rate of 40 Hz and the guiding sensor data are fed at the rate of 1 Hz to correct gyro drift.

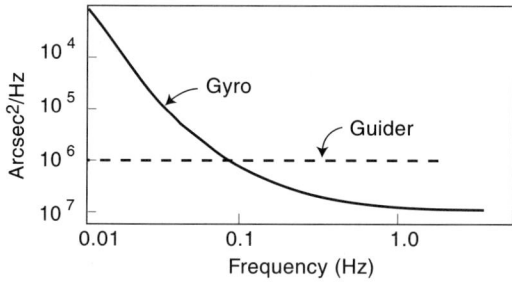

Fig. 7.11. PSD of HST's guider and gyroscope noise equivalent angles. The signals of the two sensors are blended so as to minimize the overall noise error in attitude, with preponderance given to gyro data above 1 Hz and to the guider for frequencies below that.

The second scheme makes use of an active optical element, such as a fine steering mirror, and is really part of active optics techniques, which will be examined in more detail in the next chapter. The main advantage of this scheme is that its bandwidth can be much larger than that of the main telescope because the driven element, typically a small mirror, is much lighter. It thus becomes possible to correct for higher-frequency disturbances, such as those due to wind or internal vibrations, which are beyond the reach of the main control system. The bandwidth is thus only limited by the brightness of the available guide star, which is, itself, constrained by the size of the field.

Indeed, the size of the guiding field can be severely limited. In space, optical, physical, and detector constraints limit the field to something between

---

[2] Because of the large number of optical elements in HST's fine guidance system, guide stars of magnitude 13.5 produce only 250 photons/s at the detector.

2 and 10 arcminutes in radius. On the ground, a tighter constraint arises from the atmosphere's isoplanatism, which limits the guiding field to about 1 arcminute. In practice then, correction is possible only up to a few Hertz. Beyond that, there is no choice but to make sure that disturbances will be negligible, either by construction or by isolation (Fig. 7.12).

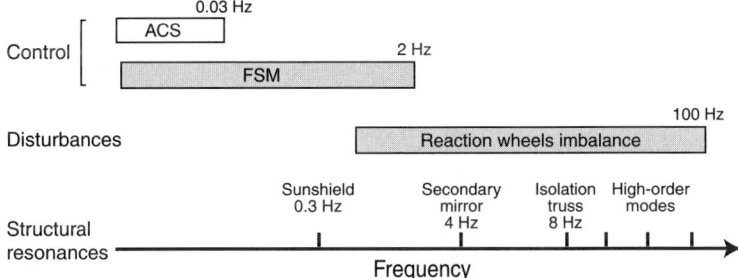

Fig. 7.12. Bandpasses of the attitude control system and line-of-sight stabilization (with fast steering mirror) compared to the range of various disturbances on NGST. The bandwidth of the attitude control system (ACS) is set at 0.03 Hz, a decade lower than the frequency of the first structural mode, which is that of the sunshield. The role of the ACS is to maintain pointing, but it has essentially no authority in correcting disturbances. The bulk of the disturbance rejection is entrusted to the fine steering mirror (FSM). The bandwidth of that active optics system is, itself, limited to 2 Hz because of the maximum field available for guiding. Beyond 2 Hz, disturbances must be made negligible by design or isolation.

Control-law structural filters

When the lowest resonant frequencies of a structure are few and far between, it is possible to avoid exciting them by setting the bandwidth of the control loop low enough compared to the first structural frequency and using a low-pass filter in the control law to dampen subsequent ones. This is a poor solution for telescopes because their structures are stiff, which results in a large number of resonant frequencies close to one another. To avoid exciting all of these frequencies would limit the bandwidth enormously. The solution consists of adding, in the control law, narrow-band filters centered around the most pronounced resonant frequencies. These filters are called "notch filters." To maximize efficiency, they have to be narrow, which implies that the matching of notch frequency to structure frequency is critical. The final center frequency of the filters is, therefore, best set after modal analysis of the assembled telescope. The width of the filter is chosen such that small variations in telescope structural frequencies, due to instrument exchange for instance, do not require reconfiguration of the servo controller. As an example, the position loop transfer gain for the VLT control system is shown in Fig. 7.13.

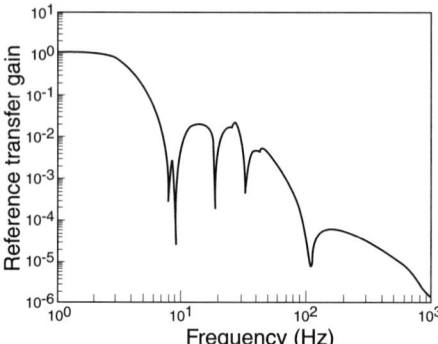

Fig. 7.13. Transfer gain of the VLT position loop v. frequency. Five notch filters are used to suppress the most pronounced modal frequencies in the structure up to 100 Hz. The -3 dB bandwidth of that loop is 3 Hz. Without notch filters, it would have been about 1 Hz and the disturbance rejection would have been much worse [6].

Sensing noise and filtering

Just as disturbances perturb the operation of a control system, so does noise in the sensing signals: it spoils the error measurements used in the feedback loop. Noise can be inherent to the sensor or be introduced by interference from other electrical sources. Noise can also be caused by wear and tear on the sensor or some physical obstruction causing the sensor to send inaccurate readings. Finally, sensors may introduce errors owing to their own dynamics and to distortions that cannot always be perfectly calibrated.

The effect of noise in the sensed data can be drastically reduced by use of a data "filter." Such a filter is a hardware or software device that processes noisy data to produce the best possible estimate. One of the best of these is the Kalman filter, which has many applications but is ideally suited to process sensor data. The Kalman filter may be implemented as a recursive algorithm to optimally process stochastic digital data using estimation by least squares [13, 14]. It is optimal because it uses all available measurements and prior knowledge of the measurement system to determine the best estimate of the current value of data. The recursive implementation does not require all previous data to be stored and reprocessed each time new data are acquired and hence is appropriate for real-time control. Kalman filtering requires that the sensing data satisfy three conditions: the system producing the data must be linear (i.e., describable by a linear mathematical model), the data noise must be Gaussian (i.e., with Gaussian probability density), and the data noise must also be "white." White data means that there is equal power at all frequencies and that the data is not correlated in time. Attitude sensing and guiding data will normally meet these conditions. Kalman filters are in standard use in space mission control laws and could probably also be used to advantage in ground telescopes.

7.3.3 Disturbance rejection

"Disturbances" are those undesirable forces and torques acting on the telescope system that perturb its operation because they are not under the direct influence of the controller the way the drive motor torque is. They include (1) external torques such as those due to wind or the space environment and (2) internally generated effects such as friction in bearings, motor ripple, and mechanism motion in the telescope and instruments. The origin and analysis of such disturbances will be investigated in detail in Sections 7.6 and 7.7. From the point of view of the control system, disturbances fall into two categories: linear and nonlinear.

Linear disturbances

Linear disturbances are those that create a parasitic force or torque such that the system is still governed by linear differential equations with constant coefficients (such as those of Section 7.2.1). The problem with nonlinear differential equations is that they are not solvable in closed form. Some nonlinear disturbances can still be treated in the traditional fashion, however, because their behavior is near-linear; that is, they are essentially linear over a small range of input values. This is the case, for example, for motor ripple and wind gusts. To minimize sensitivity to such linear disturbances, one must

- increase the bandwidth and servo gains of the telescope control system to cover the frequency spectrum of the disturbance with sufficient rejection power;
- decrease the magnitude of the disturbance by protecting the telescope (e.g., placing it in an enclosure and behind a windscreen to reduce wind effects), by reducing the cross sections of telescope members, or by passive or active isolation; or
- add a special control loop to actively compensate for the effect of the disturbance.

The first approach only works if the telescope can be stiffened. The second approach will be examined later in this chapter and in the chapters on enclosures and site selection.

The third approach has not yet been implemented, but there have been proposals. Since wind is, by far, the main disturbance on ground-based telescopes, a "wind feed-forward loop" was proposed for the control system of the VLT. Wind speed was to have been measured with a fast-response sensor immediately upstream of the telescope and a correction torque injected into the servo system at the proper time to counteract gusts. It is estimated that if this kind of ad hoc correction loop were adopted, the line-of-sight jitter due to wind could be reduced in the VLT by a factor of 2 [15].

272 7. Pointing and Control

Nonlinear disturbances

Nonlinear disturbances are more problematic in the sense that they cannot generally be eliminated by the control system. This is the case with friction. In Section 7.2.1, we assumed that damping was linear with the angular velocity. In reality, this is only true when velocity is not close to zero. At near-zero velocity, friction creates abrupt changes in velocity, a phenomenon referred to as "stick-slip" (Chapter 6), resulting in cyclic position errors, as shown schematically in Fig. 7.14. For sophisticated analysis, this effect can be represented mathematically using the "Dahl model" [16].

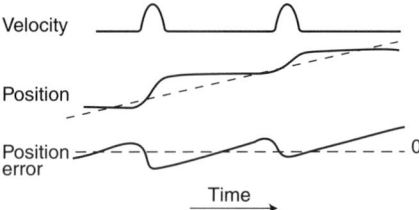

Fig. 7.14. Stick-slip near zero velocity creates abrupt changes in velocity (top) that result in cyclic position error (bottom).

To minimize stiction or any other nonlinear discontinuous torque source, the control system must have a high enough bandwidth for the servo to detect a rapid change in the mount response and quickly increase or decrease torque to overcome it. If the telescope relies only on a low bandwidth position loop (and all position loops wrapped around the largest system inertias are low-bandwidth), then the motion at around zero velocity is going to be jittery. On the other hand, if a high-bandwidth velocity loop is wrapped around the motor with tachometer feedback, this loop can detect an instantaneous change in speed resulting from a step in friction and quickly respond to correct the problem. If the bandwidth of the velocity loop is so high that the position loop is insensitive to it, then the result in position control will be an almost flawless transition through the zero-velocity regime. This type of control incorporating a separate velocity loop is essential for combating nonlinearities in ground telescopes.

In addition to solid friction due to bearings, gears, seals, motor brush drag, etc., several other nonlinearities may affect smooth operation of the control system. They include

- gear backlash,
- nonlinear spring rates of bearings, gears, and gearboxes,
- sudden torque changes in cable wraps other than those due to solid friction,
- quantization of digital commands,
- encoder and tachometer quantization,

- time delays, usually associated with sample-and-hold data,
- operational discontinuities such as step commands.

Methods to reduce the effect of nonlinearities include

- avoiding solid friction altogether by using hydrostatic bearings and minimizing the impact of other sources of nonlinearities by design,
- maximizing the inertia/friction ratio,
- using a separate drive to move cable wraps,
- avoiding driving the servo gains higher than needed, especially the integrating gain in the position loop (the drawback being an increase in wind sensitivity),
- avoiding zero velocity by incorporating a continuous mechanical motion stage in the drive with no effect on pointing, or by "biasing" the velocity of the drive system to avoid zero velocity altogether, as can be done with reaction wheels in space telescopes (see Section 7.4.2),
- "dithering," which consists of feeding a constant sine wave torque command to keep the drive system restless (with the drawback of increased wear).

When the nonlinear behavior is repeatable and predictable and can be characterized, its effects can be countered by introducing a command at the right time to create a motion of the telescope exactly opposite to that of the nonlinear effect. This is what was done on the Hubble Space Telescope to eliminate the jerk created by friction when the reaction wheels go through zero velocity. The time of "zero crossing" is predicted by monitoring the wheel speed as a function of time and, at the predicted time, a pulse command in the opposite direction is introduced by the pointing-control system.

As telescopes become larger, their sensitivity to disturbance increases because of sheer size, so that the value of the above solutions diminishes. Fortunately, there is a way out which consists of adding a new control layer to act directly on the optical path instead of on the structure supporting the optical elements. With this, we enter the realm of active and adaptive optics, which will the subject of Chapter 8.

7.4 Attitude actuators

7.4.1 Drives for ground-based telescopes

Until the 1970s, most ground-based telescopes used a worm gear system because its high gear ratio and excellent intrinsic accuracy permitted open-loop tracking using a constant-speed motor. The worm gear offers a large speed ratio in a single pair (e.g., 1/720), resulting in a very stiff drive [17]. Moreover, because of its intrinsic mechanical irreversibility, the worm gear is the only type of drive to offer absolute safety against telescope imbalance: should the telescope ever be operated while imbalanced, whether through human error or breakage of components, it will not "run away." Although used extensively in the past, the worm gear drive has major disadvantages. First, its mechanical efficiency is poor, on the order of 10% to 15%. Second, the system is irreversible. This makes it impossible for the servo to compensate for structural deflections or wind action with a traditional motor-mounted tachometer. Third, precise alignment of the worm with respect to the wheel is required. This alignment must be maintained regardless of structure deflection, thermal expansion, variations in oil pad film thickness, and gear eccentricity or wobbling. The solution is to "float" the worm carriage while guiding it with side rollers that move along races machined onto the gear. But this is expensive, complex, and difficult to maintain. And finally, worm gears are costly and limited in size. As a result, these drives have been abandoned for large telescopes.

Spur gears avoid the irreversibility problem of the worm gear while maintaining positive control: gears cannot slip under abnormal torque, as friction drives do. Spur gears have been used on several 4-meter class telescopes, but because of their complexity and relatively high cost, they have now been abandoned in favor of roller or direct drives.

Roller drives

Roller drives (also called friction drives), when properly made, are inherently smoother and more accurate than gear drives. A roller drive is simply a driving cylindrical roller which is pushed against a cylindrical journal of the telescope axis to be driven (Fig. 7.15).

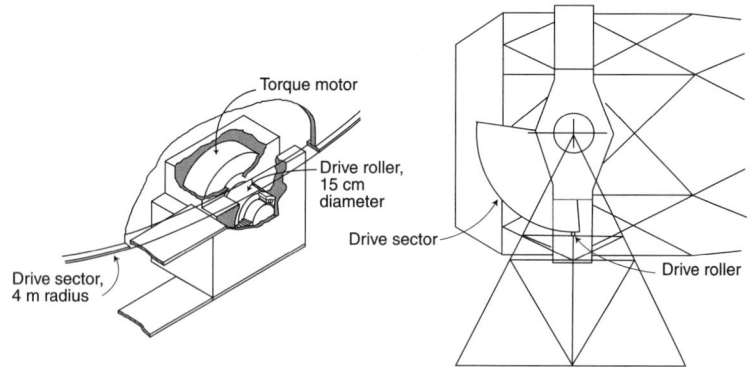

Fig. 7.15. Roller drive used for the altitude axis of the Keck telescopes. The detail of the roller is shown on the left. The roller drives a cylindrical sector mounted on the altitude axis, as shown on the right.

The magnitude of the force applied to the roller is determined by the torque requirement and the coefficient of friction. Maximum torque occurs when accelerating the telescope to slewing velocity.

The maximum compressive stress that will occur between the roller and the journal is given by

$$\sigma_c = 0.59\sqrt{\frac{W}{L}\frac{1/r_1 + 1/r_2}{1/E_1 + 1/E_2}}, \qquad (7.15)$$

where W is the preload forcing the roller against the journal, L is the length of the line of contact perpendicular to the rolling direction, E_1 and E_2 are the Young moduli of the two materials in contact, and r_1 and r_2 are the radii of the two rolling surfaces [18].

As an example, the characteristics of the Gemini 8-meter telescope azimuth and altitude axis drives are shown in Table 7.3.

Table 7.3. Roller drive of the Gemini telescopes

Parameter	Value
Diameter of the drive journal	12 m
Diameter of the roller	300 mm
Width of the roller	75 mm
Reduction ratio	40:1
Motor size	200 N m
Preload force on rollers	13 500 N

To avoid damaging the journal in case of slippage, the roller should be made of a softer material than the journal rim (e.g., brass).

Roller drives are fine devices but rely on high-quality surfaces, not a trivial matter for large driven journals. They always exhibit some eccentricity error, are intrinsically prone to stick-slip, and, more important, require extremely accurate alignment. If the alignment is not perfect, the drive will "jump" after enough tracking error has accumulated, and the roller or journal may be degraded fairly rapidly.

Direct drives

A direct drive, which eliminates all mechanical systems, is the ultimate choice. For smaller telescopes (2-meter class), direct drive systems employ commercial open-frame motors mounted directly on the telescope axes. For larger telescopes, custom fabrication is required, as was the case for the ESO VLTs and the Subaru and GTC telescopes. In all of these, the rotor and stator of each axis motor were integrated in the structure. Large-diameter direct drive motors can be viewed as slightly curved linear motors (Fig. 7.16). The magnetic race, generally manufactured in small segments (around 1 m in length), is bolted to the main bearing journal, and a number of small motors (winding pads) are installed facing the magnetic race at strategically stiff areas of the telescope structure. The air gap, that is, the distance between the race (rotor)

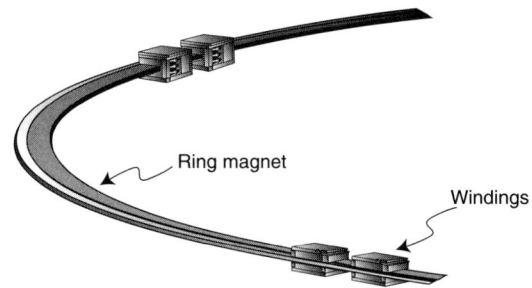

Fig. 7.16. Perspective view of the GTC direct drive. The motor is composed of a ring-shaped magnet assembly mounted on the rotating part and winding segments mounted on the fixed part. The winding segments have a U-shape in order to balance the magnetic forces inside the motor. The winding segments are "staggered" (i.e., not at angles regularly repeated with respect to the magnet poles) so as to reduce motor ripple and cogging. (Courtesy of Phase Motion Control Inc.)

and windings (stator), is on the order of a few millimeters. Each linear motor provides a tangential thrust on the telescope structure, and the sum of all thrusts produces the torque to move the telescope.

Direct drive motors are structurally simple and have no moving parts. They are intrinsically free from friction and stick-slip effects and need no maintenance. When combined with hydrostatic bearings and a separately driven cable wrap to eliminate friction, they provide an essentially perfect drive system free of nonlinearities.

In contrast to all other drive mechanisms, in which the force is concentrated on a pinion or a wheel, direct drives distribute the thrust along the structure, thus minimizing localized deformation and maximizing structural stiffness. This results in the highest drive stiffness possible and does not further degrade the intrinsically low stiffness of very large telescopes. In the case of the VLTs, the lowest locked rotor frequency in altitude is about 8 Hz, whereas it would have been about 4.4 Hz with a conventional gear drive.

Another key advantage of direct drives is their insensitivity to mechanical misalignment (on the order of a few millimeters), dramatically reducing telescope installation time. Finally, as direct drives are generally realized as multiple linear motors on the same magnetic race, they are intrinsically redundant. A drive section can generally be removed for maintenance while still leaving the telescope completely operational.

The disadvantages of direct drives are relatively high cost and slight torque ripple and nonlinearities associated with electromagnetic "cogging." Cogging is a condition, most pronounced at low motor speeds, in which motor rotation is jerky as a result of magnetic forces that develop between the motor stator and the motor's permanent magnets [6]. But with the present state of technology, torque ripple and cogging average much less than 1%, and the motor can be designed such that the ripple spatial frequency is located where the corresponding angular error is small and easily correctable by the control loop.

Reduction ratio

When choosing the reduction ratio for gear or roller drives, the traditional engineering practice is to match the inertia of the drive to that of the load. If J_T and J_M are the moments of inertia of the telescope and motor, respectively, the reduction drive ratio, N, would be such that $J_T = N^2 J_M$ (Fig. 7.17). This ensures maximum energy transfer efficiency. But there is no point in doing this for telescopes because they have such enormous inertia that the high N required to achieve this optimum would mean unacceptably low gear stiffness. And besides, the energy required to move well-balanced, low-friction, low-acceleration telescope systems is so small that power consumption is not an issue.

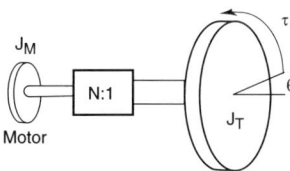

Fig. 7.17. Telescope coupled to a motor via a reduction drive.

The right approach is to select the reduction ratio so as to minimize disturbance sensitivity. Sensitivity to disturbance can be measured in terms of disturbance response defined as $\Delta\theta_D/\tau_D$, where τ_D is the disturbance torque and $\Delta\theta_D$ is the resulting change in angular position. For the sake of argument, let us assume that the telescope is coupled to the driving motor via a perfectly stiff reduction drive mechanism. Then, one has simply

$$\frac{\Delta\theta_D}{\tau_D} \propto \frac{1}{J_T + N^2 J_M}. \qquad (7.16)$$

The drive motor moment of inertia is always very small compared to that of the telescope, so its effect is significant only if N is large, say in the 1000 range [19]. High drive ratios can only be obtained by gear systems which are mechanically complex and expensive. But a greater drawback of gearboxes is that they have low stiffness. This reduces the natural frequency of the overall system and thus prevents achieving high servo loop gains, with the result that disturbance rejection actually worsens.

In conclusion, it is generally best to select the lowest reduction ratio possible. This will maximize overall system performance.

Countertorque preloading

Loaded rolling elements have significant amounts of internal deformation (Fig. 7.18, left). If the load direction is reversed, the local deformation must recover and move in the opposite direction before the roller will generate a full opposing reaction force. But this behavior is not linear and, at the moment of torque reversal, the spring rate approaches zero. This nonlinear behavior is illustrated in Fig. 7.19 (left). The effect is not specific to roller drives: gears, which are inherently composed of rolling elements, exhibit the same behavior.

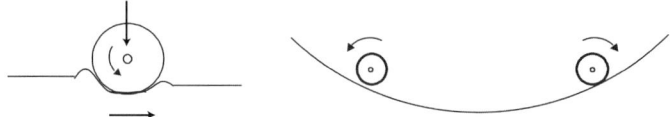

Fig. 7.18. At left, local deformation due to roller loading (exaggerated for clarity). The asymmetric deformation is a source of nonlinear behavior which can be mitigated by using two rollers preloaded against one another (right).

Since drive-system stiffness is one of the terms in the closed-loop control system gain, this effect amounts to a region of unstable, low-gain operation whenever an external disturbance approaches zero. For example, wind turbulence would excite the system and this would result in a low-frequency servo oscillation, an effect described as "hunting."

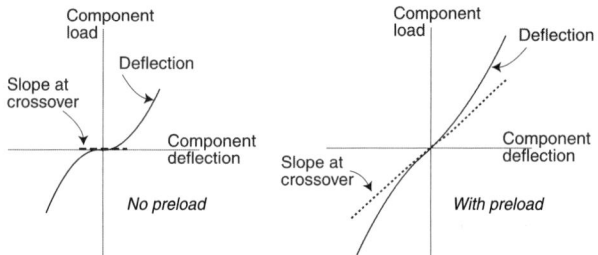

Fig. 7.19. Effect of countertorque preloading. With a single drive, the slope of the torque/rotation curve is close to zero at near-zero torque, resulting in little stiffness (left). Stiffness is increased by adding a second drive to counteract the first (right).

It is therefore essential to reshape the fundamental system-stiffness curve of the drive system. This can be accomplished by introducing a permanent countertorque to "bias" the system. This, in turn, is usually achieved by means of a second identical drive applying a torque opposite to that of the first one (Fig. 7.18, right). To minimize power requirements, the two drives act together to slew the telescope when performance is not important, but are rearranged as a pair of countertorqued drives during tracking. Figure 7.19 (right) shows the much improved stiffness profile.

7.4.2 Space telescope attitude actuators

Space telescopes cannot be driven by reaction against the ground; a torque must be applied, either by mass ejection or momentum transfer. Mass ejection systems have limited lifetimes, so space telescopes typically use flywheels which are accelerated or decelerated to transfer momentum from the wheels to the telescope. These flywheels are called "reaction wheels" (also "reaction wheel assembly" or RWA), and they typically rotate with speeds of up to 3600 rpm in either direction. A sectional view of a typical reaction wheel is shown in Fig. 7.20.

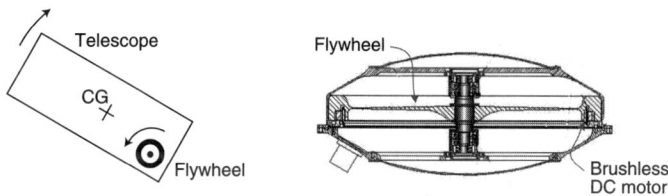

Fig. 7.20. At left, principle of using momentum transfer to slew a telescope: accelerating the rotation of a flywheel moves the telescope in the opposite direction. At right, cutaway view of a typical reaction wheel (used with permission of Goodrich Corporation).

A space telescope is normally equipped with a set of four such wheels, one for each axis plus a spare. To slew the telescope, the spin speed of the reaction wheels is increased or decreased depending on the desired direction. Because overall momentum remains constant, the telescope moves in the opposite direction. To stop the telescope as it arrives at its target, the reaction wheel speeds are returned to their original values (Fig. 7.21). These speed changes are commanded by sending current pulses to the wheel motors. The pulses, called "jerks," are shaped so as to avoid exciting resonances in the telescope.

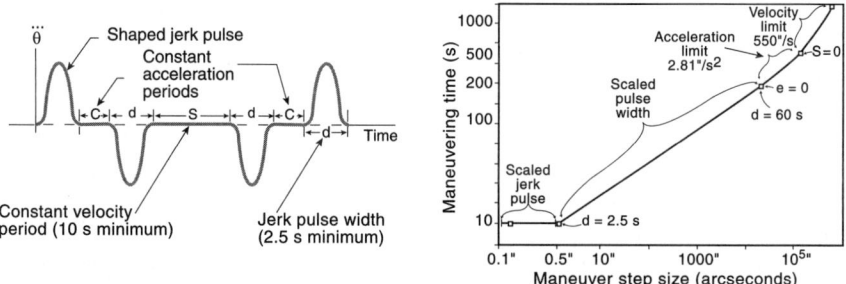

Fig. 7.21. At left, use of spin velocity pulses to slew a telescope (θ being the commanded angle). The slew profile for HST is shown on the right.

280   7. Pointing and Control

When the wheel rotation has to reverse, there will be an undesirable pulse on the telescope as the wheel speed goes through zero and torque is suddenly increased to overcome friction. This effect can be compensated by introducing an opposite torque at the exact time the wheel speed passes through zero, as was done on HST (see Chapter 6). Another approach is to simply bias the wheel speed so that the rotation sense never does reverse. For example, instead of operating within a range of −1000 rpm to +1000 rpm, one can center wheel speed on +1100 rpm and operate between 100 and 2100 rpm.

Momentum dumping

Space telescopes have to combat an external torque due to imbalanced solar radiation pressure (and residual atmosphere in near-Earth orbit) which, although small, does accumulate to significant values after several hours. This torque is continuously counterbalanced by the reaction wheels, but their spin speed will progressively build up, and control of the spacecraft will be lost when the wheel speeds reach their maximum values. Accumulated momentum must be eliminated before this happens. This is generally accomplished by use of a mass ejection device such as a gas thruster or magnetic torquers for near-Earth orbits.

7.5 Attitude sensors and guiding system

Pointing a telescope at a given target requires the ability to determine the absolute direction of the optical axis. For ground-based telescopes, the position of each axis is measured by position encoders mounted on each of the moving axes. For space telescopes, the main attitude sensors are generally gyroscopes, the drifts of which are corrected by star trackers.

7.5.1 Position encoders

In early telescopes, the polar axis and declination axis shaft angles were determined visually using "hour and declination circles." These were simply large circles mounted on each axis, with scaling marks every 10 arcminutes or so. These circles were later replaced by a pair of transmitting synchros mounted on the drive systems, with repeater synchros at the console. Today's telescopes employ position encoders (more commonly called simply "encoders"), which are the modern equivalents of hour circles and nearly always of the optical type. Optical encoders consist of a series of reference marks on some sort of support and a counting head which reads these marks. The support can be a tape wrapped around a cylindrical surface coaxial with the telescope axis or a disk mounted perpendicular to the axis. The marks are read by a noncontacting, self-adjusting optical head, which means no wear and low maintenance. The reference marks have a linear accuracy which can be as good as 3 μm for linear tapes and 1 μm when a very stiff special glass or quartz substrate is used.

Encoders fall into two categories: incremental and absolute. An absolute counting encoder "knows" the absolute position upon being turned on, whereas an incremental encoder does not. Incremental encoders simply supply a differential position measurement. However, if referenced to an absolute position, an incremental encoder will also provide an absolute measurement (potentially with even higher accuracy than that of an intrinsic absolute encoder), although miscounts can happen due to dust, grease or vibration.

An example of an incremental optical tape encoder is shown on the left in Fig. 7.22. A grating on a tape is imaged through a set of four gratings, each being phase-shifted from the others by one-quarter of the grating period. As the tape moves with respect to the reading head, the light-dark modulation produced by each of the scanning gratings is read electronically. The four signals are then combined to produce an incremental measure of the tape's displacement. For absolute encoding, a number of tracks are added to allow unique determination of the position by the use of some encoding technique. The principle is shown on the right in Fig. 7.22 for a rotary encoder.

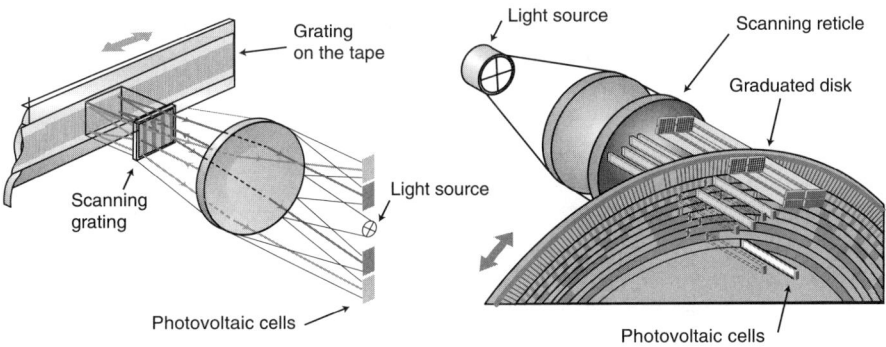

Fig. 7.22. Examples of an incremental optical tape encoder (left) and of an absolute disk encoder (right). In these implementations, the tape is metallic and is measured by reflected light, whereas the absolute encoder uses a glass disk and transmitted light. (Courtesy of Heidenhain Corporation.)

Absolute encoders are generally mounted directly on a large journal and used for absolute pointing determination. They have an intrinsic absolute accuracy of about $1''$. Incremental encoders can have a higher angular resolution (e.g., $0.1''$) and are typically used for tracking. They are sometimes mounted on a gear or friction system for increased magnification. Tape encoders should be mounted on as large a lever arm as possible to maximize resolution [20]. In the case of the VLT, which uses such encoders, tapes have grid spacings of 40 μm and a diameter of 1.6 m for the altitude axis and 7 m for the azimuth axis.

7.5.2 Tachometers

Tachometers are used to supply velocity information to the velocity loops of ground-based telescopes. They should be mounted close to the drive in order to achieve dynamic stability. It is important to understand that the performance of the control loop at medium to high frequency (i.e., the domain where the position loop is insufficient) depends entirely on, in fact mimics, the tachometer signal. The choice of origin, processing, and placement of the tachometer system is consequently critical to the performance of the whole system.

A traditional solution consists of using conventional brushed DC tachometer generators on a roller mounted on the journal used in driving the telescope axes. The roller mechanism should be totally separate from that of the drive, to avoid corrupting the velocity information by torsion stress, and it should have a flexible mount in order to follow any runout in the driven journal. But as a result, this flexible mount will allow small rotations of the whole tachometer assembly at rotation reversals. This means that the tachometer will not be able to measure correctly through the zero-speed range of the telescope, that it will have backlash there, and that this will result in the telescope oscillating across zero speed.

In geared systems (not in direct drives), an alternative is to mount the tachometer on the drive motor shaft, but care must be taken to prevent the tachometer from being sensitive to commutation spikes and mechanical deformation induced by the motor. Furthermore, this solution would close the loop on motor speed, not telescope speed, thus allowing the transmission inaccuracies to be passed along to the optical system.

To avoid the problems associated with rollers, the tachometer can be of the "direct drive" type, working in reverse mode to that of a direct drive motor. This solution is exempt from nonlinearities, but a direct tachometer for a telescope must be endowed with enormous sensitivity to magnetic field gradients in order to produce a meaningful signal at speeds well below 1 turn/day. This sensitivity is feasible but also means that such a direct tachometer would be sensitive to magnetic flux variations in or near the telescope's steel structure. Such direct, analog tachometers were originally implemented on the VLTs, but had to be abandoned for this reason.

All systems described so far involve a brushed, mechanically or electronically commutated DC generator which produces an analog voltage proportional to telescope rotation speed. Apart from the intrinsic limitations in the device itself, the analog signal is awkward because of the extremely wide speed range required for telescope pointing and tracking: during tracking, the tachometer signal must retain a good signal-to-noise ratio although the signal is at the submillivolt level. This is clearly impractical. Recent designs have therefore resorted to obtaining tachometric feedback by processing the digital signal of the main telescope encoders [21].

Using the encoder data is an ideal solution from mechanical and rational viewpoints. All ground-based telescopes must have quality encoder systems, intrinsically free from backlash and stick-slip, and with subarcsecond accuracy. The needed signal is therefore readily available. This apparently simple solution faces two main problems however:

- Encoders read position, so the speed signal must be obtained by differentiating the encoder readings at two consecutive readouts. This must be done rapidly enough to avoid compromising the dynamic performance of the system. Usually, the raw encoder data needs extensive processing, so there may be a computing-time problem.
- Although the encoder error is always lower than the telescope pointing accuracy, typically less than $0.1''$ in large telescopes, this error may have a very high spatial frequency. Consequently, the "ghost" speed ripple caused by encoder error, amplified by the differentiation process, may impair the performance of the speed loop.

These problems require careful design and programming, but this new "digital tachometer" approach is in line with the evolution of servo technology in industrial applications, from which analog tachometers disappeared in the early 1990s.

7.5.3 Gyroscopes

A gyroscope consists of a rapidly spinning flywheel used to sense and measure changes in the orientation of the spin axis (the name is often abbreviated to "gyro"). The most common type of gyroscope for spacecraft attitude sensing is the "single-degree-of-freedom rate gyro." Single degree of freedom refers to the fact that the flywheel is mounted on a single gimbal and only the corresponding axis is sensed. Three gyros with different orientations are thus needed to measure all three angles of attitude. Rate gyros are so called because they measure the angular rate of change in the direction of the spin axis (by opposition to rate integrating gyros, which provide the total spacecraft rotation from an inertial reference).

A single-degree-of-freedom rate gyro is represented schematically in Fig. 7.23. The gimbal is restrained by viscous damping and a restoring spring, and the rotation of the output axis is proportional to the spacecraft's angular rate about the input axis. In actual high-accuracy gyros such as those used on space telescopes, the spin axis is supported by gas bearings and the output axis rotation is balanced by an electromagnetic torque. The applied torque is the measure of the angular rate.

The HST gyros, despite their technology dating from the 1970s, are still state of the art. They have an angular resolution of 0.25 mas and drift at the rate of 1 mas/s [22]. They need to be corrected every few seconds with absolute positional data supplied by star trackers or fine guiding systems.

Other types of gyro, such as ring-laser and fiber-optic gyros, may be simpler, less expensive, and more reliable, but they are far from offering the same precision.

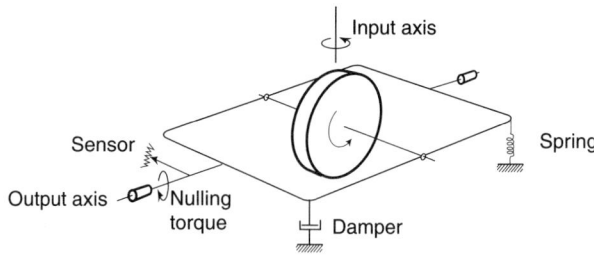

Fig. 7.23. Schematic geometry of a single-degree-of-freedom rate gyro.

7.5.4 Star trackers and Sun sensors

Star trackers are devices with their own optics that measure the locations of stars within their fields of view. Star trackers should not be confused with "fine guiding systems," which centroid on stars but use the telescope's main optics to do so. The fine guiding systems, which will be examined in the following section, have high accuracy but a much smaller field of view and usually do not provide absolute attitude determination. Star trackers have a large field of view, typically about 8×8 degrees, permitting the measurement of several star positions. These observations are then compared with the coordinates from a star catalog to determine the spacecraft's attitude. Star trackers have a sensitivity of about magnitude 6, with an update rate of 10 Hz. The best star trackers have an absolute accuracy of about $2''$ rms [23].

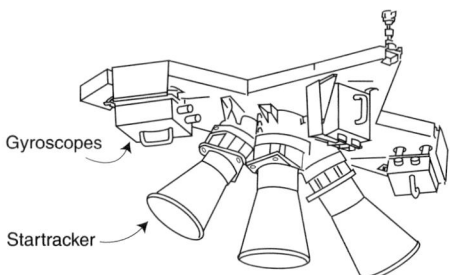

Fig. 7.24. Fixed-head star tracker system used for HST. The three star trackers, located on the side of the telescope opposite the Sun, provide complete determination of the attitude whenever the Earth is not in the way. Baffles protect the optics from the Sun when the spacecraft is rolled with respect to its nominal position.

In addition to the gyros and star trackers, a number of coarser sensors are used to determine the attitude of the spacecraft following launch, during the initialization phase, or to monitor the proper operation of the main attitude system. These additional sensors, which include Sun sensors and magnetic sensors (when close to the Earth), have an accuracy on the order of arcminutes.

7.5.5 Guiding system

By itself, telescope tracking is sufficiently accurate for durations of only a few minutes at best. On the ground, the limit stems from errors in the gravity compensation systems, encoder imperfections, atmospheric refraction, and atmospheric turbulence. In space, the limit is due to gyroscope drift and to sensing noise in the star trackers.

This problem is solved by monitoring the position of a relatively bright star in the vicinity of the field being observed and then correcting for detected errors. This procedure is referred to as "guiding." Early telescopes used guiding telescopes mounted on the side of the telescope tube. But it is difficult to ensure that the line of sight of a guiding telescope remains parallel to that of the main telescope, regardless of its inclination. It is much better to guide using a star in the very field of the telescope. This eliminates differential line-of-sight errors and the guiding system benefits from the full aperture of the main telescope, thus improving its sensitivity.

In most cases, a single guide star will be enough because "roll," the angle of rotation around the line of sight, is usually well controlled either by attitude sensors for space telescopes or by calculation for alt-az telescopes on the ground. If the accuracy of this roll determination is not sufficient, a second guide star, far enough from the first one to create a sufficient lever arm, will be required. This is the case with HST, whose ultimate tracking accuracy of 7 milliarcseconds is reached by using two guide stars at least 13 arcminutes apart [24].

Guiding system implementations fall into two categories: passive and active. In the passive (no moving parts) type, the guiding field is "paved" with large array detectors. Once a guide star has been identified, the subarea of the detector containing that star is read at a rapid rate to supply the two-dimensional position signal. In the active type, a probe equipped with a small array or a four-quadrant detector is positioned at the expected location of a guide star within the guiding field. If required, the guide star is searched for by moving the probe in a spiral fashion. Once the star signal has been confirmed, the centroiding process is initiated. The probe is typically mounted on an x-y platform. Alternatively, the detector can remain fixed and the guide star light be directed to it via a rotating optical system capable of exploring the entire guiding field. The latter design was employed for HST (a system referred to as a "star selector system") [2].

The accuracy of a guiding system as a position sensor is limited by two effects:

286   7. Pointing and Control

- A systematic error due to differential motion between the image of the guide star and the line of sight defined, for example, as the center of the science field. These systematic errors have two possible origins: internal to the telescope or external. Internal errors stem from mechanical or thermal effects which modify the physical distance between the science detector and the guiding probe. An external effect is one that affects the apparent relative positions of the guide star and the science field. This effect is due to atmospheric refraction on the ground and to differential velocity aberration in space. Although such effects are fully predictable, they are by nature correctable only in "open-loop" fashion and may exhibit residual errors in their correction.
- A random error due to noise in the detector and photon noise in the guide star signal.

The systematic error should be minimized by proper mechanical, structural, and thermal design of the guider and its mount and by adequate analysis of the atmospheric and velocity aberration correction. The random error is typically expressed in terms of the "noise equivalent angle" (NEA), which is the rms error in one's knowledge of the guide star centroid due to sensor noise. When sky background is negligible, the NEA is given by

$$\text{NEA} = \frac{\sqrt{1 + R_n/N_s}}{k\sqrt{N_s}}, \qquad (7.17)$$

where N_s is the total number of detected photons, R_n is detector readout noise over the total area of the detector being read out, and k is the slope of the centroiding function, that is, the constant of proportionality between the excursion angle and the error signal. N_s is a function of the incoming flux and optical characteristics of the optics and detector and is given by the formula

$$N_s = \eta \mathcal{A} \cdot \text{QE} \cdot \text{BP} \cdot \Phi_0 \cdot 10^{-0.4m} \, t, \qquad (7.18)$$

where QE is the quantum efficiency of the detector, $\mathcal{A}$ is the telescope's collecting area, η is the throughput of the optics, BP is the wavelength bandpass, Φ_0 is the guide star photon flux per unit wavelength bandpass for a zero magnitude star, m is the star's magnitude, and t is the integration time between readouts. To the first order, Φ_0 is equal to $9.6 \cdot 10^{10}$ and $4.5 \cdot 10^9$ photons/(m² μm s) in the V-band and K-band, respectively.

In the case of diffraction-limited images, where centroiding is done using the central peak of the image, k is given by

$$\frac{16}{3\pi} \frac{D}{\lambda}, \qquad (7.19)$$

where D is the pupil diameter (assumed to be circular and without central obstruction) and λ is the wavelength [25]. The effects of pixel size and background have been analyzed by Hardy [26].

An example of the noise equivalent angle as a function of integration time is shown in Fig. 7.25. To the first order, the noise equivalent angle is inversely proportional to the square root of the integration time: the longer photons are collected, the greater the signal-to-noise ratio. But, past the knee in the curve, there is little benefit to increased integration time. Before the knee, on the other hand, noise increases rapidly so that, in practice, the optimal point will be near the knee.

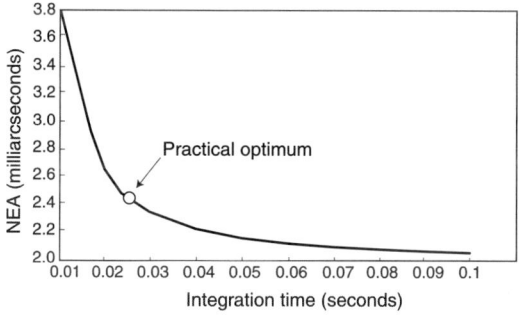

Fig. 7.25. Example of the noise equivalent angle for a guider on an 8 m diffraction-limited telescope in space working on 16.5 magnitude stars in the K-band. For the guider NEA to be compatible with the resolution potential of the telescope, the integration time needs to be at least 0.025 s, which limits the overall telescope control bandwidth to a few Hertz.

For a given telescope aperture size, optical throughput, and detector characteristics, the field required for guiding is set by the minimum star brightness needed to satisfy the control system bandwidth. Star density is minimal at the galactic poles, and that minimal density is given in Table 7.5.5 as a function of magnitude and wavelength band. These data are based on actual star counts obtained from the HST guide star catalog for the north galactic pole; the counts are slightly higher for the south galactic pole [27].

Table 7.4. Average cumulative number of stars per square degree at the north galactic pole

Magnitude brighter than	V-band	J-band	K-band
14	57	182	293
15	109	344	540
16	189	590	868
17	303	929	1286
18	459	1377	1863

Typically, one will want to ensure a 95% to 99% probability of finding at least one guide star of a given magnitude or brighter for any pointing. To guarantee this, the average number of stars in the field available for guiding,

N_{gs}, must be, using Poisson statistics,

$$N_{gs} = -\log_e(1 - P(0)), \qquad (7.20)$$

where $P(0)$ is the desired probability of finding at least one guide star in the field. For $P(0)$ of 0.95 and 0.99, the average number of guide stars in the field must be 3 and 4.6, respectively. The guiding system for HST, which was designed to satisfy this condition, is shown in Fig. 7.26 [28].

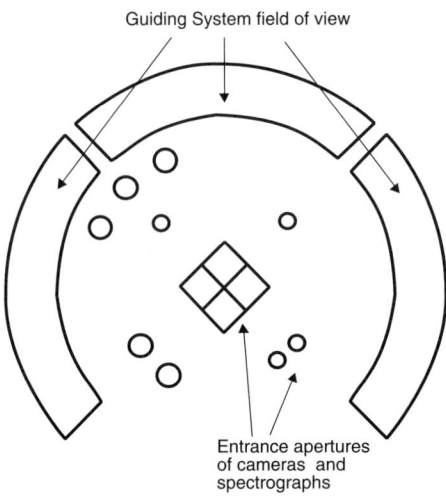

Fig. 7.26. Focal plane of the Hubble Space Telescope showing the three zones used for guiding surrounding the science field. These zones total 64 square arcminutes, which guarantees finding at least two guide stars of magnitude 14.5 or brighter to control the line of sight (pitch/yaw) as well as the roll of the telescope.

The error signal supplied by the guider can be fed directly into the pointing system, but the bandwidth of the correction can be improved by "fine steering" the optical beam itself, instead of the entire telescope. This is accomplished by the tip-tilt of a mirror in the optical train, such as the secondary mirror or, better yet, by using a dedicated, low-inertia, flat mirror located at a real pupil in front of the focal plane (see Chapter 8).

7.6 Ground-based telescope disturbances

Disturbances affecting ground-based telescopes arise from internal excitation (i.e., generated within the telescope itself) and from wind.

In the days of visual guiding, observers themselves were a potential source of internal excitation when they were in the prime focus or Cassegrain "cage." Today, the only sources of internal excitations are those due to the movement of mechanisms in the instruments (e.g., filter wheels), torque ripple in the drive motors, and friction in the telescope axes and in the cable wraps. Mechanism motion is generally insignificant because of the large moment of inertia of the telescope itself, whereas torque ripple and friction sources can usually be dealt with by proper design of the control system, as discussed in Sections 7.4.1 and 7.5.

By far the largest source of disturbance in ground-based telescopes is wind. Its effects are particularly detrimental because its power spectrum contains considerable energy at low frequencies (0.1 to 1 Hz), which are relatively close to resonance frequencies in the structure and active mirror systems.

Most astronomical sites are in isolated locations at high altitude and, as a result, are quite windy. To maintain observing efficiency, observatories are usually designed such that observation remains possible for 95% of the "clear sky" time. This results in a requirement to operate in fairly high wind conditions (e.g., up to 20 m/s at the Mauna Kea observatories).

7.6.1 Effects of wind: Generalities

Wind effects on any structure are of two sorts: static and dynamic. The static effect is simply due to the pressure created by a constant wind on surfaces impinged upon. Dynamic effects stem either from the turbulence created by an obstructing surface in a flow with constant speed or from turbulence in the incoming flow itself.

Wind static effects

Wind load on any structure is given by

$$F = C_D \rho \frac{V^2}{2} A, \qquad (7.21)$$

where C_D is the drag coefficient, ρ is the density of air, v is the velocity of air, and A is the cross-sectional area normal to wind direction. The drag coefficient depends on the geometry of the body, the turbulent state of the incident wind, and the wind velocity. Assuming that the incident flow is laminar, the drag coefficient can be described as a function of geometry and of the Reynolds number.[3] The drag coefficient for flat plates is equal to 2, whereas for cylindrical shapes, it varies between 0.4 and 1.2, depending on the Reynolds number [29].

---

[3]The Reynolds number is given by $\mathcal{R} = VL/\nu$, where L is the characteristic dimension of the object normal to the flow, V is the flow velocity, and ν is the kinematic viscosity of the fluid ($\nu = \mu/\rho$, where μ is the dynamic velocity). At Mauna Kea (4200 m altitude, -10 °C temperature), $\rho = 0.82$ kg/m^3, and $\nu = 2.0 \cdot 10^{-5}$ m^2/s.

Dynamic wind effects: Vortex shedding

In addition to the direct drag force examined above, airflow around individual members can generate forces normal to the wind direction due to "vortex shedding" (also called Von Karman vortices). Vortex shedding can excite natural resonances in the member and result in large-amplitude oscillations. Karman vortices are shed at the characteristic frequency, f, given by

$$f = \frac{SV}{L}, \tag{7.22}$$

where V is the wind speed, L is the characteristic transverse dimension of the element, and S is a dimensionless quantity called the Strouhal number, which depends on the Reynolds number but is typically around 0.2. Hence, the vortex shedding frequency for cylinders is approximately given by

$$f = \frac{0.2V}{L}. \tag{7.23}$$

Dynamic wind effects: Wind gustiness

The turbulence content in a wind flow is characterized by the power spectral density (PSD) of the wind speed. Two models are commonly used for wind speed PSD near the ground: the Von Karman spectrum and the empirical Davenport model. The Von Karman spectrum is defined as

$$S_V(\nu) = \frac{4\,I^2\,V^2 L}{[1 + 70.8(fL/V)^2]^{\frac{5}{6}}}, \tag{7.24}$$

where ν is the frequency, I is the turbulence intensity in percent, V is the mean speed in m/s, L is the outer scale of turbulence in meters, and S_V is the PSD in $(m/s)^2/Hz$. In open air, I is about 15% and L is on the order of 80 m. Inside the telescope enclosure or downstream from a porous wind screen, the incoming wind vortices are broken down into smaller ones, with size driven by the dimension of the obstruction (dome slit, louvers, etc.), and L is set by that dimension. For a telescope enclosure with a 10 m wide slit for example, L has been found to be 3.2 m. A comparison of the power spectrum inside and outside an enclosure is shown in Fig. 7.27, left.

The other frequently used power spectrum density model is the Davenport formula which was derived empirically [30, 31] as

$$S_V(z,\nu) = \frac{4.0\,k\,x^2\,V^2}{\nu\,(1+x^2)^{4/3}} \quad \text{with} \quad x = \frac{1200\,\nu}{V_{10}}, \tag{7.25}$$

where $S(z,\nu)$ is the PSD at height z in $(m/s)^2/Hz$, ν is the frequency in Hertz, V is the mean wind velocity at height z, V_{10} is the mean wind velocity at a height of 10 meters in m/s, and k is a roughness coefficient, which is about 0.08 for open terrain. As shown on the right in Fig. 7.27, this formula has proved to match data obtained at typical astronomical sites.

7.6 Ground-based telescope disturbances

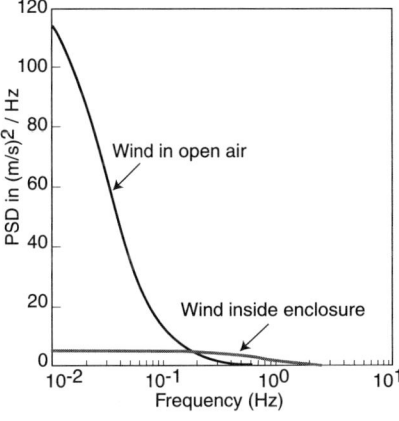
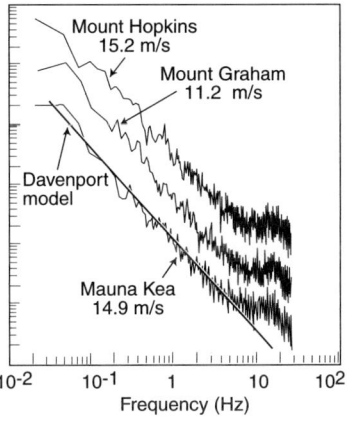

Fig. 7.27. At left, power-spectrum density inside a telescope enclosure compared to that of the outside wind. Although much reduced in amplitude, the wind inside the enclosure might still be damaging to image quality because its power spectrum is shifted to higher frequencies which are more prone to excite the telescope structure or active mirror systems. At right, power-spectrum density of the wind at three representative astronomical sites. The power spectra have been displaced vertically for clarity, but have the same magnitude. The straight line is the Davenport model, which is valid for undisturbed flow in open terrain. The higher amount of energy between 8 and 30 Hz compared to the Davenport model is thought to be an artifact of the measurement.

When wind acts on a large structure such as a telescope, one must take into account the partial decorrelation of the wind speed over the telescope area facing the wind. This is done by applying an aerodynamic attenuation factor, χ_a, which is given by

$$\chi_a^2(\nu) = \frac{1}{1 + (2\nu\sqrt{A}/V)^{\frac{4}{3}}}, \tag{7.26}$$

where A is the characteristic area facing the wind. For very large structures, the attenuation factor is quite significant, so that the influence of wind turbulence is negligible.

7.6.2 Effects of wind on telescope structure

Static and dynamic wind loads

Static wind load can be readily estimated from equation 7.21. As for dynamic wind load, the PSD of the wind torque around a given axis, S_τ, is given by

$$S_\tau(\nu) = \tau^2 S_V(\nu) \chi_a^2(\nu), \tag{7.27}$$

where τ is the static wind torque around that axis, $S_V(\nu)$ is the wind speed power spectrum density from equation 7.24 or 7.25, and $\chi_a(\nu)$ is the aerodynamic attenuation given by equation 7.26. As an example, Fig. 7.28 shows the PSD of wind torque on the altitude axis of the VLT. Multiplying this PSD by the square of the control system's rejection sensitivity supplies the PSD of angular rotation around that axis, and integrating this over frequency gives the rms squared of the pointing error [6].

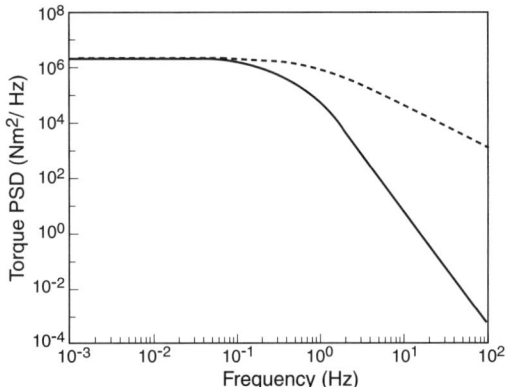

Fig. 7.28. Predicted PSD of wind torque on the altitude axis of the VLT. The dashed line represents the torque PSD without aerodynamic attenuation, and the solid line represents it with that factor taken into account. The plot shows a clear decrease of disturbance at frequencies higher than 1 Hz. The aerodynamic factor reduces disturbance even more.

Vortex shedding

The dynamic effects of vortex shedding on structural members can be estimated from equation 7.23. For cylindrical beams, the characteristic length is simply the diameter of the beam. In a telescope tube, the typical characteristic length (L) of the structural elements varies from centimeters to meters. Hence, vortex shedding frequencies for 10–20 m/s winds will typically be in the range of 1 to 100 Hz. This is also the range of the natural resonance of these elements, so that a careful check of this problem must be made to avoid undue resonances. The transversal force is given by

$$F = C_L \frac{\rho V^2}{2} A \sin(2\pi f t), \qquad (7.28)$$

where V is the wind speed, f is the vortex shedding frequency, and C_L is the lift coefficient, which varies from 0.8 for a Reynolds number of $\mathcal{R} = 10^4$ to 0.4 for $\mathcal{R} \geq 10^6$. Compared with equation 7.21, it can be seen that the oscillatory forces are on the same order as the static wind forces.

7.6.3 Effect of wind on primary mirror

A rough order of magnitude of the wind load on the mirror can be obtained from formula 7.21, where $C_D \simeq 1$ for a flat, circular disk and where V is the wind velocity in the dome at the level of the primary mirror. Inside a dome, the wind velocity is about an order of magnitude lower than that of the wind outside. As for the characteristic period of the wind force, it can be roughly estimated from the time it takes a wind gust to pass over the mirror. With a wind velocity of around 2 m/s at the level of the mirror, the main effect should be at a frequency of V/D, or about 0.2 Hz for a 10-meter telescope.

More precise data must be obtained from flow modeling or *in situ* measurements in order to take into account the attenuation by the dome and the scale of turbulence generated by the dome slit. Such measurements were done for the VLT using a 3.5-meter dummy mirror placed in the NTT dome (Fig. 7.29).

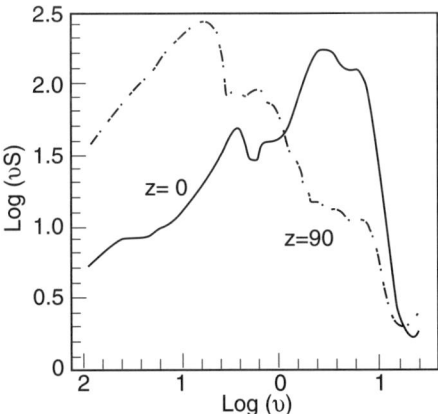

Fig. 7.29. Wind pressure power spectrum on a 3.5 m test mirror inside the NTT dome; mirror pointing to zenith (z=0) and at the horizon (z=90). The power spectrum depends strongly on the inclination of the mirror: the energy is shifted to higher frequencies when pointing to the zenith.

7.6.4 Effect of wind on telescope pier

Under the action of wind on the observatory building, the foundation soils deform and cause tilting of the telescope pier, thus affecting pointing performance during observations (Fig. 7.30). The telescope enclosure presents a very large projected area normal to the direction of the wind, resulting in wind forces in the 10^5–10^6 N range for an enclosure 30 meters in diameter. The associated overturning moment deforms the foundation soil and propagates to the telescope pier, causing image motion in the telescope. The effect can be estimated very roughly by assuming that the foundation soil behaves elastically [32, 33], but it is best determined by representing the soil as an elasto-plastic semi-infinite solid supporting the enclosure and pier foundation, then solving the problem by finite element computer analysis. Measures to reduce this effect are discussed in Chapter 11, Section 11.7.

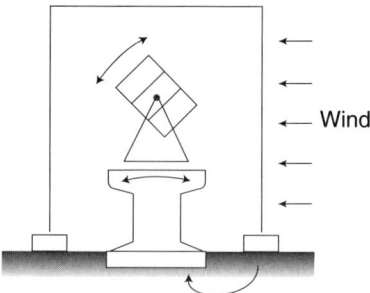

Fig. 7.30. Wind acting on the observatory building creates vibrations which are transmitted to the telescope via the soil and the pier foundations.

7.7 Disturbances in space

A space-borne telescope must contend with a variety of disturbances that potentially degrade both line-of-sight and imaging performance. These may be loosely classified as external disturbances and internal disturbances. External disturbances are the result of interaction with the space environment, and include gravity gradient torque, aerodynamic torque, solar pressure (which generally produces both a force and a torque disturbance), and magnetic torque. Internal disturbances arise from mechanisms aboard the spacecraft such as momentum wheels, thrusters, gyroscopes, filter wheels, and tape recorders, or from the release of strain energy at structural interfaces (joints, latches, hinges) during "thermal snap" events.

External disturbances will not normally degrade either the line-of-sight or the imaging performance of a space telescope. These disturbance torques are either constant or periodic with very low frequencies and are thus well within the bandwidth of a practical attitude control system. Such torques act on the spacecraft, but the resulting momentum is absorbed into the reaction wheels rather than into the body.

Internal disturbances are another matter. These are broadband excitations resulting from impulsive phenomena or, in the case of reaction wheel disturbances, high-frequency and sinusoidal (or multitone) in nature. In either case, there will generally be disturbance components with frequencies outside the bandwidth of the attitude control system that will affect the optical performance of the telescope. These disturbances must be mitigated by using active optics or some form of vibration isolation or suppression.

7.7.1 Gravity gradient torque

Any nonsymmetrical object of finite dimensions in orbit around Earth is subjected to a gravitational torque because of variations in the Earth's gravitational force over the size of the object. This effect is illustrated in Fig. 7.31, in which a "dumbbell spacecraft," consisting of equal masses m_1 and m_2 connected by a rigid element, orbits the Earth at a mean radius of R_0.

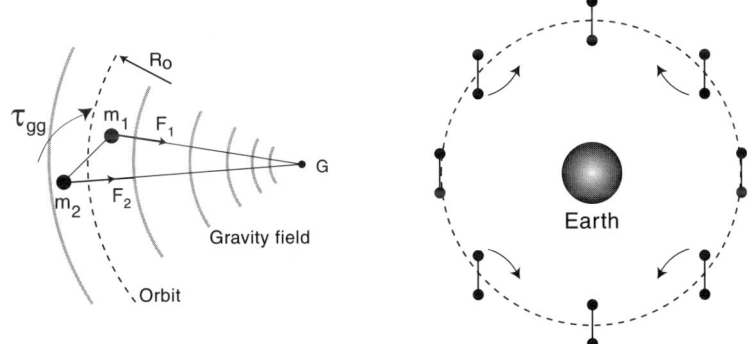

Fig. 7.31. The gravity gradient torque due to a spherical gravity field (left) creates an oscillatory torque on an orbiting spacecraft (right).

The force acting on mass m_1 is greater than that acting on mass m_2, the former being closer to the Earth's center of mass. This force differential produces a torque about the geometric center of mass of the dumbbell. In the absence of any external damping or attitude control system, this torque would result in a continuous oscillation of the dumbbell as it rotates around the Earth (Fig. 7.31, right). The gravity gradient torque is given by

$$\tau_{gg} = \frac{3\mu}{R_0^3}\left[(\vec{R}_0 \times (\mathbf{I}\cdot\vec{R}_0)\right], \qquad (7.29)$$

where μ is Earth's gravitational constant ($\mu = GM_\oplus$), $\vec{R}_0$ is the position vector from the Earth's center to the spacecraft's geometric center, and $\mathbf{I}$ is the moment of inertia tensor for the spacecraft. From this equation, we note that the torque is normal to the local vertical, vanishes for a spherically symmetric spacecraft, and is inversely proportional to the cube of the geometric distance from Earth. The gravity gradient torque is a major torque for spacecraft in low Earth orbit, but it becomes insignificant for high orbits.

7.7.2 Aerodynamic torque

Aerodynamic disturbance torques in spacecraft in low Earth orbit are caused by drag in the Earth's upper atmosphere. The magnitude of this effect varies widely since atmospheric density varies with the seasons, altitude, latitude, and solar activity (11-year cycle). Typically, this disturbance is only a factor at altitudes lower than 500 km, where it is generally the dominant external disturbance. For example, HST has to be reboosted ever few years, especially during the peak of the solar cycle.

The drag produces a torque due to the offset of the center of gravity and the aerodynamic center of pressure:

$$\tau_{\text{aero}} = -\frac{1}{2} C_D \rho V^2 \int \vec{r} \times (\vec{N} \cdot \vec{V}) \vec{V} dA, \qquad (7.30)$$

where C_D is the drag coefficient, ρ is the atmospheric density, V is the scalar velocity, $\vec{r}$ is the vector from the spacecraft's center of mass to the elemental area dA, $\vec{N}$ is the outward normal of the elemental area, and $\vec{V}$ is the velocity vector.

Closed-form solutions to this equation exist for simplified geometric shapes such as spheres, planes, and right cylinders. For complex geometric shapes, the aerodynamic torque must be evaluated by resorting to numerical solutions.

7.7.3 Solar radiation torque

The surface of a spacecraft is subjected to radiation pressure from sunlight, both direct and reflected from Earth. The force per unit area is equal to the vector difference between the incident and reflected photon momentum fluxes. Integrated over the area, the resultant force acts to perturb the spacecraft's trajectory, but this has no effect on the pointing control system. What does matter, however, is that this force will not generally pass through the center of mass of the spacecraft (Fig. 7.32). Consequently, a torque is produced about the center of mass that must be compensated by the control system. For spacecraft in high orbit, where gravity gradient and atmospheric drag are negligible, solar radiation is the dominant disturbance torque.

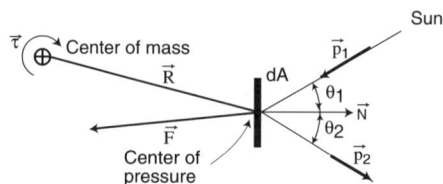

Fig. 7.32. Solar radiation pressure model.

The mean momentum, p, acting on a surface normal to the impinging radiation is given by

$$p = \frac{F_p}{c}, \qquad (7.31)$$

where F_p is the radiation flux and c is the velocity of light. F_p is the sum of the radiation flux from the Sun and from Earth (either reflected or emitted). Near Earth, the radiation flux from the Sun is the so-called solar constant, which is 1358 W/m². The contribution of Earth radiation is only important in low Earth orbits: it is about 700 W/m² at 500 km altitude but is negligible above 15 000 km.

The effect of solar pressure on a spacecraft is calculated as follows. Photons strike an elementary unit area dA at an angle of incidence θ_1 with momentum p_1. Some photons reflect specularly but others are absorbed or scattered. Denoting the coefficients of absorbed, specularly reflected, and diffusely reflected radiation as C_a, C_s, and C_d, respectively ($C_a + C_s + C_d = 1$), the forces for each of these three cases are

$$d\vec{F}_a = -p\, C_a \cos\theta_1 \vec{S} dA, \qquad (7.32)$$
$$d\vec{F}_s = -2p\, C_s \cos^2\theta_1 \vec{N} dA, \qquad (7.33)$$
$$d\vec{F}_d = -p\, C_d \cos\theta_1 \left(\frac{2}{3}\vec{N} + \vec{S}\right) dA, \qquad (7.34)$$

where $\vec{S}$ is the vector from the unit area to the Sun and $\vec{N}$ is the normal to the surface (in the third equation, diffusion is assumed to be Lambertian) [34]. The net force on the spacecraft is then the integral of the sum of these terms over the area of the spacecraft. The corresponding torque must be counteracted by the spacecraft's attitude system on a continuous basis. The effect is somewhat averaged out as the telescope is pointed from target to target, but will generally need to be canceled from time to time to avoid reaction wheel saturation, a procedure referred to as "momentum dumping" (see Section 7.4.2).

7.7.4 Magnetic torque

Magnetic disturbance torques are only a factor in low Earth orbit. They result from the interaction between the magnetic field of the Earth and that of the spacecraft. The spacecraft's magnetic field results from permanent or induced magnetism in its structure and from current loops generated in the electronics.

The Earth's magnetic field, $\vec{B}$, can be represented by a series of spherical harmonics and is determined using well-established computer codes. The magnetic torque acting on the spacecraft is given by

$$\tau_{\text{mag}} = \mathbf{m} \times \vec{B}, \qquad (7.35)$$

where $\mathbf{m}$ is the magnetic moment of the spacecraft.

7.7.5 Reaction wheel disturbances

With the possible exception of cryocoolers and instrument mechanisms, the only parts moving in a space telescope during an observation are gyroscopes and reaction wheels. The disturbance created by gyroscopes is totally negligible, but that of the reaction wheels is not. Indeed, as a rule, the reaction wheels are the dominant source of disturbance.

Although extremely well balanced and rotating on magnetic bearings, reaction wheels have rotors with masses of several kilograms turning at speeds up to 3000 rpm, so that even minute rotor imbalances can create disturbances on the order of milli-gs. As the wheel speed sweeps through its operational range, wheel vibrations will be amplified by the resonant dynamics of the structure and cause jitter in the line of sight. The effect is temporary, lasting only as long as the exciting frequency matches a resonant frequency in the structure, but the jitter level may still be unacceptable. Figure 7.33 shows the line-of-sight jitter predicted for NGST as a function of wheel speed. This preliminary analysis suggested that, unless some isolation was provided, jitter would reach values 5 to 10 times higher than the acceptable level.

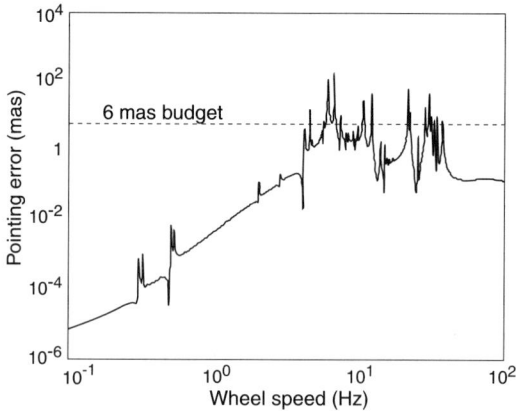

Fig. 7.33. Prediction of NGST's line-of-sight jitter due to reaction wheel disturbances with reaction wheels identical to those on HST. To meet a 6 mas maximum jitter specification at all wheel speeds, the reaction wheels must be about 10 times quieter than those of HST. This can be achieved by passive isolation.

Wheel imbalance can be modeled to the first order by small lumped masses located on the outer rim of the rotors (Fig. 7.34) [35, 36]. As the wheels rotate at a given angular velocity ω, the centrifugal forces developed in these small masses produce reaction forces and torques which are, in turn, transferred to the spacecraft structure through the wheel mounts.

These forces and torques are sinusoidal, with frequency ω and amplitude proportional to ω^2. The proportionality constants for the forces and torques corresponding to the two types of imbalance shown in Fig. 7.34 are referred to

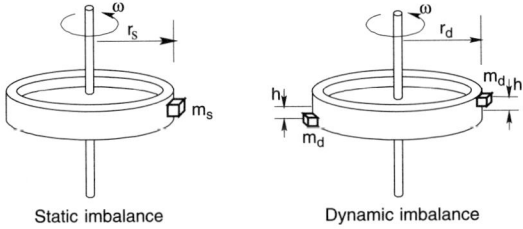

Fig. 7.34. Imbalance of a spinning reaction wheel: static (left) and dynamic (right).

as the "static imbalance" and "dynamic imbalance" coefficients. The disturbances may be modeled as summations of sinusoidal terms [37]. For a wheel spinning about the z axis, the radial force vector is given by

$$F = m_s r_s \omega^2 (x + iy) e^{i\omega t} . \qquad (7.36)$$

The wobble moment due to the principal moment-of-inertia axis not being aligned with the spin axis is given by

$$M = m_d r_d h \omega^2 (x + iy) e^{i(\omega t + \phi)} . \qquad (7.37)$$

The static and dynamic imbalances are both once-per-revolution terms, although they can be offset from each other by some phase angle ϕ, typically unknown. In addition to these imbalance terms, numerous other repetitive forces arise at various subharmonics and superharmonics of the fundamental spin rate. Imperfections in the bearings, the bearing race, and the cage that holds the bearings in relative spacing all contribute to these harmonics. As an example, Fig. 7.35 shows the axial and radial disturbance forces of an HST wheel measured at each wheel speed harmonic [38].

These disturbances are centrifugal in nature, hence their amplitudes vary as the square of the spin rate. Thus, the force in a given harmonic, m, is of the form

$$F_m = c_m \left(\frac{\omega}{2\pi}\right)^2 (x + iy) e^{im\omega t} . \qquad (7.38)$$

One additional source of disturbance is internal resonance in the reaction wheel itself. Should the spin rate or an excited harmonic of the spin rate coincide with a natural resonance frequency of the wheel, transmitted vibrations can be amplified. The first mode is typically in the range of 60–80 Hz and may increase to over 100 Hz once mounted on an isolator. Thus, for spin rates below about 70 rps, the fundamental imbalances will not, by design, excite this resonance. The higher harmonics will, however, and they can be very strong, as was the case with HST. Vibration isolation then becomes necessary.

300   7. Pointing and Control

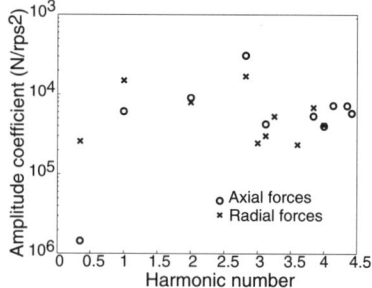

Fig. 7.35. HST reaction wheel force coefficients.

7.7.6 Other internally generated disturbances

Appendages

Appendages such as solar arrays, boom-mounted antennas, and sunshields are important sources of disturbance because of their large size and high leverage. Any disturbance generated within them due to inertial effects, mechanism motion, or thermal effects can excite fundamental modes in the telescope structure and result in sizable line-of-sight jitter. As an example, HST's original solar arrays used to vibrate at each entry into, or exit from, the Earth's shadow due to thermal snaps, causing up to 30 mas peak-to-peak of jitter lasting several seconds (Fig. 7.36).

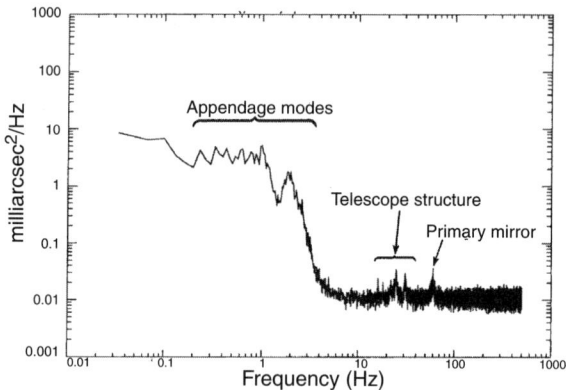

Fig. 7.36. Power spectrum of the line-of-sight jitter of HST with the original solar arrays during day/night transitions. Most of the power is under 5 Hz and is due to appendages vibrating (solar arrays, aperture door, and antennas). This problem was corrected by the installation of a new set of less sensitive solar arrays during the first servicing mission. The modes in the 15–30 Hz range are due to vibration of the telescope structure, and the mode at 60 Hz results from vibration of the primary mirror on its support points.

Clearly, the best way to avoid exciting these appendages is to design their deployment hinges and mechanisms so they will not be a source of disturbance. When this is not entirely possible, two basic approaches can be taken to mitigate the effects. The first is to the design the structure so that its resonance frequencies are high enough above those of the disturbances to avoid being excited.

The second approach is, on the contrary, to lower the structural modes so that they fall within the controller bandwidth, and use the controller to attenuate the effect.

Mechanism motion

Even minute movement of mechanisms inside the telescope can generate substantial line-of-sight jitter. Common culprits are tape recorders and filter wheels. As an example, Fig. 7.37 shows the disturbance created by the filter carousel of one of the science instruments and by the capstan mechanism of the tape recorder on HST. As a rule, it is best to avoid instrument mechanism motion during observations.

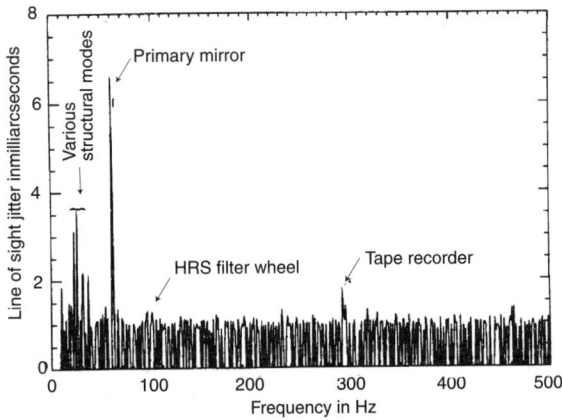

Fig. 7.37. Spectrum of disturbances in HST's line of sight measured during a dedicated test where specific disturbances were triggered. The several vibration modes of the telescope structure in the 10 to 20 Hz range and the primary mount rocking at 60 Hz are the same as in Fig. 7.36. Also visible, albeit at a lower level, are the vibrations created by the rotation of the filter wheel of the high-resolution spectrograph and the capstan of the tape recorder. (From Ref. [39].)

Fuel slosh

When a spacecraft is equipped with liquid fuel for orbital station keeping, the fuel can be a source of disturbance as it sloshes in its tank following a major telescope slew. This effect can be much reduced by installing baffles in the tanks. Slosh then typically damps out within a few minutes.

Cryocoolers

Telescopes operating in the infrared typically require a focal plane chilled well below the temperature of the light-gathering optics. If one is willing to accept a limited operating life, the cooling can be provided by a dewar with a consumable cryogen such as liquid helium. A longer-lived alternative is a continuously operating cryocooler employing one of a number of thermodynamic cycles such as turbo-Brayton, pulse-tube or Sterling. The drawback is that the compressor and other moving parts are sources of vibration. Coolers are typically designed with this in mind. For example, a piston compressor could be designed with two pistons back-to-back, thus canceling the fundamental imbalance force. But active control can enhance vibration cancelation, for example, by sensing the residual in-line acceleration or force, then adjusting the amplitude and phase of the stroke of one piston to minimize residual disturbances. With active force cancelation, the residual disturbance can be as low as 0.1 N in each of the harmonics of the cooler drive rates, which are typically in the 40–60 Hz range. An alternative is to use turbine compressors. They operate at much higher rates (e.g., 300 000 rpm), so that residual vibrations will generally be beyond the sensitivity range of the telescope.

7.8 Active and passive vibration control

Although the combination of modern feedback control systems and active optics techniques effectively rejects most disturbances, their bandwidth is limited. Disturbances with frequencies beyond their reach must be dealt with by other methods. The solutions fall into five categories [40]:

- **Tuning** of the telescope structure so that resonances fall outside of the excitation spectrum; this is normally done as part of the design.
- **Absorption** of the disturbance by a proof mass absorber. This technique has been used in tall buildings to minimize sway from wind and has been proposed for reducing the motion of secondary mirror towers in very large telescopes.
- **Compensation** of the disturbance at the source by moving a mass in the opposite direction to cancel the effect. This method is used to minimize the disturbance created by chopping secondary mirrors and mechanical coolers,
- **Damping** of the motion created by the disturbance by means of passive damping material or active structural elements. These techniques have been proposed to minimize motion in large space structures such as those

proposed for interferometry missions; this is a solution of last resort, as it involves correcting a widely distributed effect.
- **Isolation** of the source of the disturbance from the rest of the telescope. Passive isolation is systematically used in space telescopes to minimize vibration created by reaction wheels; active isolation has been proposed for additional improvement.

When feasible, this last solution is the most logical because it solves the problem at the source instead of correcting unwanted effects downstream. It has received a lot of attention and we now examine it in more detail.

7.8.1 Passive isolation of the vibration source

Passive isolation is typically accomplished by introducing a soft spring and damper between the source and the host. The transmissibility function for such a simple isolator at a frequency f is

$$T(f) = \frac{F_{\text{transmitted}}}{F_{\text{input}}} = \frac{1 + 2i\zeta_0 f/f_0}{1 - (f/f_0)^2 + 2i\zeta_0 f/f_0}, \quad (7.39)$$

where f_0 is the resonant frequency of the isolator and ζ_0 is the damping ratio. The squared term in the denominator drives down transmissibility at frequencies greater than the isolator's resonant frequency. In this inertia-dominated regime, the force generated by the disturbance is resisted by the mass of the vibrating part, such that transmissibility rolls off as $1/f^2$. The damping ratio plays a role by limiting force amplification by any disturbance acting at the isolator's resonant frequency. One typically designs the isolator so that its resonance is well below the anticipated frequency content in the disturbance. The drawback of heavy damping is that the viscous term at the numerator acts to stiffen the isolator at high frequency. Thus, at high frequencies, roll-off becomes only $1/f$. This viscous damping lockup is avoided by constructing a compound spring with a second elastic flexure in series with the damper. The compound spring approach was employed in both HST and the Chandra telescope and are described below.

HST and Chandra isolators

The HST wheel isolator is shown in Fig. 7.38. Its function is to block undesirable high-frequency vibrations from the wheel while allowing the low-frequency attitude control torques to pass through. The isolator consists of six elements with viscous damping fluid constrained to move between two bellows [41]. A third bellows is provided for thermal expansion of the fluid, as well an outer skirt to prevent fluid from escaping in the event of a leak. The springs are tuned to give a 20 Hz axial isolation frequency. They are arranged to be compliant for axial motions, but stiff in the spin direction to preserve reaction wheel command authority. Using the compound spring approach, the

304   7. Pointing and Control

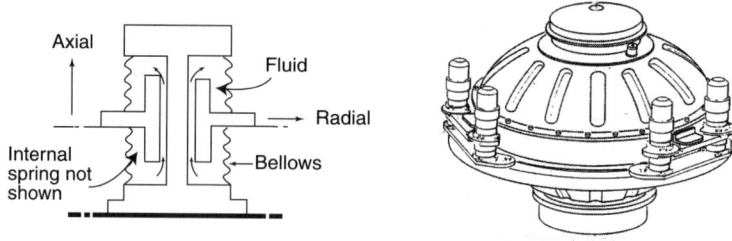

Fig. 7.38. At left, principle of the reaction wheel isolator used on HST. At right, a view of the reaction wheel mounted on its isolators.

damping in the isolator, $Q=1/2\zeta$, is about 5. The transmitted axial force is reduced from 15 N at the worst wheel speed in the hard-mount case to 0.1 N with the isolator. In addition to attenuating low-level vibrations on-orbit, the isolator was useful in attenuating launch vibration transmitted to the wheel by nearly an order of magnitude.

The Chandra reaction wheel isolator consists of six dampers in a hexapod configuration [42], as shown on the left in Fig. 7.39. Damping is provided by wafers of a viscoelastic material which is sheared between two plates. The advantage of the viscoelastic damper is that, being solid, it obviates the need for the complex bellows used on HST to prevent leaks. Since this isolator is significantly softer than HST's, compliant urethane bumpers were used to limit stroke during launch vibrations. Thus, an effective 40 Hz isolation was achieved during launch, softening the ride for the reaction wheel. The performance of the isolator is shown on the right in Fig. 7.39.

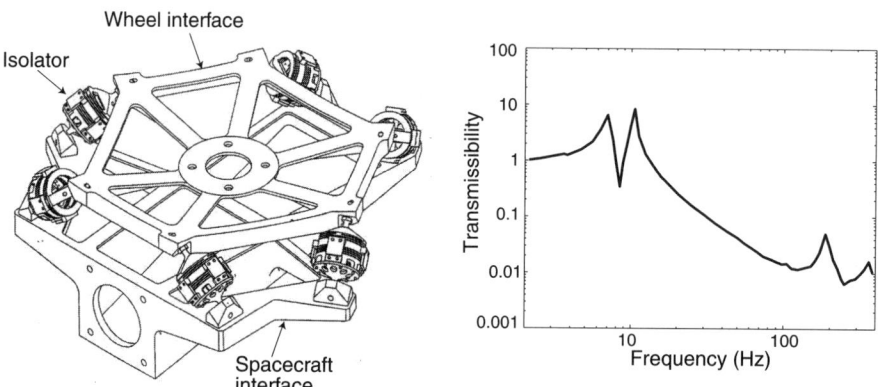

Fig. 7.39. At left, a general view of the Chandra wheel isolator. The radial force transmissibility of the system is shown on the right.

Two-stage isolation

When a single isolator is not sufficient, two can be stacked in series. But care must be taken to employ a mass ratio so that the first and second stages will not couple (i.e., coalesce into a single compliance). The isolated mass of the second stage, onto which the first isolator stage is mounted, should be at least 5 to 10 times greater than the first isolated mass. Figure 7.40 shows the transmissibility curve for a single 4-Hz isolator stage compared to that of dual 4-Hz stages. It is apparent that considerable advantage is gained by the two-stage approach at frequencies over roughly twice the isolation frequency, where a dual 4-Hz isolator is equivalent to a single 2-Hz isolator. To implement two-stage reaction wheel isolation, one could mount each wheel on an individual isolator, then place a cluster of wheels on a further isolated pallet.

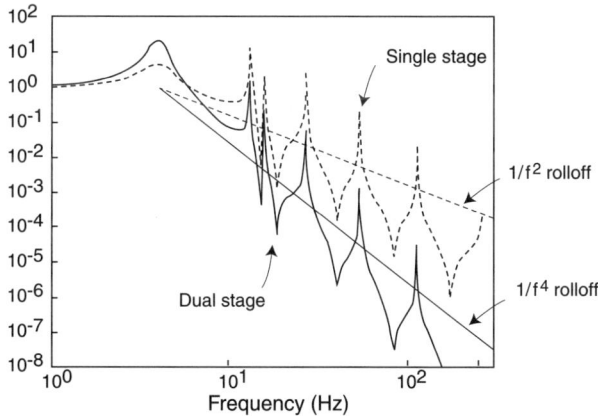

Fig. 7.40. Transmissibility of single and dual 4-Hz isolator stages.

7.8.2 Active isolation

Active isolation consists of sensing the disturbance force and applying an opposite force to counteract it. Figure 7.41 shows a prototype of an active isolation system designed to isolate a vibration source in all six degrees of freedom. It employs six active struts arranged in a mutually orthogonal hexapod configuration. Each strut is composed of a voice coil actuator in parallel with a soft spring element. The performance of the device, shown in Fig. 7.41 (right), illustrates the large improvement afforded by active isolation over passive systems.

An additional advantage of active isolation systems compared to passive ones is that they can be turned off when isolation is not needed. This is a useful feature for reaction-wheel isolation, as it avoids impeding the transmission of attitude torques during repointing. It also allows the wheels to be locked during the spacecraft's launch phase.

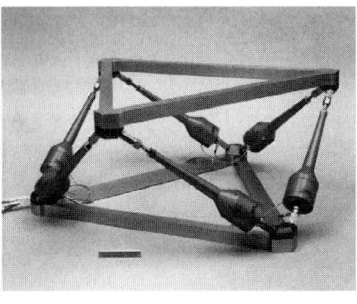 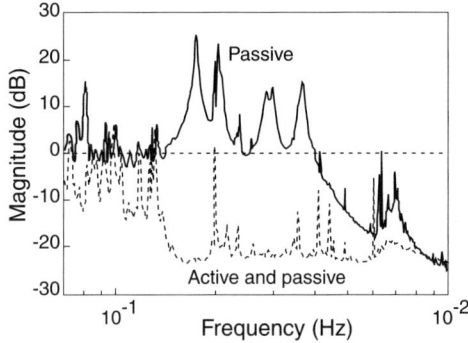

Fig. 7.41. At left, a prototype for a six-axis active isolation mount designed and built by JPL. At right, experimentally measured six-axis effective transmissibility of the mount. In the 7–100 Hz range, isolation improves by a factor of 10 when the active system is turned on.

7.9 Observatory control software

The pointing control system in charge of telescope slewing and tracking is now implemented in software and is part of a control system of much broader scope that manages the entire observatory.

For space observatories, the observatory control system is a fully integrated and extremely complex software system used to manage all scientific and engineering operations of the observatory in a seamless manner: scientific observation proposals, observation scheduling, uplinking of commands, downlinking of scientific and engineering data, scientific data calibration and archiving, and so forth. Figure 7.42 gives a schematic view of such a system.

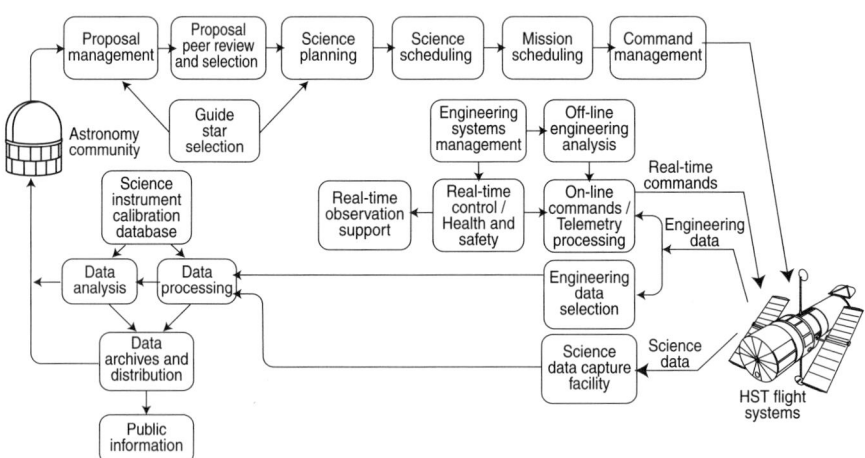

Fig. 7.42. Hubble Space Telescope operations flow.

The software itself is divided into two subsystems: "flight software," which resides onboard the spacecraft, and "ground software." Space observatories are generally operated in a "preplanned" fashion, where the sequence of observations is prepared in full detail on the ground and then uplinked to the spacecraft on about a weekly basis. The observatory then runs autonomously and is capable of working around missed observations and coping with anomalies without intervention from the ground. In case of a serious failure, the spacecraft can also direct itself into a dormant secure state (called "safe mode") to await investigation and reprogramming from the ground.

Without quite reaching this level of complexity and autonomy, control systems for large ground-based observatories are becoming increasingly sophisticated. A number of tasks that were formally executed by the "night telescope operator," such as opening the dome shutter and ventilation louvers, rotating the dome, mirror cover opening, turning on oil pumps, telescope slewing, and guide star acquisition, are now fully automated or semiautomated.

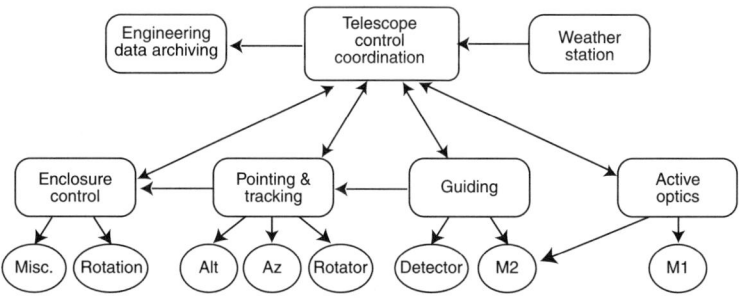

Fig. 7.43. Ground-based telescope control coordination.

The main coordination tasks of a modern ground-based telescope control system, illustrated in Fig. 7.43, are as follows:

- *Enclosure control.* This consists of rotating the enclosure and moving the shutter and windscreen up or down to follow telescope tracking. This subsystem is also typically in charge of airconditioning the telescope chamber during the day and automatically adjusting louvers and windscreen porosity at night as a function of wind and temperature conditions.
- *Telescope pointing and tracking.* This core observatory control task consists of slewing the telescope to the desired target and tracking it for the required exposure time. It involves converting the astronomical coordinates of the target into values for the altitude, azimuth, and focus rotator axis, and then applying the necessary corrections such as those for atmospheric refraction, gravity deflections, misalignment, and so forth.

- *Guiding.* This involves acquiring and locking on guide stars, applying the necessary corrections for flexure and atmospheric refraction, and interacting with the pointing control system as required.
- *Active optics.* This deals with the control of actuators on the primary mirror and the alignment of the secondary mirror to compensate for gravity and thermal effects.
- *Monitoring and engineering data archiving.* This task consists of monitoring the telescope for "health and safety" and logging pertinent data both during normal operation and during testing.
- *Weather.* This consists of monitoring weather data such as wind speed, wind direction, temperature, and humidity to serve as input to the observatory thermal control and pointing corrections.
- *Time.* Accurate pointing and tracking of the telescope requires a time source with an accuracy approaching the millisecond level. Time is typically acquired through the GPS system.

References

[1] Sirota, M.J. and Thompson, P.M., Azimuth/elevation servo design of the W. M. Keck telescope, SPIE Proc., Vol. 887, p. 168, 1988.
[2] Nurre, G., Anhouse, S.J., and Gullapalli, S.N., Hubble Space Telescope fine guidance sensor control system, SPIE Proc., Vol. 1111, p. 327, 1989.
[3] Sirota, M.J., Thompson, P. M., and Jex, H. R., Azimuth/elevation servo performance of the W. M. Keck telescope, SPIE Proc., Vol. 2199, p. 126, 1994.
[4] Campbell, M.F. and Elsaie, A.M., Structural optimization and modeling of large dynamic structures for controls simulation, SPIE Proc., Vol. 4004, p. 320, 2000.
[5] Quattri, M., Zago, L., and Plötz, F., Design evolution and performance evaluation of the VLT telescope structure, ESO Conference on *Very Large Telescopes and Their Instrumentation*, Ulrich, M.-H., ed., 1988, p. 127.
[6] Ravensbergen, M., Main axes servo systems of the VLT, SPIE Proc., Vol. 2199, p. 997, 1994.
[7] Huang, E., *Line-of-sight sensitivity equations*, Gemini Report TN-O-G0017, 1992.
[8] Bay, J.S., *Fundamentals of Linear State Space Systems*, McGraw-Hill, 1998.
[9] White, C.V. and Levine, M.B., *Experiments to measure material damping at cold temperatures*, Report JPL D-22047, Jet Propulsion Laboratory, 2001.
[10] Landau, L.D. and Lifshitz, E.M., *Theory of Elasticity*, Pergamon Press, 1986, Chap. 4.
[11] Friedland, B., *Control Systems Design*, McGraw-Hill, 1985.
[12] Dougherty, H. et al., Space Telescope pointing control system, J. Guidance, Control Dynam., Vol. 5, No. 4, p. 403, 1982.
[13] Kalman, R.E., A new approach to linear filtering and prediction problems, Trans. ASME, J. of Basic Eng., Vol. 82, p. 35, 1960.

7.9 Observatory control software 309

[14] Brown, R.G. and Hwang, P.Y.C., *Introduction to Random Signals and Applied Kalman Filtering*, John Wiley & Sons, 1992.

[15] Ravensbergen, M., ESO unpublished proposal.

[16] Dahl, P.R., Solid friction damping of spacecraft oscillations, AIAA Guiding and Control Conference, Paper 75-1104, 1975.

[17] Bertin, B., Driving the French 2 m and 3.60 m telescopes from the horseshoe, ESO Conference on *Large Telescope Design*, p. 405, 1971.

[18] Young, W.C., *Roark's Formulas for Stress and Strain*, McGraw-Hill, 1989, p. 647.

[19] Ellington, S., Disturbance rejection of the WTYN telescope position control servosystem, SPIE Proc., Vol. 2479, p. 278, 1995.

[20] Wilkes, J., Fisher, M., Selection of a tape encoding system for the main axis of the Gemini telescopes, SPIE Proc., Vol. 3112, p. 30, 1997.

[21] Erm, T. and Gutierrez, P., Integration and tuning of the VLT drive systems, SPIE Proc., Vol. 4004, p. 490, 2000.

[22] Dougherty, H., Rodden, J., Reschke, L.F., Trompetini, K., Weinstein, S.P., and Slater, D., Performance characterization of the Hubble Space Telescope rate gyro assembly, Preprints 10th IFAC Symposium, Toulouse, p. 217, 1985.

[23] Cassidy, L. W. and Abreu R., Star trackers for spacecraft application, SPIE Proc., Vol. 1304, p. 58, 1990.

[24] Eaton, D.J. and Abramowicz-Reed, L., Acquisition, pointing and tracking performance of the Hubble Space Telescope fine guidance sensors, SPIE Proc., Vol. 1697, p. 236, 1992.

[25] Tyler, G.A. and Fried, D.L., Image-position error associated with quadrant detector, J. Opt. Soc. Am., Vol. 72, No. 6, p. 804, 1982.

[26] Hardy, J.W., *Adaptive Optics for Astronomical Telescopes*, Oxford Books, 1998, p. 398.

[27] Spagna, A., *Guide star requirements for NGST*, Space Telescope Science Institute Report STScI-NGST-R-0013B, 2001 (available on the STScI website).

[28] Bradley, A., Abramowicz-Reed, A., Story, D., Benedict, G., and Jeffrys, W., The flight hardware and ground system for Hubble Space Telescope astrometry, PASP, Vol. 103, p. 317, 1991.

[29] Murdock, J.W., *Fluid Mechanics and Its Applications*, Houghton Mifflin, 1976.

[30] Davenport, A. G., The spectrum of horizontal gustiness near the ground in high winds, Quart. J. Roy. Met. Soc., Vol. 87, p. 194., 1961.

[31] Simiu, E and Scanlan, R.H., *Wind Effects on Structures: An Introduction to Wind Engineering*, John Wiley & Sons, 1978.

[32] Mikami, I. et al., Enclosure of Subaru telescope, SPIE Proc., Vol. 2199, p. 430, 1994.

[33] Medwadowski, S.J., *UC telescope pier rotations due to wind action on the observatory dome*, Keck Report No. 53, 1984.

[34] Wertz, J. R., ed., *Spacecraft Attitude Determination and Control*, D. Reidel, 1986, p. 570.

[35] Masterson, R.A., Miller, D.W., and Grogan, R.L., Development of empirical and analytical reaction wheel disturbance models, AIAA *40th Structures, Structural Dynamics and Materials Conference*, AIAA-99-1204, 1999.

[36] Bialke, B., Microvibration disturbance sources in reaction wheels and momentum wheels, Proc. ESA Conference on *Spacecraft Structures, Materials & Mechanical Testing*, 1996.

[37] Melody, J.W., *Discrete-frequency and broadband reaction wheel disturbance models*, Jet Propulsion Laboratory Interoffice Memorandum 3411-95-200 csi, 1995.

[38] Neat, G.W., Melody, J.W., and Lurie, B.J., Vibration attenuation approach for spaceborne optical interferometers, IEEE Trans. Control Syst. Technol., Vol. 6, No. 6, p. 689, 1998.

[39] Bely, P.Y., Lupie, O.L., and Hershey, J.L., The line-of-sight jitter of the Hubble Space Telescope, SPIE Proc., Vol. 1945, p. 55, 1993.

[40] Baier, H. and Locatelli, G., Active and passive microvibration control in telescope structures, SPIE Proc., Vol. 4004, p. 267, 2000.

[41] Davis, L.P., Wilson, J.F., Jewell, R.E., and Rodden, J.J., Hubble Space Telescope reaction wheel assembly vibration isolation system, Proc. NASA Workshop on *Structural Dynamics and Control Interaction of Flexible Structures*, Marshall Space Flight Center, N87-22702 16-15, p. 669, 1986.

[42] Nye, T.W., Bronowicki, A.J., Manning, R. A., and Simonian, S.S., Applications of robust damping treatments to advanced spacecraft structures, Proceedings of the *19th Rocky Mountain Guidance & Control Conference, American Astronautical Society*, Advances in the Astronautical Sciences, Vol. 92, p. 531, 1996.

Bibliography

Den Hartog, J.P., *Mechanical Vibrations*, McGraw-Hill, 1957.

Germann, L.M., Gupta, A.A., and Lewis, R.A., Precision pointing and inertial line-of-sight stabilization using fine-steering mirrors, star trackers, and accelerometers, SPIE Proc., Vol. 887, p. 96, 1988.

Kaplan, M. H., *Modern Spacecraft Dynamics and Control*, John Wiley & Sons, 1976.

Katsuhiko O., *Modern Control Engineering*, Prentice-Hall, 1997.

Kuo, B.C., *Automatic Control Systems*, Prentice-Hall, 1985.

Wertz, J. R., ed., *Spacecraft Attitude Determination and Control*, D. Reidel, 1986.

8
Active and Adaptive Optics

For telescopes much larger than 4 meters, traditional measures for keeping the figures of mirrors undisturbed and the optical trains aligned become increasingly inadequate. Instead of relying on passive means to compensate for gravity and thermal effects, it becomes advantageous to use an "active system" to control the optical quality. Extending this concept to a higher degree of correction, wavefront distortion due to atmospheric seeing can also be corrected. This chapter is devoted to these techniques, which have applications both on the ground and in space.

8.1 Fundamental principles

8.1.1 Respective roles of active and adaptive optics

Active control of optics consists of automatically correcting defects in the optics by (1) monitoring the respective positions of the individual optical elements or measuring errors in the final wavefront and then (2) applying corrections by adjusting the position and figure of the optical components. This has several advantages.

Active control can be applied to maintain proper focus and optical alignment in spite of thermal and gravitational effects. In addition, active control can be applied to correct deformations of the primary mirror, thus permitting the use of thinner, hence lighter, mirrors supported on lighter, hence less expensive, mechanical mounts.

A related advantage is the ability to correct one of the most common optical defect in telescopes: spherical aberration. Testing a fast, large primary mirror at the center of curvature in an optical shop requires canceling a 2000-wave spherical aberration to an accuracy of one-tenth of a wave. This is extremely challenging. But if the primary mirror is equipped with active control, then its conical constant can be adjusted on the sky. It even becomes possible to correct a primary mirror for slight differences in f-ratio between the Cassegrain focus and the Nasmyth focus.

Active control of optics can also be used to coalign and cophase an array of mirror segments such as those of the Keck telescopes and NGST. In fact, it is active control that makes such segmented telescopes feasible.

A further advantage is the ability to correct for the effects of wind buffeting [1]. As will be seen in Chapter 11, the modern approach to dome-seeing control consists of letting wind flow across the mirror surface to maximize heat exchange and thus reduce temperature differences. The drawback is the effect of wind buffeting the telescope, an effect that can be compensated by active control if the frequency bandwidth is high enough. But why stop there? If the wavefront surface can be corrected quickly enough, then the effects of atmospheric turbulence can also be reduced or eliminated.

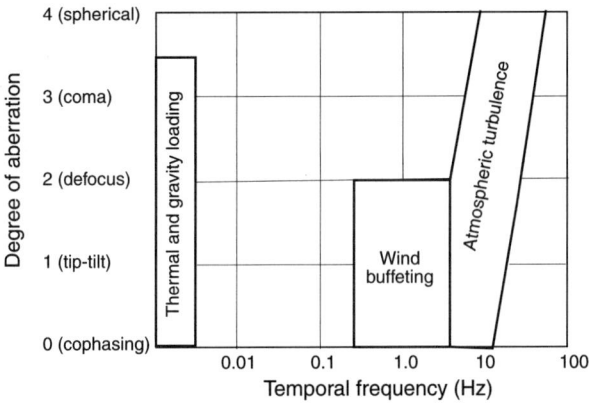

Fig. 8.1. Effects compensated by active and adaptive optics. For each effect, the range of aberrations to be compensated and the required frequency bandwidth are indicated.

Figure 8.1 sketches the range in spatial and temporal frequencies associated with each of these effects. Note the wide difference in frequency range between thermal and gravity effects and the others. Clearly, the different domains in this diagram will call for different correction techniques.

The terminology is still fluid, but *active optics* generally refers to the figure control of optical elements at low bandwidth (DC to a few Hertz) to correct residual aberrations and gravity and thermal effects, whereas *adaptive optics* refers to the correction of high-frequency wavefront disturbances (above a few

Hertz) such as those created by atmospheric turbulence. Wind disturbances straddle these two frequency domains, with the lower end potentially falling within the regime of active optics.

8.1.2 Active and adaptive optics architectures

The architecture of active control systems depends on the type and bandpass of the disturbance to be corrected.

When disturbances can be accurately predicted or measured, as is the case for gravity and thermal effects, "open-loop" control can suffice. Random or hard-to-measure disturbances such as mechanical excitations and atmospheric effects, on the other hand, require the ultimate control capability of "feedback" control systems. Feedback loops can be closed at the local level by using metrology on the element to be controlled, or at the system level by using starlight. These three approaches are schematically shown in Fig. 8.2.

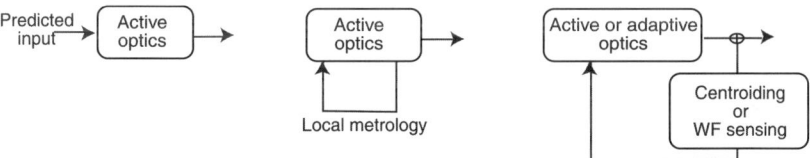

Fig. 8.2. The three basic approaches for active and adaptive optics: at left, open-loop control; at center, correction with local metrology; at right, system-level feedback control. Most active optics systems use a combination of the three methods. By nature, adaptive optics has to use system-level feedback.

The case of seeing correction is special because the "disturbance" is external to the system and embedded in the incoming light. In this case then, the only solution is a system-level correction using measurement of the wavefront errors on a reference star.

System-level feedback would seem to be the definitive answer. Why then bother with other types of correction? The answer is that the bandpass of the system level is limited by the flux of in-field stars or atmospheric turbulence, and thus may be too low. Local metrology, on the other hand, has no such limitation and can have a very high bandpass. But if a locally closed loop is used, why would a system-level loop be required? Because locally closed loops suffer from drift in time and need to be periodically reinitialized. Finally, why bother with open-loop correction when a feedback system must be used in any case for other reasons? The reason here is because open-loop correction easily removes large correction terms which could overload the more accurate feedback loops. In practice then, active optics systems will often consist of two or three layers:

- open-loop correction for predictable low-frequency disturbances such as gravity and temperature effects,

- a local-level closed loop on an internal metrology system for the correction of high-frequency disturbances such as wind or mechanical excitations,
- a system-level loop closed on starlight for periodic correction of errors or drift in the previous systems and for correction of seeing in ground-based telescopes.

Implementations differ depending on the frequency of the correction to be made and whether the main optics are passive or active. For low frequencies (a few Hertz at most), it is generally possible to make the correction directly on the main optics. High-frequency corrections can only be made on very thin optics and usually require a dedicated optical element. Two typical implementation approaches are shown schematically in Fig. 8.3.

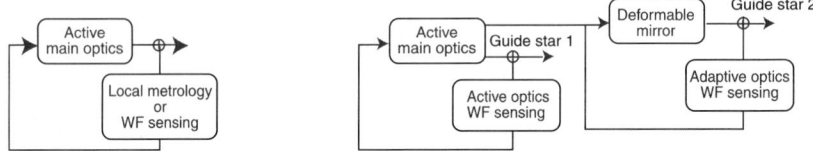

Fig. 8.3. Two basic implementations of active and adaptive optics. At left, the main optics are used for correction. At right, the main optics are active and corrected at low frequency with a wavefront sensor using a bright off-axis star, whereas an adaptive optics system using a star within the isoplanatic patch corrects for atmospheric seeing.

Actual implementations can incorporate additional loops depending on the frequencies of the correction and the type of sensing and correcting hardware. As an example, the line-of-sight jitter in systems affected by atmospheric turbulence extends over a wide range of frequencies, from DC to several tens of Hertz. Compensation here is best handled with two or more nested loops, allowing one to separate the correction of small-amplitude high-frequency disturbances from that of slow, large-amplitude drifts. As shown in Fig. 8.4, a deformable mirror corrects for high-frequency small-amplitude jitter, whereas a separate tip-tilt mirror handles mid-frequencies and the slow drifts are removed by the telescope pointing system.

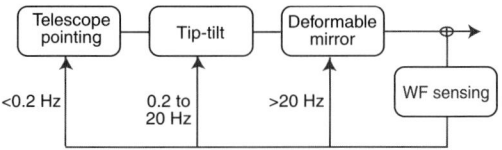

Fig. 8.4. Example of nested loops used for the compensation of line-of-sight jitter in a ground-based telescope.

The choice between local metrology and wavefront sensing methods depends on the frequency range of the disturbances to be corrected.

Wavefront sensing using the typically faint stars found in the field of the telescope can only provide error signals for slow corrections. This is adequate to correct for deformations affecting ground telescopes, since changing gravity and thermal effects have time scales on the order of several minutes.

This is also adequate for space telescopes, in general, because of the very benign environment they inhabit: no gravity, no wind, slow thermal transients. In the best cases, recalibration may be required only at intervals of several weeks.

On the other hand, optical systems sensitive to wind or to relatively fast thermal transients (e.g., segmented mirrors on ground telescopes), will need a higher correction bandpass. This can only be supplied by an internal metrology system.

Internal metrology systems measure the respective positions of optical elements in the optical train using position sensors or laser-based devices. They do not suffer from the flux limitation of the sky-based system and their bandwidth can be set higher than the expected disturbance bandwidth. However, as indicated earlier, they are not absolute, can only measure changes from a start position, and must, therefore, be initialized and calibrated by observation of a bright star.

8.2 Wavefront sensors

Wavefront sensing methods for active control of optics need to be very sensitive due to the flux limitations imposed by sky sources. These methods include derivatives of techniques used in optical shop testing described in Chapter 4 and a new technique, phase retrieval.

8.2.1 Shack-Hartmann sensor

The Shack-Hartmann sensor is an evolution of the Hartmann test. In order to benefit from the collecting area of the full telescope aperture, the telescope entrance pupil is reimaged on a lenslet array instead of a mask [2]. Each lenslet produces a star image forming the equivalent of a Hartmann pattern (Fig. 8.5) which can be conveniently recorded on a CCD detector. The position of the centroid of each lenslet image compared to a reference supplies the slope of the wavefront (or wavefront tilt) at the location in object space corresponding to the lenslet. Calibration of the system is done with a reference plane wave.

Shack-Hartmann sensors, being both compact and rugged, have become standard devices for measuring wavefront slope errors. But since they measure only tilts, not phase errors, they cannot be used to measure wavefront errors in a segmented mirror system, at least not directly. Shack-Hartmann sensors can

316   8. Active and Adaptive Optics

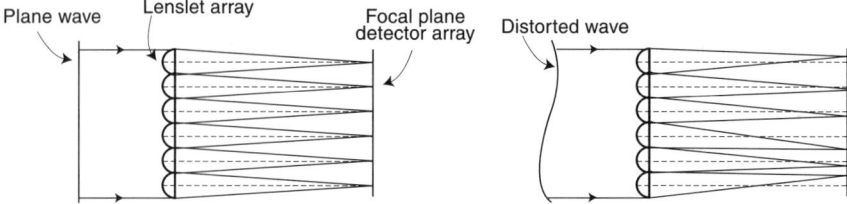

Fig. 8.5. Shack-Hartmann wavefront sensor. The lenslet array is in a plane conjugate to the telescope entrance pupil. Each lenslet forms a star image on the detector array. By comparison with a reference plane wave (left), a distorted wave produces a lateral shift of the images (right). The shift is proportional to the wavefront slope averaged over the lenslet area.

be used for both active and adaptive optics applications, the main difference being in the frequency at which the wavefront must be sampled. For adaptive optics, the sampling interval must be on the order of a millisecond instead of a minute.

These sensors have the advantage of working with broadband light, thus making use of a large number of photons. This is particularly important for adaptive optics. Wavefront distortions introduced by the atmosphere are, to a good approximation, achromatic, so they can be measured over a wide bandwidth. Indeed, most adaptive optics systems measure the wavefront distortion in the visible to compensate images in the infrared. Furthermore, since this is an incoherent light-sensing technique, extended sources such as small nebulosities, galaxy cores, small planets, or asteroids can be used. The main drawback of Shack-Hartmann sensors in adaptive optics applications is that their wavefront tilt sensitivity is fixed by design and cannot be changed to accommodate different seeing conditions.

8.2.2   Curvature sensing

The curvature sensing method is yet another substitute for the Hartmann test. It consists of recording the illumination in defocused stellar images [3]. The procedure is essentially the same as in the classical Hartmann test, but no mask is used. Also, instead of recording a single long exposure, either intrafocal or extrafocal, two exposures are made to improve accuracy, one on either side of the focal plane. The distance to focus must be large enough for the detector to be outside the caustic zone.[1] Intensity variations observed in the defocused images are opposite on each side of the focal plane (Fig. 8.6). By taking intrafocal and extrafocal images, one doubles the measurement accuracy while reducing the deleterious effects of nonuniform pupil illumination.

---

[1]The caustic zone is the zone where rays near focus intersect; see glossary.

To a first-order approximation, these intensity variations reflect variations in the wavefront's total curvature (Laplacian) [4].[2]

This method is referred to as *curvature* sensing, in contrast with the Shack-Hartmann *slope* sensing method.

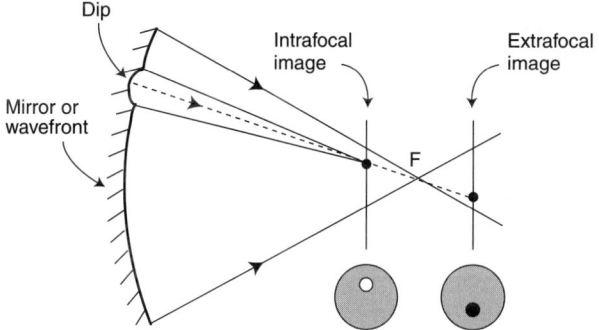

Fig. 8.6. Principle of the curvature sensor. A change in local curvature in the mirror surface or wavefront (e.g., more concave) makes the outgoing pencil converge more than nominal, resulting in an increase of illumination in the intrafocal image and a decrease in the extrafocal image.

The wavefront surface can be reconstructed from its Laplacian, provided that boundary conditions are available. These are given by the location of the beam edge. Deviations of the edge, measured in the radial direction, map the wavefront radial slopes at the pupil edge and provide appropriate boundary conditions to solve the Poisson equation describing the wavefront surface:

$$\rho(x,y) = \nabla^2 \Phi, \qquad (8.2)$$

where x and y are two perpendicular coordinates, $\rho(x,y)$ is the local wavefront curvature, and $\Phi(x,y)$ is the wavefront phase.

Like the Shack-Hartmann method, curvature sensing benefits from the utilization of the full telescope aperture. But unlike the Shack-Hartmann sensor, curvature sensing retains the advantages of the original Hartmann method, in the sense that no critical additional optical element is used. It also has the advantage that the spatial sensitivity can be adjusted at will. The greater the defocus, the higher the spatial resolution on the wavefront (at the price of

---

[2] The propagation of the electromagnetic field in the near field is defined by the transport equation

$$\nabla I \cdot \nabla \Phi + I \nabla^2 \Phi + \frac{\partial I}{\partial z} = 0, \qquad (8.1)$$

where I is the intensity, Φ is the phase, and z is the coordinate along the propagation direction and where $\nabla = \frac{\partial}{\partial x} + \frac{\partial}{\partial y}$ is the gradient and $\nabla^2 = \frac{\partial^2}{\partial x^2} + \frac{\partial^2}{\partial y^2}$ is the Laplacian. $\nabla^2 \Phi$ is the second derivative of the wavefront, the *wavefront curvature*. $\nabla I \cdot \nabla \Phi$ is the wavefront tilt, which is only important at the edge of the pupil.

lower sensitivity to aberrations). A typical implementation for adaptive optics is shown in Fig. 8.7.

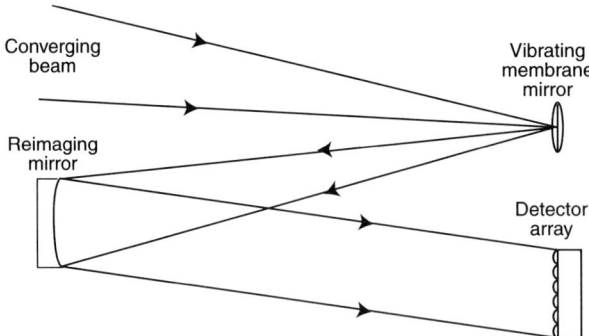

Fig. 8.7. Adaptive optics implementation of wavefront curvature sensing. An image of the telescope entrance aperture is formed on the detector array by a concave reimaging mirror. The pupil image is sequentially defocused in and out at a rate of about a kilohertz by a vibrating membrane mirror located in the stellar image plane.

Compared to Shack-Hartmann sensing, curvature sensing is less sensitive to sky background. One disadvantage of curvature sensing is error propagation in the wavefront reconstruction. But wavefront distortions shrink when the loop is closed and the sensitivity of the curvature sensor can be increased accordingly, whereas that of the Shack-Hartmann cannot. As shown analytically and by computer simulations, this increase in sensitivity can more than compensate the effect of error propagation [5].

8.2.3 Phase retrieval techniques

In the two wavefront sensing techniques examined so far, the wavefront error is determined from slope or curvature measurements made in the pupil or far from focus. An alternate method consists of extracting the wavefront phase error information directly from an image. The basic problem in working in the image plane is that one only has access to the intensity distribution. The phase information has been lost in the image formation process and it is not possible to analytically reconstruct the phase errors in the incoming wavefront from the intensity pattern in the image. Recall from Chapter 4 that image intensity is the absolute value squared of the Fourier transform of the complex amplitude of the incoming wave and that one can then derive the shape of the point spread function (PSF) from a knowledge of the wavefront. One could then make a guess at the phase error distribution in the incoming wavefront, derive the intensity distribution in the PSF, compare it to that of the actual image, and adjust the guess until the match is satisfactory. The problem with this approach is that the solution is not unique.

This fundamental difficulty can be overcome by taking a second image in which a known phase change has been deliberately introduced, a method referred to as "phase diversity." This phase change can be obtained by a slight defocus, changing the field position, or using a different wavelength. The most common method consists of taking one or several images with a slight defocus. These images are then used to constrain the estimate of the wavefront error. An initial estimate is made and iterated until the derived PSFs and the measured PSFs are acceptably close. The uniqueness and convergence of the solution is not well understood, but, for images of reasonable complexity, the solution is believed to be unique as long as the phase errors are not too large ($< 2\pi$). This method, which allows reconstruction of the phase errors in the incoming wavefront from an image in the focal plane, is referred to as "phase retrieval."

Compared to other wavefront sensing techniques, phase retrieval is the simplest, as it does not require additional sensors and introduces a minimum of new error sources. It is particularly advantageous for space applications, as it is intrinsically redundant; the same basic algorithms can be run on images obtained with all installed cameras detectors. Its disadvantage is that the required processing is massive. For space applications, it is best performed on the ground. The processing time also makes it impractical for adaptive optics applications.

The phase retrieval technique was very successfully applied to measure the wavefront error in HST from images taken by the onboard cameras. The information was then used to design corrective optical systems. This technique is also planned for the initial and periodic phasing of the NGST mirrors.

8.3 Internal metrology devices

Many internal metrology systems have been proposed, from simple maintenance of the overall shape of a segmented mirror to systems aimed at controlling the entire optical train. Two representative examples that will be discussed in detail later are shown in Fig. 8.8. A metrology system should be as "direct" as possible, meaning that it should sense the position of all optical elements in the same way that starlight does. If so, then piston, tilt, and focus errors, not only of a segmented primary mirror but also of every subsequent optical surface, will be accounted for and monitored. In this section we will survey the various solutions that have been used or proposed.

8.3.1 Edge sensors

In a segmented primary mirror, the position of each segment must be controlled with respect to the others to better than a tenth of the wavelength. At this accuracy level, it is not possible to rely on the mirror's backup structure

320   8. Active and Adaptive Optics

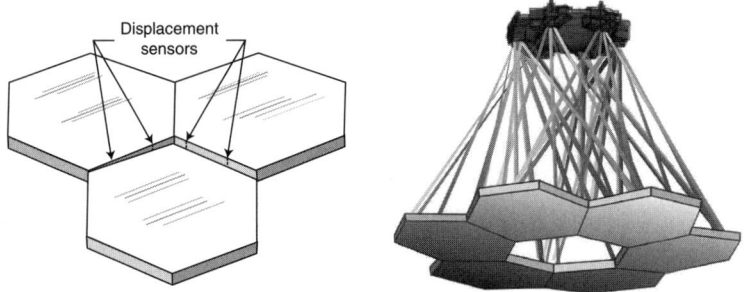

Fig. 8.8. Two types of internal metrology system. At left, electrical devices are used to sense the relative positions of neighboring segments in a segmented mirror. At right, a complex laser metrology system is used to sense the relative position of each optical element.

as an "optical bench" to measure the relative position of each segment: deformations of that backup structure due to gravity or thermal effects can be orders of magnitude larger. But if the segments are very close to each other, as they normally are in a compact aperture system, the obvious solution is to sense the position of each segment edge with respect to those of its neighbors.

Typical edge sensors measure only one direction of displacement. Hence, two such sensors are required per edge to provide the relative height and twist of adjacent mirrors. If the sensors are located at a point directly on the line between the segments, they leave unmeasured one degree of freedom of the array, which corresponds to the change in tilt between adjacent segments, that is, a change in overall focus. A separate measurement of this degree of freedom is then required. This problem can be eliminated simply by locating the sensors at a point slightly offset from the edge of the segment. This breaks the degeneracy and no additional measurement is needed. It is the solution used on the Keck telescopes, as shown on the left in Fig. 8.9.

The edge sensors on the Keck telescopes are of the capacitive type. They are made of low-expansion ceramic glass to minimize thermal sensitivity and have an operating range of ± 20 μm, a bandwidth of 100 Hz (they are actually updated much more slowly, at 2 Hz), and a noise level of about 0.5 nm. They drift at the rate of about 2 nm per day due to instabilities in the electronics, so the primary mirror edge sensors must be recalibrated every few weeks. Under normal operating conditions, their total stability is better than 20 nm. Each unit has a mass of about 2 kg [6].

Edge sensors of the type used in the Keck telescopes must be accurately made and, hence, are expensive. In addition, they are interlocked with their neighboring segments, thus making segment exchange complex. A simpler system using capacitor or inductive sensors is shown on the right in Fig. 8.9. This configuration avoids the interlocking problem, and the vertical capacitor gap (as opposed to the horizontal gap in the Keck telescope sensors) may still allow for detection of the "focus mode" even though there is no offset. In any case, this mode changes only slowly and, in the worst case, could easily be sensed and controlled by a low-order wavefront sensor.

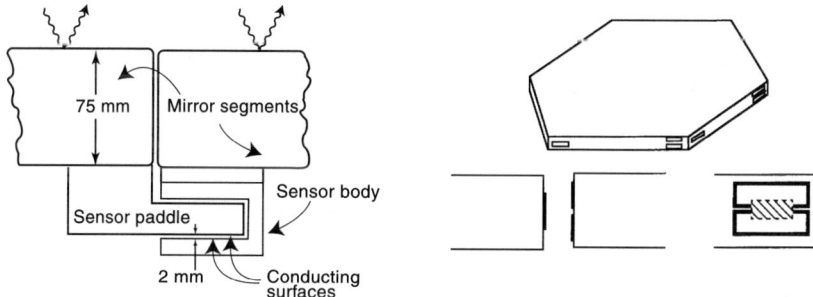

Fig. 8.9. At left, the edge sensing device used on the Keck telescopes. Each is offset with respect to the segments' center planes to provide information on tilt between adjacent segments. At right, the shearing type edge sensors proposed for the California extra large telescope (CELT), which avoid interlock between segments.

Another degree of freedom which remains unmeasured is the distance between segments in their plane. The corresponding requirement is a function of the f-ratio of the primary. This was not necessary in the case of the Keck telescopes, but would be so for faster optics.

Each side of every internal mirror segment must have two edge sensors in order to sense the twist with respect to the adjacent segment. The sensors should be located near each apex to maximize the lever arm (Fig. 8.10). For hexagonal segments, the total number of sensors, N_sensors, is then given as a function of the number of "rings," N_rings, in the overall mirror by

$$N_\text{sensors} = 6\left(N_\text{rings} + 1\right)\left(3\,N_\text{rings} - 2\right). \tag{8.3}$$

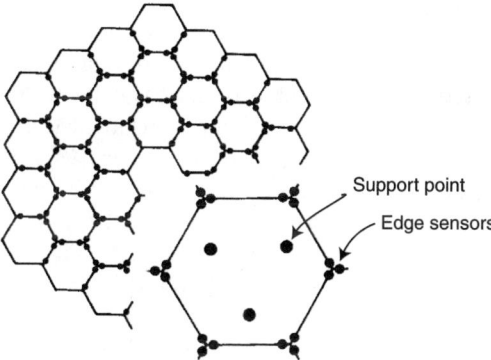

Fig. 8.10. Location of the edge sensors in the Keck telescopes' primary mirrors.

Edge sensors are best at defining the positions of segments with respect to their neighbors, but not over the scale of the overall mirror because of error propagation. This is unfortunate since the most damaging deflection modes of the supporting structure, meaning those with the largest amplitudes, are those with low spatial frequency corresponding to low-order optical aberrations such as defocus, astigmatism, and trefoil. The same is true of deformation due to temperature change. When correction of these large-scale deformation modes is required, edge sensors must be complemented with low-order wavefront sensing.

If edge sensors are an excellent means of controlling adjacent optical elements, they cannot be used to measure the relative positions of widely separated optical elements such as mirror segments in a diluted aperture system or a secondary mirror with respect to the primary. When this is required, one must turn to different methods, which we examine next.

8.3.2 Holographic grating patches and retroreflector systems

From the center of curvature, it is relatively easy to measure the angular and piston positions of segments in a segmented primary mirror system. This method is commonly used for testing optics, segmented or not, in the laboratory (see Chapter 4). Unfortunately, once a telescope is built, access to the center of curvature is rarely an option because it is so far out in front of the secondary mirror. One possibility is to glue onto, or polish into, each mirror segment three tiny spherical mirrors having their centers of curvature close to that of the secondary mirror, and then observe them through a hole in the secondary mirror.

The holographic grating patch system does just that, but in a simpler way. Holographic grating patches simulating small spherical mirrors with a radius about equal to the focal length of the primary mirror are etched directly into the front surface of the mirror segments, three patches being required for each segment. A laser source located at the segment's common center of curvature illuminates the patches, and the return beams are made to interfere with the incoming beam to detect errors in the location of the segments. (Fig. 8.11) [7, 8, 9]. This provides information on piston, tilt, and decenter errors for each segment. If the light source is located further down in the optical train, for example at the final focus, the system will sense all internal optical path differences in the optical train and will supply information on the misalignment of the secondary mirror as well [10]. The diffractive efficiency of the gratings can be adjusted so that scattered light loss can be very small. The frequency bandpass of the measuring system is a function of the laser intensity and can thus be very high. Scattered light from the laser can be filtered out in the instruments, or the wavelength can be selected such that it does not affect scientific observations.

8.3 Internal metrology devices   323

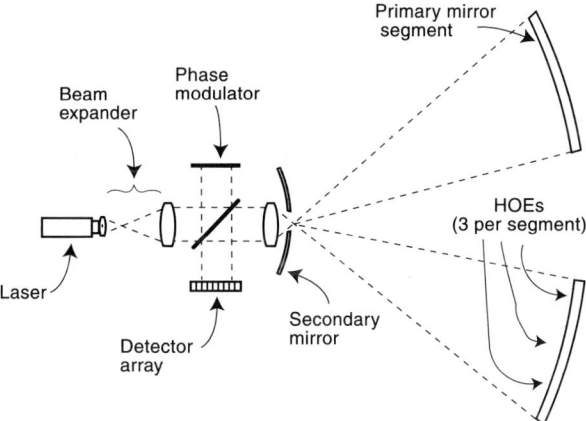

Fig. 8.11. Phasing of a segmented mirror system by the use of holographic grating patches (called holographic elements, or HOEs).

A similar solution consists of mounting retroreflectors on the primary mirror elements (three per mirror) and measuring the phase errors by interferometry. Figure 8.12 shows a possible implementation using a Dyson interferometer [11].

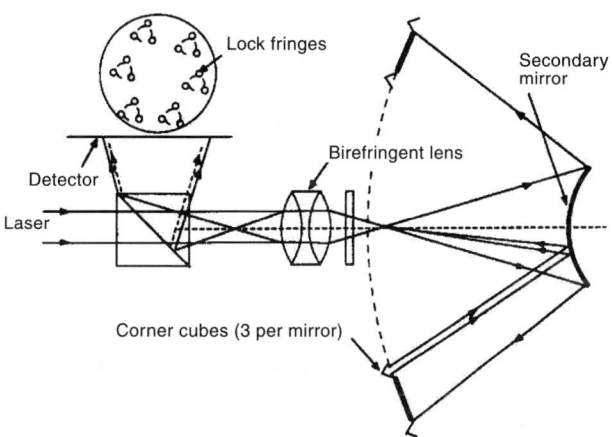

Fig. 8.12. Phasing of a segmented (diluted) mirror system by interferometric measurement of the beams returned by retroreflectors mounted on the individual primary mirror elements.

A birefringent lens is used as a beam splitter. For one polarization, the lens has no convergence and the laser points directly through the secondary mirror to one of the retroreflectors taken as a reference. Light is reflected back onto the interferometer. Going twice through a $\lambda/4$ plate, the returning beam has its polarization changed by 90° and now sees the convergence of the lens. It converges behind the lens and then expands, illuminating a detector. For the other polarization, the beam first converges and then expands, illuminating all of the primary mirror element corner cubes. The return beams, with their polarization changed by 90°, see no convergence and produce spots of light, one per reflector, which interfere with the reference beam [12].

8.3.3 Laser metrology systems

If the holographic grating and retroreflector methods make some use of the telescope as an optical system, another approach employs brute-force metrology to measure the respective positions of optical elements in a train. This can be done in a variety of ways. One scheme consists of launching multiple laser beams from the edge of the secondary mirror toward retroreflectors attached to each primary mirror segment, as shown on the right in Fig. 8.8. Not only can the position of all mirrors in the optical train be measured (e.g., primary mirror segments and secondary and tertiary mirrors), but the position of the entrance aperture of the scientific instruments in the focal plane could be measured as well, ensuring complete internal alignment of the telescope/instrument system at all times.

Although a single laser source with fiber-optic feeds can provide all of the necessary beams, the system is complex because of the large number of individual measuring channels required. On the other hand, this is the ultimate solution for maintaining the alignment of separate optical elements subjected to high-frequency disturbances [13].

8.3.4 IPSRU

Although not a metrology device per se, the Inertial Pseudo-Star Reference Unit (IPSRU) does sense the errors in the direction of the line of sight, whether they stem from internal misalignment or jitter. It is meant to mimic a sky-based guiding system, but with an internal source so as not to be limited in flux, as is the case with in-field stars. The device, commonly called "star in a box," was developed by Draper Laboratories for space-based defense applications [14]. It is composed of an inertially stabilized platform supporting a laser which feeds an alignment beam into the telescope (Fig. 8.13). The laser beam reflects off all mirrors in the train like authentic starlight, and the resulting spot in the focal plane is sensed to derive line-of-sight errors. The device uses gyroscopes which will drift eventually, but with a time scale long enough to be compensated by the normal telescope guiding system. With a 5 mW laser, the spot centroid can be measured to an accuracy of a few milliarcseconds at a rate of more than 100 Hz and might be used to correct vibrations in the optical train induced by wind or mechanical excitations.

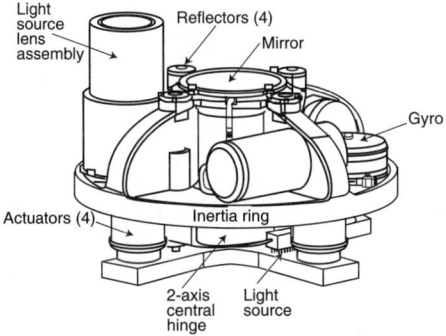

Fig. 8.13. IPSRU, a device that generates a high-power, inertially stabilized laser beam that can be used to stabilize the line of sight of a telescope at high frequencies. (Courtesy of Draper Laboratories.)

8.4 Wavefront correction systems

Wavefront correction methods can be conveniently classified according to the degree of the Zernike term one seeks to compensate. The first degree corresponds to wavefront tip and tilt. It can be compensated by tipping or tilting a mirror in the direction opposite to that of the perturbation. Slow image drifts can be compensated by the telescope pointing system, which adjusts the line-of-sight as needed. Fast response requires moving a smaller mirror such as the telescope's secondary or a dedicated flat mirror.

Three additional Zernike terms can be compensated without deformable optics: defocus and the two coma components. They can be compensated, albeit slowly, by moving the telescope's secondary mirror. Actively controlled telescopes are all equipped with remote control of the secondary mirror. The defocus and coma terms are estimated from the wavefront sensor signals and compensated by moving the secondary in piston and decenter or tilt.

Beyond these five terms, compensation of wavefront errors requires acting on a surface, which can be done on the main optics or on a dedicated mirror.

8.4.1 Fine steering mirrors

The role of a tip-tilt mirror is to compensate image motion at frequencies on the order of, and higher than, the resonance frequencies of the telescope structure (which typically range from 1 to 10 Hz). This includes effects of wind buffeting and atmospheric turbulence.

Fine steering mirrors (FSM), also called "fast steering mirrors" or "tip-tilt mirrors," are used to steer the output beam of the telescope to correct for image motion due to spacecraft jitter, atmospheric seeing, short-term errors of the telescope drive, structural vibrations in the telescope or telescope pier, and so forth. The rationale is that moving a small, light mirror to correct for image motion makes more sense than repointing a multiton telescope. The correction is usually finer and the bandwidth of the correction much higher.

Ideally, fine steering mirrors should be located at a pupil in order to minimize their size and prevent beam-walk downstream.[3] Fine steering mirrors are typically in the 5 to 20 cm size range and are motion compensated: a dummy mass moves in the direction opposite to that of the mirror so as to minimize mechanical disturbances (Fig. 8.14). Secondary mirrors can be used for beam steering if located at the entrance pupil and provided that the tilt angle stays within aberration tolerances. The drawback is that the correction bandwidth will be limited because of the mirror's larger inertia.

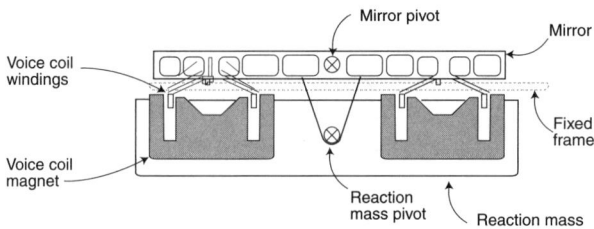

Fig. 8.14. A typical fine steering mirror.

It is important to note that FSMs correct the line of sight for only a single point in the field, typically the guide star used to supply the correction signal. This can be a problem when the telescope field is large (e.g., several arcminutes) as in the case of space telescopes. The reason is that an FSM shifts the mapping of the distortion field onto the focal plane as it rotates to compensate jitter. When distortion is not negligible, this results in a displacement of the image of all points in the field except for the point being corrected, usually the guide star (Fig. 8.15, right).

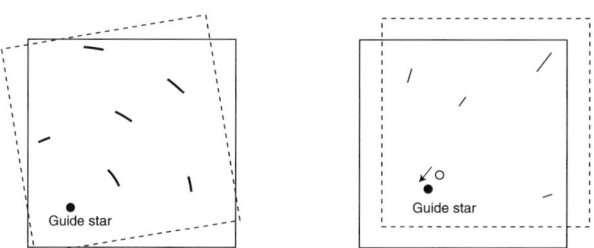

Fig. 8.15. Image blurring due to field rotation (left) and field distortion (right), induced by the use of a fine steering mirror.

---

[3]This is particularly important when a cold stop or a deformable mirror is placed at a subsequent pupil. Otherwise, as the mirror is tilted to correct line-of-sight motion, it will change the reimaging of the entrance pupil by all downstream optics.

A second independent effect is that the field rotates about the guide star as the FSM is tilted (Fig. 8.15, left). This effect is a function of the angle of incidence on the mirror and is zero for normal incidence. It can be reduced by using the smallest angle of incidence compatible with beam clearances.

8.4.2 Deforming the main optics

Compensation of any higher-order aberration beyond tip-tilt requires the deformation of an optical element; this element can be a part of the main optics (e.g., the primary mirror) or be a dedicated mirror. Deformation of the primary mirror is accomplished by applying local forces or moments to the mirror through its support system. Extensive research has been done on this subject, both for space and ground applications, and is well summarized by Wilson [15]. These studies have led to the following basic principles and guidelines:

- Moment actuators have broad influence functions, essentially extending over the full diameter of the mirror (Fig. 8.16). They correct large-scale spatial aberrations more efficiently than force actuators: fewer are needed.

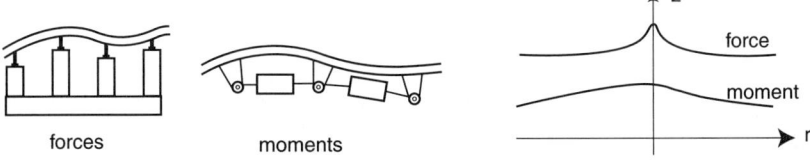

Fig. 8.16. At left and center, principle of force and moment actuators. At right, representative shapes of the corresponding influence functions.

- Moment actuators are not efficient at controlling edges. Even when most of the aberration can be controlled with moment actuators, additional force actuators are needed at the edges.
- The stiffer the mirror, the larger the actuator influence function will be. When only large spatial-scale figure errors need to be corrected, a stiffer mirror will require fewer actuators than a thin one.
- In the opposite case, if small spatial-scale figure errors must be corrected, this is best done with a more flexible mirror.
- Figure errors can be corrected up to the "Nyquist frequency": two actuators are needed per figure-error cycle.
- Correctability, the ratio of the rms of the wavefront error before correction to the rms of wavefront residual error after correction, increases with the number of actuators and decreases when the spatial scale of the figure error decreases (Fig. 8.17).

328   8. Active and Adaptive Optics

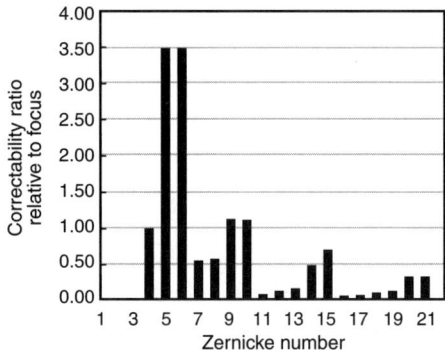

Fig. 8.17. Typical correctability ratio as a function of the Zernike coefficient for a mirror on force actuators with uniform spacing. The correctability ratio is normalized to the fourth Zernike coefficient (focus).

- Axisymmetric figure errors (focus and third-, fifth-, and seventh-order spherical aberrations) are the most demanding to correct. This is of importance in correcting segmented-mirror systems where all segments must have the same radius of curvature. This is also important when one wants to correct the residual errors typically encountered when a mirror has been characterized with a null lens from the center of curvature (Fig. 8.18).

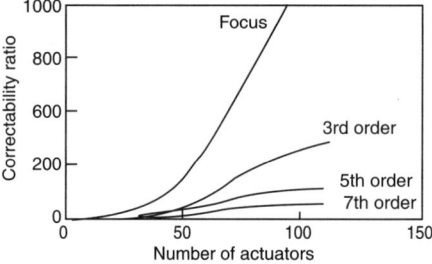 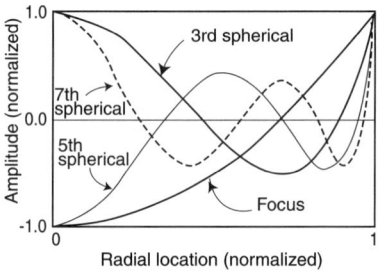

Fig. 8.18. At left, an example (based on an NGST study) of the correctability ratio for focus and third-, fifth-, and seventh-order spherical aberrations as a function of the number of actuators. For reference, the profiles of these aberrations are shown on the right [16].

- The easiest figure error modes to correct are those corresponding to the natural vibration modes of the mirror (this is understandable, since these are the modes requiring the least energy to excite). The best approach for correcting a mirror surface is then to act on those modes and derive the optimal force distribution by mapping the wavefront error measurements onto modal deformations. This approach is referred to as "modal control."

- Conversely, the natural vibration modes are also the modes that will be dominant when the mirror is subjected to thermal stress or when it undergoes backplane deformation. On space telescope mirrors, trefoil is typically the first aberration to appear because it corresponds to the natural deformation of a plate on a three-point support. Astigmatism, which is the first natural deformation mode of a circular disk, is frequent on ground-based telescope mirrors. High correctability of trefoil and astigmatism figure errors is usually essential.
- If the figure errors are steady state and well characterized, the number of actuators can be minimized by optimal placement. The drawback is that, with the correction being fairly sensitive to actuator placement, any subsequent change in the figure error may not be correctable. As a rule, when designing actuator placement, it is best to assume that the geometry of the figure error is random, even though this requires a greater number of actuators.

An excellent example of active control of the main optics in a ground-based telescope is the VLT primary mirror. The VLT axial active support system is composed of 150 active support points on six rings [17, 18]. The number of support points was determined to enable the correction of six aberration terms and to limit the sag between individual supports. Each support is a two-layered system composed of a *passive* hydraulic system carrying the axial component of the total weight of the mirror, combined with an *active* electromechanical system which is responsible for the active optics corrections. (Fig. 8.19).

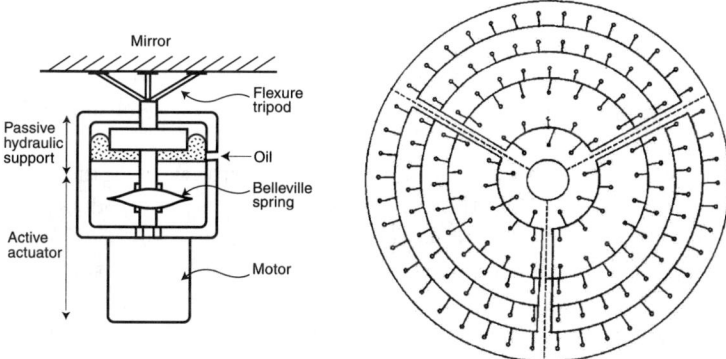

Fig. 8.19. Compound hydraulic/electromechanical active support for the VLT (left). The passive supports are connected hydraulically in three groups, thereby defining three virtual fixed points (right).

The primary mirror of the military prototype telescope ALOT is a good example of an active space-based mirror. The 4-meter diameter segmented mirror was composed of 40 mm-thick meniscus mirror segments supported on a total of 400 electrostrictive actuators.

8.4.3 Dedicated deformable mirror

Compensation of atmospheric turbulence requires a much faster response than can be provided by the supports of the primary mirror. Even when response time is not an issue, it is often advantageous to correct the wavefront on a small optical element with the corresponding small amount of power, rather than correct the main optics. Deformable mirrors (DMs) are located at a conjugate image of the region to be corrected. For atmospheric seeing correction, this should be a conjugate image of a dominant atmospheric layer. For correction of the primary mirror, this should be at a pupil. Deformable mirrors are made of thin plates actuated by electrical means [19]. The two most common systems used in astronomy, piezostack and bimorph mirrors, are shown in Fig. 8.20.

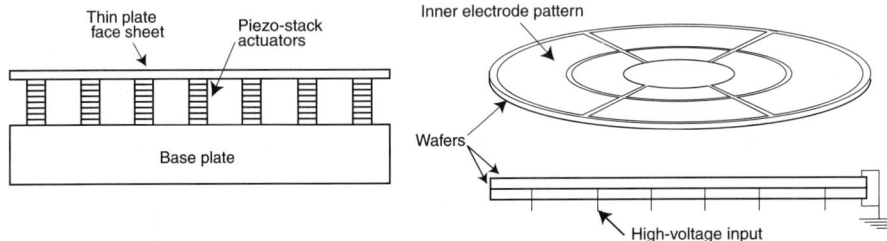

Fig. 8.20. Deformable mirrors used in astronomy: piezostack type at left and bimorph at right.

Piezostack mirrors consist of a thin glass or silicon plate supported by an array of piezostack actuators.[4] Actuators pull or push on the plate, producing local bumps or dips. This technology was developed for defense applications in the 1970s.

A bimorph mirror consists of a pair of piezoelectric wafers glued together, with electrodes in between and on the outside surfaces. The wafers are polarized in a direction perpendicular to the surface. When voltages are applied to the electrodes, electric fields are produced inside the wafers so that one wafer contracts while the other one expands, locally bending the mirror. Bimorph wafers are widely used as acoustic transducers. Although proposed long ago for deformable mirrors [20], the technology has only recently been developed specifically for astronomical application.

Compared to piezostack mirrors, bimorph mirrors are easier to fabricate and therefore less expensive. But unlike piezostack mirrors, their stroke decreases steeply with the spatial frequency of the deformation (as the square

---

[4] Alternatively, piezostacks can be replaced by electrostrictive actuators.

of it). Bimorph mirrors can compensate quite well for the low-order aberration terms, those produced not only by turbulence but also by telescope optics. They can even compensate wavefront tip-tilt motion. They may lack the necessary stroke for high-order compensation, however.

Piezostack mirrors, on the other hand, can compensate well for small-scale wavefront errors, but they often lack the stroke to compensate low-order aberration terms. They always require the use of an additional fast-steering flat mirror for image stabilization. Even small deformable mirrors may limit the frequency response of the control loop.

To provide a safe margin, it is prudent to have the first mirror's resonance frequency at least 10 times higher than the desired loop bandwidth. Current bimorphs typically resonate around 3 kHz. Piezostack mirrors are stiffer; they resonate around 10 kHz. Both types of mirror are commercially available.

Deformable mirrors impose a limit on the size of the field that can be corrected. This is because angles in the deformable mirror space are magnified by the magnifying ratio of the optical system (i.e., by the ratio of the diameter of the primary mirror to that of the deformable mirror). If the deformable mirror is adjusted to correct the optical wavefront for the center of the field, the magnification will cause it to overcorrect the light beam from a target at the edge of the field (Fig. 8.21). This is generally not a problem for the small fields encountered in natural guide star adaptive optics, but can be the limiting factor for the fields of space telescopes.

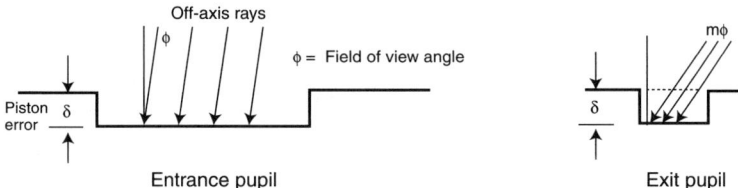

Fig. 8.21. Phase error created by a deformable mirror for large fields. In the deformable mirror image space, field angles are multiplied by the optical magnification of the system, m (the ratio of the primary mirror diameter to that of the deformable mirror). Because of this effect, wavefront corrections done for the center of the field are insufficient to correct images far from the axis. If δ is the piston error for on-axis rays that the deformable mirror corrects, rays at the edge of the field will remain uncorrected by $\delta(1 - 1/\cos m\phi)$, where ϕ is the field angle.

8.5 Control techniques

Control techniques used for active optics differ from those used for adaptive optics. In the case of active optics, one can afford the time to oversample the wavefront. A wavefront surface is generally reconstructed from the sensor measurements and expanded in Zernike aberration terms. The coma and defocus terms are subtracted from the wavefront surface and used to control the secondary mirror. The residual wavefront errors are used to control the primary mirror. In both cases, a control matrix is used to convert wavefront distortion parameters into actuator parameters.

In the case of adaptive optics, computations are usually performed in two steps. First, a fast, dedicated "wavefront computer" converts the sensor outputs into a smaller number M of signals, usually estimates of the wavefront's first- or second-order derivatives. Then, a "control computer" converts these M signals into P voltages to be applied to the deformable mirror. In most systems, the number M of wavefront parameters is larger than the number P of actuators to be controlled. These systems are overdetermined and the accuracy of the compensation is mainly limited by the mirror. However, especially when measurements are affected by readout noise, one may wish to minimize the number of wavefront measurements and take $M = P$. The sensor must then be precisely matched to sense those particular deformations that the mirror can correct. This is the case when wavefront curvature measurements are used to apply bending moments and create a corresponding curvature in a bimorph mirror. The mirror then behaves like an analog device which solves the Poisson equation and reconstructs the wavefront.

8.6 Typical active optics system implementations

8.6.1 The VLT active optics system

The use of active optics for ground-based telescopes was pioneered by ESO on the NTT and then applied to the VLT [18]. In addition to enabling the use of lightweight, low-rigidity meniscus mirrors, the active optics system is capable of correcting all of the steady-state and low (time)-frequency wavefront errors in the system. These include design and manufacturing errors in the optics, mechanically induced deformation of the optics due to mount and gravity, deformation of the support of the secondary mirror due to gravity, and at least the lowest-frequency errors due to wind buffeting the primary mirror and telescope tube.

The basic principle is shown in Fig. 8.22. Wavefront errors are measured using a bright star off-axis, then compensated by adjusting the position of the secondary mirror and deforming the primary mirror.

The off-axis reference star must be several arcminutes from the axis so as not to interfere with the science field. This is too far for the reference star to be in the isoplanatic patch, the region of the sky where phase errors remain essentially constant and which extends over about one arcminute at most. It is therefore necessary to average out the seeing effects on the reference star by integrating for a long enough time, typically on the order of 30 s. This limits

8.6 Typical active optics system implementations   333

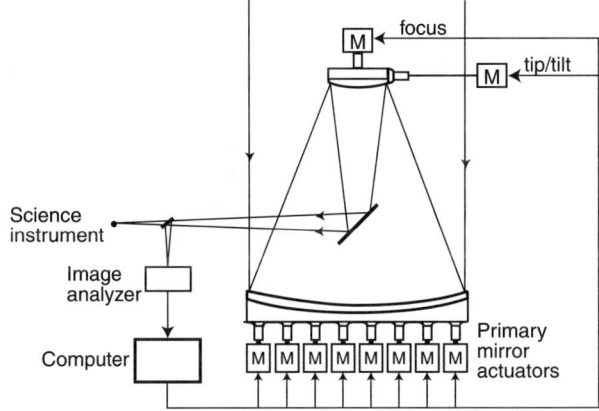

Fig. 8.22. Active optics system of the NTT and VLT.

the bandpass of the active system to about 0.03 Hz, too low for correcting wavefront tilt errors; a separate guiding system has to be used. On the other hand, the advantage is that fairly faint stars can be used, meaning that finding a reference star is not generally a problem.

Wavefront errors are measured on stars of magnitude 14 or brighter using a Shack-Hartmann sensor with a 20 × 20 array. This provides sufficient sampling of the wavefront error for correction by the primary mirror actuators, which are on a ∼ $D/12$ grid. Defocus and third-order coma are corrected by repositioning the secondary mirror. All remaining modes are corrected by adjusting the figure of the primary mirror. The system is capable of acting on 18 elastic modes of the mirror, permitting correction of wavefront errors within the bandpass of the system up to 50 nm rms [21].

8.6.2 Coaligning, cofocusing, and cophasing segmented systems

For a segmented primary mirror to be equivalent to a single monolithic mirror, each segment must be coaligned, cofocused, and cophased (Fig. 8.23). Coaligning consists of both stacking the images produced by the individual segments and properly locating the segments laterally on the parent mirror. Cofocusing ensures that all of the individual images are of the same size (i.e., the focal length of each segment is the same) and cophasing ensures that there is no piston discontinuity between the edges of neighboring segments. When these three conditions are satisfied, the mirror segments will match the figure of the ideal full-sized parent mirror.

Cofocus is usually obtained by proper figuring of the mirror segments or by using a segment-based active system. Coaligning is done by observing a star on-axis and stacking all the images from the individual segments (Fig. 8.23, left). When this is accomplished, the primary mirror acts as a "light collec-

334   8. Active and Adaptive Optics

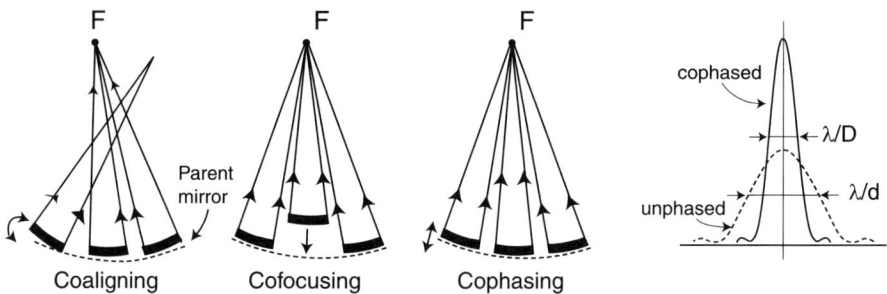

Fig. 8.23. Coaligning, cofocusing, and cophasing a segmented system (left). At right, the PSF of a segmented system where the mirror segments are unphased or phased. The width of the PSF is λ/d in the first case, where d is the diameter of the mirror segments, and it is λ/D in the second case, where D is the diameter of the full primary mirror.

tor," but not yet to its full potential. This is because the image of a point source is simply the incoherent superposition of the PSFs of all the individual mirror segments. For the mirror to provide the sharper PSF of the full-sized mirror, all of the segments must be phased. The resolution of the perfectly phased telescope is then better than the resolution of the corresponding completely unphased telescope by a factor of $\sqrt{N_{\text{seg}}}$, where N_{seg} is the number of segments (Fig. 8.23 right).

Two examples of coaligning and cophasing methods, one for a ground-based telescope and the other for a space telescope, are briefly described below.

Keck telescopes

The mirror segments of the Keck telescopes are rigid enough to maintain their figures without the need for active control. The active system only has to adjust the positions of the segments with respect to each other. Of the six degrees of freedom affecting each segment, only three are actively controlled: piston, tip, and tilt. The other three, rotation about the segment axis and decenter in two dimensions, require no adjustment after installation because the optical performance is less sensitive to these errors and dimensional changes in the support system due to gravity and thermal effects are small.

Cofocusing of the mirror segments (i.e., ensuring that they all have the same radius of curvature) is obtained by manufacture and by correction of residual errors with a warping harness (Chapter 6). Coaligning is a simple matter of stacking the individual images. For the telescope to reach its resolution and sensitivity potential, the mirror segments must be phased.

As shown in Fig. 8.24, phasing can have an enormous impact even for imaging through the turbulent atmosphere. Segment phase errors begin to limit image quality when the atmospheric coherence length (r_0) is about equal to the diameter of the segments. For a typical seeing at Mauna Kea of 0.5" (in V), this condition occurs at a wavelength of 1.8 μm. To reach diffraction-

8.6 Typical active optics system implementations   335

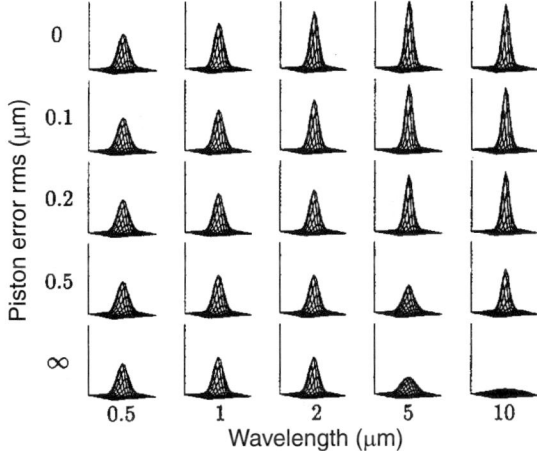

Fig. 8.24. Theoretical PSF for the Keck telescopes for a variety of wavelengths and piston errors of the primary mirror segments. Seeing is assumed to correspond to an r_0 of 20 cm at 0.5 μm. Note the devastating effects of phase errors at longer wavelengths. (From Ref. [22].)

limited image quality at that wavelength, the rms wavefront error must be less than about $\lambda/14$, that is to say, a piston error between segments of half that value, or 60 nm.

Phasing of the Keck telescopes is accomplished by a special camera that utilizes a wave-optics variation of the Shack-Hartmann test. The standard Shack-Hartmann method provides information on the slope of the wavefront but not on its phase or phase discontinuities. To provide information on the phase error between segments, the primary mirror is reimaged onto a pupil mask with 84 small (12 cm) apertures *straddling* the intersegment edges (Fig. 8.25, right).

When two segments are in phase, the image given by the subaperture is the usual circular Airy pattern. As piston error increases, a second diffraction peak appears. Its relative intensity grows with increasing piston error. The two peaks become equal at a piston error of $\lambda/4$ (total path difference of $\lambda/2$). As the piston error continues to increase, the second peak continues to grow at the expense of the first until, at a piston error of $\lambda/2$ (total path difference of λ), it has replaced the original peak (Fig. 8.25, left).

The method therefore consists of extracting the segment-to-segment piston error by correlating the observed diffraction pattern against a predetermined sequence of diffraction patterns with known piston errors. Once the segment-to-segment piston errors are known, the optimal correction is applied to the segment actuators, and the mirror system then maintains its shape using the edge sensors for error sensing. Typically, the phasing process requires only a single exposure of about 60 s on a bright star (fourth magnitude). Averaged over all the segments, the phase error is about 30 nm rms [22, 23].

336   8. Active and Adaptive Optics

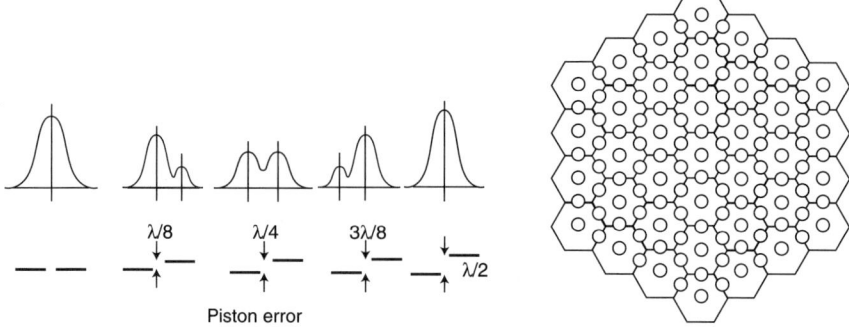

Fig. 8.25. Principle of the phasing method at the Keck telescopes. At left, schematic sequence of PSFs supplied by a single Shack-Hartmann aperture as a function of the piston error between adjacent segments. At right, primary mirror configuration of the Keck telescopes showing the 84 edge-sampled spots which provide information on the segment phases.

Cophasing NGST

NGST will need to be aligned and cophased upon launch, and then again at periodic intervals, to correct for drifts, mid-term thermal changes, and long term mechanical effects. The original baseline procedure was as follows.

Instead of the dedicated wavefront error sensor used on ground-based telescopes, NGST uses an image-based coarse phasing method followed by phase retrieval. The advantage is that wavefront sensing can then be performed with any of the science cameras, no additional equipment being required. This ensures maximum redundancy and reliability and saves on cost and mass as well. In addition, phase retrieval provides complete coverage of the phase errors, both over the mirror segments and along all segment boundaries, which cannot be obtained by other methods such as the Shack-Hartmann test. The massive computation required for phase retrieval needs to be done on the ground, but since the optics are expected to remain stable for weeks, the practical impact is negligible.

Wavefront control proceeds in three phases, as illustrated in Fig. 8.26. The first phase, coarse alignment, is used only when the optics are badly misaligned, as they are likely to be following launch and deployment. It consists of observing a bright star and coaligning the primary mirror segments by stacking up the images. When this is accomplished, all segments except two are tilted away.

The second phase consists of cophasing two mirror segments at a time by observing their interference fringes, which are dispersed as described in Fig. 8.27; this is the so-called "dispersed fringe sensing method." The capture range is very large, on the order of several mm, and the method can serve to align the optical train and phase the primary mirror segments to within

8.6 Typical active optics system implementations   337

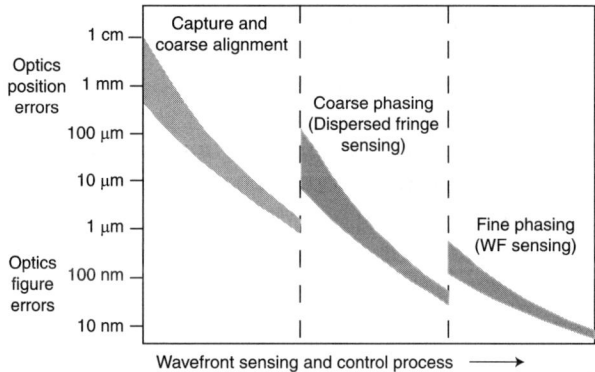

Fig. 8.26. Evolution of the optics alignment and figure errors during the wavefront control procedure proposed for NGST.

a wavefront error of about 1 µm rms. Use of this technique provides paired segment phasing to within $\lambda/4$ in just a few actuation steps [24].

The procedure is repeated for the remaining pairs of segments until all are cophased. This phasing approach is in routine use at the Palomar Interferometer, where it is used to phase up widely separated small apertures. The phasing can be further improved, to below the diffraction limit at 2 µm, by using image sharpening in white light. Image sharpening consists of maximizing the peak of the PSF by making small adjustments to the segments in a trial-and-error fashion.

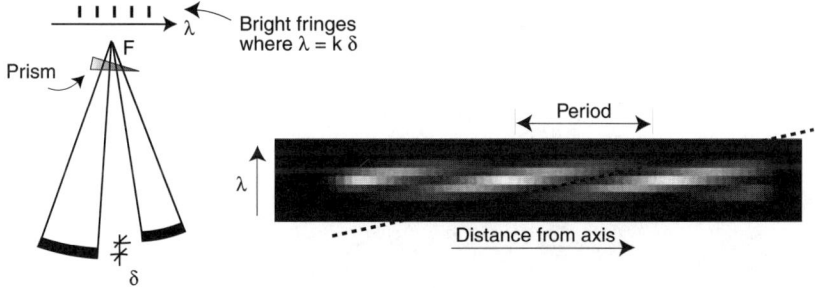

Fig. 8.27. Principle of the dispersed fringes sensor proposed for coarse alignment and phasing of NGST. The image formed by two segments is dispersed by a prism (actually a grism) and observed in broadband with a camera (left). Phase differences between the two segments cause the overlapping images to produce interference fringes which are differently spaced at each wavelength band (right). The period of the fringes provides the amplitude of the piston error between the two segments (the multiple wavelengths allowing complete determination of the piston error when it is larger than the wavelength), and the orientation of the dark fringes (dashed line) supplies the sign of that error. After correction, the two wavefronts add coherently at all wavelengths and the fringe modulation disappears.

The last phase, fine figure control mode using phase retrieval, is invoked next. A sequence of images is taken close to focus, typically at ±25 mm, then ±15 mm from focus, and then again at focus. A pupil image is also taken by "flipping in" a pupil-imaging lens. The images are taken in white light with narrow-band filters. The images are then processed in a modified Gerchberg-Saxton [25] iterative transform algorithm which leads to an estimate of the wavefront error map in about 10 iterations. With the wavefront error estimated, the control commands are readily computed using linear optimal control calculations. At the end of the process, wavefront errors are reduced to about $\lambda/20$ in the visible and $\lambda/60$ in the near infrared.

Once the telescope is aligned and phased, the wavefront control system is turned off. Passive structural stability is relied upon to hold figure and alignment during observations (no edge sensors are used). Wavefront quality is checked periodically and the wavefront control system reactivated as required.

8.7 Correction of seeing

The first-order effect of atmospheric turbulence is image motion. A compensation system composed of a sensor and fast steering (tip-tilt) mirror is therefore an effective means of improving turbulence-degraded images. But the efficiency of that correction depends heavily on the ratio D/r_0 of the telescope diameter over Fried's seeing parameter r_0. The Strehl ratios ideally achieved with perfect tip-tilt compensation are approximately described by curve 1 in Fig. 8.28, as a function of D/r_0.

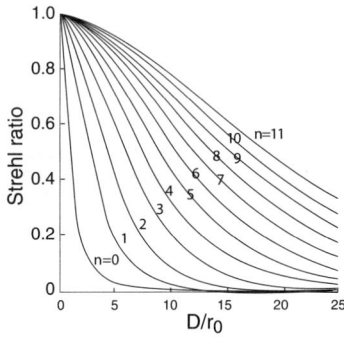

Fig. 8.28. Strehl ratios theoretically achieved by perfectly compensating the first N atmospheric Karhunen-Loeve modes (courtesy of M. Northcott). N is related to the curve number n by the relation $N = n(n+3)/2$, n being the degree of the corresponding Zernike polynomial. Strehl ratios are plotted as a function of the ratio D/r_0 of the telescope D over Fried's seeing parameter r_0.

By comparison, curve 0 shows the Strehl ratios obtained without compensation. These are theoretical estimates based on the Kolmogorov turbulence spectrum with an infinite outer scale. The Strehl ratio improvement reaches maximum when D/r_0 is on the order of 4, then decreases as D/r_0 further increases. Hence, with a 4-meter telescope, tip-tilt compensation is effective in the near thermal infrared where r_0 can be as large as 1 m. But as the telescope diameter increases, the gains diminish quickly. Further improvement requires the compensation of atmospheric aberration terms higher than the tip-tilt.

8.7.1 Historical developments

Atmospheric seeing is corrected by measuring the distortion of the wavefront coming from a point source within the isoplanatic angle and then applying the opposite distortion to a deformable mirror within the atmospheric coherence time τ_0. This idea was first proposed by Babcock in 1953 [26]. For wavefront correction, Babcock suggested using an oil film, its thickness to be controlled by electrical charges. Technology was not ready, however, and the idea was not pursued. Research efforts for the imaging of satellites were started by the U.S. Department of Defense (DoD) in 1973, with the first successes of adaptive optics achieved in 1977 [27]. This technology was first used for astronomy in the ESO 3.6 m telescope in the late 1980s [28]. Many other systems followed and all large telescopes are now or will soon be equipped with adaptive optics systems. A basic implementation is illustrated in Fig. 8.29.

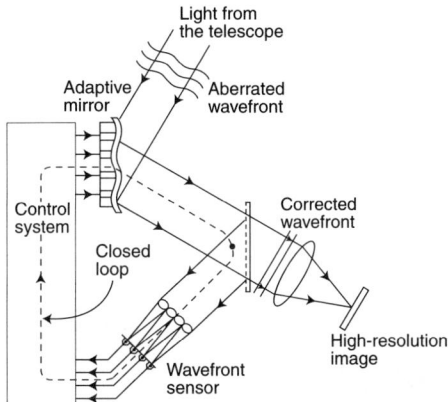

Fig. 8.29. Basic principle of adaptive correction of atmospheric seeing. The wavefront error is measured with a wavefront sensor and corrected with a deformable mirror.

Wavefront correction requires the availability of a bright source near the object of interest. Unfortunately such situations are rare on the sky. An alternate to natural stars was needed. Synthetic laser beacons were proposed in a military context by the U.S. Department of Defense as early as 1982, with their first prototypes dating from 1984. This idea was independently proposed and documented in the open literature by Foy and Labeyrie in 1985 [29]. The DoD work was declassified in 1991.

Experimentation with both natural and artificial guide stars for application in astronomy has been very active since then. We will now briefly review the corresponding techniques. Detailed coverage of the subject can be found in books by Hardy and by Roddier listed in the bibliography at the end of the chapter.

8.7.2 Adaptive optics using natural guide stars

It is possible to use natural guide stars over a reasonable portion of the sky to compensate for the atmospheric wavefront error when working at 1 μm and above. Figure 8.30 gives the "Strehl seeing angle" (the radius of a "top hat" image containing the same total energy as the observed image and having the same peak intensity) as a function of wavelength for an 8-meter telescope and HST. One notes that an 8-meter telescope with adaptive optics will have a resolution superior to that of HST beyond about 1 μm. But the sky coverage achievable with natural guide stars is only on the order of 1% because of the limited size of the isoplanatic patch around the desired target.

As one goes into the visible, the gain diminishes dramatically, as does the availability of guide stars that are bright enough. At 0.5 μm, sky coverage is much too small to be of general use: one typically needs an eighth-magnitude star less than 5 arcseconds away. Seeing compensation in the visible requires artificial stars.

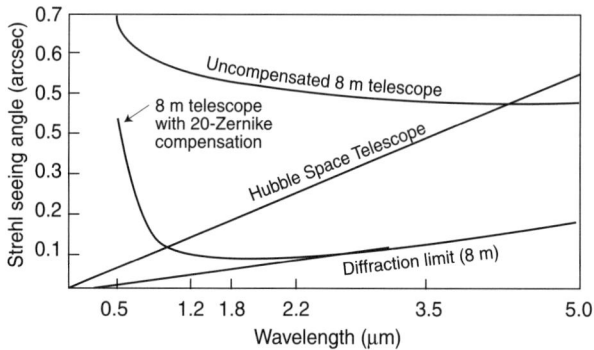

Fig. 8.30. Strehl seeing angle as a function of wavelength for an 8-meter telescope and HST.

8.7.3 Adaptive optics with laser stars

Laser stars have the potential to considerably improve the degree of correction and sky coverage of adaptive optics systems, both in the visible and the near infrared. There are two methods of creating an artificial star: Rayleigh scattering and bouncing from the sodium layer.

The **Rayleigh scattering** method involves focusing a powerful laser beam to a point at an altitude of 10 to 20 km, above most atmospheric turbulence. Only backscattered photons from the height at which the laser is focused then contribute to the wavefront estimate. Unfortunately, because the Rayleigh laser star is necessarily well inside the atmosphere (some air must be present for it to backscatter from), it misses the wavefront disturbances above it.

The **sodium laser star** is created by illuminating the sodium layer at an altitude of about 90 km.[5] Because of its higher altitude, it provides a more complete sampling of atmospheric turbulence than the Rayleigh star, but as the sodium density is low, the return signal is weak.

The main problem with laser stars, both Rayleigh and sodium, is that they do not supply information on tip-tilt because the outgoing and return beam follow the same path (Fig. 8.31). For tip-tilt correction, a natural guide star is still required. It can be fairly faint, however, because it is only needed for the tip-tilt information, not the full wavefront error determination. Still, the more correction one seeks with the adaptive system, the more precise the tip-tilt correction will need to be, and this imposes limitations on how faint the guide star can be.

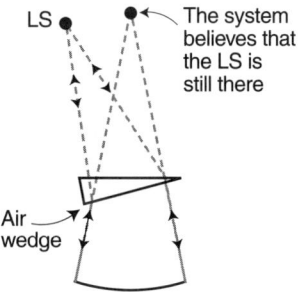

Fig. 8.31. Unlike natural guide stars, artificial laser stars (LS) do not provide tilt information. Any air wedge in the atmosphere generating line-of-sight tilt is traversed *twice* and is therefore not sensed.

Another problem with laser stars is that they project a conical rather than a parallel beam through the turbulent layers (Fig. 8.32, left). This means that the wavefront perturbation that they sense is not the same as that of a target at infinity, even if it is in the same line of sight. This effect is called the "cone effect" or "focus anisoplanatism."

---

[5] The reason behind the presence of sodium at this altitude is still debated; it may have been deposited by meteorites. There is very little of it, only 300 kg distributed over the entire Earth.

342   8. Active and Adaptive Optics

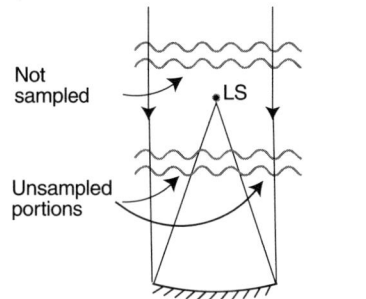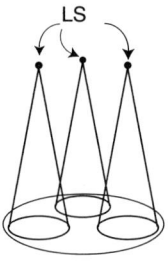

Fig. 8.32. A single laser guide star does not fully sample the atmosphere above the aperture (left); this requires a multiple laser star system (right).

The severity of this effect depends on the altitude of the laser star, Rayleigh or sodium, the telescope diameter, and the intensity of turbulence. Because of this effect, sodium laser stars cannot be used when D/r_0 is larger than 20. This corresponds to a wavelength of about 1 μm on an 8-meter telescope under fair seeing conditions.

Several solutions have been proposed to overcome the cone effect. All involve the use of several laser stars to probe the entire cylindrical beam, as shown on the right in Fig. 8.32. One method involves the use of dedicated laser stars for subpupil areas, effectively decreasing the cone effect for each individual area. The phases determined in each area need to be "stitched" to produce an estimate of the phase over the entire pupil. A more promising method is "multiconjugate adaptive optics" (MCAO), which uses several wavefront sensors and associated laser stars to invert the turbulence profile. MCAO implies the use of two or more deformable mirrors, optically conjugated to different altitudes, to compensate for the phase distortions over an extended field of view. The goal of MCAO is a uniform PSF over a relatively large field of view, on the order of 1 to 2 arcminutes. A secondary advantage of MCAO is that it eliminates the cone effect altogether, holding out the promise of using laser star adaptive optics in future extremely large telescopes.

References

[1] Wilson, R.N. and Noethe, L., Closed-loop active optics: its advantages and limitations for correction of wind-buffeting deformations of large flexible mirrors, SPIE Proc., Vol. 1114, p. 290, 1989.

[2] Shack, R.V. and B. C. Platt, B.C., Production and use of a lenticular Hartmann screen, J. Opt. Soc. Am., Vol. 61, p. 656, 1971.

[3] Roddier, C. and Roddier, F., Wavefront reconstruction from defocused images and the testing of ground-based optical telescopes, J. Opt. Soc. Am., Vol. A-10, p. 2277, 1993.

[4] Roddier, F., Curvature sensing and compensation, a new concept in adaptive optics, Appl. Opt., Vol. 27, p. 1223, 1988.

[5] Roddier, F., Error propagation in a closed-loop adaptive optics system: a comparison between Shack-Hartmann and curvature wavefront sensors, Optics Commun., 113, p. 357, 1995.

[6] Mast, T.S., Gabor, G., and Nelson, J.E., Edge sensors for a segmented mirror, SPIE Proc., Vol. 444, p. 297, 1983.

[7] Wissinger, A.B., Sensing and control for large optical systems, Proceedings of a workshop on *The Next Generation Space Telescope*, Bely, P.Y., Burrows, C.J., and Illingworth, eds., Space Telescope Science Institute, 1989.

[8] Bruning, J.H. et al., Digital wavefront measuring interferometer for testing optical surfaces and lenses, Appl. Opt., Vol. 13, No. 11, p. 2693, 1974.

[9] Kishner, S.J., High bandwidth alignment sensing in active optical systems, SPIE Proc., Vol. 1532, p. 215, 1991.

[10] Bely, P-Y, Ford, H.C., Burg, R., Petro, L. and White, R., Post: polar stratospheric telescope, Space Sci. Rev., Vol. 74, p. 101, 1995.

[11] Dyson, J., Common-path interferometer for testing purposes, J. Opt. Soc. Am., Vol. 47, p. 386, 1957.

[12] Bely, P.Y., Burrows, C.J., Roddier, F., and Weigelt, G., HARDI: a high angular resolution deployable interferometer, Proc. ESA colloquium on *Targets for Space-based Interferometry*, 1992.

[13] Lau, K., Breckenridge, W., Nerheim, N., and Redding, D., Active figure maintenance control using an optical truss laser metrology system for a space-based far-IR segmented telescope, SPIE Proc., Vol. 1696, p. 60, 1992.

[14] Gilmore, J., et al., Inertial pseudo star reference unit, IEEE Proc. of the *Plans Symposium*, 1992.

[15] Wilson, R.N., *Reflecting Telescope Optics II*, Springer-Verlag, 1999, p. 274.

[16] Stier, M., Metha, P., and Rockwell, R., *Beryllium mirror study for the NGST*, Raytheon Report, PR-E26-0001, 1999.

[17] Stanghellini, S., Design and construction of the VLT primary mirror cell, SPIE Proc., Vol. 2871, p. 314, 1996.

[18] Schneermann, M., Cui, X., Enard, D., Noethe, L., and Postema, H., ESO VLT III: the support system of primary mirrors, SPIE Proc., Vol. 1236, p. 920, 1990.

[19] Ealey, M.A., Deformable mirrors: design fundamentals, key performance specifications and parametric trades, SPIE Proc., Vol. 1543, p. 36, 1991.

[20] Steinhaus, E. and Lipson, S.G., Bimorph piezoelectric flexible mirror, J. Opt. Soc. Am., Vol. 69, p. 478, 1979.

[21] Guisard, S., Noethe, L., and Spyromilio, J., Performance of active optics at the VLT, SPIE Proc., Vol. 4003, 2000.

[22] Chanan, G., Troy, M., and Ohara, C., Phasing the primary mirror segments of the Keck telescopes: a comparison of different techniques, SPIE Proc., Vol. 4003, p. 188, 2000.

[23] Chanan, G., Mast, T. and Nelson, J., Phasing the mirror segments of the W.M. Keck telescope, SPIE Proc., Vol. 2199, p. 622, 1994.

[24] Shi, F., Redding, D. et al., DCATT dispersed fringe sensor: modeling and experimenting with the transmissive phase plates, PASP Proc. Vol. 207, p. 510, 2000.

[25] Gerchberg, R.W. and Saxton, W.O., A practical algorithm for the determination of phase from image and diffraction plane pictures, Optik, Vol. 35, p. 237, 1972.

[26] Babcok, H.W., The possibility of compensating astronomical seeing, PASP, Vol. 65, p. 229, 1953.

[27] Hardy, J.H., Active optics: a new technology for the control of light, Proc. IEEE, Vol. 66, No. 6, p. 651, 1978.

[28] Rigaut, F, Rousset, G., Kern, P. et al., Adaptive optics on a 3.6 m telescope: results and performance, Astron. and Astrophys., Vol. 250, p. 280, 1991.

[29] Foy, R and Labeyrie, A., Feasibility of adaptive telescopes with laser probe, Astron. Astrophys., Vol. 152, L29, 1985.

Bibliography

Active optics

Ealey, M.A., ed., *Active and Adaptive Optical Components*, SPIE Proc., Vol. 1543, 1991.

Roddier, F., ed., *Active Telescope Systems*, SPIE Proc., Vol. 1114, 1989.

Wilson, R.N., *Reflecting Telescope Optics II*, Springer-Verlag, 1999.

Adaptive optics

Alloin, D.M. and Mariotti, J.-M., eds., *Adaptive Optics for Astronomy*, NATO ASI Series, Kluwer Academic Publishers, 1994.

Hardy, J.W., *Adaptive Optics for Astronomical Telescopes*, Oxford Books, 1998.

Roddier, F, ed., *Adaptive Optics for Astronomy*, Cambridge Univ. Press, 1999.

Tyson, R.K., *Principles of Adaptive Optics*, Academic Press, 1991.

9
Thermal Control

Thermal issues permeate all aspects of telescope design. Minute temperature variations in the optics or supporting structure may affect mirror figures and alignment. Small temperature differentials between the telescope or enclosure and the ambient air degrade the image quality of ground-based telescopes. Thermal emission from the optics and their environment creates instrumental background that degrades the sensitivity of ground and space observatories in the infrared. Many of these issues have already been treated in the chapters dealing with the pertinent systems. In this chapter, we concentrate more specifically on "thermal control"; that is to say, ways of reducing residual thermal effects to an acceptable level in systems that have already been designed to minimize them.

9.1 General requirements

The purpose of the thermal control system is to control the temperature of critical systems so as to keep them within their design temperature range and maximize the observatory's scientific performance. More specifically, the function of the thermal control is to

- maintain the temperature of optics and supporting structure within their design operational range (a condition specific to space telescopes),
- minimize seeing, in the case of ground telescopes, by maintaining the temperature of systems near the light path, particularly the primary mirror, close to ambient air temperature,

- maintain adequate operating temperatures for instruments and detectors,
- minimize instrumental background in the infrared.

In the following sections, we examine various techniques for fulfilling these requirements.

9.2 Thermal environmental conditions

The thermal environment encountered on the ground is well known and will not be examined here. The thermal environment of near-Earth space is summarized in Fig. 9.1. For higher orbits, the heat input from Earth is negligible and the Sun is the only significant source. Radiation out to space is controlled by the temperature of space, which is 7 K near Earth (not 3 K, due to dust particles in the inner solar system).

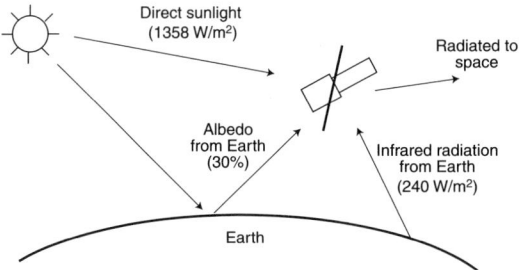

Fig. 9.1. The thermal environment of space observatories in low Earth orbit.

9.3 Temperature control techniques

Thermal control can be classified as passive or active. Passive measures consist of using coatings, insulation, or radiators to control the amount of external heat input or to dump waste heat. Active measures consist of using heaters and, on the ground, ventilation and coolants.

The thermal properties of typical materials used as thermal coating or insulation are listed in Table 9.1. In this table, κ is the thermal conductivity, C_p is the specific heat, α is the solar absorptance (defined with respect to the solar spectrum), and ϵ is the emittance (defined at ambient temperature). Thermal conductivity and specific heat data are not shown for coatings: since they are thin with respect to the material they are applied to, conductive heat transfer properties are dominated by the properties of the substrate. VDA Kapton is a Kapton film with a vapor-deposited aluminum coating on one side. Its thermo-optical properties are given separately for the two surfaces, as Kapton is relatively transparent to visible wavelengths and somewhat opaque to infrared wavelengths. The thicker the Kapton layer, the higher the infrared emittance. VDA-Kapton properties are also given for the "beginning of life" (BOL) and after 5 years on geosynchronous orbit (GEO), as the material degrades over time with exposure to the space ultraviolet and energetic particle environment.

Table 9.1. Typical material properties

Material	κ W/m K	C_p W s/kg K	α	ϵ
Maxorb	–	–	0.90	0.10
Aluminum paint (typical)			0.30	0.31
Black paint (typical)	–	–	0.90	0.85
White paint (Chemglaze A276)	–	–	0.25	0.88
Beryllium (polished)	204	1925	0.02	0.02
VDA-Kapton (VDA side)	0.15	1005	0.12	0.03
VDA-Kapton (Kapton side, BOL*)	0.15	1005	0.36	0.61
VDA-Kapton (Kapton side, 5 yr GEO*)	0.15	1005	0.66	0.61
Black Kapton	0.15	1005	0.92	0.88
Carbon-fiber composite	1 – 200	880	0.60	0.85

* for 25 μm thick Kapton.
Space material data from Ref. [1].

A common means of insulating space hardware is the use of multilayer insulation (MLI). MLI blankets consist of several layers of aluminized plastic sheets. The inside layers are made of Mylar and the outer one is Kapton, a material with greater resistance to ultraviolet exposure. Contact between layers is avoided by using separator nets.

A heater is a simple device composed of an electrical resistance sandwiched between two sheets of insulating material. Most electronic, electrical, and mechanical equipment can operate within a typical temperature range of −10 to +40 °C. This is generally not a problem for ground observatories, but thermal control is definitely needed in space observatories for reliable operation of computers, reaction wheels, gyros, and transponders. In space, temperature-controlled equipment will generally be "cold biased," meaning that it will run cold unless its heaters are on. Thermostatically controlled heaters act to bring temperatures up to the desired level.

In space systems, radiators are used to radiate away waste heat. Heat-generating equipment is conductively connected to radiators via solid copper bars or heat pipes. A heat pipe uses fluid phase change and capillary force to transfer heat from one end of the pipe to the other (convection would not work in zero gravity). Such a device can transfer more than 100 times as much energy as a copper bar of the same cross section.

348   9. Thermal Control

Radiators pose a nontrivial problem for cryogenic observatories. Taking the temperature of space as 0 K, the radiated energy is calculated by the Stefan-Boltzmann equation

$$E = \epsilon \sigma T^4 A, \tag{9.1}$$

where A is the radiator's area, T is its temperature, ϵ is its emissivity (about 0.8), and σ is the Stefan-Boltzmann constant, $5.67 \cdot 10^{-8}$ W/ (m^2 K^4). At cryogenic temperatures, radiators become extremely inefficient and have to be huge. For example, at 60 K, dissipating the mere 250 mW generated by the NGST telescope and instruments requires a radiator no less than 4 m^2.

9.4 Thermal control for dimensional control

The goal is to maintain the figures of mirrors and their relative positions within required tolerances in spite of changes in the thermal environment. This can be achieved either passively, by the adoption of a particular optical design or the use of low-expansion materials, or else actively, via some sort of thermal control. Passive means have been described in Chapters 4 and 6. We now examine the active solutions.

9.4.1 Mirror figure control

With the use of ultra-low-expansion materials such as ULE or Zerodur, thermally induced deformation of mirrors has essentially disappeared in ground-based telescopes. However, borosilicate glass is still used for large mirrors because it can be cast in complex shapes to produce honeycomb blanks (e.g., LBT 8-meter mirrors). Besides improving stiffness, lightweighting blanks helps reduce mirror seeing by lowering thermal inertia. Bulk temperature changes[1] are not a real problem because they are slow and only affect focus, not wavefront quality. Diametric and nonuniform axial temperature gradients, on the other hand, deform the mirror surface in ways that cannot be corrected by adjusting focus. This effect can be avoided by ventilating the honeycomb cells with temperature-controlled air, as shown in Fig. 9.2.

Thermally induced mirror figure deformation is a nonproblem for large meniscus mirrors with active optics. Even low-order active optics can correct for thermal effects since these will generally affect only the lowest structural modes (focus, astigmatism, trefoil, etc.). Also, due to the large thermal inertia involved, changes are slow and well within the bandpass of active optics. The problem is solved to the point where the use of low-expansion material for the meniscus is no longer required. The VLT and Gemini 8-meter mirrors are

---

[1]Bulk temperature change is defined here as a *uniform* change of temperature throughout the entire blank.

9.4 Thermal control for dimensional control

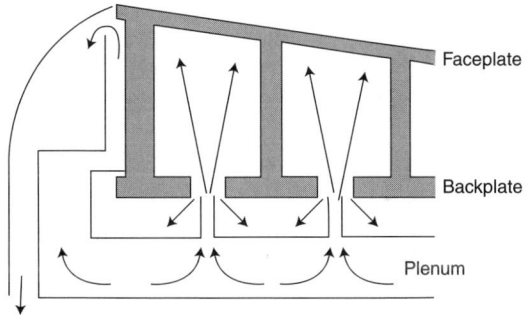

Fig. 9.2. Ventilating a honeycomb mirror to minimize temperature gradients. (From Ref. [2].)

made of low-expansion material, but could just as easily have been made of borosilicate glass or metal.

On the other hand, segmented-mirror systems still face thermally induced figure deformations, even though active optics may be used. The reason is that the active optics system usually corrects for errors between segments but not for wavefront errors within the segments themselves. In such cases, mirror blanks with a low coefficient of thermal expansion (CTE) are essential.

Space mirrors may or may not require temperature control, depending on environmental conditions and the mirror blank's CTE. Although made of low-CTE material (ULE), HST's primary mirror is temperature controlled to 21 ± 1 °C for two reasons. First, because this was the temperature at which it was figured and tested, and its temperature in orbit could not deviate from that by more than 3 °C without affecting image quality; second, because the mirror is exposed to strong and variable diametric gradients depending on Sun orientation and orbital day/night conditions.

Being far from Earth, NGST's primary mirror will not suffer from strong heat input variations. However, according to current designs, thermal transients due to large-angle slews will not be negligible (Fig. 9.3) and may require either slow bandpass active optics control or minimal thermal control of the mirrors.

9.4.2 Controlling optics separation and alignment

Older telescopes were infamous for their focus variations. As temperature dropped during the night, focus would change significantly on a time scale of an hour or less. This was partly due to deformation of the mirror itself, but also to changes in the length of the steel structure supporting the Cassegrain mirror. The effect was corrected manually by observers between (or sometimes during) exposures. Schmidt telescopes, whose focuses are inaccessible to visual checking, were equipped with very simple passive focus controls consisting of low-expansion-coefficient rods (Invar or low-expansion ceramic) resting on the

350   9. Thermal Control

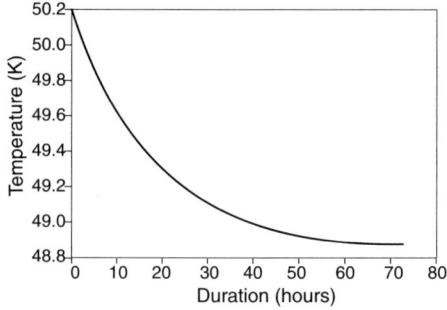

Fig. 9.3. Predicted transient profile for NGST's primary mirror trailing edge following a worst-angle slew. Although relatively small, the temperature drop may be excessive for very long observations.

primary mirror and supporting the focal plane (Fig. 9.4). This method was also recently used to maintain focus on the Sloane telescope.

A simpler solution is to measure the temperature of the tube struts and use a computer look-up table to determine focus offsets.

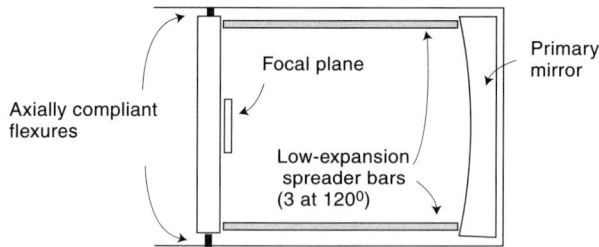

Fig. 9.4. Schmidt telescope focus control using three low-expansion spreading bars.

For space applications, one shies away from mechanisms. They are difficult to build with redundancy and, although generally reliable, their failure can endanger the entire mission. Passive solutions, or at least those having low duty-cycles, are therefore preferable. One possibility is to adopt an athermal design (see Chapter 6), using the same material for both the mirror and the Cassegrain supporting structure. When temperature increases, the expansion of the tube automatically compensates for the change in radius of curvature (hence focus) of the optics. This is the solution adopted for the SIRTF telescope.

The next best scheme is to apply temperature control to the entire system. Heater/sensor systems are highly reliable, have no moving parts, and redundancy can easily be incorporated by adding completely independent sensing/heating loops.

Then, there is the hybrid approach: controlling the temperature of the most critical elements actively and using passive control (i.e., insulation and low-expansion-coefficient material) for less critical elements. This was the solution adopted for HST: the primary and secondary mirrors are temperature controlled, and the structure holding the secondary is made of ultra low-expansion graphite epoxy (Fig. 9.5).

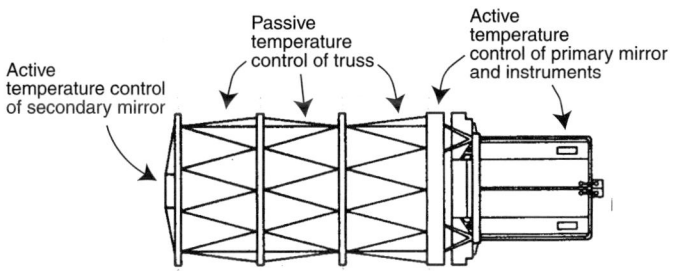

Fig. 9.5. Thermal control of HST. The primary and secondary mirrors are thermally controlled to 21 °C. The "metering truss" is not, but is made of low-expansion material to minimize variations in the primary/secondary distance.

Due to the number of thermal sources involved and possible amplification mechanisms (e.g., "bimetallic effects"), thermal effects can be subtle. Examples are numerous. In spite of careful design, HST does experience a focus change on the order of ±5 μm on an orbital time scale [3]. This periodic focus change, commonly referred to as " breathing," is clearly thermally induced, although the exact mechanism remains something of a mystery. In the case of ground telescopes, the struts at the top of the tube, which face the cold sky on one side and the warmer interior of the dome on the other, can introduce lateral shift of the secondary mirror and, hence, coma. A significant advantage of actively controlled telescopes is that all such problems disappear.

9.5 Avoiding locally induced seeing

In ground-based observatories, thermal instabilities generated by the very presence of the telescope and enclosure can be a source of image degradation, a phenomenon referred as "local seeing" or "dome seeing."

In free convection, temperature gradients and fluctuations are greatest close to the heat-exchange surface, whereas the more distant regions experience the largest velocities [4]. Thus, seeing effects, being caused by temperature fluctuations, decrease rapidly with the distance from the heat-exchange surfaces.

The theory of atmospheric turbulence summarized in Chapter 1 (Section 1.3.5) is also applicable to the turbulence encountered in the local telescope environment, albeit with an outer scale in the range of a centimeter to a few meters (size of dome slit, louvers, free-convection plumes, etc.) rather than tens of meters as found in the atmosphere. According to this theory, the image spread

is given by
$$\theta = k \left[\int C_T^2 dl \right]^{3/5}, \qquad (9.2)$$

where l is the distance along the light path, k is a constant, and C_T^2 is the temperature structure coefficient, which is proportional to the temperature gradient across the turbulent layer. For a single layer of thickness l over which the temperature varies by ΔT, the angular image spread will be

$$\theta = k \frac{\Delta T^{6/5}}{l^{3/5}}. \qquad (9.3)$$

The sources of free convection in the telescope's local environment fall into three categories:

(1) free convection throughout the entire telescope chamber generated by a floor warmer than the ambient air, or by telescope parts that are colder than the ambient air,

(2) convection caused by a temperature difference between the surface of the primary or secondary mirror and ambient air, and

(3) localized free convection generated by heat sources on the telescope or inside the enclosure.

From equation 9.2, we note that when combining the contributions of several sources to the image spread, the individually calculated or estimated image spreads will add up according to a three-fifths power law. For example, the image spread due to the combined effects of the enclosure (θ_{encl}), mirror (θ_{m}), and local heat sources (θ_{hs}) will be

$$\theta = (\theta_{\text{encl}}^{5/3} + \theta_{\text{m}}^{5/3} + \theta_{\text{hs}}^{5/3})^{3/5}. \qquad (9.4)$$

In the following sections, we will look at each of these sources of seeing, quantify them, and discuss ways of mitigating their effects.

9.5.1 Thermal control of the enclosure during the day

Clearly, locally induced seeing will be much reduced if, when night falls, the telescope and telescope chamber are close to being in thermal equilibrium with the outside air. To that end, the first line of defense is the protection afforded by the enclosure during the day in rejecting solar heat and insulating the air inside from the warmer air outside (see Chapter 11). To reduce radiative coupling in daytime between the inner skin of the enclosure and the telescope, the enclosure should be well insulated and the inner skin should have low emissivity. But heat leaks through the enclosure walls are inevitably too large to keep the telescope at nighttime temperatures. Some form of active cooling of the telescope chamber is always required during the day, generally in the form of air conditioning.

The air conditioning system is usually set to the outside air temperature predicted for the coming evening. To the first order, this temperature can be assumed to be equal to that at the beginning of the previous night. However, this estimate can be improved by tracking the exterior temperature during the day, comparing it to that of the previous day, and making corresponding adjustments. Naturally, the air conditioning system is turned off as soon as the enclosure is opened. To avoid the risk of condensation on the optics, a dew-point sensor is used to control the temperature of coolant so that it never falls below the ambient dew point.

9.5.2 Seeing caused by a warmer floor

Using the basic law of equation 9.2, Zago [5] has derived an estimate of enclosure seeing as

$$\theta_{\text{encl}} \approx 20.9 \, D_d^{-1/5} q_s^{4/3}, \tag{9.5}$$

where D_d is the enclosure's diameter and q_s is the surface heat flux. This relationship is plotted in Fig. 9.6

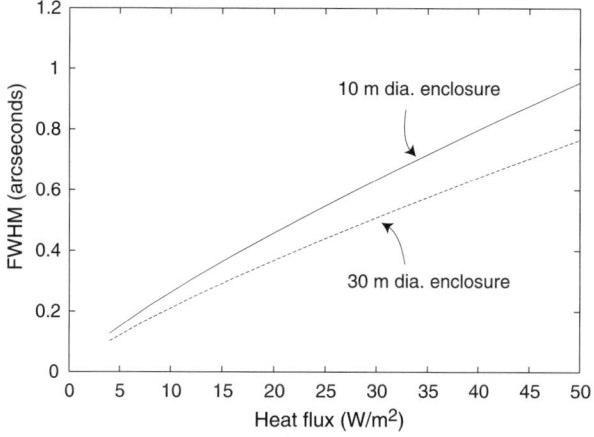

Fig. 9.6. Seeing caused by a warmer enclosure floor for an enclosure diameter of 10 and 30 m.

This relationship indicates that natural convection in the enclosure volume will begin causing significant seeing effects (≈ 0.4 arcsec) at a heat flux on the order of 20 W/m². Considering that a typical free-convection heat transfer rate is 3 W/m² K at the floor, a warm floor will contribute a seeing of about 0.06″ to 0.08″ per °C of floor–air temperature difference. This effect can be mitigated by actively controlling the temperature of the floor to keep it close to that of the ambient night air or, more simply, by insulating it.

Cooling floors were successfully used in several 4-meter class observatories in the 1960s and 1970s. A cooling floor consists of coils embedded in the

354   9. Thermal Control

telescope chamber floor in which a cooling fluid, generally glycol, circulates. The idea was to absorb any heat from below and force stratification in the telescope chamber. This system is extremely effective in blocking heat from the lower floors and also cools the telescope chamber as a whole by radiative coupling. The drawback is relatively high construction cost and high thermal inertia: errors made in the prediction of the nighttime air temperature take several hours to correct. A more economical and potentially better solution is to absorb heat from the lower floors by ventilation.

But if care has been taken to eliminate all inessential heat sources from the building, a "passive" solution, simply relying on insulation, may be better in the end. The reason is that the telescope pier, typically made of concrete, then becomes the single largest source of heat transfer to the telescope chamber. Having a ventilation space between the pier and the telescope chamber is not always practicable; one is better off insulating the entire telescope chamber floor. To minimize thermal inertia, the structural floor covering the insulation should be made of a low-thermal-capacity material such as wood.

9.5.3 Seeing due to heat sources or sinks in the telescope chamber

Naturally, all heat-generating equipment that does not need to stay near the telescope should be located elsewhere in the building or outside of the observatory (see Chapter 11). But some heat sources cannot be avoided. This is the case with electronic cabinets, pumps, and motors that, by function, must be on or near the telescope. An example of the effect of such a source is shown on the left in Fig. 9.7. Hunting for heat sources and cold sinks in an existing observatory can be conveniently done with an infrared camera [6].

Equation 9.5, which relates seeing caused by a warm floor to heat flux, is applicable to other potential sources of free convection located inside the enclosure. However, in view of the smaller exchange surfaces with respect to the inner air volume, the seeing rate per °C of surface–air temperature difference will be lower than for the floor. The practical requirement adopted by ESO is that no heat source in the telescope chamber should generate more than 10 W/m^2. For ease of contractual verification, this condition was translated into the requirement that the surface temperature of all equipment located in the telescope chamber should be within 1.5 °C of the ambient temperature in calm air (no wind). This condition also applies to cold equipment such as cooled instruments, cooling pipes, and hydrostatic pad systems. However, when the equipment is located below the primary mirror (e.g., Cassegrain focus), this condition is relaxed to the range $+1.5/-5$ °C. These are rule-of-thumb specifications and are on the conservative side. Although detailed verification after installation may not be warranted, having such guidelines forces designers to be vigilant and is an incentive to take preventive measures during the design phase rather than dealing with thermal problems after the fact.

9.5 Avoiding locally induced seeing

 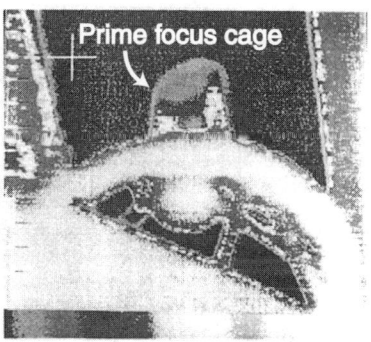

Fig. 9.7. Thermographs of the CFH telescope mount (left) and top of tube (right). At left, the lightest shade corresponds to areas heated by the hydrostatic bearing oil: pads, portion of the supported horseshoe, and oil pipes. This problem was later solved by cooling the oil. At right, the top of the prime focus cage is seen to be cooler (darker on the image) than the rest of the tube due to its wide exposure to the night sky.

General principles for minimizing the effects of local heat sources are as follows.

Small or intermittent heat sources can be dealt with passively. This is done by enclosing the heat source in an insulated shell to dilute the areal density of heat transfer to ambient air and by evacuating the heat generated to a large thermal sink such as the telescope structure (Fig. 9.8, left).

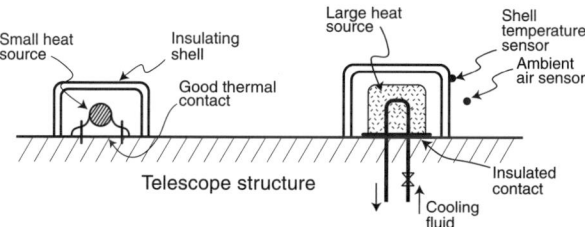

Fig. 9.8. Heat generated by small or intermittent heat sources can be dissipated by conduction to large thermal sinks (left). Large heat sources require active cooling (right).

Large heat sources such as electronics cabinets, telescope and enclosure drive motors, and hydrostatic pad systems require active cooling. This is accomplished by insulating the source from ambient air and supporting structures, and by removing heat via an active cooling system (Fig. 9.8, right). The temperature of the outer shell is maintained equal to that of ambient air by regulating coolant flow.

Continuously activated secondary mirrors are a critical case because they are located directly in the incoming beam. Electromechanical actuators used to adjust the secondary mirror rapidly, whether for fast beam-switching in infrared observations or for the correction of guiding errors due to wind loading, will produce heat which may be detrimental to seeing unless removed by a cooling circuit. The magnitude of the effect can be estimated from the following formula which was established empirically, based on experimental measurements at the ESO 2.2-meter telescope [5]:

$$\theta = 0.018 \, Q^{4/5} D^{-9/5}, \tag{9.6}$$

where θ is the FWHM image spread in arcseconds, Q is the total heat flow from the secondary unit in watts, and D is the secondary mirror's diameter in meters.

9.5.4 Seeing due to telescope structure cold areas

Steel structures used in telescopes generally have a small thermal inertia with a time scale on the order of half an hour.[2] They are not a source of local seeing, with the exception of the top of the tube, which has a wide view of the sky and hence cools down (Fig. 9.7, right). As a result of this radiative cooling, the upper tube ring, spider, and secondary mirror or prime focus cage generate a slow flow of cold air which falls down the optical beam. This effect can be mitigated by coating these areas to reduce radiative coupling.

Metal parts not located directly in the light beam (e.g., upper ring) can be coated with low-emissivity paint, an aluminum flake pigment paint for example, which is gray in the visible but has low emissivity in the infrared. Metal parts located in the optical beam (spider, secondary mirror) should be coated with a product that is black in the visible to minimize scatter, and has low emissivity in the infrared (e.g., nickel foil such as Maxorb[3] or one of the more exotic products used in space applications — see Chapter 5). A potentially better solution consists of covering the telescope structure with insulation topped with aluminum foil, as was done on the Subaru telescope.

9.5.5 Mirror seeing

Mirror seeing is caused by natural convection over the optical surface whenever that surface is warmer or colder than ambient air. The effect is generated in the region just above the viscous-conductive layer, where temperature fluctuations are the largest and most intermittent. Most of the degradation occurs in a thin but very turbulent layer floating a few millimeters above the surface.

---

[2]From this point of view, a thin-walled tube of large diameter is preferable to a thick-walled tube of small diameter with similar structural performance.

[3]Maxorb is a product of Special Metals.

Mirror in still air

The results of experiments performed by various researchers [7, 8, 5, 9] have led to the widely accepted relationship valid for still air:

$$\theta_m = 0.4 \, \Delta T_m^{6/5}, \tag{9.7}$$

where ΔT_m is the difference between the optical mirror surface and ambient air temperatures in degrees Celsius and θ_m is the resulting angular image spread (FWHM) in arcseconds.

Mirror in forced flow

Mirror seeing diminishes when the mirror is ventilated. As suggested by a number of laboratory experiments [5], seeing seems to vary as a function of air velocity above the mirror according to

$$\theta_m = 0.18 \, \mathrm{Fr}^{-0.3} \, \Delta T_m, \tag{9.8}$$

where Fr is the densimetric Froude number[4] defined as

$$\mathrm{Fr} = \frac{TV^2}{\Delta T g D}, \tag{9.9}$$

in which V is the wind velocity, D is the diameter of the mirror, and g is the acceleration of gravity (Fig. 9.9). This law indicates that larger mirrors should produce more seeing than smaller ones under the same conditions of ventilation.

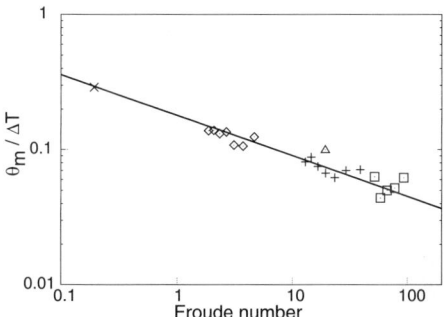

Fig. 9.9. The ratio of image spread to temperature difference between the mirror surface and ambient air ($\theta_m/\Delta T$) as a function of the Froude number for a ventilated mirror. The line represents the function $\theta_m/\Delta T = 0.18 \, \mathrm{Fr}^{-0.3}$.

To benefit from natural flushing by wind entering the enclosure, the primary mirror should be as free from surrounding structures as possible. The

---

[4]The densimetric Froude number, which takes thermal effects into account, is a modification of the usual Froude number.

configuration of the Gemini telescopes is excellent in this respect. Because the telescope has no Nasmyth focus, it was possible to locate the primary mirror high up near the altitude axis, with little surrounding obstruction, instead of burying it as usual in the tube "cavity."

If wind flushing a mirror will decrease seeing, it will also, unfortunately, increase dynamic effects. As shown in Fig. 9.10, there is an optimal wind speed at which flushing is beneficial while the wind's dynamic effects on the mirror remain acceptable.

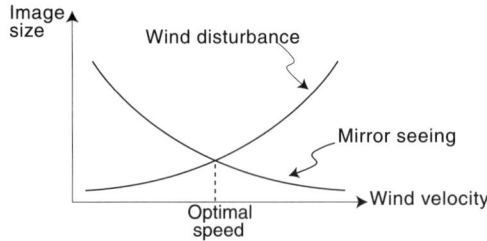

Fig. 9.10. Notional plot showing how mirror seeing improves with wind velocity while dynamic effects worsen. Optimal wind speed can be maintained by adjusting the enclosure's louvers and windscreen.

In the case of the VLT mirror, wind will not perturb it significantly up to about 1.5 m/s (a wind of that speed produces astigmatism of 150 nm rms, which is considered acceptable). Since this value is sufficient to avoid mirror seeing, one tries to maintain this wind speed near the mirror by opening or closing the enclosure's louvers and windscreen. But this may mean a much higher wind speed at the top of the tube, which faces the enclosure slit, and thus cause increased tube shake. In the case of the VLT again, this problem is solved by measuring wind speed at the mirror and at the top of the tube, and automatically opening and closing the louvers and windscreen with an algorithm that attempts to minimize an expression of the form [10]

$$(V_\mathrm{m} - 1.5)^2 + kV_\mathrm{tr}, \tag{9.10}$$

where V_m and V_tr are wind velocity in meters per second near the primary mirror and at the tube top ring, respectively.

Active mirror cooling

General thermal control of the telescope chamber during daytime and wind flushing of the primary mirror at night may not always suffice to ensure low mirror seeing. This is the case when the actual nighttime air temperature differs from the prediction used for daytime temperature control, for example, due to the passage of a front. The mirror temperature may then take several hours to come into equilibrium with the ambient. This is also the case at sites where average temperatures systematically drop during the night by more than about 2 °C.

Lightweighted mirrors have relatively small thermal inertia, and active inner ventilation as shown in Fig. 9.2 will cope well in such situations. Ventilation works poorly with solid mirrors because of the smaller heat-exchange area.

The solution adopted for the VLT and Gemini solid meniscus mirrors uses radiative cooling of the mirror back [11], [12], [13]. As shown in Fig. 9.11, a cooling plate is placed close to the back surface of the meniscus mirror. The plate is insulated on its rear face and is cooled by a glycol-water mixture circulating in a dense system of coils. In the case of the VLT meniscus mirror (17 cm thick), the radiative cooling system enables the front mirror surface to track nighttime ambient temperature changes as high as 0.5 °C/h. During the day, fans behind the mirror are used to increase the convective exchange, which makes it possible to cool the mirror by 2 °C in 6 hours when necessary.

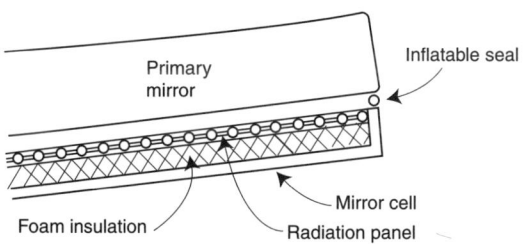

Fig. 9.11. Radiative cooling of the back of a solid meniscus mirror (Gemini).

Mirror surface heating

Mirrors made of low-expansion materials do not suffer from temperature gradients inside the blank. Their one problem is mirror seeing. Since only the temperature of the front surface affects mirror seeing, why not simply control the temperature of the optical surface and forget about the bulk temperature? This is the solution that was proposed and tested for the Gemini telescopes [13], [14, 15].

The idea is, during the day, to cool the rear surface of the mirror to below the expected temperature of the coming night and, at night, to heat the front surface by feeding a current into the aluminum optical coating, so as to match ambient air temperature. By increasing or decreasing the heating current, the temperature of the mirror surface can be brought to track air temperature changes. The current must be very uniform, which requires a large number of electrodes located on the outer edge of the mirror and at the central hole. The energy required is estimated at about 1 kW.

References

[1] Gilmore, D.G., ed., *Satellite Thermal Control Handbook*, Aerospace Corp., 1994.
[2] Cheng, A.Y.S. and Angel, J.R.P., Steps toward 8 m honeycomb mirrors VIII: Design and demonstration of a system of thermal control, SPIE Proc., Vol. 628, p. 536, 1986.
[3] Hasan, H. and Bely P.Y., Effect of OTA breathing on Hubble Space Telescope images, in *The Restoration of HST Images and Spectra II*, Hanish, R.J. and White, R.L., eds., Space Telescope Science Institute, 1994, p. 157.
[4] Townsend A. A., *The Structure of Turbulent Shear Flow*, Cambridge University Press, 1976, pp. 381–392.
[5] Zago L., An engineering handbook for local and dome seeing, SPIE Proc., Vol. 2871, p. 726, 1996.
[6] Williams, J.T., Beckers, J.M., Salmon, D., and Kern, P., IR thermography and observatory thermal pollution, SPIE Proc., Vol. 628, p. 30, 1986.
[7] Iye M. et al., Evaluation of seeing on a 62 cm mirror, PASP, Vol. 103, p. 712, 1991.
[8] Lowne C. M., An investigation of the effects of mirror temperature upon telescope seeing, Monthly Notice Royal Astron. Soc., Vol. 188, p. 249, 1979.
[9] Racine, R., Salmon, D., Cowley, D., and Sovka, J., Mirror, dome and natural seeing at CFHT, PASP, Vol. 103, p. 1020, 1991.
[10] Cullum, M. and Spyromilio, J., Thermal and wind control of the VLT, SPIE Proc., Vol. 4004, p. 194, 2000.
[11] Wilson, R.N., *Reflecting Telescope Optics II*, Springer-Verlag, 1999, pp. 332–335.
[12] Bäumer, V. and Sacré, P., Operational model for VLT temperature and flow control, SPIE Proc., Vol. 2871, p. 657, 1997.
[13] Greenhalgh, R.J.S., Stepp, L., and Hanson, E, The Gemini primary mirror thermal management system, SPIE Proc., Vol. 2199, p. 911, 1994.
[14] Hanson, E., Hagelbarger, D. and Pearson, E., Prototype testing of a surface heating system for the Gemini 8 m telescopes, SPIE Proc., Vol. 2871, p. 667, 1997.
[15] Bohannan, B., Pearson, E.T., and Hagelbarger, D., Thermal control of classical astronomical primary mirrors, SPIE Proc., Vol. 4003, p. 406, 2000.

Bibliography

Gilmore, D.G., ed., *Satellite Thermal Control Handbook*, Aerospace Corp., 1994.

Schember, H. and Rapp, D., *Key issues in the thermal design of spaceborne cryogenic infrared instruments*, in *Optomechanical Design*, Yoder, P., ed., SPIE CR43, SPIE, 1992, p. 111.

Siegel, R., Howell, J.R., *Thermal Radiation Heat Transfer*, McGraw-Hill, 1972.

10
Integration and Verification

Integration and verification is the process by which, after manufacture, the various elements of the observatory are assembled to form a complete functional system, then verified to confirm that the observatory complies with the scientific requirements. The process culminates with the commissioning of the observatory on the sky. As illustrated in Fig. 10.1, integration and verification (I & V) is a "stair-step" process. Each major phase must be fully validated before the next step can be taken.

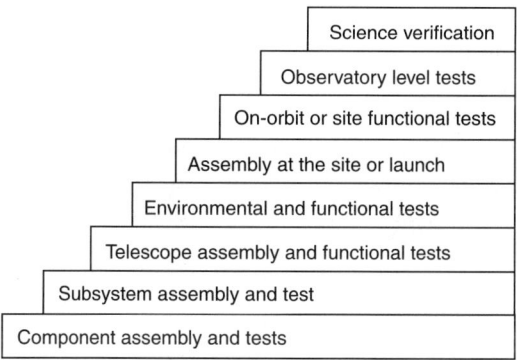

Fig. 10.1. Stair-step process for integration and verification. (Adapted from Ref. [1].)

In our discussion of this process, we will use the following hardware nomenclature:

- **Parts** are individual parts such as optical elements, electronic integrated circuits, and mechanical bearings.
- A **component** is a complete unit such as an electromechanical actuator or an electronic "black box."
- An **assembly** is a functional group of parts and components such as a mirror on its active mount.
- A **subsystem** is all the components and assemblies comprising a major functional element of the observatory, such as the telescope, a science instrument, or the enclosure.

The integration and verification procedure varies depending on agency policies, contractual arrangements, availability of test equipment, and the size and expertise of the observatory staff. For ground observatories, integration and verification are generally the responsibility of the contractor only up to component or subsystem level, whereas the commissioning of the entire observatory is performed by the observatory staff. Space observatories, on the other hand, are generally under the responsibility of a prime contractor who is in charge of the entire process, including on-orbit verification. In all cases, however, the observatory or space agency staff must take part in the verification process; they must supervise the work, perform independent verifications when possible, and, most importantly, become familiar with the idiosyncrasies of the hardware they will have to deal with later.

10.1 Integration and verification program, methods, and techniques

The I & V program should be developed early, in conjunction with the design, as it has strong implications on cost, risk, schedule, and often on the design itself. For example, a given design may be so novel or fine-tuned that it will require exhaustive testing to ensure its validity, whereas another approach may be so traditional and conservative that no verification is required at all. As another example, it might, in some cases, be more economical to do a detailed structural analysis of a given piece of equipment than to perform a modal test after it is built.

In order to establish a rational verification program, it will thus be important to perform detailed trade-off studies to support decisions about verification methods and the selection of test facilities and to assess how testing might be used to mitigate programmatic risks.

For ground telescopes, it is important to determine the optimal level of assembly and verification to be done in the shop. Observatories are generally in remote, high-altitude sites. This is not the place to perform functional tests requiring complex equipment and specialized personnel; nor is it the ideal place to rework parts. Although preassembly functional testing of a telescope in the shop is expensive, it does minimize unpleasant surprises during final assembly at the site, and it will usually pay for itself. Some shortcuts can be taken, however. In general, shop assembly is done without the optics, which are tested on their mirror mounts separately in a cleaner, safer environment.

For space telescopes, similar decisions must be made concerning end-to-end testing. Because of the difficulty in mimicking the zero-gravity and thermal environment of space, true end-to-end testing of the optics is generally not practical. But extensive functional testing of the fully assembled telescope is the norm.

10.1.1 Verification methods

Methods used for verification fall into three broad categories, namely, in increasing order of confidence: inspection, analysis, and test. Deciding which one is best for each type of equipment and system has a strong bearing on risks, cost, and schedule, and so should be made with care.

Verification by **inspection** is the lowest level of verification. It consists of a physical evaluation of the equipment to verify its dimensions, construction features, and workmanship.

Verification by **analysis** consists of verifying that the design of the equipment meets the applicable environmental and functional requirements. This can be done by examination and critique of the design, computer modeling, or hardware simulation.

Verification by **test** provides the ultimate level of confidence. It consists of subjecting the equipment to actual tests to evaluate its compliance with requirements. These tests fall into two categories: functional and environmental. Functional testing comprises a series of electrical and mechanical tests conducted to establish the satisfactory performance of the equipment. For space equipment, functional testing is generally carried out under ambient conditions. Environmental testing is a series of tests conducted to assure that the equipment can sustain the environment it will be subjected to and can perform satisfactorily in that environment. These tests, generally applicable only to space hardware, include vibration, acoustic, vacuum, and thermal testing to verify that the equipment will survive launch and on-orbit conditions. Environmental testing may be combined with functional testing to verify the performance of critical equipment and subsystems in the space environment.

10.1.2 Incremental verification

A key concept in the verification process is that of "incremental verification." Integration and verification should proceed gradually and in concert, with every assembly step being verified before the next level of assembly is initiated. If not, problems that could have been resolved at the component or assembly level with minimal cost and schedule impact will crop up later, when integration is well advanced, resulting in costly work stoppage and delays. This is sometimes referred to as the "marching army syndrome": as integration progresses, technical complexity and the number of people involved increase strongly, so that the further along one is in the integration process, the more disruptive an otherwise trivial mishap will be.

This is especially true of space observatories: in the months preceeding launch, when a large number of contractors are involved and very expensive testing equipment is being used, the cost per day is in the hundreds of thousands of dollars range, whether or not any work is actually accomplished. This is not the time to spot a faulty component, which stops the assembly or testing process and requires extensive dismantling. On the other hand, the qualification of parts and assemblies can be expensive; one must not "overtest" when risks are small. Determining the optimal set of validation procedures is therefore a matter of judgment and must take into consideration the program risks involved.

The verification flow is the reverse of the requirements decomposition flow which led to final design and manufacture (Fig. 10.2). The successive verifications of components, assemblies, subsystems, and complete systems correspond, at least conceptually, to the successive levels of the requirements. An exact one-to-one match is not always applicable, but this general approach will help guide the definition of the verification steps and the quantification of performance.

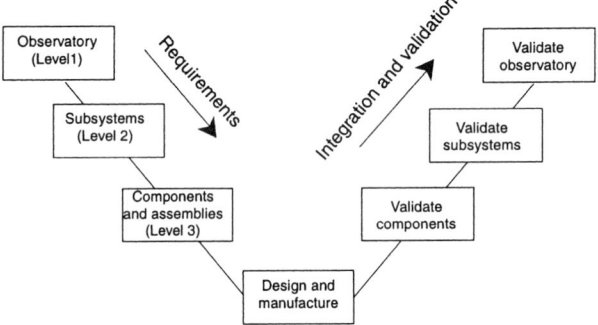

Fig. 10.2. The "V-process model" in which each verification step is the counterpart of a requirements decomposition level.

As an example, a simplified diagram of the verification flow for NGST is shown in Fig. 10.3 [2].

10.1.3 Verification requirements matrix

The process described above is best handled with a "requirements matrix." This is a document that describes how each of the design requirements (e.g., levels 1, 2, 3, etc.; see Chapter 3) is to be verified and at what stage in the manufacture or integration and verification phases it is to occur. This

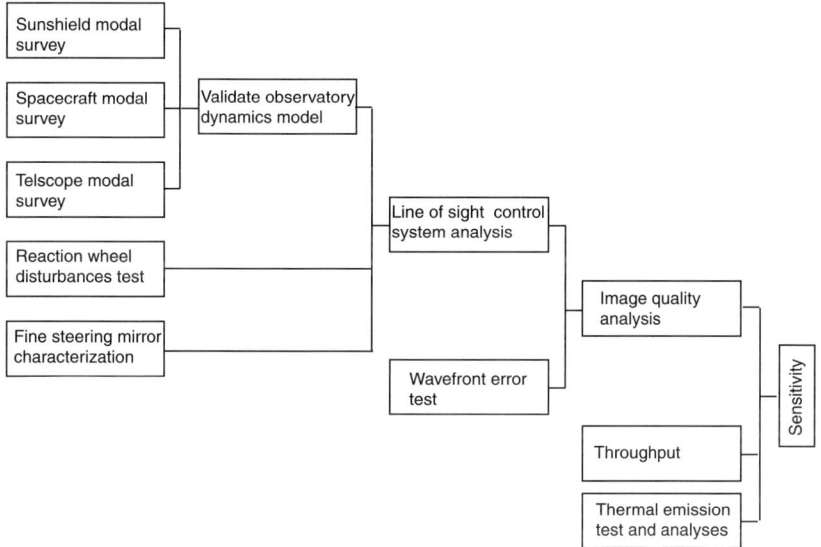

Fig. 10.3. Flow of tests and analysis to verify NGST sensitivity (breakdown is shown only for line-of-sight jitter).

document should be prepared early in the design phase, typically for the preliminary design review, and forms the basis of the verification program.

10.1.4 Verification based on end-to-end computer modeling

A refinement of the verification requirements matrix consists of using an end-to-end mathematical model of the observatory. As discussed in Chapter 3, such end-to-end computer models, when detailed enough, can determine the final performance of the observatory with high fidelity. The verification process then consists of gradually validating the assumptions in the mathematical model as the hardware is being produced and its performance evaluated by actual tests.

Compared to the verification matrix method, this has the advantage of putting the spotlight on the performance of the complete observatory instead of on details which may or may not be critical. This method allows rational decisions to be made concerning acceptance or rejection of tested hardware. When hardware does not quite meet the allocated performance budget, the final performance of the observatory can be quickly evaluated to see if one can mitigate the corresponding impact by tightening requirements on other hardware. Conversely, if a piece of hardware meets its requirements with ample margin, the model can be exercised to see if this permits the requirements for the remaining hardware to be relaxed.

10.2 Observatory validation

We are now at the stage where the observatory has been fully assembled on site or launched in the case of space telescopes. The final verification consists of confirming that the observatory is functionally operational and can accomplish its intended scientific purpose. Verification at the system level is referred to as *validation* This validation is realized in two steps: engineering verification and science verification.

10.2.1 Engineering verification

Engineering verification consists of (1) repeating the major functional tests which were performed during the construction and assembly phases and (2) performing tests at the system level in order to verify the overall *engineering* performance of the observatory. These performance tests should be tailored to verify the essence of the high-level requirements (e.g., level 1). As an example, the series of tests performed to verify observatory performance of the Hubble Space Telescope after deployment and functional checkout were as follows:

- Target acquisition
- Pointing stability over 24 hours
- Line-of-sight jitter
- Moving target tracking
- Optical image quality
- Optical throughput at various wavelengths
- Stray light due to earthshine and moonshine and to spacecraft glow
- Noninterference during parallel observations

10.2.2 Science verification

The purpose of "science verification" is to ensure that the observatory is operational as a scientific tool and meets its original requirements. As opposed to observatory-level tests, which encapsulate observatory performance in engineering terms, science verification is the acid test, taking real scientific programs as "test particles." These tests are defined by astronomers and executed as if they were true science proposals in order to exercise the system from the user's viewpoint.

In this final I & V phase, every observation mode should be exercised and tested. Observing modes are defined as combinations of telescope and instrumental parameters (e.g., spectral resolution, field of view, or focus used), that lead to significantly different types of observation. For example, imaging and grism spectroscopy with a given instrument constitute two modes, whereas observations made with different filters or grisms belong to the same mode.

This can be a long process: the HST science verification lasted 6 months. And there is a fine line to walk between keeping the reins on a new astronomical tool long enough to make sure that all the knobs turn smoothly and turning it over early to observers, who may become frustrated by capabilities that are untested and still unreliable.

References

[1] Lewis, K.D., and Arthur, L.C., Spacecraft integration and test, in *Spacecraft Structures and Mechanisms — From Concept to Launch*, Sarafin, T.P. and Larson, W.J., eds., Kluwer Academic Publishers, 1998, p. 819.

[2] Menzel, M., Triebes, K., Leary, D., and Krim, M., *A Strawman Verification, Integration and Test Program for NGST*, NGST Monograph 4, NASA, Goddard Space Flight Center, 2000.

Bibliography

Geballe, T. and Gillett, F., Gemini system verification, Gemini Newsletter No. 18, 1999.

Gilmozzi, R., The first six months of VLT Science operations, ESO Messenger, No. 98, p. 25, 1999.

Gray, P.M., Assembly and integration to first light of the four VLT telescopes, SPIE Proc., Vol. 4004, p. 1, 2000.

Menzel, M., Triebes, K., Leary, D., and Krim, M., *A Strawman Verification, Integration and Test Program for NGST*, NGST Monograph 4, NASA, Goddard Space Flight Center, 2000.

Sarafin, T.P. and Larson, W.J., eds., *Spacecraft Structures and Mechanisms — From Concept to Launch*, Kluwer Academic Publishers, 1998.

Wallander, A., Spyromilio, J., and Wirenstrand, K., *Commissioning VLT unit telescopes: methods and results*, SPIE Proc., Vol. 4004, p. 234, 2000.

11
Observatory Enclosure

Ground-based optical telescopes contain many components that must be protected from moisture, wind, lightning, and contamination by atmospheric particles. This protection is provided by the telescope "enclosure."[1] The enclosure and associated building can account for 20–30% of the total cost of the observatory. Recent trends have therefore been to keep cost down by using primary mirrors with faster f-ratios to reduce telescope tube length and by reducing the clearance between the telescope and the enclosure. Figure 11.1 shows how dramatic the reduction of enclosure size compared to telescope size has been over the last 25 years, starting with the MMT.

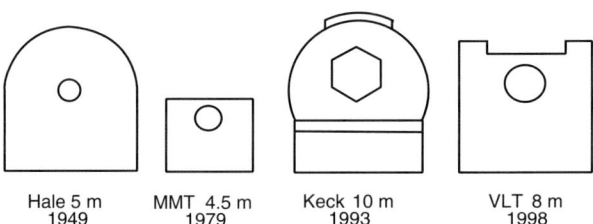

Fig. 11.1. Evolution of enclosure size compared to telescope diameter. At left, a representative of traditional design. The three examples of modern design shown on the right exhibit a much reduced enclosure-size/aperture-size ratio.

---

[1] The term "enclosure" is now preferred to the nonspecific "building" and to the overly specific "dome" that, in the strict sense, refers only to more or less spherical shapes.

Strictly speaking, the term "enclosure" refers only to that part of the observatory that houses the telescope. Although this chapter will mostly deal with the requirements and design issues pertinent to the enclosure proper, we will also briefly cover some of the issues related to associated buildings, piers, foundations, and handling equipment.

11.1 Enclosure functions and requirements

In the closed position, the enclosure should protect the telescope and its instrumentation against sun, rain, snow, strong winds, dust, and lightning. In the open position, the enclosure should allow the telescope free access to the sky in all directions, typically down to 10° above the horizon. It should also be designed to minimize thermal effects that cause image degradation. In addition to these three main requirements, enclosures generally also provide:

- housing for the telescope and instrument control room,
- housing for supporting equipment (coating tanks, hydraulic pumps, pneumatic systems, control and data processing systems, air conditioning),
- laboratory space for scientific instrument setup and calibration,
- storage areas for spare parts and scientific instruments not currently in use,
- structural support for handling equipment (cranes, carriages, lifting equipment),
- office space and rest areas for observers and technical staff,
- general access and circulation (personnel and freight elevators),
- stray light control (from the Moon and artificial lighting).

11.2 Overall enclosure configuration

Enclosures can be classified into three general types: traditional, corotating, and retractable (Fig. 11.2). The main advantages and drawbacks of each of these types are as follows.

Traditional dome

In this common approach, the enclosure fully clears the telescope in all directions. Although this requires the enclosure to be larger than a corotating one, the large volume has the advantage of reducing wind speed inside. This design also has numerous operational advantages. Since the enclosure can be rotated independently of the telescope, cranes or hoists can be mounted on the enclosure to service the telescope and instruments. The telescope can be

370   11. Observatory Enclosure

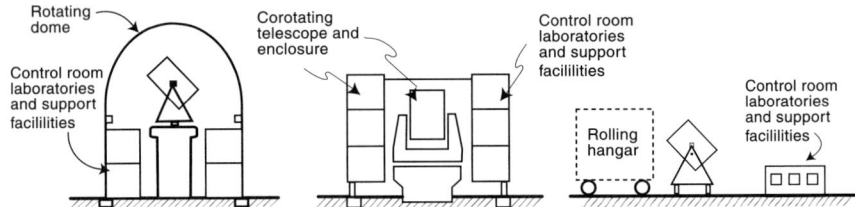

Fig. 11.2. The three types of enclosure: traditional dome (left), corotating enclosure (middle), and complete open-air operation (right) with a shelter rolled over the telescope during the day or in inclement weather.

moved during the day for maintenance or setups without the need to rotate the enclosure with it. Finally, the free area surrounding the telescope can be used for storage or handling. The large volume of traditional domes, on the other hand, fosters stagnant air pockets and vorticity which can occur when wind blows in through the slit at an angle (Fig. 11.3) [1]. These effects can be mitigated by having a relatively large slit and incorporating louvers and fans in the enclosure shell to ensure adequate flushing.

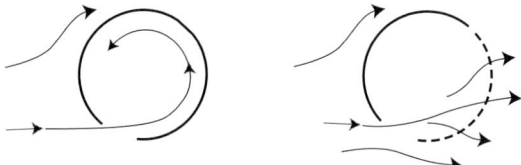

Fig. 11.3. With the slit at about 40° from wind direction, a traditional dome can experience internal vorticity with air velocity close to that of the external wind (left). The effect disappears with the use of louvers. Louvers also significantly improve flushing of the enclosure for all wind angles (right).

Corotating building/enclosure

This is the approach that the MMT [2] pioneered and that was also used for NTT and the Subaru telescope [3]. The office, laboratory, and storage areas are located on either side of the telescope and separated from the telescope chamber by well-insulated, low-thermal-capacity walls. The overall size of the building/enclosure is thus minimized and cost is reduced accordingly. The two walls sandwiching the telescope guide the air stream smoothly, ensuring good flushing (Fig. 11.4).

Air funneling, on the other hand, increases the wind speed around the telescope and this leads to higher structural vibrations. Laboratory and office space also suffer from vibrations induced by rotation and wind action. Other drawbacks of the corotating system are that the rotating mass is significantly increased, and electrical and fluids feedlines must have relatively complex

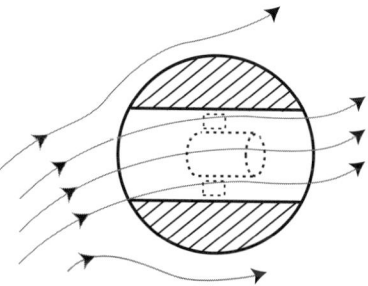

Fig. 11.4. Air funneling in a narrow, corotating enclosure ensures good ventilation but increases wind shake at high wind speeds.

wrap-up systems or slip joints. Handling and storage space in the telescope chamber are also seriously restricted.

Rolling hangar or retractable enclosure

In principle, this approach eliminates all enclosure-induced thermal effects. In the case of a rolling hangar, the structure must be stored far enough away from the telescope (about six times its scale length) and preferably on the side opposite prevailing winds so that its wake will not affect the telescope. All auxiliary equipment must also be housed far enough away, or with a sufficiently insulated barrier, to limit heat generation near the telescope. This solution was used for the Sloan Digital Sky Survey 2.5-meter telescope. With the enclosure rolled away, the telescope is protected by a wind baffle driven in altitude and azimuth independently of the telescope [4].

An alternative to the rolling hangar is the retractable dome, a solution which was originally proposed for the VLT (Fig. 11.5).

Fig. 11.5. Early VLT concept for open-air operation. Enclosures made of fabric were to protect each telescope during the day and be retracted at night. A windscreen made of tiltable louvers was to protect the telescopes from prevailing winds [5, 6, 7]. This approach was not implemented because telescope and mirror wind shake was determined to be excessive [8].

Although the open-air solution is considered best from the thermal point of view, it has not been used for large telescopes because of the difficulty in implementing a windscreen effective in strong winds and for all wind directions. Windscreens also transfer the energy in the wind pressure from its natural low spatial frequency into the higher spatial frequency corresponding to the windscreen gap length, which makes figure control of large, active, thin mirrors more difficult. This effect is also present in traditional enclosures, but with lower amplitudes because of the protection they afford.

11.3 Height of telescope chamber above the ground

Since the surface layer can be an important contributor to seeing (see Chapter 12, Section 12.2.1), it is important that the telescope be located above this layer and that the enclosure not interact with the corresponding flow and risk enhancing turbulence in and around the enclosure.

At the best astronomical sites, the depth of the surface layer can be as little as 5–10 m overall, but this depends strongly on local topography and the statistics of wind direction. This height can be determined by microthermal measurements [9, 10] or, more economically, by computer modeling [11]. The purely aerodynamic characteristics of this layer can be determined by wind-tunnel studies or, *in situ*, using a vertical pole with streamers at appropriate intervals (but this will provide no information on the thermal effects, which can be predominant). In general, it is found that the telescope chamber floor should be located at a height of 10–20 m to be well above the surface layer at night. Locating the telescope chamber relatively high above ground also minimizes the problem of dust blowing in.

11.4 Wind protection and flushing

11.4.1 Basic principles

During observations, the most important role of an enclosure is to protect the telescope from wind disturbances. Thin active mirrors are perturbed when exposed to winds in excess of 2 m/s and telescope shake usually begins to be detrimental to pointing when wind much exceeds 5 m/s. Good observatory sites are, by nature, unprotected and windy. Since, in the interest of observing efficiency, it is desirable to be able to observe in winds up to 20 m/s, an attenuation of an order of magnitude is required. As shown in Fig. 11.6, this is roughly what a typical enclosure will provide (see also Fig. 7.27).

Unfortunately, this reduction of wind speed is accompanied by a modification of the wind's power spectrum. In general, the added turbulence will shift energy from lower to higher frequencies, moving low-frequency power into the 1–5 Hz range. This can cause problems for telescope control systems and ex-

11.4 Wind protection and flushing

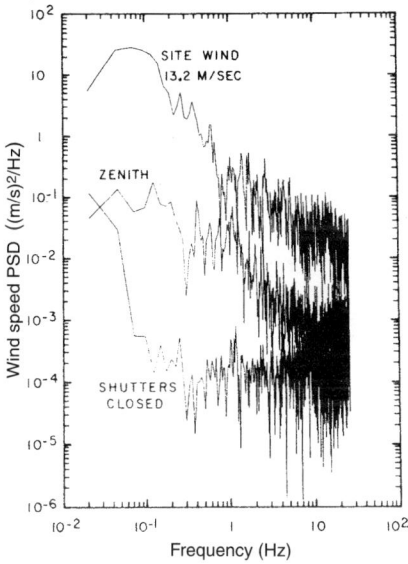

Fig. 11.6. Power spectrum density of the wind inside a typical enclosure with the telescope pointing at zenith (mid-curve) compared to that of the wind outside (upper curve). The power spectrum density for the closed enclosure is shown for reference (lower curve). (From Ref. [12].)

cite resonances in the telescope structure. Fortunately, however, the effect is significant only if most of the volume inside the enclosure is characterized by the presence of well-developed turbulence. Otherwise, the total power shifted to higher frequencies is not critical.

On the other hand, some degree of wind-forced flushing of the telescope chamber is beneficial. It removes air that has been heated or cooled by interior surfaces before that air can pass through the telescope's light path and degrade image quality. The air velocity required for well-managed flushing is modest. A uniform flow of 1 m/s flushes a telescope chamber 30 m in diameter 120 times per hour. In practice, flow is likely to be somewhat nonuniform, but a high flushing rate and good shielding from wind can still be achieved through proper design of the openings.

Extensive wind- and water-tunnel experiments have been done to study airflow in and around enclosures. The results of these studies can be summarized as follows.

- Flushing rates of more than 50 times per hour can be achieved for all orientations of the telescope enclosure and for the median wind speed at a typical telescope site (5 to 8 m/s).
- The walls of the enclosure should be at least 20% porous, and up to 50% in sites with low median wind speed.

- Ventilation openings on flat surfaces are more effective over a larger range of angles with respect to wind direction than those on round surfaces (e.g., spherical domes). This is because the negative pressure induced by flow over round surfaces prevents air from entering. Consequently, only openings that are directly upstream accept air.
- Openings must be provided in all major wall surfaces. This may affect the location of support areas and large equipment. Unventilated corners, stairwells, etc. should be avoided.
- The wake of a telescope enclosure can affect other telescopes at the same site. The wake is turbulent and extends for about six enclosure dimensions downstream.

11.4.2 Windscreens and louvers

Since ventilation helps minimize dome seeing, one will want a fully open slit at low wind speed. A reduced slit opening will be preferable when wind speed is high, however, in order to avoid telescope shake. This is accomplished by moving the shutter down (if of the up-and-over type) and using a windscreen. A windscreen usually consists of a series of panels which are raised along the lower part of the slit. In the VLT enclosure, windscreen porosity can be adjusted from 10% to 60% by rotating the windscreen flaps in order to optimize the "vibration versus ventilation" compromise (see Chapter 9, Section 9.5.5). Natural ventilation can be effective even in moderate winds, provided that the openings are large enough, making fan-forced ventilation unnecessary. Controllable openings should be provided over the full height of the enclosure, with larger ones at the level of the primary mirror (Fig. 11.7).

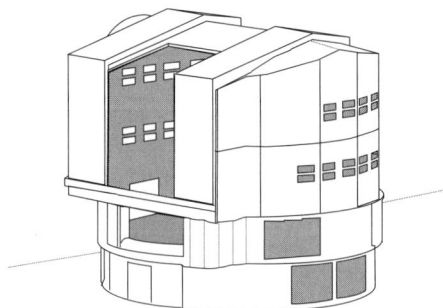

Fig. 11.7. The openings and louvers on the VLT enclosures. A variable-porosity windscreen can be raised to protect the lower part of the telescope.

11.4.3 Wind- and water-tunnel studies and numerical modeling

The flow in and around enclosures can be rather easily studied in wind and water tunnels. This empirical technique is commonly used to optimize enclosure shape and flushing [14, 15, 16, 17].

In water-tunnel studies, water flows at about 10 cm/s around 1/100 to 1/200 scale models. Dye is injected into the water to trace the flow around and through the model. The Reynolds number[2] for the model system is much smaller than for the full-scale enclosure. However, as long as the Reynolds number is above 4000 and edges and surfaces are not rounded, flow is similar for both systems. Small-scale round surfaces can be modeled if roughness is added to the surface to cause the boundary layer to become turbulent. Examples of enclosure types that have been subjected to water-tunnel studies are shown in Fig. 11.8.

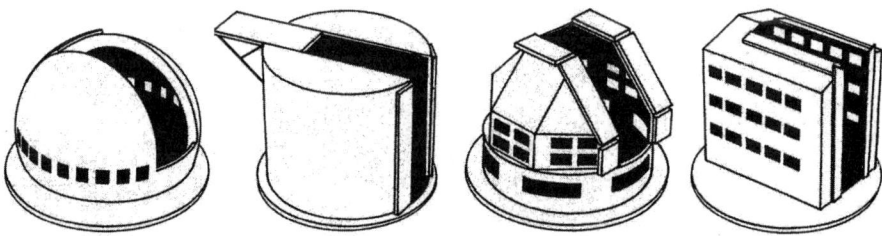

Fig. 11.8. Typical enclosure shapes that were used to study aerodynamic effects and flushing in water-tunnel experiments. The octagonal and rectangular shapes, third and fourth from left, were found to be the most efficient for flushing. (From Ref. [14].)

Numerical simulations are inexpensive alternates to wind- and water-tunnel studies and they also permit the study of thermal effects. These simulations, referred to as "computational fluid dynamics" (CFD), can be nonlinear and thus can examine the role of flushing, turbulence, and the resulting effect on seeing [18]. They are useful for comparing various enclosure designs and for optimizing the size and location of vents. Numerical simulations are a valuable complement to wind- and water-tunnel tests. Although these tests can incorporate more complex geometries, it is difficult to reproduce flows at the proper Reynolds numbers because of the small scale of the models.

---

[2]The Reynolds number is a dimensionless parameter used for assessing whether a flow is laminar or turbulent; see glossary.

11.4.4 Acoustic modes in the enclosure

It is worth noting that an enclosure, with the shutter open and ventilating vents closed, can act as a "Helmholtz resonator" when excited by incoming wind. The natural frequency is given by

$$f \simeq \frac{c}{2\pi}\sqrt{\frac{a}{V}}, \tag{11.1}$$

where c is the speed of sound, a is the characteristic size of the opening, and V is the volume of air inside the enclosure [13]. The frequency is independent of the shape of the enclosure. For a 30 m diameter enclosure, the frequency is on the order of a few Hertz, a frequency low enough to be a potential problem for thin active mirrors. However, this effect has so far never been identified in existing enclosures.

11.5 Thermal design

11.5.1 Basic principles

If the enclosure protects the telescope and telescope chamber from the Sun's heat during the day, it can be a source of image degradation at night because air heated or cooled by the surfaces of the enclosure may be transported into the line of sight. The problem is almost exclusively due to the mixing of air parcels at different temperatures from the ambient; pressure fluctuations due to aerodynamic effects are negligible because they do not significantly affect the refraction index of air (see Chapter 1).

For visual observers of the nineteenth century, it was common knowledge that refractors can have better seeing than reflecting telescopes due to the stability of the air column inside the tube. This led Bernard Lyot in the 1940s [19] to propose enclosing reflectors in a tube with a glass plate at the top and sealing the space between the tube and the enclosure by means of bellows (Fig. 11.9). This solution was successfully implemented at the Pic-du-Midi Observatory on small telescopes used for solar observations [19, 20] but, unfortunately, it cannot be extrapolated to those much larger than 1 meter in diameter because of the difficulty in manufacturing large, optical-grade glass plates.

Still, this same basic philosophy was applied to the large telescopes of the 1960s and 1970s: efforts were made to (1) isolate the inside of the enclosure from the exterior by using small slits and windscreens and (2) control the temperature of both the telescope and the inside of the enclosure so as to minimize the temperature difference with the outside air at night. Temperature control of the telescope and telescope chamber during the day was typically obtained by means of a "cooling floor" set to the nighttime temperature (Fig. 11.10)

11.5 Thermal design   377

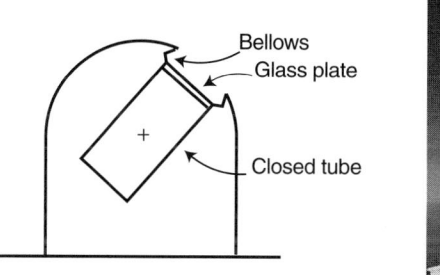

Fig. 11.9. Isolating the telescope from the inside of the enclosure and from the outside eliminates enclosure seeing in principle, but this requires a large glass plate to close off the top of the telescope tube. This solution was only partially implemented for the Pic-du-Midi 2-meter telescope shown on the right, as the glass plate was never installed. (Photo by Robert Futaully, courtesy of Observatoire Midi-Pyrénées.)

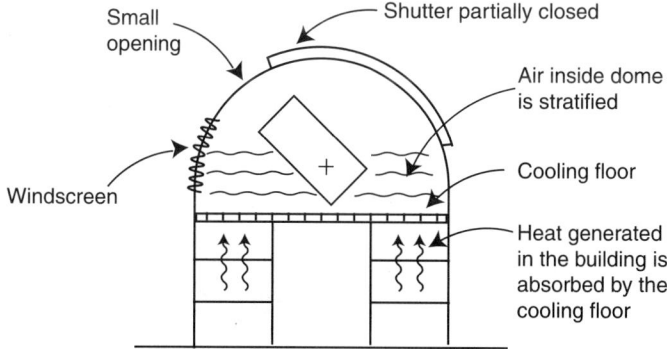

Fig. 11.10. Thermal design of enclosures in the 1970s.

These measures also created a stable, "stratified" environment inside the enclosure which persisted at night, at least under low-wind conditions. The approach worked relatively well because good astronomical sites have low variations in nighttime air temperature after sunset and because variations from one night to the next are also small. In spite of the large thermal inertia of the telescope structures and mirrors of the time, the difference between the interior and exterior nighttime temperatures rarely exceeded a few degrees Celsius [21]. Although this approach was quite successful overall, as exemplified by the excellent image quality systematically obtained at CFHT [22], the large thermal inertia of the entire observatory prevented complete elimination of locally generated seeing. Attempts to minimize local seeing by using fans or by fully opening enclosure slits under moderate wind conditions were unsuccessful. As has often been observed [23, 24], slow convection cells in a quiet, confined interior are better left undisturbed; actively blowing them away only makes things worse.

378   11. Observatory Enclosure

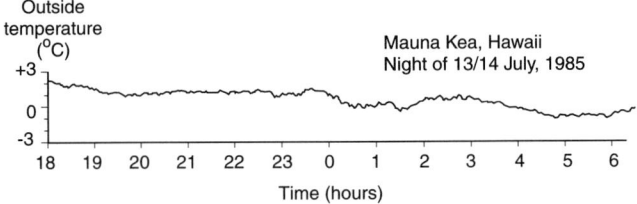

Fig. 11.11. Typical nighttime outside temperature at Mauna Kea. Although the standard deviation is only 0.5° C, the average time derivative is 0.6 °C/h.

Examination of the external temperature profile in Fig. 11.11 shows that, although the temperature drop at night is small, (e.g., annual mean of 1.3 °C from sunset to sunrise at Mauna Kea), there are frequent short-time-scale fluctuations that a large-thermal-capacity system just cannot track. A very short time constant is essential. The new philosophy, first implemented on the innovative MMT [25], calls for a two-prong approach:

(1) The telescope structure and optics should be designed for a very low thermal inertia (lightweighted or thin mirrors, "open" structures).

(2) The enclosure should be wide open at night to maximize exposure of the telescope to ambient air and allow it to reach thermal equilibrium.

In this new approach, it is important, however, that the entire enclosure be well flushed. If pockets of air at different temperatures persist, then turbulence can mix these different regions and contribute to image degradation. The more open the enclosure, the better, with open-air operation being the ultimate solution. A widely open enclosure develops an airflow which is generally fully turbulent with a Reynolds number of $\sim 10^6$–10^7, but turbulence per se is not a source of degradation if the flow is isothermal. The drawback of widely open enclosures is that, under windy conditions, telescope shake will begin to dominate, so that a compromise between thermal effects and dynamic effects has to be found. This can be done by using variable-porosity openings to adjust the degree of natural flushing to wind speed and direction. This makes it possible to always operate at the point where image degradation from the sum of thermal and dynamic effects is minimal (see Chapter 9, Section 9.5.5). In practice, this philosophy translates into the following design and operating principles (Fig. 11.12):

– Prevent air from the surface layer from entering the enclosure or flowing above the telescope by locating the telescope at sufficient height and by proper aerodynamic design of the enclosure.

– Minimize heat input during the day by designing the enclosure with high external reflectivity, high insulation, and, if possible, natural flushing of the skin. Complement this with air conditioning of the telescope chamber and temperature control of optical elements with high thermal

11.5 Thermal design

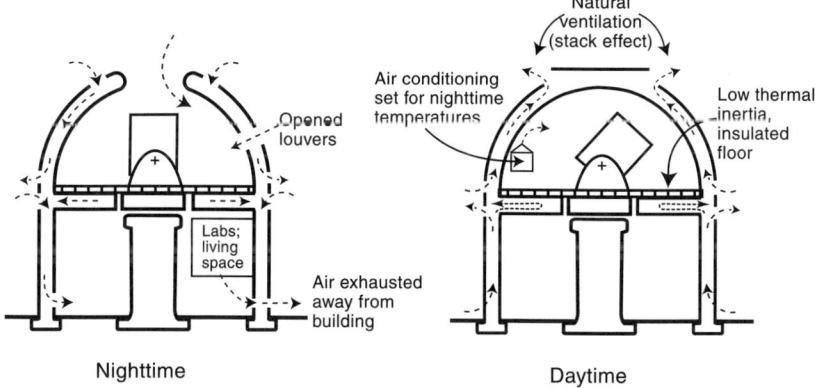

Fig. 11.12. General principles of enclosure thermal control at night (left) and during the day (right).

inertia (primary mirror). Use meteorological forecasts for the coming night to adjust all active thermal control set-points.
- Keep heat-generating equipment inside the telescope chamber at an absolute minimum. Locate hydraulic plants, air compressors, chillers, etc. some distance away from the building or below the observing-floor thermal barrier. Incorporate cooling systems to remove heat from essential equipment such as oil-bearing systems and instruments which must be near the telescope.
- Keep heated offices and laboratories in the telescope building to an absolute minimum.
- Design the enclosure (external skin, inside walls) to minimize temperature differences with the ambient air (low thermal inertia, proper emissivity).
- Locate air exhausts at some distance from the building and downwind from prevailing winds. Provide a second exhaust facing a different direction for use when winds are not from the prevailing direction.
- When the enclosure is open, allow wind to flow smoothly and with little resistance through the enclosure. Shapes of enclosure edges and windscreen should be designed to avoid putting energy in the high frequencies, which can excite structural modes in the telescope and especially the primary mirror.
- When very large openings are not practical, use fan-forced air ventilation to flush the enclosure when external wind speed is low (e.g., less than 4 m/s).
- Create a thermal barrier under the telescope chamber to prevent heat from the lower floors from penetrating into the telescope chamber (insulated and ventilated floor, seals around the telescope pier).

- Design the telescope chamber floor for low thermal inertia (e.g., wood).
- Keep external air from infiltrating into the enclosure during the day by installing proper seals at the enclosure wheel/track and prevent leakage from the lower floors of the building. This can be accomplished by keeping the telescope chamber at a slight overpressure ($\sim$ 5 mm water column) and the lower building in slight negative pressure.

11.5.2 Enclosure external skin emissivity

Many telescope enclosures have been coated with white titanium dioxide paint. This paint has a low solar absorptivity which reduces heating of the enclosure skin during daytime. However, it also has high thermal emissivity and, at night, the enclosure skin cools by radiating to the cold sky. Hence, air passing over the enclosure skin is cooled and pockets of this colder air fall into the enclosure opening, thus increasing local seeing (Fig. 11.13).

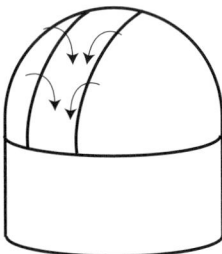

Fig. 11.13. Cooling of the enclosure's exterior skin at night can create a downflow of cold air falling onto the telescope.

It is better to use a thin external skin made of unpainted aluminum. At night, such a skin will track the ambient air temperature to better than 0.3 °C even with little wind. However, using aluminum (or other metal with a low-emissivity coating) implies that the internal surface of the skin be well ventilated during the day, as the high conduction of aluminum considerably increases heat transfer through the enclosure wall.

External steel surfaces of existing enclosures have been successfully covered with aluminum Mylar tape to reduce thermal emissivity and improve ambient temperature tracking.

11.6 Structural and mechanical design

11.6.1 Loading cases

Telescope enclosures, like ordinary buildings, should be designed to comply with local building codes. However, observatories are usually in remote, mountainous areas and, as it happens, where seismic risks and maximum wind speeds are often higher than the norm. In such cases, building codes may not give sufficient guidelines and it will often be necessary to establish design criteria by consultation with meteorological and geophysical institutions. As an indication, the specifications to which the VLT were designed are as follows [26]:

- Survival wind speed: 50 m/s
- Survival earthquake: 8.5 Richter
- Maximum operational wind: 20 m/s with gusts up to 30 m/s
- Maximum "operational" earthquake: 7.75 Richter
- Maximum shutter opening and closing time: 2 min
- Maximum enclosure rotation speed: 2 °/s
- Enclosure emergency stop from maximum rotation speed: <5 s

At high-altitude sites, snow and ice loads can be significant. At Mauna Kea, ice can build up on all surfaces due to freezing rain. The Gemini enclosure was designed for a 150 mm ice buildup and a snow-loading of 150 kg/m^2.

In the low latitudes, where most observatories are built, local architects may have little experience with some of the design requirements of a high-elevation site. These include issues such as exits that cannot be blocked by snow, snow removal, prevention of ice dams, and so forth. It is important to ensure that the design team includes expertise in such matters.

11.6.2 Enclosure shape

A spherical enclosure allows the telescope to be moved around inside the enclosure; it also has a smaller volume than the corresponding cylindrical shape. The spherical shape permits the use of an up-and-over shutter which, during high winds, can be lowered to minimize the size of the opening. The spherical shape is also advantageous in case of snow, which tends to accumulate less than on a flat horizontal surface. And finally, axisymmetric enclosures have the advantage of being free from "windvaning" in strong winds.

As indicated earlier (Section 11.4.1), cylindrical enclosures permit better ventilation. They are also thought to be less expensive because straight structural members can be used, although this saving may be offset by the larger surface and volume and the less efficient structural shape.

The overall shape of the enclosure and building taken together has an influence on the flow of air up and around it. As illustrated in Fig. 11.14, hemispherical enclosures have a tendency to "lift" the airstream near the ground up above the telescope [14], with potential detriment to seeing. This effect can be eliminated by extending the dome sphere down below its equator ("horseshoe" shape) and setting it on a cylindrical base, as was done for the CFHT and Keck observatories. Cylindrical enclosures do not suffer from this effect.

An enclosure with a small or largely open base may even lower the boundary layer height.

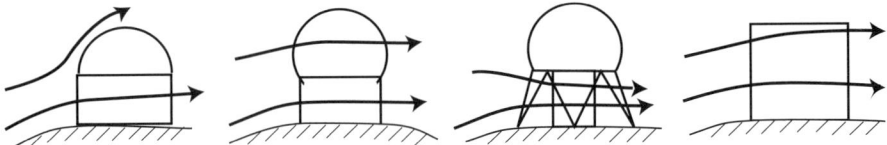

Fig. 11.14. Traditional domes tend to lift the surface layer up over the dome (left). This effect can be counteracted by extending the dome below its equator (second from left). A better solution is shown third from left, where the dome is supported by an open truss structure forcing the surface layer down [21]. Cylindrical shapes also eliminate the effect (right).

11.6.3 Shutter

Shutters fall into two categories, "up-and-over" and "biparting" (Fig. 11.15). The up-and-over type has many advantages. It minimizes wind drag and does not perturb the flow around and above the dome when opened. In high winds, it can be brought back over the top of the dome to minimize the size of the opening and reduce telescope wind shake (in combination with a windscreen in the lower part of the slit).

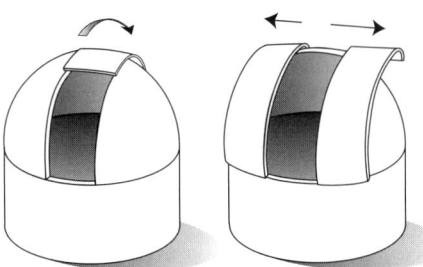

Fig. 11.15. Enclosure shutters. "Up-and-over" at left, "biparting" at right.

The up-and-over shutter can even be designed with multiple separable sections to serve as windscreens (Fig. 11.16). This type of shutter is also well suited to snow conditions, as it will push down snow accumulation on the back of the dome when it is being opened. In addition, the risk of snow falling in through the slit is minimized.

The drawback of the up-and-over shutter is that it is only applicable to spherical enclosures. It is also relatively complex to fabricate, as it needs to be made of several sections and requires circular tracks. The biparting shutter is simpler to fabricate and less costly.

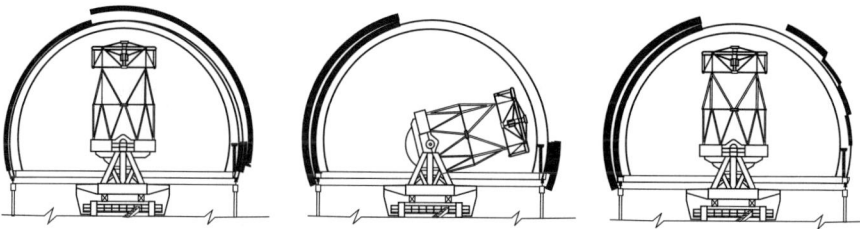

Fig. 11.16. The Gemini shutter and windscreen arrangement. The shutter is made of two independently driven sections. It is shown closed at left. When winds are low, the upper shutter is completely raised to provide maximum air flushing (center). In high-wind conditions, the lower shutter section is raised, drawing the windscreen up with it (right). (From Ref. [11].)

A drawback of cylindrical enclosures in snowy sites is that snow accumulation on the flat top must be removed before the shutter is opened. One solution is to melt the snow by circulating hot air under the roof [27].

11.6.4 Bogies and drive

Enclosures rotate on track/wheel systems. Wheels are generally grouped by pairs and mounted on a spring-suspension cart called a bogie or truck. Springs absorb unevenness in the track and deformation of the enclosure caused by snow or wind. Bogies should be fitted with side rollers to keep the enclosure centered and resist lateral wind loads. These side rollers should also be spring-mounted to absorb runout error and deformation in the rotating enclosure.

The bogies can either be mounted on the building (they are then stationary) or on the rotating enclosure. Several factors influence this choice.

One of these is structural and involves discrete load-bearing capability. The structure with the strongest girder should carry the track, whereas the one with mostly discrete support points (e.g., columns or arches) should carry the bogies.

A second factor involves load uniformity. If the enclosure load is strongly nonuniform along its periphery, it is better to mount the bogies on the enclosure so they can be variably spaced to carry approximately the same weight. This is the case for domes with an up-and-over shutter: the shutter arches carry a heavier load than the others and can then be supported by more bogies. More generally, the major advantage of mounting bogies on the rotating enclosure is that this keeps the structural load path constant. Variable loads result in constant flexing of the enclosure, which is detrimental to both structural elements (e.g., shutter tracks) and nonstructural ones (e.g., metal siding).

Another factor involves the way the enclosure is driven. When the bogies are also used as drives, it is better to have them fixed to the building to avoid the need for slip rings.

The enclosure can be rotated using motors mounted on the supporting wheels or by the use of separate roller drives. At least three equally spaced drive units should be used to avoid imbalance torque on the enclosure and possible wheel binding. To avoid shocks, the drives should be capable of operating continuously at slow speed rather than in a stop-and-go fashion.

Very smooth rotation is required in order to minimize vibrations that could propagate to the pier and from there to the telescope. It is particularly important that the wheels roll without skidding. To satisfy this condition, the wheel axes must intersect at the center of the plane of the track, as shown in Fig. 11.17. Another useful precaution is to grind the track flat after fabrication. This can be done in the shop or after assembly at the site. In the latter case, grinding can be performed by simply mounting a grinding wheel on the track and rotating it for several hours.

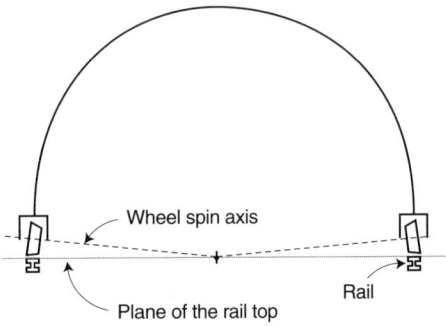

Fig. 11.17. Rail and bogie geometry to eliminate wheel-skidding.

A lightweight enclosure may have insufficient mass to resist overturning if subjected to high wind loads. In that case, hold-down clips or spring-loaded rollers may be required.

11.6.5 Weather seals

The rotating enclosure and shutter and all other openings in the telescope chamber must be sealed to minimize air leaks when the enclosure is closed.

Positive sealing of the enclosure can be achieved by the use of liquid seals or inflatable seals. Liquid seals consist of a vertical ring-shaped blade running along the bottom edge of the dome and partly submerged in a trough filled with a nonfreezing liquid. Inflatable seals are simply pneumatic seals; they are deflated whenever the dome is being rotated.

As a rule, compression seals should be preferred to sliding seals. In cold climates, heat tracing of seal joints is necessary to melt ice that may form on the seal. Attempting to open frozen seal joints will usually lead to damage.

11.7 Telescope pier

The function of the pier is to provide a stable platform for the telescope. As indicated earlier, its height is set by the need to have the telescope above the atmospheric ground layer. The pier must be extremely stable and not be affected by wind disturbance or enclosure rotation. Some designs take advantage of the fact that telescope pointing is unaffected by horizontal motion, and minimize tilt at the expense of some lateral motion (e.g., parallelogram action).

Piers are generally made of concrete because of its low cost and excellent damping properties. Ideally piers should rest on bedrock. When this is not possible, two types of foundation are used, depending on the soil characteristics of the site. In the first approach, long columns (piles) are either poured or driven into the soil and the pier is built on these columns. Friction along the sides of the columns provides the support for the pier. In the second approach, a large slab is poured and the support comes from the bearing-pressure of the soil under the slab.

Care must be taken to prevent mechanical vibration from machinery and elevators from being transmitted to the pier. It is critical to have no physical contact between the building and the pier. All cables and pipes leading from the building to the pier and then to the telescope should have flexible loops.

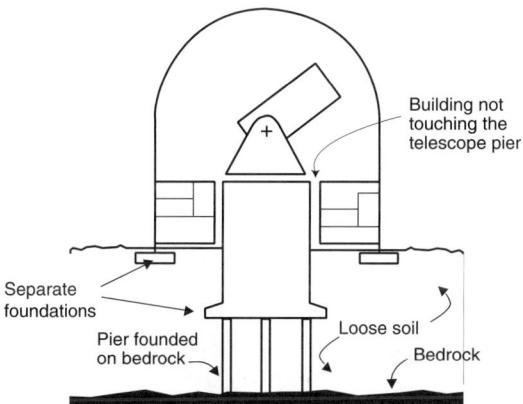

Fig. 11.18. Separating the pier from the building in order to minimize transmission of vibrations.

The building and dome present a very large wind-attack area. The associated overturning moment causes additional forces on the building foundations which can be expected to propagate through the soil and into the pier. These effects can be roughly estimated by hand calculations [3, 28] but are better calculated by computer analysis using wind models and soil mechanics methods.

11. Observatory Enclosure

The results are strongly dependent on the type of soil, whether it is rock or particulate. As schematically shown in Fig. 11.18, transmission is generally reduced if:

- the building foundation is structurally separate from the pier foundation,
- the building foundations are built over a damping layer (sand, lava cinder), while the pier is mounted on bedrock, and
- the pier foundation is at a lower level than the building foundation.

For additional isolation, the building supporting the telescope enclosure can be mounted on absorbing pads as shown in Fig. 11.19.

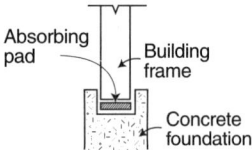

Fig. 11.19. Absorbant material placed at the base of the building to prevent transmission of vibrations to the telescope pier.

11.8 Handling equipment

The telescope chamber should be provided with equipment for

- installing instruments at the telescope focuses,
- exchanging top units on telescopes, when applicable,
- removing the primary mirror for realuminizing,
- accessing various parts of the telescope for maintenance.

A notional handling arrangement is shown in Fig. 11.20. Traditional enclosures (noncorotating) facilitate handling, since they provide rotation around

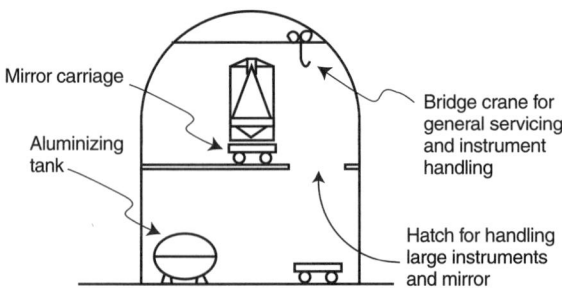

Fig. 11.20. Typical dome-mounted crane for general servicing.

the building axis: by installing a bridge crane with some radial travel, most of the telescope floor can be served. From this point of view, cylindrical enclosures are better than spherical ones because the bridge crane can extend to the full radius and provide access to the entire floor area. Since overhead cranes cannot reach the lower part of the telescope, the primary mirror and Cassegrain instrumentation are usually handled with a special carriage traveling on the observing floor.

References

[1] Schneermann, M., VLT enclosures wind-tunnel tests and fluid dynamic analyses, SPIE Proc., Vol. 2199, p. 465, 1994.

[2] Beckers, J., Ulich, B.L., and Williams, J.T., MMT – the first of the advanced technology telescopes, SPIE Proc., Vol. 332, p. 2, 1982.

[3] Mikami, I., Yamauchi, N., Itoh, N., Kawahara, S., Cocksedge, C.P.E., Ando, H., Karoji, H., Kodaira, K., Noguchi, T., and Hayashi, S., Enclosure of Subaru telescope, SPIE Proc., Vol. 2199, p. 430, 1994.

[4] Hull, C., Limmongkol, S., and Siegmund, W.A., Sloan digital sky survey telescope enclosure: design, SPIE Proc., Vol. 2199, p. 1178, 1994.

[5] Bonneau, A. and Zago, L., Design and construction of the inflatable dome prototype, ESO Conference on *Progress in Telescope and Instrumentation Technologies*, p. 867, 1992.

[6] Zago, L., The design of telescope enclosures for the VLT, ESO Conference on *Progress in Telescope and Instrumentation Technologies*, p. 235, 1992.

[7] Zago, L., Design and performance of large telescopes operated in open air, SPIE Proc., Vol. 628, p. 350, 1986.

[8] Esnard, D., The European southern observatory Very Large Telescope, J. Opt., Vol. 22, No. 2, p. 33, 1991.

[9] Erasmus, D.A. and Thompson, L.A., Ground turbulence at Mauna Kea observatory: location and ground height for future telescopes, SPIE Proc., Vol. 628, p. 148, 1986.

[10] Cayrel, R., Knowledge acquired during the site testing for the Canada-France-Hawaii telescope, Proc. ESO Workshop on *Site Testing for Future Large Telescopes*, p. 45, 1983.

[11] Raybould, K., Ford, R., Gillett, P., Hardash, S., and Pentland, G., Gemini enclosure and support facility design philosophy and design description, SPIE Proc., Vol. 2199, p. 452, 1994.

[12] Forbes, F., Wind loading of large astronomical telescopes, SPIE Proc., Vol. 332, p. 198, 1982.

[13] Fletcher, N.H., *The physics of musical instruments*, Springer-Verlag, 1990.

[14] Siegmund, W.A., Wong, W.-Y., and Forbes, F., Flow visualization of four 8 m telescope enclosure designs, SPIE Proc., Vol. 1236, p. 567, 1990.

[15] Zilliac, G.G. and Cliffton, E.W., Wind-tunnel study of an observatory dome with a circular aperture, PASP, Vol. 103, No. 669, p. 1211, 1991.

[16] Forbes, F., Wong, W.-Y., Baldwin, J., Siegmund, W., Limmongkol, S., and Comfort, C., Telescope enclosure flow visualization, SPIE Proc., Vol. 1532, p. 146, 1991.
[17] Ando, H., Barr, L., Miyashita, A., Sakata, K., and Shindo, S., Some airflow properties of telescope enclosures estimated from water-tunnel tests, PASP, Vol. 103, p. 597, 1991.
[18] De Young, D.D. and Charles, R.D., Numerical simulation of airflow over potential telescope sites, Astron. J., Vol. 110, p. 3107, 1995.
[19] Rösch, J., Solutions against man-made seeing, Proceedings of Flagstaff Conference on *Identification, Optimization and Protection of Optical Telescope Sites*, p. 146, 1986.
[20] Rösch, J., Aerodynamic and thermal problems around and inside a dome and telescope, in *Instrumentation for Astronomy with Large Optical Telescopes*, Humphries, C.M., ed., D. Reidel Publishing, p. 79, 1982.
[21] Bely, P. and Lelievre, G., Seeing control in domes and telescopes, Proceedings of the Flagstaff Conference on *Identification, Optimization and Protection of Optical Telescope Sites*, p. 155, 1986.
[22] Racine, R., Salmon, D., Cowley, D., and Sovka, J., Mirror, dome and natural seeing at CFHT, PASP, Vol. 103, p. 1020, 1991.
[23] Texereau, J., *How to Make a Telescope*, Interscience, 1962, p. 235.
[24] Steavenson, W.H., *Air disturbance in reflectors*, Vistas Astron. I, p. 473, 1955.
[25] Beckers, J. and Williams, J.T., Seeing experiments with the MMT, SPIE Proc., Vol. 332, p. 16, 1982.
[26] Schneermann, M., Marchiori, G., and Dimichino, F., The VLT enclosures – design and construction, SPIE Proc., Vol. 2871, p. 650, 1996.
[27] Neff, D.H., et al., The structural design of the corotating enclosure for the Large Binocular Telescope, SPIE Proc., Vol. 4004, p. 135, 2000.
[28] Medwadoski, S., *UC telescope pier rotations due to wind action on the observatory dome*, Keck Report No. 53, 1981, revised 1984.

12
Observatory Sites

The selection of the site for a ground-based observatory or the orbit for a space mission is one of the most important decisions in the development of an astronomical facility. It impacts scientific performance, design, fabrication, and operations at the highest level. In this chapter, we review the range of observatory sites available both on the ground and in space, site-testing methods for ground sites, selection criteria, and the specific environmental conditions found in space.

12.1 Ground versus space

One rarely has the choice between building a new facility on the ground or in space. This decision is made a priori by the agency to which one belongs or offers the proposal. Still, with the performance of ground observatories equipped with adaptive optics approaching that obtainable in space, and with the costs of space observatories converging with those of the largest ground-based facilities, it is important to tailor the respective scientific and operational goals to minimize potential overlap. Understanding what is best done from the ground and from space is essential.

12.1.1 Advantages of ground-based facilities

The advantages of building telescope facilities on the ground are obvious. The Earth's environmental conditions are familiar to designers and builders; access and construction, even in remote sites, is well within modern capabilities; also, human access allows for the correction of initial difficulties, tuning up, repairs, and the upgrading of the facility as technology evolves and science requirements change.

Although premium sites are rare, there are still many good locations to fit the needs of individual institutions. And even if the cost of a large facility is significant, it does not necessarily require the level of funding that only governments and international organizations can afford, but remains within reach of university consortiums and private funding. This relative ease of funding and corresponding minimal red tape permits novel designs to be tried out and makes niche applications possible.

12.1.2 Advantages of space-based facilities

From the scientific point of view, observing from space presents four compelling advantages over ground-based observations:

(1) Image quality is not affected by the atmosphere and is thus limited only by the optics. Optics can be made diffraction limited, dramatically improving sensitivity and spatial resolution. Although, for ground telescopes, atmospheric turbulence compensation techniques promises close to diffraction-limited image quality, for the near future the corrected field will be limited by the isoplanatic patch, which is typically only a few arcseconds in diameter. In space, field is restricted only by the optical design and can typically reach several arcminutes.

(2) Access to a wide wavelength region is totally unimpeded. In particular, the OH lines between 1 and 2 μm are avoided and the "cosmological window," the wavelength region around 3.5 μm where zodiacal light is at a minimum, is fully open.

(3) Sensitivity is enhanced in the infrared by elimination or reduction of background emission from the atmosphere and telescope. Space allows for the use of telescopes actively or passively cooled to cryogenic temperatures, something impossible on the ground because frost would collect on the optics. A significant reduction in atmospheric obscuration can be achieved by observing from airplanes and balloons, but the thermal emission of the telescope remains an overwhelming limitation beyond 2 μm.

(4) Instrumental stability is excellent. Because the space environment is benign, space telescopes and instruments experience very little variation in thermal and dynamic conditions. Optical alignment, point spread function, throughput, and detector characteristics remain unchanged for months or years, resulting in highly stable observation properties and calibrations.

As illustrated in Fig. 12.1, the advantages of observing from space are so substantial in the infrared that it is difficult for ground facilities to compete, regardless of telescope size.

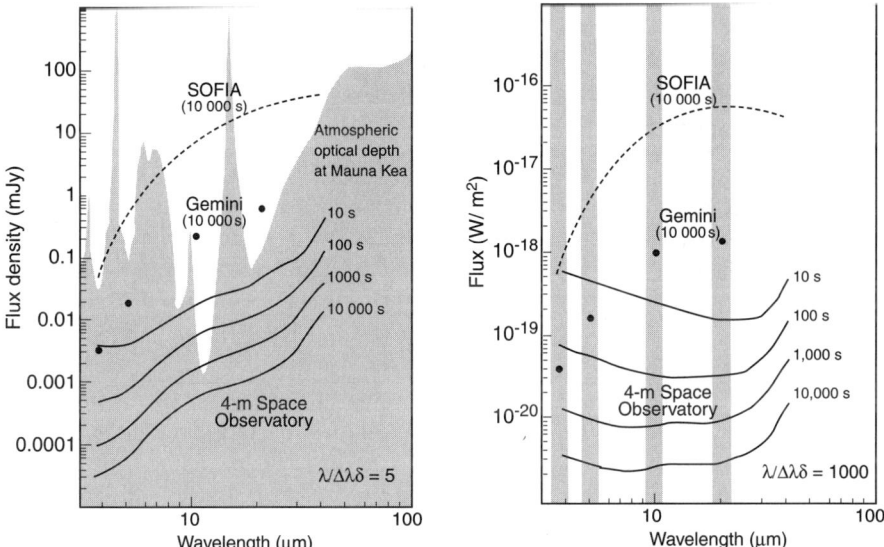

Fig. 12.1. Broadband imaging (left) and spectroscopic (right) point-source sensitivities for a 4-meter infrared space telescope (solid lines), SOFIA (dashed line), and an 8-meter ground-based telescope such as Gemini (dots). Integration times for point sources, calculated for a S/N of 5, are shown beside each curve or data point. The shaded areas show the relative optical depth at left and, at right, the approximate regions of good transparency from Mauna Kea. (From Ref. [1]).

In addition to scientific superiority, space also offers engineering advantages thanks to the low levels of mechanical and thermal disturbances and absence of gravity. Gravity sag and wind buffeting make it increasingly difficult to build ever larger telescopes on the ground, but there would seem to be no limitation placed on the size of space telescopes by the fundamental laws of physics.

The major drawback of space telescopes is cost, which is 10 to 100 times higher than the cost of ground-based telescopes of comparable size. Hence, space missions are only justified if it is essentially impossible to obtain the desired scientific data from a ground-based facility. And since it is difficult to predict the gains that future technological developments could bring to ground telescopes, it is good practice to assume that a space telescope should be at least two orders of magnitude more powerful than its ground-based competition at the time the space mission is proposed.

Another disadvantage of space facilities is that, in general, their lifetimes are short and instruments cannot be upgraded as new technology appears or science requirements change. The Hubble Space Telescope is an exception in this regard: on-orbit maintenance was made possible by the use of low Earth orbit and the NASA Space Transportation System. This maintenance is extremely expensive, however, and since future space telescopes will likely be in much higher orbits, on-orbit maintenance will not be an option for them for the foreseeable future.

12.1.3 Aircraft and balloons

Bridging the gap between ground-based facilities and true space platforms, airplanes and balloons benefit from the lack of atmospheric turbulence and absorption that space affords, and at a fraction of the cost of satellites. Sounding rockets played a critical role in the early days of space astronomy, but are rarely used now because of their short flight duration. Astronomical observations from the upper atmosphere, and in particular the stratosphere, now rely on aircraft and balloons.

Observing from the stratosphere is especially advantageous in the infrared. Being very dry, atmospheric opacity and radiance are minimal there, an advantage magnified by the low pressure which reduces spectral line broadening. Ambient temperature is also low, which reduces telescope optics emissions. Finally, the stratification hampers the generation of turbulence and associated seeing effects. In the near infrared (up to 3 μm) and, to a lesser extent, in the longer infrared, the resulting conditions approach those of space. Extensive observations from the stratosphere have been made with the highly successful Kuiper Airborne Observatory (KAO), and a larger airborne observatory, SOFIA, is in the making. For the visible and near infrared, however, observations from aircraft suffer from image degradation due to air turbulence and vibration. Image quality on the KAO was on the order of 1" [2], and this is also the image quality expected from SOFIA.

Balloons can offer greater altitude and longer observing times, although, until recently, observations were limited to a few days and reflights were infrequent. But long-lasting balloons can now provide weeks of observations. Observatories based on airships and tethered aerostats have also been proposed [3]. Such airborne platforms have the potential to offer reliable launch and recovery of payloads, long on-station time, and a lower cost per hour than aircraft.

12.1.4 Capabilities of various observatory platforms

In addition to the purely scientific differences discussed above, the various possible sites and observatory platforms also differ according to their logistics and telescope size and mass capabilities. Telescopes of great size can be installed on the ground, whereas launchers, airplanes, and balloons impose stringent limits on payload mass and dimensions. Table 12.1 is an attempt to summarize in broad terms the capabilities of these various site possibilities.

It must be noted in particular that the maximum payload sizes and masses shown are those imposed by current technology, not fundamental limits.

Table 12.1. Comparison of capabilities of ground-based and space facilities

	Ground based	Airplane	Balloon	Sounding rockets	Spacecraft
Altitude (km)	< 5	14	30	100	>500
Max exposure time	8 h	10 h	days	15 min	days
Avail. time (h/yr)	2000	1200	–	–	7000
Max telescope dia. (m)	10–30	2.5	4	0.4	~ 8
Max payload mass (ton)	–	1.5	1	0.1	5–10
Lifetime (yr)	40	20	–	–	10

12.2 Desirable characteristics for ground-based sites

Based on the astronomical requirements described in Chapter 1, the general requirements for optical/infrared ground-based observing sites are as follows [4, 5, 6].

- **Minimal cloud cover.** Clearly, cloud coverage is detrimental to astronomical observation, but it must be emphasized that even partial cloud cover is deleterious because the passage of clouds can affect photometric observations. In addition, clouds emit at infrared wavelengths, creating an additional background. Clouds cease to be a problem above the tropopause, which is at about 8 km in the arctic regions and 18 km in the tropics. Very high-altitude clouds are composed of ice crystals, producing little opacity.
- **Low water vapor.** Water vapor is the primary atmospheric absorber and emitter, especially at wavelengths over 10 µm.
- **Low temperature.** In the near infrared, background radiation emitted by the overlying atmosphere and telescope optics typically dominates scattered light and thermal emissions from the zodiacal background. This local background radiation decreases as an exponential function of temperature, so that sites with low temperatures are desirable (e.g., the Antarctic).
- **Low atmospheric pressure.** Aside from atmospheric spectral lines and bands, the opacity and emissivity of the overlying air is determined by pressure broadening of spectral lines; in the line wings, higher altitude results in less opacity for a given absorption-column density.
- **Dark sky.** Low levels of city light pollution are of particular importance. Measurements have shown that the intensity of artificial illumination of

the night sky due to city lights varies as the inverse 2.5 power of the distance from the city and that the amount of light produced by cities in regions of homogeneous economic development is proportional to their populations [7] (Fig. 12.2).

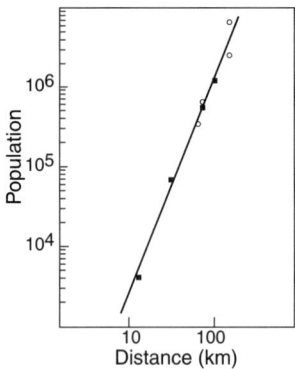

Fig. 12.2. Variation with urban population of the distance at which the light of a California city produces an artificial night sky illumination of 0.2 mag at 45° altitude in the direction of the city. More refined predictions can be made using the Garstang model [8].

- **Low optical turbulence (good seeing).** This is a critical factor in maximizing telescope sensitivity and spatial resolution.
- **Low nighttime wind velocity.** Reduces the risk of telescope wind shake.
- **Low nighttime temperature variation.** Minimizes the effects of dome seeing.
- **Low nighttime relative humidity.** Minimizes the risk of frost on optical elements.
- **Low level of radio wave and microwave radiation.** Avoids perturbation of detectors and electronic equipment.
- **Low dust pollution.** Minimizes contamination of the optics.
- **Low enough latitude.** Maximizes sky coverage and access to significant celestial regions (i.e., galactic center, Large Magellanic Cloud).
- **Good site accessibility.** Reduces infrastructure and operation costs.
- **Low seismicity.** Reduces structural requirements.

12.2.1 Seeing

Of all the factors listed above, low optical turbulence is arguably the most important, since sensitivity and resolution are critically affected by actual image quality. As explained in Chapter 1, seeing is directly related to microthermal activity; that is to say, high-frequency temperature fluctuations associated with atmospheric turbulence.

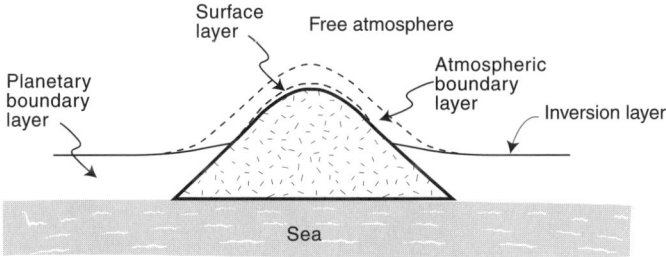

Fig. 12.3. Schematic representation of the various layers where seeing occurs (case of an island shown).

As illustrated in Fig. 12.3, the atmosphere above an astronomical site can be divided into four main layers: the surface layer, the planetary boundary layer, the atmospheric boundary layer, and the free atmosphere:

- In the **surface layer**, also called the ground layer, turbulence is generated by wind shear due to frictional and topographic effects at ground surface. The height of this layer is strongly influenced by the geometry of the site and the large-scale "roughness" of the ground. Boulders and crags will increase the height of this layer; so will trees. The surface layer can range in depth from a few meters at the best sites to as much as several tens of meters. The height of the surface layer is about 5 meters at Paranal [9] and 6 to 10 meters at Mauna Kea [10]. Whenever possible, the enclosure slit and telescope and, in particular, the primary mirror should be located above the surface layer.

    The ideal mountain shape is an isolated conical peak. With such a configuration, impinging airflows tend to divide and flow around the peak on either side, rather than being forced up-slope and over the top. There is some evidence that the slope of this conical peak must be greater than 7° but less than 18° to avoid up-slope motion of the airflow [11]. When the top of the mountain is flat, the observatory should be placed as close as feasible to the ridge so as to sit in the unperturbed flow (Fig. 12.4). When several telescopes are clustered near each other, they should be laid out in the direction perpendicular to the prevailing wind so as to avoid interference among themselves. Mountain ridges have been used as observatory sites, but are less than ideal. Abrupt or not, they disturb the airflow, forcing air up-slope. This action mixes free air with air cooled by contact with the surface of the mountainside, causing significant variation in the refractive index above the ridge top [12].

- The **planetary boundary layer** is the layer in the atmosphere where frictional dissipation due to the Earth's surface is still significant and

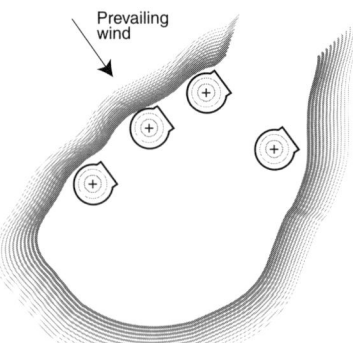

Fig. 12.4. Layout of the four VLT telescopes at the Paranal site. Three of the telescopes are located close to the edge facing prevailing winds. The fourth one is not, in order to improve imaging capability when the four telescopes are combined interferometrically.

in which there is a continuous vertical transfer of air mass due to the diurnal cycle. Air in close proximity to the ground is continually heated or cooled as it passes over areas heated by the Sun or cooled at night by radiating to the sky. Microthermal activity results from pockets of air at different temperatures moving up or down due to buoyancy and aerodynamic effects. The upper limit of the planetary boundary layer is the "inversion layer," typically located at about 1000 m altitude. Most astronomical sites are above this layer.

– The **atmospheric boundary layer** is the layer immediately above the planetary boundary layer. Although free of the large-scale convectional effects found below the thermal inversion layer, it is still affected by the ground, both mechanically and thermally. This may be due to the presence of mountain ranges that emerge above the planetary boundary layer or to gravity waves generated by abrupt changes in the topography (Fig. 12.5). To avoid the effect of gravity waves, it is best to be near the sea on the side of the prevailing winds. In the finest astronomical sites, the atmospheric boundary layer is reduced to a minimum, but at some existing observatory sites, the atmospheric boundary layer can be the dominant source of seeing. At the ESO site in La Silla, Chile, this layer extends up to 500 m above the site and 80% of the seeing is generated there.

– The **free atmosphere** is essentially unaffected by the ground. The activity in this region is synoptic (i.e., of very large-scale origin such as trade winds in the tropics, westerly winds in the mid-latitudes, and the jet streams in the higher layers). Microthermal activity in the free atmosphere results from wind-induced mechanical turbulence in zones where vertical temperature gradients exist. This generally occurs in the

12.2 Desirable characteristics for ground-based sites

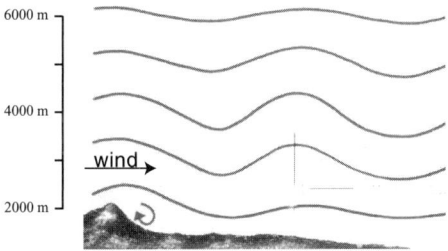

Fig. 12.5. Gravity waves generated by abrupt topography changes.

shear zones above and below the jet streams at altitudes of 12 km and higher. As shown in Fig. 12.6, jet stream velocity is maximal in the mid-latitudes, implying that seeing due to the free atmosphere is likely to be better in the tropics and near the poles. On average, seeing attributable to the free atmosphere is about 0.4" [13].

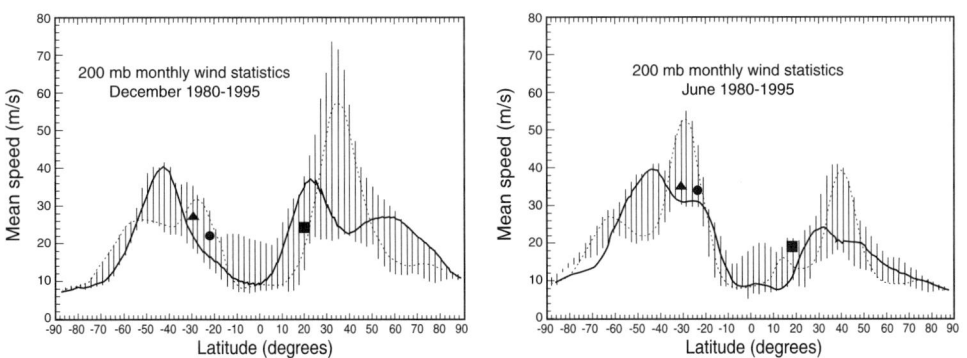

Fig. 12.6. Sixteen-year statistics for the wind velocity at an altitude corresponding to a 200-millibar atmospheric pressure as a function of latitude in December (left) and June (right). The vertical bars represent the variation in longitude, the solid line is for longitude 0, and the dashed line for longitude 180°. The small triangles, dots and squares show the 200 mb wind velocity above La Silla, Paranal, and Mauna Kea, respectively.

12.2.2 Criteria for extremely large telescopes of the future

Although the 4 to 10 meter class telescopes built so far have required excellent seeing in order to take full advantage of their larger apertures, this may be less true for the extremely large telescopes of the future. There are two reasons for this. The first is that the emerging atmospheric compensation techniques have the potential to correct for mediocre seeing. What counts then is not seeing in itself, but how well seeing can be corrected. One important factor affecting correction is the atmospheric coherence time, τ_0. The smaller the value for τ_0, the more difficult it is for the adaptive optics system to compensate. The coherence time is linked to wind velocity in the upper atmosphere and is expected to be longer at low latitudes and at the poles, where wind speeds are lower (Fig. 12.6) [14].

The second reason involves wind speed at ground level: preferred sites will be those with low average wind speed because image degradation due to wind buffeting will be less.

As a result, the sites of choice may shift from those with primarily excellent seeing to those with low wind speed and seeing characteristics that are easier to correct.

12.3 Location and characteristics of the best observing sites

The areas of the world having more than two octas of cloud cover at least 50% of the time throughout the year are shown in Fig. 12.7. In general, minimum cloud cover occurs in two latitude zones around the world, extending from roughly 10° to 35° north and south.

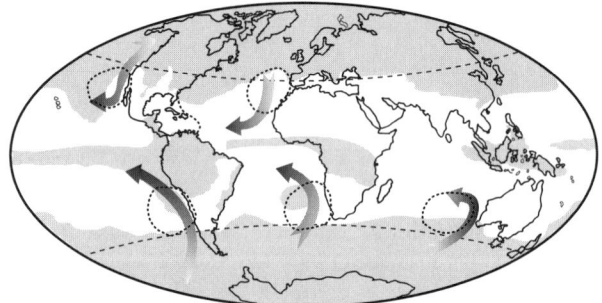

Fig. 12.7. World map showing areas with more than two octas of cloud cover 50% of the time annually (shaded), cold ocean currents (arrows), and approximate boundaries of the regions of stable tropical maritime air (dotted circles). Latitudes 40° north and south are indicated by dashed lines.

Applying the criteria defined in the previous section, we can now identify those locations that are likely to have the best observing conditions.

Island sites.

The finest observing conditions will be found at island sites which meet the conditions stipulated above. Ideally, the island should be in a tropical region to benefit from low cloud cover and lower upper wind velocity. It should be far from any land mass to benefit from unperturbed airflow and possess a single peak high enough to be above the inversion layer to guarantee good conditions regardless of wind direction. Additionally, the island should have low light pollution. The number of such locations is extremely small and include Mauna Kea in Hawaii and the Canary Islands.

Coastal sites

Conditions at coastal sites will be similar to those at island sites when the prevailing winds come from the direction of the sea. A "cold sea" is beneficial because it lowers the inversion layer. These conditions are found along the California and Chilean coastal ranges, which are next to cold seas (California and Humboldt-Peru currents) and on the western sides of continents in latitudes where winds are from the west (Fig. 12.8).

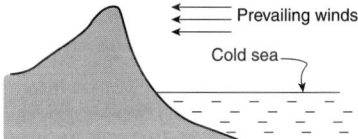

Fig. 12.8. In the best cases, coastal sites can benefit from the same near-ideal conditions found on some islands, i.e., a mountain high enough to be above the inversion layer and facing unperturbed airflow coming from the sea.

Inland sites

Contrary to traditional wisdom, inland sites can be as good as coastal or island sites as long as they are high enough and face into unperturbed airflows. This situation may occur in mountain ranges downwind from a large flat plain or desert [14]. However, this will mean altitudes of 3000 m or higher, at which height problems of low temperature and snow accumulation can be serious and where clear weather may be infrequent, as cloud caps tend to form around such peaks. Thus, only high-altitude peaks in arid climates facing flat plains and deserts are likely to be acceptable.

Antarctica

The two poles potentially offer exceptional conditions for astronomical observations. The global wind circulation induced by Earth's rotation has "singularities" there which results in a very calm upper atmosphere. The atmosphere at the poles is also thermally stable and stratified, especially during winter night, thanks to the cold surface and absence of solar heating. This is particularly true of Antarctica, with its large land mass that is a thermal sink in winter. Low mechanical turbulence and thermal stability there should combine to create excellent seeing conditions. Another important advantage of the poles is the low level of precipitable water in the atmosphere, which is favorable to infrared observations. The low temperatures at ground surface are also beneficial because the thermal emission of the optics is reduced.

Of the two poles, the Antarctic is the more attractive because of the "solid ground," the more stable atmosphere, the colder ground temperature, and existing infrastructure. Infrared telescopes are already operating at the South Pole station but, unfortunately, seeing is not very good there due to microthermal activity in the first 1000 m of altitude. Other sites at higher altitude on the continent appear more promising.

12.3.1 Characteristics of the major observatory sites

The main characteristics of the major observatory sites are listed in Table 12.2. In this table, photometric nights are defined as those having more than 6 continuous hours with no clouds over 18° above the horizon; wind velocity is measured 10 m above the ground and is the median velocity, meaning that the wind velocity is lower than this value 50% of the time; seeing is the mean seeing; and the amount of precipitable water in the atmosphere is that occuring at nighttime only. The data in Table 12.2 were compiled from various published reports and personal communications. It must be emphasized that the listed values are subject to caution because measuring equipment and interpretation is not uniform and the time spans over which data were obtained may be different. In addition, there are long-term trends which can change these data over time scales of 5 to 20 years.

Table 12.2. Comparison of the best observatory sites

Site	Latitude	Altitude (m)	Photom. nights (%)	Wind vel. (m/s)	Mean seeing (″)	Precip. water (mm)
Mt. Palomar, California	33°N	1706				6.0
Mt. Graham, Arizona	33°N	3300	41			
Kitt Peak, Arizona	31°N	2130				7.1
Mt. Hopkins, Arizona	31°N	2600	48	6.7	0.70	
Mt. Locke, Texas	30°N	2100				
Mt. Fowlkes, Texas	30°N	2100	33	5.0	1.03	
La Palma, Canaries	28°N	2400	79	5.5	0.73	
Tenerife, Canaries	28°N	2400	79	6.0	0.71	3.8
Mauna Kea, Hawaii	19°N	4200	60	5.6	0.45	1.6
Paranal, Chile	24°S	2660	77	6.3	0.76	2.3
Cerro Pachon, Chile	30°S	2700				
Las Campanas, Chile	29°S	2280			0.76	
La Silla, Chile	29°S	2400	62	4.6	0.92	3.9
Cerro Tololo, Chile	30°S	2200	65		0.68	4.8
Siding Springs, Australia	31°S	1100				
South Pole	90°S	2800			>1	<1

Data shown where available. Values subject to caution; see text.

12.4 Evaluation methods for ground-based sites

The general approach to site selection for a new observatory has changed dramatically over the last few decades. Originally, the aim was to find a suitable mountain that had a favorable climate, was easily accessible, and was a reasonable distance from the home institution. With the very large, expensive telescopes of today, it has become critical to extract the maximum from the observatory. Since the quality of the site is such an important factor in observatory performance, finding the best possible site for a new facility is crucial.

Scientific priorities vary, but, in general, a site investigation will need to provide data on site climatology, infrared sky background, and image quality.

Site climatology (local temperature, wind velocity, and cloud cover) is obtained by traditional meteorological means. Temperature and wind velocity are generally measured at various heights above the ground up to the level of the future telescope. Cloud cover can be determined from visual observations, wide-angle radiometric monitoring (at 10 μm wavelength), and satellite observations.

Infrared sky background is measured with well established techniques such as the use of an Infrared Sky Radiance Monitor [15].

The third aspect of the site investigation, the measurement of image quality, is, by far, the most critical and difficult. We will now examine the techniques used in making these measurements in some detail.

12.4.1 Methods for testing image quality

The relationship between image quality and the atmospheric turbulence parameters was not fully understood until the 1980s. One could only rely on measuring image quality directly (e.g., the coherence length of the atmosphere, r_0) with a seeing monitor telescope. The problem there is that, because seeing fluctuates with atmospheric conditions, r_0 is itself a random variable. Its time variations cover a broad spectrum of frequencies, with periods extending from a few minutes to days or even years. As for most other atmospheric fluctuations, the lower the frequency, the larger the magnitude of the fluctuation. Such random variables are said to be "nonstationary." Nonstationarity strongly limits the accuracy of any statistical estimate. If one wants to compare different sites, seeing must be measured at all of them over a long time span (at least a year), but one will have no assurance of the predictive value of the results, since prevailing meteorological conditions may differ appreciably from one year to the next. Moreover, seeing measurements do not indicate why one site is superior to another or suggest where to look for better ones.

The reliability of the prediction is greatly improved by understanding the origin of the image degradation at each potential site. This requires measurement of the temperature fluctuations in the atmosphere and was originally done by balloon soundings using microthermal sensors. However, such probings of the atmosphere are by nature too spotty. Today, continuous monitoring of atmospheric turbulence is available thanks to *in situ* and remote sensing techniques. Additionally, computer modeling can now provide a reliable picture of atmospheric turbulence on both local and global scales. A well thought out site-testing campaign will thus rely on overall optical measurements combined with direct probing and modeling of atmospheric turbulence to obtain an understanding of the various sources of image degradation.

Overall seeing is the summation of image broadening due to several zones along the optical path:

Seeing source:	Telescope	Surface layer	Boundary layer	Free atmosphere
Height range:	< 10 m	1–30 m	30–1000 m	> 1 km

Each of these regions is best investigated with specialized techniques: microthermal probes near the ground, acoustic sounders for the surface layer, radar sounding for the free atmosphere, and a dedicated seeing-monitor telescope to provide a measure of the overall seeing.

In addition, flow visualization by wind-tunnel study or computer modeling can be of use in determining the best location at a given site. We now examine each of these techniques in detail.

12.4.2 Microthermal sensors

The most direct way to measure the temperature-structure coefficient (C_T^2) is to use sensitive differential thermometers to determine the temperature difference between two points. The exact separation between the two temperature sensors is not critical but is traditionally set to 1 meter so that the $r^{2/3}$ term is equal to 1 in SI units (Chapter 1, Section 1.3.5, Equation 1.10). The sensors must have a fairly wide bandpass, between 0.5 and 100 Hz, for example, in order to capture the entire spectrum of temperature fluctuations, and this requires very thin sensors. These are generally made of platinum wire 10 μm in diameter, and temperature is determined by measuring their resistance. The temperature-structure coefficient is obtained by squaring the temperature difference between the two sensors.

To probe the surface layer, the paired sensors can be mounted at various heights on a mast typically 30–40 m high (Fig. 12.9) [10]. This direct method can also be used to probe the optical path in the atmosphere up to 20 km, by mounting a pair of sensors on a radiosonde [13].

12.4.3 Acoustic sounder

Acoustic sounders, also known as "sodars" (for "sound detection and ranging"), measure the temperature-structure coefficient, C_T^2, in the atmospheric

12.4 Evaluation methods for ground-based sites   403

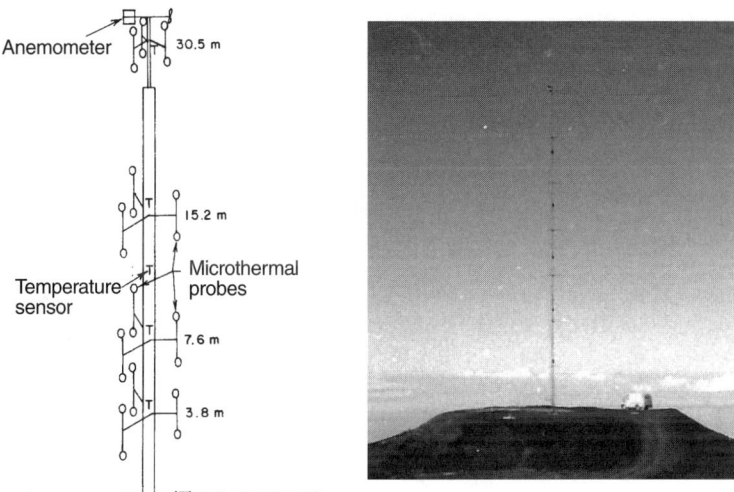

Fig. 12.9. Schematic view of a mast equipped with microthermal probes (left) and an actual example: the 40 m mast installed at the site of the CFH telescope.

boundary layer based on sound scattering from small-scale thermal and velocity fluctuations [16]. An acoustic pulse of constant frequency is beamed into the atmosphere and the reflected acoustic energy, known as backscatter, is analyzed to determine echo intensity versus altitude (Fig. 12.10). Scattering strength is proportional to C_T^2, such that echo-intensity plots give information about the vertical distribution of turbulence layers in the atmosphere.

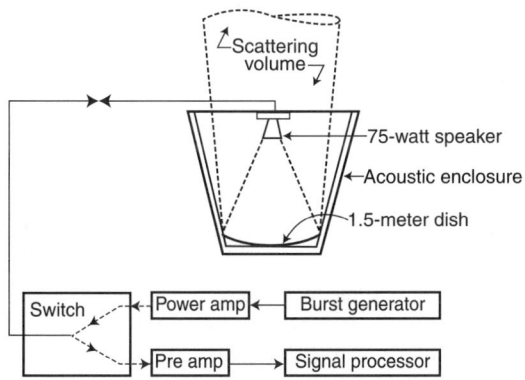

Fig. 12.10. Schematic of an acoustic sounder system.

Sodars typically operate at frequencies of about 2 kHz. They have a spatial resolution on the order of a few meters, ensuring that small turbulence layers do not go undetected, and they are capable of reaching heights of 800 m, albeit with reduced spatial sensitivity. In the best astronomical sites, nighttime atmospheric turbulence is so low that, in practice, the requirements for spatial resolution and sensitivity limit the vertical range to about 200 m. In addition, the accuracy of C_T^2 measurement is strongly limited by calibration, which is difficult to make and is only good to within a factor of 2.

Acoustic sounders have been used successfully for theoretical investigations [17, 18] and would certainly be useful under conditions of strong turbulence (e.g., solar observatories), but their application to the very low levels of turbulence found at the best nighttime astronomical sites is limited.

12.4.4 Site flow visualization

When the site under investigation is not an isolated mountain peak but an extended area with several possible sites, it may be useful to perform flow-visualization studies to determine the best potential locations and where to install the seeing-monitoring equipment. Such studies make it possible to determine the approximate height of the surface layer at various locations as a function of wind direction and to check on the wakes of surrounding peaks or existing domes. Flow visualization can be performed in a wind-tunnel using scale models of the site [19] or be carried out by computer simulation (Fig. 12.11).

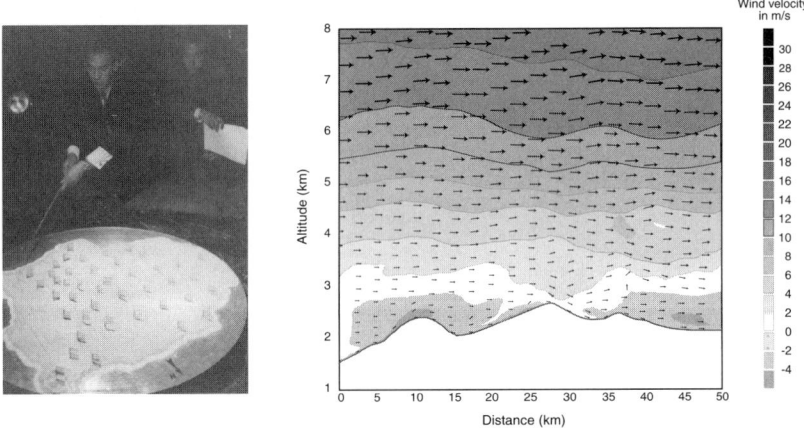

Fig. 12.11. Two methods of flow visualization. At left, a wind-tunnel study of the Mauna Kea summit area. At right, a computer simulation of the flow over Paranal, Chile (courtesy of M. Cure, University of Valparaiso, Chile, and Institute for Meteorology of Munich Technical University, Germany).

12.4.5 Radiosondes

Meteorological radiosondes are launched routinely from numerous places in the world to measure temperature and wind velocity and direction in the atmosphere up to 30 km altitude. When radiosondes are launched near an observatory site under consideration, the data collected can be used for a preliminary evaluation of upper-atmosphere seeing. An empirical model proposed by Hufnagel [20] uses wind speed as a parameter, the index of refraction structure parameter being given by

$$C_n^2 = 2.7 \, 10^{-16} \left(3u^2 \left(\frac{z}{10} \right)^{10} e^{-z} + e^{-z/15} \right), \tag{12.1}$$

where z is the altitude in kilometers and u is the rms wind speed in meters per second in the range of 5 to 20 km of altitude.

A more refined model making use of temperature, wind speed, and wind-direction profiles has been developed by VanZandt [21] and calibrated against radar measurements of atmospheric turbulence. The rationale behind the model is as follows.

The stability criterion for spontaneous growth of small-scale disturbance waves in a stably stratified atmosphere can be expressed in terms of the Richardson number. This dimensionless number is a measure of the ratio of the work done against gravity by the vertical motions in the waves to the kinetic energy available in the shear flow. Turbulence occurs when the Richardson number is lower than 0.25.

Microthermal-fluctuation balloon soundings of the atmosphere have shown that turbulence is confined to thin horizontal layers on the order of a few meters to a few tens of meters thick, separated by nonturbulent layers [22, 23]. Meteorological radiosonde samplings, typically hundreds of meters apart, are thus too infrequent to permit direct detection of turbulence. However, a statistical distribution of the wind shear and stability parameter can be derived, and the fraction of the layer that is probably turbulent can be calculated. Assuming steady-state conditions, the well established theory of optical propagation through turbulence can then be applied, and the average C_n^2 determined for each radiosonde sampling layer.

Although the validity of the method has been demonstrated on theoretical grounds, attempts at correlating radiosonde weather measurements to observed seeing has had mixed results [24]. Application to astronomical site investigation is also limited unless a regular launch site happens to be close by. However, the same atmospheric modeling principles may be used with the emerging technique of numerical atmospheric modeling, an approach we examine next.

12.4.6 Numerical modeling of the atmosphere

A very promising method for investigating lower-atmosphere effects at a given site consists of using the recently developed "meso-scale" meteorological models. These models are initialized with conventional meteorological analysis or radiosoundings and provide a three-dimensional description of wind, temperature and moisture on a horizontal grid with spacings of 500 to 1000 meters. Using the approach sketched above, the microscopic optical turbulence is estimated from these macroscopic meteorological parameters to provide C_n^2 profiles above any grid point on the ground [25].

If these models prove to be reliable, they could be used with advantage to choose between several sites in a given general area. Since conventional large-scale meteorological data is typically available over several past decades, these models could also be very useful in determining whether site testing, which is typically done on a time scale of a few years only, is representative over the long term.

12.4.7 Optical seeing monitors

Optical seeing monitors, which measure atmospheric turbulence integrated along the entire path from the telescope to a star, are clearly the most important tools for investigating seeing. These instruments are typically composed of a 15–30 cm telescope equipped with a photoelectric focal plane detector to measure image size. They operate in an automatic fashion, minimizing the need for human presence.

Polaris telescopes, which are telescopes solidly mounted on a concrete pier and pointing at the north celestial pole, were extensively used in the 1960s for the 4-meter class telescope site testings. Thanks to their fixed mounts, wind shake is reduced and tracking errors are eliminated. On the other hand, they cannot be used in the southern hemisphere (no pole star), nor at lower latitudes since seeing is too degraded at large zenith angles.

One alternate is the "star trails" method. In this case, a traditionally mounted telescope is used but tracking is stopped during observations to eliminate tracking errors. Star images leave "trails" on the focal plane, the width of which is a measure of seeing. The problem with this method is that wind shake corrupts the measurements and also that seeing is only measured in one direction.

Shearing interferometers have also been successfully used to measure wavefront coherence [26]. Their advantage is that the measurement is not affected by the telescope optics, but they do suffer from tracking errors and are difficult to operate in adverse environments.

These various problems have led to the development of the Differential Image Motion Monitor (DIMM), which has now become the workhorse instrument for site testing [27]. The method consists of measuring the differential image motion of the same star seen through two different paths in the atmosphere. This differential image motion is directly related to r_0, which is a measure of seeing. The advantage of the differential measurement is that it cancels out telescope tracking errors and wind shake. This method was first used in a qualitative manner by Stock [28], but had to wait for the theory of differential time of arrival developed by Fried [29] and analytical solutions by Roddier [26] to become a quantitative tool.

The instantaneous position of the image of a star seen through a small hole is a function of the slope of the wavefront above that hole. But that image position will also depend on the motion of the hole itself. By collecting starlight through *two* holes in the same mask, the differential motion of the two images is then only a function of the difference of the wavefront slopes above the two holes and is independent of the motion of the mask (Fig. 12.12, left). According to atmospheric turbulence theory, this slope difference is a function of r_0 and wavelength. More precisely, the differential longitudinal and transversal image motions, σ_l^2 and σ_t^2, respectively, are given by

$$\sigma_l^2 = 2\lambda^2 r_0^{-5/3} \left[0.18 D^{-1/3} - 0.097 d^{-1/3} \right], \tag{12.2}$$

$$\sigma_t^2 = 2\lambda^2 r_0^{-5/3} \left[0.18 D^{-1/3} - 0.145 d^{-1/3} \right], \tag{12.3}$$

where D is the diameter of each aperture, d is their separation, and λ is the wavelength [27].

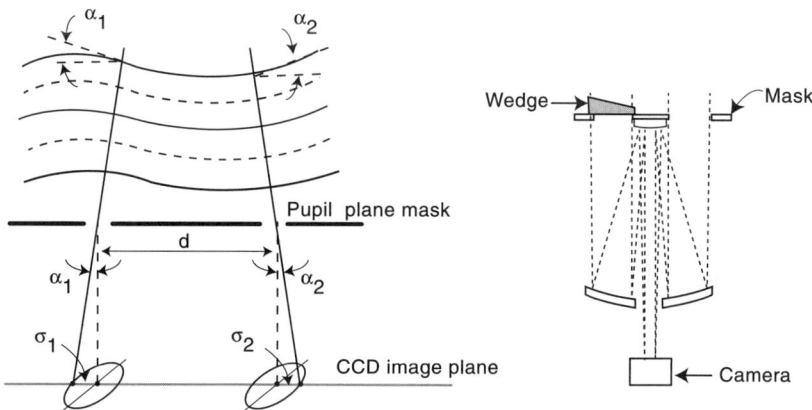

Fig. 12.12. Differential image motion monitor. A star is observed through two apertures, and the differential motion of the two images is a measure of the wavefront distortion (left). An example of an implementation with an entrance-aperture mask is shown on the right.

An example of such a setup is shown on the right in Fig. 12.12. Light from a bright star is collected in a telescope through two holes in an aperture mask. One of the two beams is deviated by a prism so as to separate the two images in the focal plane. For convenience in matching pupil characteristics to seeing conditions, a better solution is to reimage the pupil and use a beam splitter to separate the two beams. Typically, DIMMs are built around a 30 cm telescope located in open air to avoid dome seeing and at least 5 m above the ground so as to be above most of the surface layer. The holes in the mask are a few centimeters in diameter and separated by about 20 cm. DIMMs operate on stars of magnitude V≃ 3 at less than 30° from zenith, and image centroids are measured every 200 ms and averaged to produce a statistical estimate of seeing every minute. The accuracy of the measurement is limited by photon noise and is about 10% for seeing larger than 0.2″.

Seeing measured with DIMMs is well correlated with the seeing observed in telescopes up to about 2 m in diameter [27]. Larger telescopes, when not affected by "dome seeing," exhibit better seeing than DIMM measurements indicate. This phenomenon is due to the fact that the relation between r_0 and observed seeing assumes an outer scale of turbulence, L, which is infinite (or very large compared to telescope size). In reality, L varies between 10 and 100 m and averages 25 m, a dimension commensurable with the aperture of the largest telescopes, so that a correction must be applied. In general, 8- to 10-meter class telescopes have seeing in the visible that is about 10% better than that determined by DIMMs. The effect is larger in the infrared.

12.5 Space orbits and the moon

Most astronomical satellites have been placed in near-equatorial low Earth orbit where they enjoy the perfect transparency of space and cosmic ray shielding by the Earth's magnetic field, without having required much more than minimal launch capability.

Higher orbits, which are energetically more expensive to reach, can have significant advantages, however. These range from geosynchronous orbits which are accessible within hours of travel, to the second Lagrange point, which is about 3 months away.

But in parallel with purely observational considerations, the choice of orbit is strongly tied to the available launch systems, their cost, and their performance. Even with the advances of over half a century of rocketry, the current and foreseeable launch vehicles place substantial financial and physical constraints on observatory design. To place a payload in high orbit, for example, requires approximately 100 times the payload mass in fuel, sophisticated engines, and staging systems. Moreover, the high cost and risk of developing new launch systems result in a limited menu of launch and orbit options. For large telescopes, the available fairing dimensions are also critical and essentially require that the telescope be deployable in orbit.

In this section, we will look at the main orbits offering particular advantages for astronomical observations. Environmental conditions and launcher capabilities will be presented in the two subsequent sections.

12.5.1 Low-inclination low Earth orbit

Low-inclination low Earth orbits (LEO) are orbits ranging from 300 to 1000 km in altitude with inclinations up to about 30°. Altitudes below 300 km are not possible because of atmospheric drag. Altitudes above 1000 km are not desirable because of the dense population of high-energy particles trapped in the Van Allen belts. Low inclinations are advantageous because of the gain in launch velocity contributed by the Earth's rotation. The orbit's inclination is typically that of the latitude of the launch site (28° for NASA launches from Cape Kennedy, 5° for ESA launches from Kourou in French Guiana), since this maximizes payload mass capability. According to Kepler's third law, the period of the orbit is equal to

$$P = \frac{2\pi}{\sqrt{GM}} a^{\frac{3}{2}}, \qquad (12.4)$$

where a is the semimajor axis of the orbit, G is the constant of gravitation ($G = 6.67 \cdot 10^{-11}$ N m^2/kg^2), and M is the mass of the Earth ($M = 5.98 \cdot 10^{24}$ kg). For an observatory on a circular orbit with an altitude of 600 km (such as HST), the period is about 96 min.

A specific feature of low Earth orbits is that the plane of the orbit rotates with time. This is due to the oblate shape of the Earth, which results in its center of gravity not being coincident with its center of mass (Fig. 12.13, left). Far from the Earth, the difference is negligible, but for low Earth orbit the effect is significant. In the case of HST for example, the orbital plane rotates by a little more than 6° per day (56-day period).

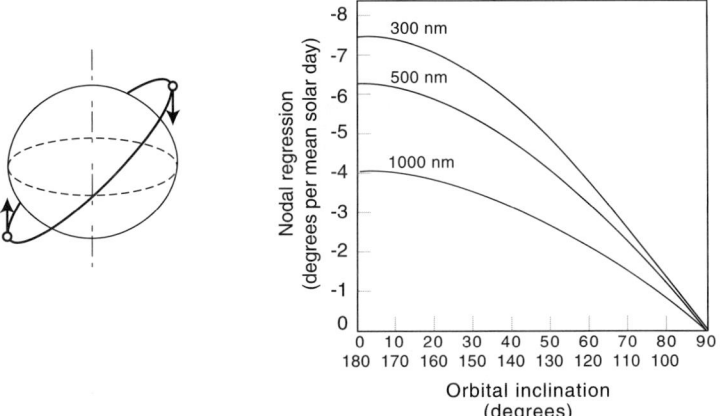

Fig. 12.13. Perturbative torques caused by the Earth's equatorial bulge (left) result in a regression of the orbital nodes. When the satellite is in one of the two positions shown, the net effect of the equatorial bulge is to produce a slight torque about the center of the Earth. This torque will cause the plane of the orbit to precess. The effect is a function of the orbit's inclination and altitude as shown in the plot at right. (After Ref [30].)

The vast majority of scientific satellites are placed in low Earth orbit, either by a dedicated launcher or by the U.S. Space Transportation System (STS, commonly known as the "Space Shuttle"). This is an obvious choice for Earth-observing instruments but is much less so for astronomy. The major advantage of a low Earth orbit is that the mass that can be placed in orbit is much greater. When the Space Shuttle is used, the facility can also be "maintained," that is, repaired or upgraded. Another advantage of low Earth orbits is that, being inside the Van Allen belts of trapped particles, the cosmic ray level is low.

On the other hand, observing efficiency is poor because of frequent target occultation by the Earth (Fig. 12.14). Within a typical 96 min orbital period, almost half the time (about 40 min) is lost. Long exposures require that target and guide stars be "reacquired" at every pass.

For infrared telescopes, another difficulty is thermal heating of telescope optics by the warm Earth. The heat input from Earth is significant (240 W/m^2) and prevents passive cooling of the optics to low temperatures. The telescope must be well baffled to minimize stray light from both the Sun and bright Earth, and the optics must be actively cooled.

Periodic eclipsing of the Sun by the Earth also produces large temperature swings, which can result in optical misalignment, image degradation, and pointing errors. For example, the temperature of HST's lightshield varies from about −80 °C on the dark side to about +50 °C on the sunside of the orbit. Thus, even though the primary mirror is made of low-expansion material, it had to be extensively insulated and temperature controlled to avoid figure changes.

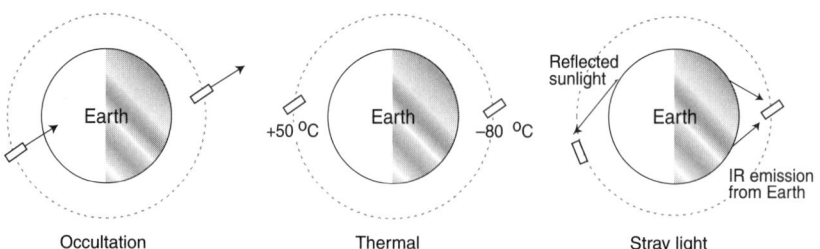

Fig. 12.14. Low Earth orbits are not ideal for optical astronomy. As shown schematically, they suffer from frequent target occultation (left) and large temperature swings (center), and require extensive baffling for protection against reflected light and infrared radiation from the Earth (right).

12.5.2 Sun-synchronous orbits

As explained above, low Earth orbits precess, with the effect diminishing as inclination increases. By selecting a near-polar orbit of appropriate altitude, it is possible to make the precession rate equal to 1 day per year eastward, thus keeping the orbital plane in a fixed direction relative to the Sun. Such orbits are called "Sun synchronous." If, in particular, the orbital plane is selected to be perpendicular to the Sun (i.e., in the plane of Earth's terminator) the Sun always remains in the same half of the sky as viewed from the spacecraft. If viewing is limited to the anti-Sun region, radiation from the warm Earth and the Sun can be blocked by simple thermal and stray light shielding (Fig. 12.15). Large radiators can then be mounted on the cold side of the spacecraft, allowing relatively low temperatures to be reached passively. The viewing limitations is quite constraining, obliging one to wait a full year to cover the entire sky. Nevertheless, a Sun-synchronous orbit can be a good choice for all-sky survey telescopes such as IRAS.

Polar orbits require specific launch sites with uninhabited areas down range (Vandenberg Air Force Base in California for U.S. launches) and are less efficient in launch velocity since they do not benefit from Earth's rotational velocity.

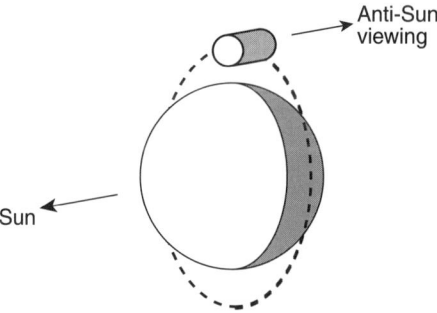

Fig. 12.15. Sun-synchronous orbit with the orbital plane perpendicular to the Sun's direction.

12.5.3 Geostationary and geosynchronous orbits

If a satellite is on a circular, equatorial, direct (westward) orbit with a period exactly equal to a sidereal day (23 h 56 min), it will appear stationary from Earth. As per equation 12.4, such an orbit is at an altitude of about 36 000 km, or 5.6 Earth radii. This so-called "geostationary Earth orbit" (GEO) is of great utility for communication and weather satellites but is also useful for astronomy satellites, as it permits continuous contact with a single ground station, thus allowing real-time, ground-observatorylike operations. Another advantage of this orbit is that it is relatively far from Earth (the Earth subtends an angle of only 16°), so that the portion of the sky blocked by Earth is small and the heat input from Earth is minimal. The drawback is that it is in the Van Allen belts, which increases detector background and the risk of permanent damage to electronics.

Variations of this orbit are the "geosynchronous" orbits, which have the same property of synchronicity with Earth's rotation but are noncircular and nonequatorial. As seen from the Earth, the satellite is no longer stationary but describes a "figure eight," more or less large depending on orbit parameters. Communication contacts with satellites in such orbits can still be very long. The International Ultraviolet Explorer (IUE) had such an elliptic orbit (27 000 km × 44 000 km), which afforded long periods of visibility from both U.S. and European ground stations on a daily basis. However, these orbits are still for the most part in the Van Allen belt, which is anything but ideal for infrared detectors.

12.5.4 High Earth orbits

To avoid the effects of the Van Allen belts, one must go to much higher altitudes. These high-Earth orbits (HEO) can be circular or highly elliptical. One example is the 100 000 km altitude circular orbit with a period of about 4 days shown in Fig. 12.16, which avoids the Van Allen belts completely. The drawback of this orbit is that the circularization is energetically costly and the allowable payload mass is reduced accordingly. Highly elliptical orbits (e.g., 1000 × 200 000 km) are less demanding and allow weeks of observing time far from Earth, but the satellite traverses the radiation belts twice at each perigee pass.

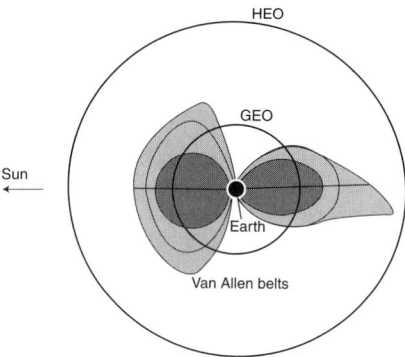

Fig. 12.16. High Earth orbit at 100 000 km altitude, which avoids the Van Allen belts. The geostationary orbit is shown for comparison.

12.5.5 Sun-Earth Lagrangian point 2

When a satellite leaves Earth's immediate vicinity, the gravity pull of the Sun is no longer negligible and the satellite's motion cannot be described in simple ways. This is the so-called three-body problem. The trajectories are not closed curves like the ellipses of the two-body problem and cannot be calculated from simple equations like those expressing Kepler's laws. There is, however, a special case when the third body is of negligible mass compared to that of the other two, and one of two large masses is in circular orbit around the other. In this case, there are five positions in the orbital plane where the small-mass object, once inserted into that plane, will move in a circular orbit, "locked in" relative to the other two objects. These points are known as the "Lagrangian points" associated with the two large masses and are referred to as L1 to L5 (Fig. 12.17). At these points, the gravitational pull from the two main bodies is balanced by the orbital centrifugal force (Fig. 12.18, left). Each pair of large masses in the solar system has such Lagrangian points, in particular the Earth-Moon system and the Sun-Earth system.

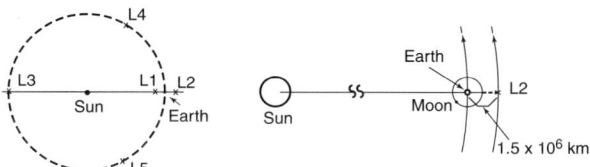

Fig. 12.17. The Lagrange points of the Sun-Earth system are shown on the left. An enlarged view of the Earth/Moon/Lagrange point 2 is shown on the right.

Two of the five Lagrangian points, L4 and L5, are stable, and the other three are metastable, meaning that an object placed there will not return to it if perturbed. And perturbing forces are, of course, always present, namely solar pressure and pull from the Moon and other planetary bodies. However, a spacecraft can be kept in orbit around these two metastable points by periodic station-keeping maneuvers [31]. In essence, the spacecraft orbits the Lagrange point rather than orbiting a celestial body. Such orbits are called "halo orbits" (Fig. 12.18, right).

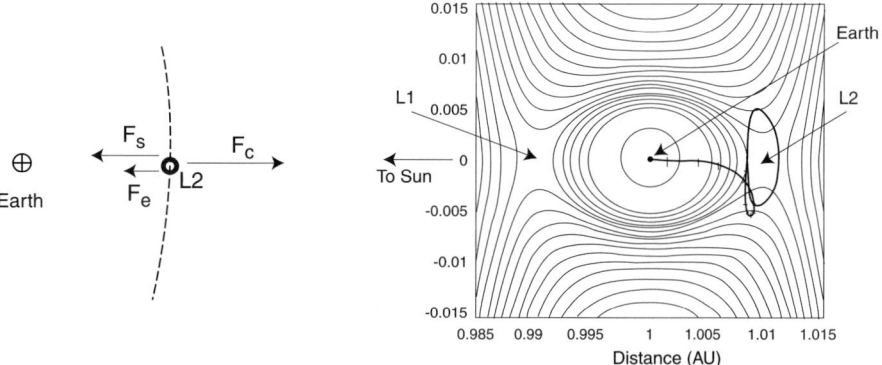

Fig. 12.18. At a Lagrange point, here L2, the orbital centrifugal force, F_c, exactly balances the gravity pull from the Sun, F_s, and the Earth, F_e (left). The gravity potential around Earth and the L1 and L2 points of the Sun-Earth system is shown on the right. The launch trajectory and halo orbit around the L2 point are also shown. The duration of the voyage to L2 is about 3 months.

The L2 point of the Sun-Earth system is ideal for astronomical viewing, since the Sun, Earth, and Moon are always on one side of the telescope maximizing sky coverage. This is also an ideal orbit for passively cooling a telescope without escaping the Earth altogether and requiring long-distance communication links. A single shield can protect the observatory from the Sun and, with minimal viewing constraints, from the Earth and Moon as well. In addition, the constant distance from the Sun (1 AU) provides a stable thermal environment and continuous solar illumination for generating onboard power. Finally, the distance to Earth is small enough for wide-band radio communication without resorting to very large ground antennas. Examples of missions launched or planned for L2 include MAP, NGST, and FIRST.

The meteoroid fluence predicted for L2 is shown in Fig. 12.19. This fluence is relatively benign compared to that of the orbital debris in low Earth orbit.

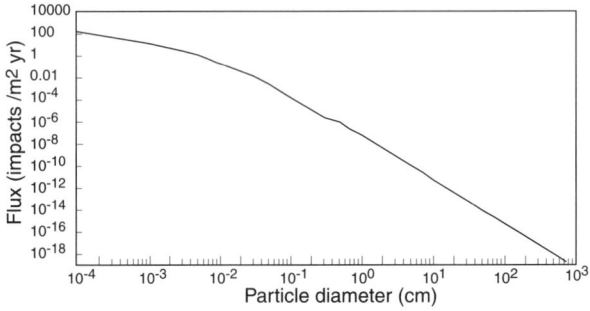

Fig. 12.19. Meteoroid fluence at L2. These meteoroids travel at 20 to 70 km/s. (Data from Ref. [32].)

12.5.6 Drift-away orbit

Instead of placing a spacecraft in a specific orbit as in the above-described options, it is also possible to launch one at the escape velocity and let it trail the Earth. This is an energetically economical solution, since no orbit insertion is required. Such an orbit provides the same stable thermal and power environment as an L2 orbit. Its advantage is that no onboard propulsion system is needed. Because of the uncertainty in the exact impulse provided by the propulsion system, however, it is necessary to launch with some impulse margin to avoid the risk of the spacecraft falling back to Earth. In practice, this results in its drifting away from Earth at about 0.1 AU per year (Fig. 12.20). After several years, the distance is such that large ground antennas (such as the Deep Space Network) are required for communication. One alternative is to *arrest* the drift at a distance where viewing restriction and thermal input due to Earth are negligible (e.g., 0.1 AU), but this requires carrying a propulsion module.

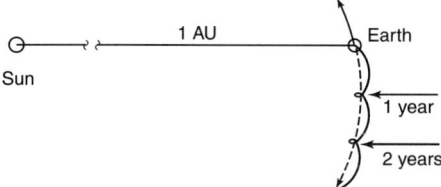

Fig. 12.20. Drift orbit

12.5.7 Heliocentric elliptical orbit

By increasing the launch energy even more, it is possible to leave the Earth-Moon environment altogether and send the spacecraft into an orbit around the Sun, for example on a 1 × 3 AU orbit (Fig. 12.21). The advantage is that the zodiacal background decreases significantly with distance from the Sun (approximately by the third power), which can lead to important gains for missions limited by the natural background (e.g., infrared spectroscopy). Although heliocentric orbits are more costly energetically and the allowed payload mass is reduced, smaller-aperture telescopes can be used because of the lower background. For example, a trade-off study for NGST showed that the performance of a 6 m telescope at 3 AU would be comparable to that of an 8 m telescope at L2. Aside from the energetic cost, the drawback of such orbits is a radio communications and power penalty occasioned by the great distance from Earth and Sun when the spacecraft is near aphelion.

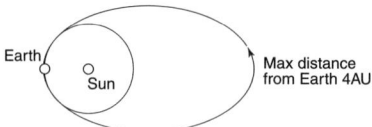

Fig. 12.21. Heliocentric 1 × 3 AU orbit with a period of about 2.8 years.

12.5.8 Moon

Ever since the Apollo mission, the Moon has been regarded as a desirable site for astronomy because it appears to combine the advantages of ground-based astronomy with those of space. A number of studies exploring the possibilities of lunar-based astronomy have illuminated the Moon's advantages as well as its problems [33, 34, 35, 36].

Like Earth, the Moon offers a stable platform on which large and potentially widely separated instruments can be installed (e.g., interferometers) and, assuming a manned lunar base, also offers the possibility of repairing and upgrading telescopes and instruments. Having essentially no atmosphere, the Moon, like space, suffers from neither seeing nor opacity effects. At night, the Moon is also relatively cold.

Technically, the main problem has to do with the alternation of day and night. Unlike a free-flying observatory in high orbit which can be protected at all times from Sun and Earth radiation, an observatory on the Moon would be swamped by stray light and be prohibitively hot during the lunar day (when the Moon's surface temperature reaches 400 K). Shielding could not adequately protect the telescope and, in practice, observations during the lunar day would be impossible. This would reduce observational efficiency by a factor of 2. And even if no observing were done during the lunar day, the observatory would still be subjected to the huge temperature variations so detrimental to optical alignment and mechanisms.

The Moon is not ideal for infrared astronomy, either. Although nighttime temperatures are relatively low (soil at 100 K), they still fall short of the very low temperatures (30 K) passively available in space. Furthermore, although gravity is only one-sixth of that on Earth, its effects are still significant for very large telescopes. Finally, it is widely acknowledged that automatic deployment of an observatory on the Moon's surface would be extremely difficult. Prior site preparation and man-assisted installation would appear to be essential, so any major lunar observatory would have to wait for a manned base.

12.5.9 Sun-Jupiter Lagrangian point 2

Arguably, the best astronomical observing site in the entire solar system may be the second Lagrange point of the Sun-Jupiter system [37]. As shown in Fig. 12.22, this location is in the "total shadow" of Jupiter. This remarkable condition, which is due to the large size and mass of Jupiter, is unmatched in the vicinities of other planets. In total darkness, an observatory located there could reach very low temperatures, potentially as low as 7 K, the equilibrium temperature in the solar system. Unimpeded by Sun avoidance constraints, the sky coverage would be close to 100%. And finally, with no need for a sunshield (always a major design complication), telescopes and interferometers with huge apertures or baselines could be located there. A drawback is that power would have to be generated by thermonuclear means.

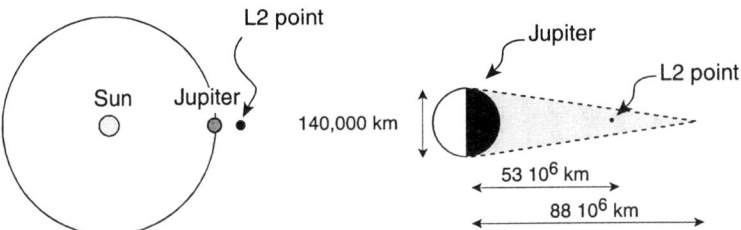

Fig. 12.22. The second Lagrange point of the Sun-Jupiter system (left), which is in the total shadow of Jupiter (right), may be the very best location for very large-aperture infrared telescopes in the solar system.

12.6 Radiation in the space environment

Space-based telescopes face radiation conditions much harsher than those encountered on the ground. Radiation effects can lead to increased background in instruments, loss of viewing time during periods of high radiation, long-term degradation and eventual failure of detectors and supporting electronics, and sudden, permanent failure of electronics from single particle strikes or discharging.

12.6.1 Sources of radiation

Radiation is a broad term covering both transmission of particles and true electromagnetic radiation. Except for ultraviolet light, which can affect some materials, radiation effects on spacecraft are not due to electromagnetic waves but to the following energetic particles:

- Protons and electrons trapped in Van Allen belts
- Heavy ions trapped in the magnetosphere
- Transient protons and heavy ions from solar flares
- Transient cosmic ray protons and heavy ions

The energy range of these particles is considerable. Trapped electrons have energies up to 10 MeV, and trapped protons and heavier ions can reach hundreds of MeV. Solar protons have energies up to hundreds of MeV and the heavier ions reach the GeV range. Galactic cosmic-ray protons have low-level fluxes with energies up to TeV. Particle levels from these sources depend heavily on the level of activity of the Sun. Figure 12.23 (left) shows the solar-cycle activity as measured by the number of sunspots for solar cycle 23. The length of a solar cycle can range from 9 to 13 years, averaging 11 years. Figure 12.23 (right) shows the solar proton fluence during active years.

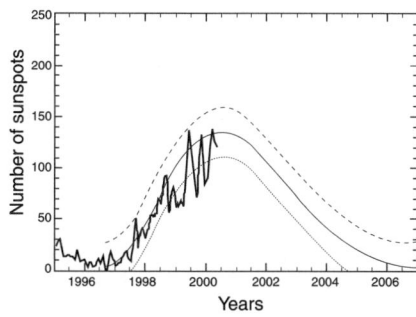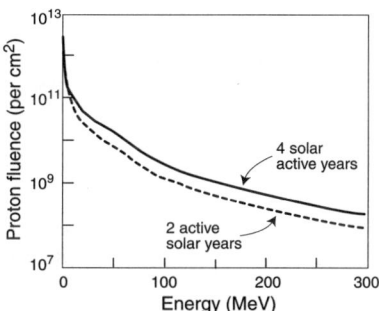

Fig. 12.23. Solar cycles are often measured by sunspot counts. The figure on the left shows the actual sunspot counts plotted with the predicted counts for solar cycle 23. The plot on the right gives the solar proton fluences for 2 and 4 solar-active years.

The Earth's magnetosphere acts as a filter for the transient particles. The levels of these particles encountered by a space telescope therefore depend primarily on the inclination and altitude of its orbit. Telescopes in geostationary and highly inclined orbits can be exposed to transient particle levels equivalent to those beyond the magnetosphere. Because of the high energies of these particles, it is not feasible to shield telescope systems against their effects.

Earth-orbiting telescopes can avoid the regions of high radiation of the trapped particle belts by flying above the peaks at geosynchronous or below the peaks in low Earth orbits. Even in these orbits, moderate levels of trapped radiation are encountered in low Earth orbits in the South Atlantic Anomaly (SAA) protrusion of the belts and, in geosynchronous orbits, in the outer edges of the belts. Shielding can help reduce or eliminate effects from electrons, but massive shielding would be required to absorb the higher-energy protons and this is not normally a design option for telescopes.

Space also contains a low-energy plasma of electrons and protons with fluxes up to 10^{12} cm^2/s. Thin layers of material easily stop this plasma, so it is not a hazard to most spacecraft electronics. However, it is damaging to surface materials, including optics coatings, and differentials in the plasma environment can contribute to spacecraft surface charging and discharging problems [38, 39].

12.6.2 Radiation effects

Radiation effects include surface erosion, dielectric charging and discharging, and damage to electronics. Surface erosion is caused by the solar-wind plasma and low-energy particles. Atomic oxygen in the ionosphere and ultraviolet exposure may also be causes of surface erosion. There is some evidence, for example, that a combination of these effects may be responsible for erosion of the external insulation on HST. Spacecraft charging and discharging is dealt with by proper grounding and adequate shielding of dielectrics. In general, the most damaging radiation effects are those affecting the electronics.

Effects on electronics can be divided into two categories: short term and long term. **Short-term effects** are due to single-particle ionization and are referred to as "single-event effects" (SEEs). SEEs are caused by a single charged particle passing through a sensitive junction of an electronic device. The net effect is that the circuit is perturbed and may lose data (an effect referred to as a "single-event upset" or SEU). Short-term effects can also be serious for detectors: the passage of a sufficiently energetic particle through a critical device region can even lead to permanent failure. Shielding is not very effective against SEEs because they are induced by high-energy particles. The prefered method for dealing with destructive failures is to use SEE radiation-hardened parts.

The **long-term effects** fall themselves into two categories referred to as total ionizing dose (TID) and displacement damage dose (DDD). TID is a long-term degradation of electronics due to the cumulative energy deposited in a material. DDD has similar long-term degradation characteristics, but it stems from a different mechanism: the displacement of nuclei in a material from their lattice position. Over time, sufficient displacement can occur to change a device's properties. Detectors are particularly susceptible to DDD. These long-term effects are primarily caused by protons and electrons trapped in the Van Allen belts and by solar-event protons. The effect of galactic cosmic rays is negligible in the presence of the other sources. Electronics can be "radiation hardened" to minimize these effects, and shielding may be used to eliminate most of the degradation (Fig. 12.24). Heavy shielding is often used on CCDs, but overshielding produces neutrons that increase DDD.

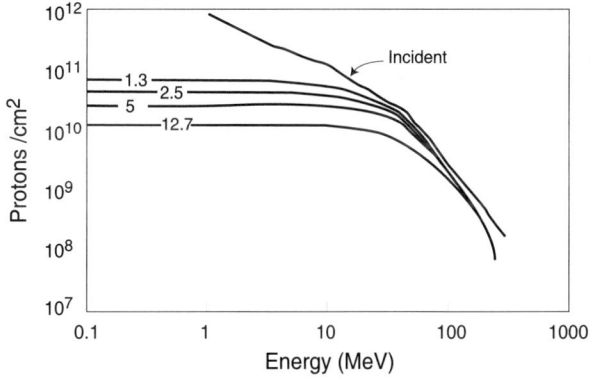

Fig. 12.24. Incident and shielded integral proton fluences for four solar-active years for various aluminum shielding thicknesses expressed in millimeters.

12.6.3 Dependence of radiation levels on observatory location

The space-radiation environment in the most common locations for space observatories (low-Earth orbits, geostationary or geosynchronous orbits, and the second Lagrangian point) is summarized below.

Low Earth orbit

The high-energy particle exposure in low-Earth orbit accumulates during passes through the South Atlantic Anomaly of the Van Allen belts. Typical spacecraft shielding absorbs lower-energy, lighter electrons, so most radiation effects in LEO are caused by trapped protons. Total ionizing and nonionizing doses below 1000 km altitude are not usually a problem. The galactic cosmic-ray and solar particle environments at LEO depend on the inclination of the spacecraft's orbit. As inclination increases, exposure to galactic cosmic-ray and solar particles also increases. In addition, the spacecraft encounters high

420   12. Observatory Sites

levels of solar protons and heavy ions over the magnetic poles where, during solar particle events, these levels increase dramatically. Figure 12.25 is a classic illustration of single-event upsets for a satellite in a polar LEO orbit. Normally, surface charging and deep-dielectric charging are not hazards for spacecraft in the LEO regions. The spacecraft passes in and out of the trapped electron regions too quickly for charge accumulation to be effective.

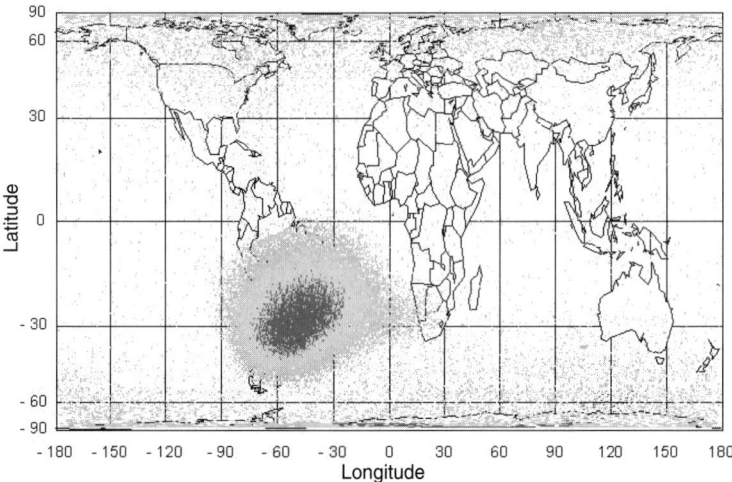

Fig. 12.25. Single-event upsets in a solid-state recorder on a polar orbit spacecraft at 700 km altitude. The area of strong radiation above Brazil is the SAA, which is due to the reduced intensity of the geomagnetic field there The black dots in this region correspond to >20 errors per 4000 km^2 cell. (Data from the SeaStar satellite.)

Geostationary and geosynchronous orbits

Outer-belt electrons dominate geosynchronous and geostationary orbit particle exposure. Although here, at 36 000 km altitude, spacecraft are beyond the peak particle level of the outer belt, they still encounter high-energy electrons and accumulate total ionizing dose effects. Shielding is only partially effective. High levels of low-energy trapped protons also occur in this region and may affect detectors. Single-event effects from high-energy cosmic rays occur at rates akin to interplanetary levels, and satellites are also periodically exposed to particles from solar events. Although the Earth's magnetosphere normally offers some protection against solar protons, it is disturbed during solar particle events and GEO spacecraft are exposed to the full effects of the storm. The nondestructive events may contaminate instrument data, whereas the protons add to the degradation of systems due to total ionizing dose and nonionizing displacement damage. The continuous exposure of GEO satellites to low- and high-energy electrons also makes spacecraft charging/discharging a serious problem, and differential levels of electrons caused by electron storms may well be lethal to systems.

Lagrangian Point 2

At Lagrangian point 2 (L2), spacecraft are basically in an interplanetary environment at 1 AU, well beyond the region of particle trapping. One important difference is that the "halo" orbit, used to maintain the L2 position, takes the spacecraft in and out of the tail of the Earth's magnetosphere. It is possible that the differential between the plasma in the magnetotail and in the solar wind will cause surface charging of the spacecraft. The magnetotail plasma can also become accelerated during solar storms. However, since storm particles do not attain energies nearly as high as those of galactic cosmic rays or solar particles, mitigation measures designed to protect the spacecraft against these will also cover magnetotail populations.

High-energy particle exposure at L2 is strongly influenced by the solar cycle. During the Sun's active phase, the likelihood that the spacecraft will encounter particles from solar events increases significantly. Figure 12.26 shows the cumulative distribution of the solar proton flux for particles with energies greater than 30 MeV, as obtained from the GOES satellite [40]. Unlike LEO and GEO orbits, where degradation effects accumulate slowly, the degradation here occurs in short, sudden bursts. The L2 single-event effects environment is very similar to that of the GEO, with a low-level continuous rate due to galactic cosmic rays and sudden increases in rates due to solar protons and heavier ions.

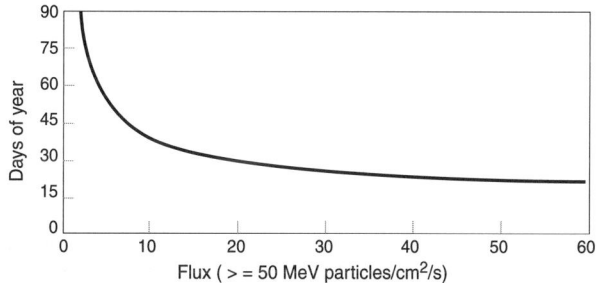

Fig. 12.26. Average cumulative distribution of solar proton flux counts >50 MeV for peak solar cycles years 1989–1991.

Finally, one must also consider the environment encountered during transfer or staging orbits out to L2. Some orbit phasing requires long passes through the trapped proton and electron belts. The risk level of SEEs in the peak of the proton belts will be comparable to that found at the peak of a solar proton event. Electron accumulation brings the risk of surface and deep-dielectric charging.

12.7 Launchers

The limitations of launch vehicles are major constraints on the design of a telescope payload. Principal among these limitations are mass and stowed volume. Figure 12.27 shows the fairing dimensions of the three major U.S. and European launchers, and Table 12.3 gives the payload mass to orbit. The height of the volume reserved for the payload depends somewhat on the choice of the payload adapter (the hardware that provides a structural interface with, and system for separating from, the launch vehicle). Items such as payload adapters, payload harnesses, and flight termination systems must also often be included in the available payload volume. Driven by commercial launch market competition, the mass-to-orbit performance of launch vehicles generally increases every few years and, consequently, potential users should consult vehicle manufacturers for their latest performance capabilities.

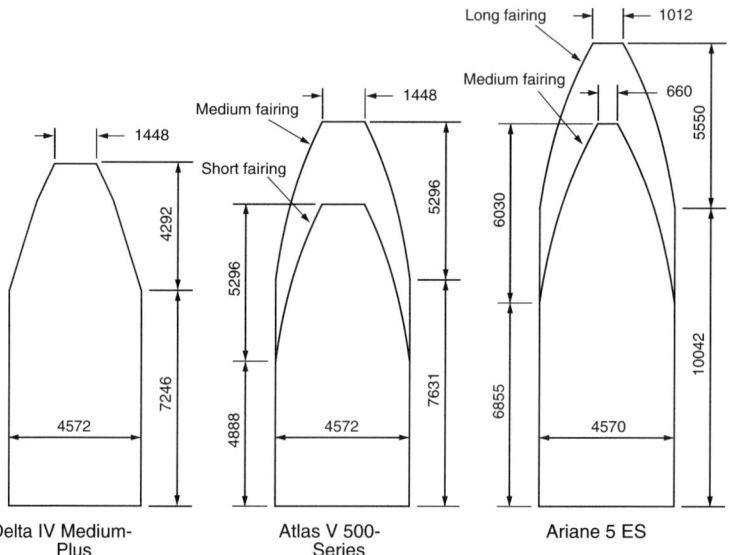

Fig. 12.27. Fairing dimensions for the major U.S. and European launchers. Dimensions are in millimeters. The payload volumes shown are the nominal "static" volumes, which are less than the available volume to allow for sway and vibration during launch. The actual volume available for the payload itself also depends on the dimensions of the payload adapter used.

In addition to limitations in mass and volume, designers of space telescopes have to contend with two other conditions imposed by launchers: launch loads and contamination. Loads occurring during launch are discussed in Chapter 6. As for contamination during launch preparations, it is usually dealt with by pumping conditioned air through the fairings. The standard air cleanliness level is Class 100 000.

Launch costs for the U.S. and European launchers listed in Table 12.3 are in the vicinity of $100M (year 2000). At about $5000/kg for LEO to $20 000/kg for high orbits, modern launching systems are still very expensive and the old saying "It costs an ounce of gold to launch an ounce of lead" is still very much true.

Table 12.3. Maximum payload mass of the major launch vehicles (kg)

Orbit	Delta IV Med+	AtlasV-500	Ariane 5ES
LEO, low inclination	>12 000	> 20 000	> 20 000
LEO, polar Sun-synchronous	9600	17 000	15 000
GTO	6100	8600	7600
Escape, C3=0	4300	6100	4300

References

[1] Dressler, A., ed., *Exploration and the search for origins: a vision for ultraviolet-optical-infrared space astronomy*, AURA Report, 1996.
[2] Elliot, J.L. et al., *Investigation of the images formed by the telescope in the Kuiper Airborne Observatory*, MIT Report, 1986.
[3] Bely, P.Y., Ford, H.C., Burg, R., Petro, L., White, R., and Bally J., POST: Polar stratospheric telescope, Space Sci. Rev., Vol. 74, p. 101, 1995.
[4] Walker, M.F., High quality astronomical sites around the world, ESO Proceedings on *Site Testing for Future Large Telescopes*, p. 3, 1983.
[5] Lynds, R. and Goad, J.W., Observatory site reconnaissance, PASP, Vol. 96, p. 750, 1984.
[6] Merill, K.M., NNTT site evaluation project: an overview, Proceedings of the Flagstaff Conference on *Identification, Optimization and Protection of Optical Telescope Sites*, p. 3, 1986.
[7] Walker, M.F., The effect of urban lighting on the brightness of the night sky, PASP, Vol. 89, p. 405, 1977.
[8] Garstang, R.H., Night-sky brightness at observatories and sites, PASP, Vol. 101, p. 306, 1989.
[9] de Baas, A.F. and Sarazin, M., The temperature structure function for complex terrain, Proceedings of the *Eighth Symposium on Turbulent Shear Flows*, 1991.
[10] Erasmus, D.A., Identification of optimum sites for daytime and nighttime observing, Proceedings of the Flagstaff Conference on *Identification, Optimization and Protection of Optical Telescope Sites*, p. 86, 1986.
[11] Walker, M.F., A comparison of observing conditions on the summit cones and shield of Mauna Kea, PASP, Vol. 95, p. 903, 1983.
[12] McInnes B. and Walker, M.F., Astronomical site testing in the Canary Islands, PASP, Vol. 86, p. 529, 1974.
[13] Barletti, R. et al., Mean vertical profile of atmospheric turbulence relevant for astronomical seeing, J. Opt. Soc. Am., Vol. 66, p. 12, 1976.

[14] Sarazin, M., The ideal site revisited for future ground based telescope projects, Proceedings IAU workshop on *Astronomical Site Evaluation in the Visible and Radio range*, 2000.

[15] Morse, D. and Gillet, F., *Water vapor monitor engineering report*, AURA Engineering Report 73, KPNO, Tucson AZ, 1982.

[16] Singal, S.P., Acoustic remote sensing, Proceedings of the *Fifth Annual Symposium on Acoustic Remote Sensing in the Atmosphere and Oceans*, 1990.

[17] Forbes F.F., Barker, E.S., Peterman, K.R., Cudaback, D.D., and Morse, D. A., High altitude acoustic soundings, SPIE Proc., Vol. 551, p. 60, 1985.

[18] Barker, E.S., Site testing with an acoustic sounder at McDonald Observatory; theory and practice, Proceedings of the Flagstaff Conference on *Identification, Optimization and Protection of Optical Telescope Sites*, p. 49, 1986.

[19] Cayrel, R., Knowledge acquired during the site testing for the Canada-France-Hawaii telescope, Proceedings of the ESO workshop on *Site Testing for Future Large Telescopes*, p. 45, 1983.

[20] Hufnagel, R.E., Propagation through atmospheric turbulence, in *The Infrared Handbook*, Office of Naval Research, Department of the Navy, Washington, D.C., 1989.

[21] VanZandt, T.E., Green, J.L., Gage, K.S., and Clark, W.L., Vertical profiles of refractivity turbulence structure constant: comparison of observations by the Sunset radar with a new model, Radio Sci., Vol. 5, No. 13, p. 819, 1982.

[22] Bufton, J. L., Comparison of vertical profile turbulence structure with stellar observations, Appl. Opt., Vol. 12, No. 8, p. 1785, 1973.

[23] Barat, J., Some characteristics of clear-air turbulence in the middle stratosphere, J. Atmos. Sci., Vol. 39, No. 11, p. 2553, 1982.

[24] Bely, P.Y., Weather and seeing on Mauna Kea, PASP, Vol. 99, No. 616, p. 560, 1987.

[25] Masciadri, E., Vernin, J., and Bougeault, P., 3D mapping of optical turbulence using an atmospheric numerical model, Astron. Astrophys., Vol. 137 (Suppl.), p. 185, 1999.

[26] Roddier, F., *The effects of atmospheric turbulence in astronomy*, Progress in Optics, Vol. 19, Wolf, E., ed., North Holland, 1981, p. 281.

[27] Sarazin, M. and Roddier, F., The ESO differential image motion monitor, Astron. Astrophys., Vol. 227, p. 294, 1990.

[28] Stock, J. and Keller, G., Astronomical seeing, in *Telescopes*, Stars and Stellar Systems Vol.I, Kuiper, G.P. and Middlehurst, B.M., eds., Chicago Univ. Press, 1960, p. 138.

[29] Fried, D.L., Differential time of arrival: theory, evaluation and measurement feasibility, Radio Sci., Vol. 10, No 1, p. 71, 1975.

[30] Bate, R.B., Mueller, D.D., and White, J.E., *Fundamentals of Astrodynamics*, Dover, 1971, p 157.

[31] Farquhar, R.W. and Dunham, D. W., Use of libration-point orbits for space observatories, in *Observatories in Earth Orbit and Beyond*, Kondo, Y., ed., Kluwer Academic Publishers, 1990.

[32] Anderson, B.J. and Smith, R.E., *Natural orbital environmental guidelines for use in aerospace vehicle development*, NASA Technical Memorandum 4527, Marshall Space Flight Center, 1994.

[33] Smith, H. J., Lunar-based astronomy, in *Observatories in Earth Orbit and Beyond*, Kondo, Y, ed., Kluwer Academic Publishers, p. 365, 1990.
[34] European Space Agency, *Kilometric baseline space interferometry*, ESA Report SCI(96)7, 1996.
[35] Burns, J.D. and Mendell, W. W., eds., *Future Astronomical Observatories on the Moon*, NASA Conference Publi. No. 2489, NASA, Washington, D.C., 1989.
[36] Bely, P.Y. and Breckinridge, J.B., eds., *Space Telescopes and Instruments*, SPIE Proc., Vol. 1494, pp. 86–233, 1991.
[37] Bely, P.Y. and Petro, L., Presentation to the Admnistrator, NASA Headquarters, 1999.
[38] Holmes-Siedle A. and Adams, L., *Handbook of Radiation Effects*, Oxford Univ. Press, 1993, p.16.
[39] Frederickson, A.R., Upsets related to spacecraft charging, IEEE Trans. Nucl. Sci., Vol. 43, No. 2, p. 426, 1996.
[40] GOES web page: www.gsfc.nasa.gov/gsfc/earth/goesl/goesl.htm

Bibliography

Observatories

Howse, D., *The Greenwich list of observatories*, Hist. Astron. Vol. 17, Pt. 4, i-iv, p.1, 1986.

Kfiscunas, K., *Astronomical Centers Of the World*, Cambridge Univ. Press, 1988.

Site testing

Ardeberg, A. and Woltjer, L., eds., ESO Workshop on *Site Testing for Future Large Telescopes*, 1983.

Millis R.L., Franz, O. G., Ables, H.D., and Dahn, C.C., eds., Proceedings of the Flagstaff Conference on *Identification, Optimization, and Protection of Optical Telescope Sites*, 1986.

Roddier, F., *The Effects of Atmospheric Turbulence in Astronomy*, Progress in Optics, Vol. 19, Wolf, E., ed., North-Holland, 1981, p. 281.

Rösch, J., ed., *Le Choix des Sites d'Observatoires Astronomiques*, IAU Symposium No. 19, Rome, 1962.

The Infrared Handbook, Chapter on *Propagation through atmospheric turbulence*, Office of Naval Research, Department of the Navy, Washington, D.C., p. 6-1, 1989.

Space orbits

Bate, R.B., Mueller, D.D., and White, J.E., *Fundamentals of Astrodynamics*, Dover, 1971.

Kondo, Y., ed., *Observatories in Earth Orbit and Beyond*, Kluwer Academic Publishers, Astrophysics and Space Science Library, Vol. 166, 1990.

Larson, W.J. and Wertz, J.R., *Space Analysis and Design*, Kluwer Academic Publishers, 1992.

Appendix A
Commonly Used Symbols

α	coefficient of thermal expansion; right ascension	f	focal length; frequency
β	backfocal to focal length ratio	g	acceleration of gravity
δ	deflection; declination	G	gravitational constant
ϵ	thermal emissivity	HA	hour angle
θ	field angle; seeing angle	h	thickness; height; altitude (angle)
κ	thermal conductivity	I	intensity; moment of inertia
λ	wavelength	J	inertia
ν	Poisson's ratio; frequency	K	structural stiffness
ρ	material density	k	proportionality constant; gain
σ	stress; image size; Stefan-Boltzmann constant	m	optical magnification; mass
τ	torque	N	f-ratio; number of counts
τ_0	coherence time (atmosphere)	n	number count
ϕ	wavefront polar angle	Q	amount of heat
φ	latitude	r	running radius on a mirror
ω	angular frequency	r_0	coherence length (atmosphere)
A	azimuth angle; area	RA	right ascension
b	backfocal distance	$\mathcal{R}$	Reynolds number
C_n^2	refraction index structure coefficient	s	mirror separation (prim.–sec.)
		S/N	signal-to-noise ratio
C_p	specific heat	T	temperature
C_T^2	temperature structure coefficient	ΔT	temperature gradient
		t	time
D	mirror diameter	v	velocity
E	modulus of elasticity	Z_n	Zernike polynomial of order n
Fr	Froude number	z	mirror surface ordinate

Appendix B
Basic Data and Unit Conversions

Name	Symbol	Value
Physical constants		
Velocity of light	c	$2.998 \cdot 10^8$ m/s
Gravitational constant	G	$6.670 \cdot 10^{-11}$ N m^2/kg^2
Planck constant	h	$6.625 \cdot 10^{-34}$ J s
Boltzmann constant	k	$1.381 \cdot 10^{-23}$ J/K
Stefan-Boltzmann constant	σ	$5.670 \cdot 10^{-8}$ W/m^2 K^4
Standard gravitational acceleration	g	9.807 m/s^2
Astronomical constants		
Solar luminosity	$L_\odot$	$3.90 \cdot 10^{26}$ W
Solar mass	$M_\odot$	$1.989 \cdot 10^{30}$ kg
Earth mass	$M_\oplus$	$5.976 \cdot 10^{24}$ kg
Mean radius of the Earth		6371 km
Eccentricity of the Earth's orbit		0.0167
Astronomical unit	AU	$1.49 \cdot 10^{11}$ m
Mean distance from Earth to Moon		$3.84 \cdot 10^8$ m
Altitude of geosynchronous orbit		36 000 km
Distance to Sun/Earth L2		$1.5 \cdot 10^6$ km
Solar constant	G_S	1358 W/m^2

Unit Conversions

Time
1 day	86 400 s
1 sidereal day	86 164.091 s
1 year	$3.1558 \cdot 10^7$ s

Angle
$1''$	$4.848 \cdot 10^{-6}$ radian
1 radian	$206\,264''$
1 steradian (sr)	$3.283 \cdot 10^3$ degrees^2
	$1.182 \cdot 10^7$ arcmin^2
	$4.255 \cdot 10^{10}$ arcsec^2
1 arcsec^2	$2.350 \cdot 10^{-11}$ sr

Length
1 statute mile	1609 m
1 nautical mile	1852 m
1 parsec	3.26 light years = $3.086 \cdot 10^{16}$ m
1 lightyear	$9.46 \cdot 10^{15}$ m
1 AU	$1.496 \cdot 10^{11}$ m
1 Angström (Å)	10^{-10} m

Energy
1 BTU	1055 J
1 eV	$1.6 \cdot 10^{-19}$ J
1 cal	4.186 J
1 kcal (cal kg)	4186 J

Pressure
1 bar	$1 \cdot 10^5$ Pa
1 torr	133 Pa
1 psi	$6.89 \cdot 10^{-3}$ Pa

Velocity
1 mph	0.447 m/s
1 km/h	0.278 m/s
1 knot	0.514 m/s

Miscellaneous
1 Jansky (Jy)	10^{-26} W/m^2 Hz

Appendix C
The Largest Telescopes

The largest ground-based, space and airborne optical telescopes are listed in Tables C1, C2 and C3. Particulars for the headings and contents are as follows:

Telescope: The common name of the telescope; refer to the Glossary for the full official name. The list is limited to astronomical telescopes and excludes reflecting telescopes used for military purposes.

Diameter: The diameter of the useful aperture of the primary mirror(s), converted to an equivalent diameter in meters if not circular; the list of ground-based telescopes is limited to those with primary mirrors 3.0 meters or larger in diameter.

Mirror material: The material of the primary mirror substrate. *Cer-Vit Sitall* (i.e., *Astro-Sitall*), and *Zerodur* are ultra-low-expansion glass-ceramics; *ULE* is an ultra-low-expansion fused silica, *Pyrex* and *E6* are borosilicate low-expansion glass products.

Mass: The mass shown for a space observatory is for its entire payload, including instruments and space support systems.

Date completed: Date of first operation of the telescope; for observatories still under construction in mid-2001, the expected completion or launch dates are shown in parentheses.

Table C.1. The largest ground-based telescopes

Telescope	Location	Diameter (m)	Primary type	Primary f-ratio	Primary material	Mount	Date completed
LBT[1]	Arizona	11.8	honeycomb	1.14	E6	alt-az	(2004)
Keck I	Hawaii	10.5	segmented	1.75	Zerodur	alt-az	1993
Keck II	Hawaii	10.5	segmented	1.75	Zerodur	alt-az	1998
GTC	Canaries	10.4	segmented	1.65	Zerodur	alt-az	(2004)
Hobby-Eberly[2]	Texas	9.5	segmented	1.8	Zerodur	fixed	1999
SALT	South Africa	9.5	segmented	1.8	Zerodur	transit	(2004)
Subaru	Hawaii	8.4	meniscus	1.8	ULE	alt-az	1999
Gemini North	Hawaii	8.3	meniscus	1.8	ULE	alt-az	2000
Gemini South	Chile	8.3	meniscus	1.8	ULE	alt-az	2000
VLT UT1	Chile	8.2	meniscus	1.75	Zerodur	alt-az	1998
VLT UT2	Chile	8.2	meniscus	1.75	Zerodur	alt-az	1999
VLT UT3	Chile	8.2	meniscus	1.75	Zerodur	alt-az	2000
VLT UT4	Chile	8.2	meniscus	1.75	Zerodur	alt-az	2001
TIM	Mexico	7.0	segmented	1.5	Zerodur	alt-az	(2004)
MMT conversion[3]	Arizona	6.5	honeycomb	1.25	E6	alt-az	1999
Magellan I	Chile	6.5	honeycomb	1.25	E6	alt-az	1999
Magellan II	Chile	6.5	honeycomb	1.25	E6	alt-az	(2001)
BTA	Caucasia	6.0	solid	4	Sitall	alt-az	1976
LZT	Canada	6.0	liquid	1.6	mercury	transit	(2001)

1. Two 8.4 m primary mirrors.
2. Fixed altitude mount; equivalent diameter.
3. Originally with six 1.8 m mirrors; 6 m mirror installed in 1999.

Table C.2. The largest ground-based telescopes (Continued)

Telescope	Location	Diameter (m)	Primary type	Primary f-ratio	Primary material	Mount	Date completed
Hale	California	5.1	honeycomb	3.3	Pyrex	equatorial	1949
WHT	Canaries	4.2	solid	2.5	Cer-Vit	alt-az	1987
SOAR	Chile	4.2	meniscus	1.75	ULE	alt-az	(2002)
Blanco	Chile	4.0	solid	2.8	Cer-Vit	equatorial	1976
LAMOST	China	4.0	segmented	5		meridian	(2004)
AAT	Australia	3.9	solid	3.3	Cer-Vit	equatorial	1975
VISTA	Chile	3.9	meniscus	2	Sitall	alt-az	(2005)
Mayall	Arizona	3.8	solid	2.8	fused quartz	equatorial	1973
UKIRT	Hawaii	3.8	meniscus	2.5	Cer-Vit	equatorial	1978
CFHT	Hawaii	3.6	solid	3.8	Cer-Vit	equatorial	1979
ESO 3.6	Chile	3.6	solid	3	fused silica	equatorial	1976
NTT	Chile	3.0	meniscus	2.2	Zerodur	equatorial	1989
MPI	Spain	3.5	solid	3.5	Zerodur	equatorial	1984
WIYN	Arizona	3.5	honeycomb	1.75	E6	alt-az	1994
TNG	Canaries	3.5	meniscus	2.2	Zerodur	alt-az	1997
ARC	New Mexico	3.5	honeycomb	1.75	E6	alt-az	1994
Shane	California	3.1	honeycomb	5	Pyrex	equatorial	1959
IRTF	Hawaii	3.0	solid	2.5	Cer-Vit	equatorial	1979

432   Appendix C.   The Largest Telescopes

Table C.3. The major space, airborne and balloon borne telescopes

Telescope	Site	Diameter (m)	Spectral range (µm)	Primary mirror material	Mass (kg)	Launch date	Lifetime (yr)
Stratoscope I	balloon	0.3	–	–		1957	
OAO 2	LEO	0.2 to 0.4	–	–	2000	1968	4
OAO 3	LEO	0.80	–	–	2200	1972	9
KAO	airplane	0.9	40–100			1976	20
IUE	GEO elliptic	0.45	0.11–0.33	beryllium	700	1978	20
IRAS	900 km polar	0.57	8–120	beryllium	1080	1983	0.9
HST	LEO	2.4	0.2–2.5	fused silica	11 000	1990	~15
ISO	GEO elliptic	0.6	2.5–200	fused silica	2200	1995	1.5
SOFIA	airplane	2.5	0.3–1600	Zerodur	~2 000	(2002)	(>20)
SIRTF	drift	0.85	3–180	beryllium		(2004)	(2.5)
FIRST	L2	3.5	60–670		~3300	(2007)	
NGST	L2	6.5	0.6–28	glass or beryllium	~3800	(2010)	(10)

Appendix D
Sharpness

A common and important problem in astronomy is the following. Given an image $I_{ij} = B + AP_{ij} + N_{ij}$ that is the sum of a background B (perhaps including a dark current contribution), a point source with total detected counts A, distributed by a point spread function P_{ij} (so that $\sum P_{ij} = I$, by definition), and noise N_{ij} with zero mean and known variance σ_{ij}, what is the best way to estimate A and what is the signal-to-noise ratio of the results?

Clearly, the answer does depend on how A is estimated. The simplest and more commonly used estimator is to choose an aperture that includes most of the flux and not too much background noise. However, this is not the best that can be done. A more general approach is to choose a set of weights W_{ij} and sum the weighted contributions to an estimate of A. The special case of aperture photometry is obtained when all weights have the same value inside the aperture so that pixels outside make some contribution (after all, they contain some signal), whereas pixels inside but near the edge of the aperture are less important than more central pixels because they individually have poorer signal-to-noise ratio.

The image noise has zero mean $< N_{ij} > = 0$, is uncorrelated from pixel to pixel $< N_{ij} N_{kl} > = \delta_{ik}\delta_{jl}\sigma_{ij}^2$, and has known variance $\sigma_{ij}^2 = B + AP_{ij} + R^2$ (the sum of Poisson noise and read noise, R). The estimator of A that we choose is the most general linear unbiased estimator (and this estimator when optimized achieves the Cramer-Rao bound, so is the best that can be done and no nonlinear estimator will give better signal-to-noise ratio).

$S = \sum(I_{ij}B)W_{ij}/\sum W_{ij}P_{ij}$ is unbiased because by construction $<S> = A$. The noise N on this signal is given by

$$N^2 \equiv (S - <S>)^2 = \frac{\sum W_{ij}^2 \sigma_{ij}^2}{(\sum W_{ij}P_{ij})^2}. \tag{D.1}$$

This result has been obtained by direct substitution and reduction using the assumed uncorrelated nature of the noise. To maximize the signal-to-noise ratio $<S>/N$, one must minimize $N^2/<S>^2$ with respect to the weights W_{ij}. To find the solution, one differentiates with respect to W_{ij} and sets the result to zero. This procedure shows that in the general case, the optimal weights are proportional to

$$W_{ij} \propto \frac{P_{ij}}{\sigma_{ij}} = \frac{P_{ij}}{B + R^2 + AP_{ij}}. \tag{D.2}$$

Two limiting cases are typically encountered. If the photon noise in the source dominates ($AP_{ij} \gg B + R^2$) over most of the image, then the expression for W_{ij} becomes constant and the signal-to-noise ratio comes out in the square root of the number of expected counts. This result is reassuring, but not usually the case for faint object imaging.

In the background- or read-noise-limited case when $AP_{ij} \ll B + R^2$ over all the image, the weights are proportional to the corresponding value of the PSF, and we can normalize them so that $W_{ij} = P_{ij}$ without loss of generality. In this case, the signal-to-noise ratio is given by

$$\left(\frac{<S>}{N}\right)^2 = \frac{A^2 \Psi}{B + R^2}, \tag{D.3}$$

where

$$\Psi = \frac{\sum P_{ij}^2}{(\sum P_{ij})^2} \tag{D.4}$$

is the quantity that we call sharpness. Notice that Ψ is always less than 1, but positive. It becomes equal to one in the sharpest case when all of the flux is in one pixel (note that one cannot just increase the pixel size to get better signal-to-noise ratio, because the quantity B generally grows with pixel size). This cleanly separates the contributions from the total detected flux A from the contribution caused by the telescope design.

This quantity has many amazing properties, which make it easy to work with. When the pixels are critically sampling the PSF or better, the sharpness is independent of pixel phase, so it does not matter where exactly the star is relative to pixel boundaries. For such pixels, the sharpness scales inversely with the pixel area (so smaller pixels give smaller sharpness, but this is exactly canceled by the corresponding change in B for the background-limited case). Thus the signal-to-noise ratio is independent of pixel size, provided the pixels are sufficiently small and the read noise is not important. The summation can then be turned to an integral, and the signal-to-noise ratio can be written in terms of the integral of the square of the PSF. By Parseval's theorem, this is the integral of the square of the MTF, and the MTF can be obtained by an autocorrelation of the aperture function. Thus, one can derive analytic expressions for the sharpness in a number of important cases (such as circular apertures or redundant and nonredundant arrays).

(*Contributed by Christopher Burrows*)

Appendix E
Derivation of the Equation of Motions

The Lagrange equations for a generalized system of n independent coordinates are

$$\frac{\partial}{\partial t}\left(\frac{\partial L}{\partial \dot{q}_i}\right) - \frac{\partial L}{\partial q_i} + \frac{\partial F}{\partial \dot{q}_i} = Q_i, \quad i = 1, \cdots n, \tag{E.1}$$

where q_i are the generalized coordinates (degrees of freedom), such as the displacement of a mass or a rotation angle for a moment of inertia, $L = T - V$ is the Lagrangian, with T and V being the system's kinetic and potential energy, respectively, both expressed in terms of q_i, F is the the Rayleigh dissipation function, and Q_i is the externally applied force (normally applied to the ith mass). The Rayleigh dissipation function is present when the frictional forces are linear and proportional to the velocities and is given by

$$F = \frac{1}{2}\sum_{i-1}^{n} B_i \dot{q}_i^2, \tag{E.2}$$

where B_i is the ith viscous friction coefficient.

Let us apply the Lagrange equations to the altitude axis of the lumped mass model shown in Fig. 7.1. The Lagrange coordinates are the rotation angles of the three lumped masses in the system: θ_m, θ_M, and θ_T for the motor, mount, and tube, respectively. The corresponding angular velocities and angular accelerations are denoted by $\dot{\theta}$ and $\ddot{\theta}$, respectively. The gearbox is a speed reducer of ratio $N : 1$, where N can be either positive or negative (negative when a clockwise motor rotation causes a counterclockwise load rotation). The motor torque, τ_m, is applied against the motor moment of

Appendix E. Derivation of the Equation of Motions

inertia, J_m, and reacts against the gear case and mount moment of inertia, J_M. The tube is assumed subjected to disturbance torque τ_d. A fourth coordinate θ_G, the rotation of the output shaft of the gearbox, is not an independent variable and can be defined by an equation that relates the relative output shaft rotation of the gearbox to the relative input shaft rotation:

$$\theta_m - \theta_M = N(\theta_G - \theta_M) \quad \text{or} \quad \theta_G = \frac{1}{N}\left[\theta_m + (N-1)\theta_M\right]. \tag{E.3}$$

The kinetic energy of the system is

$$T = \frac{1}{2}(J_m\dot{\theta}_m^2 + J_M\dot{\theta}_M^2 + J_T\dot{\theta}_T^2), \tag{E.4}$$

and the potential energy is

$$V = \frac{1}{2}\left[K_M(\theta_M - \theta_G)^2 + K_T(\theta_T - \theta_M)^2\right]. \tag{E.5}$$

Substituting θ_G from equation E.3 into equation E.5 gives

$$V = \frac{1}{2}\left[K_M\theta_M^2 + K_T\left(\frac{\theta_m}{N} + \frac{N-1}{N}\theta_M - \theta_T\right)\right]. \tag{E.6}$$

The Rayleigh dissipation function is given by

$$F = \frac{1}{2}\left[B_m(\dot{\theta}_m - \dot{\theta}_M)^2 + B_T(\dot{\theta}_T - \dot{\theta}_M)^2 + B_M\dot{\theta}_M^2\right]. \tag{E.7}$$

The terms of the Lagrange equation for the first independent coordinate, θ_m, can then be expressed as follows:

$$\frac{\partial}{\partial t}\left(\frac{\partial L}{\partial \dot{\theta}_m}\right) = J_m\ddot{\theta}_m, \tag{E.8}$$

$$\frac{\partial L}{\partial \theta_m} = -\frac{K_T}{N}\left(\frac{\theta_m}{N} + \frac{N-1}{N}\theta_M - \theta_T\right), \tag{E.9}$$

$$\frac{\partial F}{\partial \dot{\theta}_m} = B_m(\dot{\theta}_m - \dot{\theta}_M). \tag{E.10}$$

Since the forcing function for the coordinate θ_m is $Q_1 = \tau_m$, inserting the above terms into the Lagrange equation leads to the dynamic time domain equation of motion for the motor:

$$J_m\ddot{\theta}_m + B_m(\dot{\theta}_m - \dot{\theta}_M) + \frac{K_T}{N}\left(\frac{\theta_m}{N} + \frac{N-1}{N}\theta_M - \theta_T\right) = \tau_m. \tag{E.11}$$

Using a similar process, one can develop the remaining two equations of motion. For θ_T, we obtain

$$J_T\ddot{\theta}_T + B_L(\dot{\theta}_T - \dot{\theta}_M) - K_T\left(\frac{\theta_m}{N} + \frac{N-1}{N}\theta_M - \theta_T\right) = \tau_d, \tag{E.12}$$

Appendix E. Derivation of the Equation of Motions

and for θ_M,

$$J_M\ddot{\theta}_M + B_M\dot{\theta}_M + K_M\theta_M - B_m(\dot{\theta}_m - \dot{\theta}_M) - B_T(\dot{\theta}_T - \dot{\theta}_M)$$
$$+K_T\frac{N-1}{N}\left(\frac{\theta_m}{N} + \frac{N-1}{N}\theta_M - \theta_T\right) = -\tau_m. \qquad \text{(E.13)}$$

Equations E.11, E.12, and E.13 represent the simultaneous linear, time-domain dynamic equations of motion for the lumped mass system around the altitude axis. Note that the external forcing function τ_m is independent because its value varies with the electric current applied to the motor.

(Contributed by Marvin (Tim) Cornwell)

Appendix F
Glossary

This glossary covers terms and acronyms that may be encountered in the telescope building profession, whether or not they appear in this text. Definitions will be found under the most commonly used term of reference, acronym or full form, with the alternative forms cross referenced. Words used in definitions that are themselves defined in the Glossary are identified by q.v. (quod vide, i.e., "see this").

AAS American Astronomical Society. An organization of professional astronomers in the United States, Canada, and Mexico. The society has more than 6000 members and publishes the *Astronomical Journal* and the *Astrophysical Journal*.

AAT Anglo-Australian Telescope. A 4 m equatorial telescope at Siding Springs, Australia.

aberration In optics, the imperfections of an image due to wavefront errors (of geometric or chromatic origin).

aberration of starlight The apparent angular displacement of a celestial object caused by the finite velocity of light in combination with the motion of the observer.

absorption coefficient In optics, a measure of the attenuation of the intensity of light as it passes through a medium.

absorption line A narrow region of the spectrum within which the intensity of radiation is lower than in the adjoining regions; typically produced when radiation from a background source passes through cooler matter.

achromat A composite lens that does not produce noticeable chromatic aberration. It is generally composed of two lenses with different indices of refraction, their powers being selected so as to cancel out chromatic effects.

acoustic sounder A device for measuring atmospheric turbulence by the scattering of sound. Also called a "sodar."

acquisition (of a target) The action of placing a target in the aperture of a science instrument. This happens automatically when the absolute pointing accuracy is better than the field of the instrument. When such is not the case, as in the centering of a target in a spectrograph slit, special techniques such as blind offsetting (q.v.) or spiral search are required.

active optics The controlled deformation or displacement of optical elements to correct for slowly varying effects (<1 Hz), such as gravitational deflections and temperature drifts. See also **adaptive optics**.

A/D Analog to Digital. See **analog-to-digital converter**.

adaptive optics The controlled deformation or optical elements to correct for rapid fluctuations (> 1 Hz) in image quality. On the ground, this technique is used to correct for atmospheric turbulence. See also **active optics**.

adiabatic lapse rate The rate of change of temperature with altitude of a parcel of dry air which is raised or lowered in the atmosphere without exchanging heat with the surrounding air. The adiabatic lapse rate in the atmosphere is 9.8 °C/km. The actual lapse rate of temperature in the troposphere averages about 6 °C/km.

afocal Characteristic of an optical system that receives parallel rays of light from a distant source and outputs parallel rays of light at a different magnification.

afocal telescope A telescope with no final focus, both the object and image being at infinity.

air mass A measure of atmospheric extinction as a function of the path length traversed by starlight in the atmosphere before it reaches the telescope. The air mass is equal to 1 when pointing at the zenith and is about 3 for a 70° zenith angle.

Airy disk The central portion of the diffracted image of a point source formed by an optical system with a circular aperture. It contains 84% of the total energy in the case of a circular aperture with no obstruction by vanes or secondary mirror. The diameter of the disk (i.e., the diameter of the first dark ring) is $2.44\lambda/D$ measured on the sky, where λ is the wavelength and D is the diameter of the aperture. Named after George Airy, who was the first to derive the mathematical description of the PSF of a circular aperture.

albedo The fraction of incident sunlight that the surface of a celestial body (e.g., Earth, Moon, planet) reflects (identical to reflectance).

aliasing See **Nyquist theorem**.

ALOT Adaptive Large-Optics Technologies. A DoD-sponsored project that developed a lightweight, 4 m space telescope equipped with an advanced active optics system.

alt-az mount Altitude-azimuth mount. A mounting for a telescope, one axis of rotation being horizontal (altitude axis) and the other vertical (azimuth axis).

altitude The angular distance above the horizon to a celestial object, as measured along a vertical circle. Also called elevation.

altitude axis In an alt-az mount, the horizontal axis about which the tube of a telescope rotates.

aluminizing The process of coating a mirror surface with aluminum.

amorphous solid A state of solid material in which atoms are organized over short ranges but lack the recurring pattern found in crystals. Glass is an amorphous solid. The amorphous solid state is obtained by the rapid cooling of a viscous fluid (e.g., glass) or direct solidification of the vapor phase by vacuum deposition or other techniques (e.g., silicon).

analog signal A signal that continuously represent a variable.

analog-to-digital converter (ADC) An electronic device that converts analog signals to an equivalent digital form.

anastigmat An optical system that does not suffer from common optical defects such as coma, astigmatism, or spherical aberration.

Anglo-Australian Telescope See **AAT**.

Angström A unit of length used for light wavelengths or coating thicknesses, equal to 10^{-10} m.

angular size The angle over which an object appears to extend.

annealing A process of heating followed by cooling, used for softening metals or removing internal stresses. Also, slow cooling following melting in mirror blank fabrication.

ANSI American National Standards Institute. A professional organization in the United States responsible for accepting and designating the standards developed by other organizations as national standards.

antireflection coating A coating applied to a lens or optical window to minimize reflections and maximize transmission.

AO Announcement of Opportunity. A NASA announcement inviting a proposal. AOs tend to be for larger programs than NRAs (q.v.), but are not as specific as RFPs (q.v).

APART Arizona's Paraxial Analysis of Radiation Transfer. A program for analyzing stray light in optical systems, developed at the University of Arizona and commercialized by Breault Research Organization, Inc.

aperture The size of the first optical element in an optical system (e.g., the primary mirror of a telescope). The aperture diameter is the simplest measure of the light-gathering power of a telescope.

aperture stop A physical element, usually circular, that limits the light bundle an optical system will accept.

apex In orbital mechanics, one of two points on an elliptic orbit lying on the major axis.

aphelion The point at which a body (spacecraft, planet) in a heliocentric orbit is farthest from the Sun.

apogee The point at which a body in orbit around Earth reaches its farthest from the Earth.

ARC Astrophysical Research Consortium, which operates a 3.5 m telescope at the Apache Point Observatory, New Mexico.

areal density The mass per unit area (e.g., of a mirror).

Arecibo radio telescope A 305 m radio dish at the National Astronomy and Ionosphere Center in Arecibo, Puerto Rico. The dish is not movable and consists of a fixed metallic surface located in a natural valley. The receiver is supported at the focus by cables and moves to track sources being observed. The Arecibo telescope is the largest telescope of any kind in the world. It was completed in 1963 and is operated by Cornell University for the NSF.

array Short for "detector array." A two-dimensional matrix of individual electronic detectors, typically constructed on centimeter-sized wafers of silicon or other materials.

ASCII American Standard Code for Information Interchange. ASCII (pronounced "askee") is a standard developed by the American National Standards Institute (ANSI) to define how computers write and read characters. The ASCII set of 128 characters includes letters, numbers, punctuation, and control codes.

aspect ratio of mirror The ratio of thickness to diameter. A misleading characteristic of mirror flexibility. See **diameter to thickness ratio**.

aspheric An optical element (lens or mirror) that does not have a spherical surface (e.g., conic).

astatic lever A counterweighted lever used in mirror support systems.

astigmatism An optical aberration which causes off-axis rays to form an ellipse or a straight line at the focal plane instead of being brought to a point focus. An optical system designed to avoid such defects is known as an "anastigmat."

astrology The pseudoscience that treats the supposed influence of the configurations and positions of the Sun, Moon, and planets on human destiny.

astrometry The branch of astronomy concerned with the measurement of precise positions of celestial objects.

astronautics The study of celestial mechanics and engineering fields as applied to placement and control of manned or unmanned objects in space.

Astronomical Almanac A yearly publication of the U.S. Naval Observatory and the Royal Greenwich Observatory which provides the ephemerides of the Sun, Moon, and planets and other astronomical data.

astronomical unit (AU) The mean distance between the Earth and the Sun (about 149 million kilometers).

astronomy The branch of science that treats the physics and morphology of that part of the universe that lies beyond the Earth's atmosphere.

astrophysics The branch of astronomy concerned with the composition and physical properties of celestial objects.

athermal, athermalized Designed so as to ensure that system changes do not occur over a given temperature range.

atmospheric refraction The bending of light rays as they pass through atmospheric layers of varying density.

attenuation The reciprocal of gain. A dimensionless ratio defining the decrease in magnitude of a signal as it passes between two points or two frequencies. Large values of attenuation are expressed in decibels (dB).

attitude The orientation of a spacecraft with respect to a reference frame.

AU Astronomical Unit (q.v.).

AURA Association of Universities for Research in Astronomy. A consortium of universities and other nonprofit institutions that manages observatories in Arizona, Hawaii, and Chile, and the Space Telescope Science Institute (STScI).

autocollimation A technique used to test the alignment and image quality of an optical system. A source is placed at the focus of the system and the output (collimated) beam is reflected back to it by a flat mirror.

AXAF Advanced X-Ray Astronomy Facility, renamed the Chandra X-Ray Observatory (q.v.) after its launch in 1999.

azimuth The angular distance measured clockwise along the horizon from a specified reference point (usually North) to the intersection with the great circle passing through a body on the celestial sphere.

azimuth axis In an alt-az mount, the vertical axis about which the mount rotates.

back emf (back electromotive force) The voltage generated when a permanent magnet motor is rotated. This voltage is proportional to motor speed and is present whether or not the motor windings are energized.

back focal length In an optical system, the distance from the vertex of the last optical element to the focus.

background-limited observation An observation whose signal-to-noise ratio is limited by the background noise. The source of the background can be cosmological emission, zodiacal light, atmospheric emission, or thermal emission from the system itself, but not from the detector itself. When only natural background is considered (i.e., excluding thermal emission of the observatory), one refers to "sky limited" (on the ground) or "zodiacal light limited" (in space) observations.

backlash The relative movement of interlocked mechanical parts that occurs when motion is reversed (as in gears). The consequence is hysteresis in the control system.

baffle A structure in an optical system that obstructs or scatters stray light which would otherwise reach the detector.

bake out A cleaning process in which an item is heated, during or after manufacture, to outgas contaminates.

band gap The difference between the lowest energy level of the upper (conduction) band and of the lower (valence) band in an insulator or semiconductor, usually expressed in electron-volts.

bandpass In optics, the portion of the spectrum which is transmitted through an optical system. In control systems, the disturbance frequency range over which the system has control authority.

bang-bang A servo control process that uses a square-wave control. When control is needed, the controller commands the opposite extreme point. Typically used in thermal control, but also in crude mechanical systems.

baud rate The rate of a serial communication data transmission, expressed in bits per second (from Émile Baudot, an early telegraph innovator).

B.C.E. Before the Common Era. For dates, equivalent to B.C.

beam splitter An optical device for dividing an incoming beam into two separate beams, one being transmitted and the other reflected. See also **dichroic**.

beam walk The displacement of the footprint of a light beam on an optical element as a function of the change of direction in the field of view. There is no beam walk at a pupil.

bias frame The readout of a CCD detector of zero integration time with shutter closed. The number of electrons registered per pixel must be subtracted from a science exposure, since they were not created by photons from the source.

bimorph mirror A type of deformable mirror. See **DM**.

binary star Two stars in orbit about their common center of mass.

birefringent Said of an optical material whose index of refraction has a different value in different directions.

bit A binary digit. In digital computing, the smallest unit of information. A bit can either be "on" or "off," represented as a "1" or a "0." Data processed by a computer is organized into larger groups such as bytes (8 bits).

blackbody A hypothetical perfect radiator which absorbs and reemits all radiation incident upon it. A blackbody has an effective emissivity of 1.

Blanco telescope An NOAO 4 m telescope on Cerro Tololo, Chile. It is a near twin of the Mayall telescope on Kitt Peak.

blank The substrate used for a mirror after it is made into the correct size and thickness, but before the optical figure is ground.

blind offset A telescope pointing procedure used for faint targets which consists of (1) very accurately predetermining the position of the target with respect to a nearby reference star (e.g., by measurement of a previously taken long-exposure image of the field), (2) pointing the telescope to the reference star, (3) centering the telescope on that star, and (4) offsetting the telescope by the predetermined target/reference-star vector.

blind pointing Pointing a telescope in a specific direction solely by using its attitude sensors or encoders.

blind spot The region of the sky near zenith where targets cannot be tracked with an alt-az mount because the required azimuth drive velocity is too high.

BOE Basis of Estimate. The justifications for arriving at a particular cost estimate, which include the estimating methods, approach taken, and prices used.

boiling time See **coherence time**.

BOL Beginning of Life. Term used to refer to the beginning of operation of a facility (especially in space).

Bol'shoi Teleskop Azimultal'nyi See **BTA**.

bonnette A combination of guiding head and field visualization, generally mounted at a telescope focus directly in front of the instrument (from the French for "eyepiece cup").

boresight In a telescope, the mechanical axis of the tube, which is near but not necessarily coincidental with the optical axis. See also **line of sight**.

borosilicate glass A low-expansion glass such as Pyrex (a Corning product).

boule From the French, "ball." In optics fabrication, an elementary volume of raw glass, typically about one meter in diameter. Large mirror blanks are sometimes made by cutting individual boules into hexagonal segments and fusing them.

boundary layer In atmospheric physics, the air layer near the ground where wind velocity increases from zero at the surface to its full value, which corresponds to external frictionless flow. The boundary layer extends to a height of roughly 1 km and is more properly referred to as the "planetary boundary layer." The layer nearest the ground, where thermal and friction effects are strongest and which is affected by surface roughness and small-scale topography, is called the "ground" or "surface layer." The layer above the planetary boundary layer, which can be affected by large-scale thermal effects and large scale topography, is called the "atmosphere boundary layer." For more details, see Chapter 12.

brassboard, breadboard An engineering hardware mock-up used to verify a design. A breadboard is cruder than a brassboard, the latter being implemented with specific components.

BRDF Bidirectional Reflectance Distribution Function. A function that characterizes light scatter off surfaces.

bright time See **dark time**.

BTA Bol'shoi Teleskop Azimultal'nyi. A 6 m aperture telescope of the Special Astrophysical Observatory on Mt. Pastukhova, Russia.

BTDF Bidirectional Transmission Distribution Function. A function that characterizes light scatter off transmissive optical elements (e.g., lenses and windows).

burn rate The monthly rate at which a contractor's funds are expended during the period of the contract.

bus The module containing the space support systems for a space observatory. See **SSM**.

byte A unit of information used in reference to computers and quantities of data. A byte consists of 8 bits (q.v.) and generally corresponds to a single character or number. See also **MB**.

C_3 The C_3 coefficient is the square of the hyperbolic excess velocity, or velocity at infinity, and is used to describe a vehicle's orbital energy with respect to that required for escape. This is easier to understand by looking at the governing equation for a vehicle thrusting on an escape trajectory:

$$V_{\text{bo}}^2 - 2\mu/R_{\text{bo}} = V_\infty^2 = C_3$$

where R_{bo} is the distance from Earth center at burnout, V_{bo} is the burnout velocity, V_∞ is the velocity at infinity (hyperbolic excess velocity), and μ is the Earth's gravitational constant ($\mu = GM_{Earth}$). From an energy perspective, the above equation is: kinetic energy at burnout + potential energy at burnout (always negative) = kinetic energy at infinity. For trajectories that do not escape Earth, such as a transfer trajectory to L2, the value of C_3 is negative.

C_n^2 Coefficient of the structure function which describes the statistical variation of the index of refraction in the atmosphere. Seeing is a function of the integral of C_n^2 along the optical path in the atmosphere.

CAD Computer-Aided Design. Computer techniques used in the design and drawing of mechanical systems.

CAIV Cost As an Independent Variable. A design in which project cost is allowed to vary when determining the optimal architecture.

calibration The determination of the relationship between values indicated by a sensor and the actual corresponding values. For an astronomical instrument, the procedures employed to remove the instrumental signature from the scientific data.

camera In astronomy, an instrument for recording telescopic images, consisting of the optics and a photosensitive detector.

Canada-France-Hawaii Telescope See **CFHT**.

Cassegrain An optical arrangement in a two-mirror reflecting telescope in which light is reflected by a convex secondary mirror to a focus near the primary mirror.

Cassegrain focus The final focus of a two-mirror "Cassegrain" optical system.

caustic zone In an optical system, the zone in which rays approaching focus intersect. When no aberrations are present, all rays intersect at the focus. In the presence of aberrations, however, they meet at different points. The larger the aberration, the greater the spread of these intersection points and the larger the caustic zone.

CCB Configuration Control Board. A board which approves or disapproves change requests for project implementation and procedures (but not changes to the scientific requirements). The project manager is normally the board chairman.

CCD Charge-Coupled Device. A solid state light detector that has replaced photographic emulsions as the primary recording medium for astronomical images in the visible. The recording portion of the chip is divided into discrete photosensitive elements (pixels).

CDR Critical Design Review. A design review to evaluate the complete design of a project.

C.E. Common Era. In dates, equivalent to A.D.

celestial equator A great circle of the celestial sphere 90° from the celestial poles.

celestial mechanics That branch of astronomy dealing with the motions and gravitational influences of solar system bodies.

celestial poles Points about which the celestial sphere appears to rotate; intersection of the celestial sphere with the Earth's polar axis.

celestial sphere An imaginary sphere of arbitrary radius upon which celestial bodies may be considered to be located, when seen from Earth.

central obscuration In an on-axis reflecting telescope, the part of the aperture that is blocked by the secondary mirror and baffle.

centroiding An image-processing technique for determining the "center of light" of a guide star.

Cerenkov radiation A luminous emission occuring when charged particles (e.g., cosmic rays) cross a material medium (such as an optical element) at a speed higher than the speed of light in that medium.

Cer-Vit Ceramic Vitrified. An ultralow-expansion glass ceramic produced by Owens Illinois in the 1970s; no longer in production (similar to Zerodur).

CFD Computational Fluid Dynamics. A numerical technique for analyzing the thermal and dynamic effects of the air surrounding bodies such as telescopes and enclosures.

CFHT Canada-France-Hawaii Telescope. A 3.6 m aperture telescope on Mauna Kea, Hawaii jointly operated by the National Research Councils in France and Canada and by the University of Hawaii.

CFRP Carbon-Fiber-Reinforced Plastic. A family of composite materials of carbon fibers in a polymer matrix that includes "graphite epoxy" (GrEp).

CGH Computer-Generated Hologram. A hologram used in the testing of aspheric optics.

Chandra X-Ray Observatory The Chandra X-Ray Observatory, formerly known as the Advanced X-Ray Astronomy Facility (AXAF), is a NASA observatory operating in the 0.1–10 keV band. Launched in 1999, it is named after the late Indian-American astrophysicist Subrahmanyan Chandrasekhar of the University of Chicago.

characterization A process for determining a sensor's output compared to a basic input. This is similar to a calibration but is less rigorous and not completely traceable.

chopping A technique for observing faint sources in the presence of a strong, varying sky background. It consists of rapidly alternating pointing between the source and an empty portion of the sky.

clean room A room or area where temperature, humidity, and concentration of airborne particulates is strictly controlled. Clean-room specifications are defined by various national and international standards (e.g., U.S. Federal Standard 209 and ISO EN 146611-1). Particulate concentration is generally defined as the number of suspended particles of a given size in a unit volume of air. In FED-STD-209, cleanroom "class" is defined as the maximum number of particulates 0.5 µm or larger per cubic foot. A typical clean room for optics and electronics assembly, class 10 000, contains fewer than 10 000 particulates per cubic foot. See also **cleanliness level**.

cleanliness level An established maximum of allowable contaminants based on size, distribution, and quantity on a given surface area. Cleanliness levels are formally defined by standards, such as U.S. MIL-STD-1246, based on counts and sizes of particles deposited onto a unit area. See also **clean room**. Note that clean-room *class* defines the maximum number of particulates per unit volume of air in a room, whereas *cleanliness level* categorizes the maximum number of particulates deposited on a given surface area (e.g., optical surface).

closed loop A broadly applied term relating to any system in which the output is measured and compared to the input, which is then adjusted to reach the desired output condition.

COBE COsmic Background Explorer. A NASA satellite that operated from 1989 to 1993 measuring the primordial background radiation.

CODE V An optical design ray-trace program developed and commercialized by Optical Research Associates.

cogging A condition in which a motor does not rotate smoothly, but steps or jerks from one position to another during revolution. Cogging is most pronounced at low motor speeds.

coherence length A parameter, represented by the symbol r_0, introduced by David Fried in 1966 to characterize atmospheric turbulence. In a beam affected by atmospheric turbulence, r_0 is the diameter of the area in the incoming wavefront where the rms of the phase fluctuation is 1 radian (i.e., within which the beam is essentially in phase).

coherence time In a light path affected by atmospheric turbulence, the time over which, at a given aperture point, the rms of the phase error difference is 1 radian. It is also referred to as the "Greenwood time delay." Its inverse is the Greenwood frequency. The coherence time is somewhat shorter than the average lifetime of an individual speckle (q.v.), also called "boiling time." At good observatory sites, the coherence time is on the order of 10 ms.

coherent light source A light source producing radiation in which all the emitted waves vibrate in phase (such as a laser).

cold stop In an infrared instrument, a stop, generally located at a pupil, which is cooled in order to minimize thermal radiation toward the detector.

collimated beam A beam of parallel rays.

collimation The process of aligning the optical system of a telescope to minimize aberrations at the focus.

collimator An optical element in an instrument producing a beam of parallel rays. Also, a system in optical shops used to simulate a point source at infinity in order to test telescope optics.

Columbus telescope See **LBT**.

coma An optical aberration in which the image of an off-axis point source is a comet-shaped (hence the name) blur.

commissioning The phase following construction (or launch) during which the capabilities of an observatory are demonstrated in its final operational configuration. During commissioning, both verification and validation tests are performed on the complete system to ensure that the observatory meets all its science requirements and is ready for operation.

conduction band The upper energy band in a semiconductor that is not completely filled with electrons. Electrons can conduct in a conduction band.

configuration management Technical and administrative action to monitor changes to project elements, obtain the necessary approvals, and disseminate the approved changes.

controller An electronic device in a feedback control system (hardware or software) that processes a signal to regulate a controlled variable.

corner cube See **retroreflector**.

coronagraph An optical system used to block the light of a star in order to permit the observation of the star's surroundings. Developed by Bernard Lyot in the 1930s for the observation of the Sun's corona, hence the name. Also used for the detection of faint sources very close to a bright star (e.g., low-mass companion, circumstellar disk, planets). It consists of a mask located in a focal plane to reduce the light of the star by occultation, followed by a stop in a pupil plane to block the light diffracted by the edge of the entrance aperture. This second stop is called a Lyot stop.

cosmic rays Charged particles of matter (not radiation), mostly electrons, protons, and helium nuclei, moving through the Galaxy at close to the speed of light. They are produced by stars, supernovae, etc.

cosmological window The spectral region around 3.5 µm where zodiacal light is minimal, potentially allowing the most sensitive observations of the cosmos beyond the solar system.

cosmology The study of the general nature of the universe in space and time.

COSPAR Committee on Space Research. A scientific committee of the United Nations established to encourage cooperative programs of rocket and satellite research. COSPAR is only concerned with scientific research and does not address technological problems.

cost plus fee A type of contract in which the contractor is reimbursed for all allowable costs and receives an additional percentage of those costs in fee.

coudé A French term meaning "bent," used to describe the series of flat mirrors on a ground-based telescope which fold the optical beam so as to keep the focus stationary as the telescope rotates. The term is also applied to the focus itself.

CPM Critical Path Method. A mathematical technique for analyzing and optimizing project schedules. Similar to PERT (q.v.).

critical path In project scheduling, the string of connected activities requiring the longest time for completion.

cryogenic Relating to low-temperature refrigeration and/or achieving, maintaining, and experimenting with low temperatures. Generally used for temperatures lower than 100 K (or at least lower than those achievable with thermoelectric coolers, which is about 200 K).

cryo-null-figuring (improperly called "cryofiguring") A figuring technique for mirrors operating at cryogenic temperatures wherein the mirror is figured at room temperature, tested at cryogenic temperature, then refigured at room temperature to correct (null out) the surface error determined during the test.

cryostat A vessel designed to keep its contents at a low (cryogenic) temperature. The external part is a dewar (q.v.) which eliminates conduction losses. To minimize losses via radiation, the cryogenic content is surrounded by reflective radiation shields at intermediate temperatures.

CSA Canadian Space Agency. An agency of the Federal Government of Canada.

CTE Coefficient of Thermal Expansion. The proportionality factor between the relative change of dimension of a material ($\Delta l/l$) and temperature.

CTIO Cerro-Tololo Inter-American Observatory. An NOAO observatory located on Cerro Pachón, Chile, which houses the Blanco 4 m telescope (q.v.).

curvature sensing A wavefront-error-sensing method invented by François Roddier, which consists of measuring the local curvatures of the wavefront and integrating them to determine the wavefront error.

CVD Chemical Vapor Deposition. A process in which solid substances are made by deposition over a substrate in a controlled atmosphere. The substrate can then be removed to leave a free-standing piece.

dark time The period in the lunar month when the Moon is less than a quarter full. "Bright time" is when the Moon is more than half-full; other nights are classified as "gray time."

DARPA Defense Advanced Research Project Agency. A U.S. DoD agency that promotes R&D in defense technologies.

DDD Displacement Damage Dose. Radiation-induced degradation of an electronic device due to displacement of nuclei from their lattice position in a material.

deadband The range through which an input signal, although introduced into a system, does not produce an observable response.

dead time The interval between the initiation of a change in the input and the start of the resulting observable response.

Decadal Survey A report by a committee of the U.S. National Academy of Sciences which recommends priorities to NASA and NSF for all federally funded projects in astrophysics. The committee meets every 10 years to make recommendations for the following decade (e.g., 2000–2010).

DEC See **declination**.

decibel (dB) A dimensionless number expressing a logarithmic measure of the ratio of two signal levels or two powers. By definition, the number of decibels related to two amounts of power P_1 and P_2 is

$$10 \log_{10}\left(\frac{P_1}{P_2}\right)$$

For example, 3 dB represent a power ratio of about 2, and 10 dB represent a power ratio of 10.

declination (also called δ or DEC) The angular distance on the celestial sphere measured along the hour circle passing through a celestial object. Measured from zero at the equator to $+90$ ° north and -90 ° south.

deformable mirror See **DM**.

deformation Also referred to as "strain." Dimensional change of a body produced by stress. The deformation is *elastic* if the deformation disappears when stress is removed. It is *permanent* if the deformation remains when stress is removed. The least stress that causes permanent deformation is called the *elastic limit*.

depth of focus The tolerance on the axial position of the detector relative to the best optical focus.

design review A formal, systematic examination of a design to evaluate its requirements and its capability to meet those requirements. See also **PDR** and **CDR**.

detector In optical astronomy, a device for recording images or spectra.

devitrification The process by which an amorphous substance (e.g., glass) converts to crystalline form (and, in the case of glass, loses its transparency).

dewar A double-walled vacuum vessel (thermos-bottlelike) for the storage of cryogenic materials or thermal isolation of cold-temperature equipment (e.g., detectors and cold stops in an infrared instrument). Named after its inventor, Joseph Dewar.

diameter-to-thickness ratio (also "aspect ratio") The diameter of a mirror divided by its thickness. Traditionally, a mirror is considered stiff for a 6:1 ratio and flexible for 15:1 ratio or greater. As shown by André Couder, this rule of thumb is completely wrong and should never be used to judge mirror stiffness. Stiffness is inversely proportional to D^4/t^2, not to D/t.

dichroic A thin-film interference coating that separates a light beam into two separate wavelength bands, (e.g., visible and infrared). See also **beam splitter**.

diffraction The deviation of light from rectilinear propagation when an incoming wavefront passes over the edge of an obstructing body (e.g., an opaque body in the beam itself, or the edge of an aperture). This phenomenon is a characteristic of the wave nature of light and also occurs with water and sound waves, as well as with atomic particles which show wavelike properties. When the various portions of the wavefront interfere at a point beyond the obstacle, the pattern formed is called a "diffraction pattern." The point spread function (q.v.) of an optical system is a particular diffraction pattern occurring at focus. Diffraction caused by obstructions in the aperture of a telescopes, such as a secondary mirror support vane, gives rise to "spikes" in stellar images.

diffraction grating An optical surface, transparent or reflecting, ruled with parallel grooves at precisely spaced distances. The active parts are not the grooves but the flat sections between them, which act like a large number of parallel slits. Light passing through (or reflecting from) these slits diffracts and interferes in a way which depends on wavelength, causing different wavelengths to be steered in different directions. The overall effect is similar to that of a prism, but the spectral dispersion, which is a function of groove spacing, can be much higher.

diffraction limit The finest detail that a perfect (aberration-free) optical system can discern in the absence of atmospheric turbulence. This limit, which is then only due to the wave nature of light (hence, to diffraction effects), is a function of the size and shape of the aperture and of the wavelength of light.

diffraction limited Term applied to a nearly perfect optical system in which aberrations (or seeing effects) are negligible; traditionally defined as having a Strehl ratio greater than or equal to 0.8 (Maréchal condition).

diluted aperture A telescope aperture which is not complete, but has a collecting area large enough and spatially distributed so as to fully sample the uv plane (q.v.). The aperture may be diluted because of missing segments in the primary segmented mirror of a single mount telescope, or because the system is an interferometric array composed of separate telescopes. The dilution factor is the ratio of the actual collecting area to the total area of the aperture. Dilution is typically 25% or higher. An incomplete aperture which does not fully sample the uv plane is referred to as a "sparse aperture." The dilution factor of a sparse aperture is typically 5% or less.

DIMM Differential Image Motion Monitor. An automatic device for measuring seeing, used in observatory site testing.

DIRBE Diffuse Infrared Background Experiment. A COBE (q.v.) onboard experiment.

dispersion The spreading of light as a function of wavelength as it passes through a transparent medium. The effect is due to the fact that the refractive index of transparent substances varies with wavelength. It is lower for long wavelengths (e.g., red) than for short ones (e.g., blue).

dither In a mechanical system, a useful oscillation of small magnitude introduced to overcome the effects of friction, hysteresis, or clogging. In astronomical observations, a small stepwise motion of the line of sight that is introduced during an observation to reduce (1) pixelization effects in digital detectors or (2) background fluctuations when working in the infrared.

DM Deformable mirror. A mirror whose figure can be deformed to compensate for wavefront errors. The two main types used in astronomy are (1) the piezostack mirror, composed of a thin glass plate supported by an array of piezostacks acting in piston fashion to deform the plate, and (2) the bimorph mirror, composed of a pair of piezoelectric wafers embedded with electrodes that act in shear to deform them (in a manner similar to the bimetallic effect).

DoD Department of Defense of the United States of America.

DOF Degrees of Freedom. For an element in a system, the number of ways that element can move.

dog and pony show Slang term for an informative briefing presentation, often for nonexperts, as opposed to a working-level session.

drift Undesired change in an input-output relationship over a period of time.

DRM Design Reference Mission. A representative set of mission activities used in simulations to validate the hardware and operational software of space missions prior to launch.

DRP Design Reference Program. A strawman scientific program used to determine the optimal set of observatory parameters (spatial resolution, sensitivity, wavelength range, lifetime, etc.) that globally satisfies a scientific goal at its most general level.

DSN Deep Space Network. An international network of large antennas that supports interplanetary spacecraft missions and, occasionally, radio astronomy observations.

The DSN currently consists of three deep-space communication facilities placed approximately 120° apart around the world: at Goldstone, California, near Madrid, Spain, and near Canberra, Australia. This placement permits constant communication with spacecraft as the Earth rotates. The DSN is managed and operated for NASA by the Jet Propulsion Laboratory.

duty cycle For a repetitive cycle, the ratio of on time to total cycle time. Duty cycle (%) = [On time/(On time + Off time)] × 100%.

dynamic range Ratio of the largest to the smallest signal level a circuit or detector can handle (expressed in dB for electronic systems).

E6 A low-expansion glass made by the Ohara Corporation in Japan.

échelle grating From the French "échelle," ladder. A grating with short steep groove facets facing toward the light. Thus, the angle of incidence for optimum efficiency (blaze angle) is high, greater than 45°.

échelle spectrograph A spectrograph which uses an échelle grating. Associated with the use of high orders, an échelle spectrograph allows high-resolution spectra to be obtained. Since it operates in several diffraction orders, it can cover a large spectral domain, provided that a cross-dispersing element is added to separate the overlapping orders and permit stacking onto a two-dimensional detector array.

ecliptic The mean plane of Earth's orbit around the Sun.

edge sensor A device for determining the position of the edge of a mirror segment relative to adjoining segments.

effective focal length The product of the aperture diameter by the focal ratio of the converging beam at the focal position in use. For a single mirror, the effective focal length is the same as the focal length of the mirror.

eigenfrequencies From the German, eigen, "proper," "own." Characteristic vibration frequencies of a system in the absence of externally applied excitations. The lowest frequency is called natural or fundamental frequency.

elastic limit See **deformation**.

electromagnetic compatibility (EMC) The ability of communication and electronic equipment to operate together without suffering or causing unacceptable degradation because of unwanted electromagnetic radiation. This is especially important in space systems and instrument detectors.

electromagnetic waves (or **radiation**) A combination of oscillating magnetic and electric fields spreading in wavelike fashion through space at a constant speed of about 300 000 km/s. They are characterized by their wavelengths and extend from gamma-rays (very short wavelengths, $\sim 10^{-8}$ m) up to radio waves (very long wavelengths, several hundreds of meters).

electro-optics The branch of optical science dealing with the effects of applied electrical voltage on the optical properties of materials.

elevation See **altitude**.

enclosure In an observatory, the structure/building surrounding and protecting a telescope. Also called a "dome" when approximately hemispherical in shape.

encoder A measuring device that converts mechanical motion into encoded electronic signals.

engineering model An advanced prototype used during the development phase to demonstrate the maturity of a design and prepare the final specifications and drawings. See also **prototype** and **flight model**.

EOL End of Life. Term used to refer to the expected end of operation of a facility (especially in space).

ephemeris A table of predicted positions of bodies in the solar system or of a spacecraft (plural: ephemerides).

epitaxy The growth of crystals on a crystalline substrate that mimic the orientation of that substrate (used in solid-state detectors).

epoch A particular instant of time used as reference in the determination or measurement of celestial object positions. A catalog for the equinox 2000, for example, lists positions valid for that date (or epoch). To obtain positions at some other epoch, the effects of proper motion, nutation, and stellar aberration must be included in calculations. A full description of an object's position must include both the epoch of the measurement and the equinox (q.v.) to which it is referred.

equatorial mounting A mounting for a telescope, one axis of which is parallel to the Earth's axis, so that a motion around this axis can compensate for the Earth's rotation.

equinox Either of the two points (vernal, autumnal) on the celestial sphere where the ecliptic (the apparent path of the Sun on the sky) intersects the celestial equator. Due to precession, this point moves over time, so positions of stars in catalogs are usually referred to a given "equinox." Currently, the standard equinox is that of Julian year 2000 and is denoted by the prefix J (i.e., J2000). The previous common standard was for J1950 and the differences in an object's position between equinoxes 1950 and 2000 may amount to several arcminutes. Care must be taken to distinguish the *equinox* value, which relates to the position of objects in a time-dependent coordinate system, from the *epoch* value, which refers to the position of a specific object at a given date expressed in that coordinate system. The latter will be different due to effects such as stellar aberration and proper motion.

error, random In a sensor, the amount of error remaining after calibration. See also **error, systematic**.

error, systematic In a sensor, a repeatable error that either remains constant or varies according to some law. This type of error can be eliminated by calibration. The residual error is referred to as "random."

ESA European Space Agency. An intergovernmental organization with a mission to provide and promote the exploitation of space science, research and technology, and space application for exclusively peaceful purposes. It has 15 European member states: Austria, Belgium, Denmark, Finland, France, Germany, Ireland, Italy, Norway, the Netherlands, Portugal, Spain, Sweden, Switzerland, and the United Kingdom. Canada takes part in some projects under a cooperation agreement.

escape velocity The minimum velocity required to remove an object from a given point in a gravity field (e.g., the surface of the Earth) to infinity, without the imposition of a thrust at a later time.

ESO European Southern Observatory. A major observatory operated by Belgium, Denmark, France, Germany, The Netherlands, and Sweden, with sites at La Silla and Paranal in Chile.

étendue From the French, "expanse." In astronomical optics, the product of the solid angle under which a source is seen by the area of the primary mirror of the telescope. This quantity remains constant throughout the optics up to the detector, provided that all diaphragms on the light path are properly sized. The larger the étendue, the larger the field of view that can be accommodated. Commonly used as a figure of merit for the useful field of view of a spectroscopic device.

Fabry-Perot An interferometer, named for its two inventors (Charles Fabry and Alfred Perot), composed of two parallel, high-reflectivity plates separated by a gap. An incoming plane wavefront is multireflected inside, producing the equivalent of a narrow-band filter for wavelengths whose wavefronts are in phase. By slightly changing the gap, the corresponding peaks are shifted. This compact device can be placed in front of a camera for large-field imaging of emission lines.

factor of safety In structural analysis, the ratio of the load that causes failure to the service load.

failure analysis The systematic examination of an item or its diagram(s) to identify and analyze the probability, causes, and consequences of potential and real failures.

far infrared The part of the infrared spectrum from 30 μm to $\sim$ 500 μm. See also **infrared**.

fast Fourier transform A computer algorithm devised by James Cooley and John Tukey in 1967 for the numerical computation of the Fourier transform. This extremely efficient algorithm has revolutionized many fields and, in particular, optics, in that it allows the PSF of optical systems with wavefront errors and complex apertures to be readily determined.

fast optics An optical system with a small f-ratio (imported from the terminology of photography).

fatigue The weakening and eventual failure of a material due to repetitive stresses *within* its elastic range. It is caused by the gradual propagation of microcracks generated by internal defects. Fatigue should not be confused with the permanent deformation (and potential failure) that occurs when a material is stressed *beyond* its elastic limit.

fault tolerant Referring to an electronic design in which a single-event upset or the failure of a single piece of hardware does not significantly degrade the system's performance.

feedback A signal transferred from the output back to the input for use in a closed-loop system.

FEM Finite Element Model. A mathematical model of a structure made of two- or three-dimensional subdivisions called "finite elements." Used in the computer-based calculation of stresses and deflections under load. See also **NASTRAN**.

field derotator A device that compensates for field rotation at the focal plane of alt-az telescopes.

field of regard The maximum possible angular pointing ability of a telescope (as opposed to "field of view," which is the field accessible with a given pointing).

field of view The region of the sky visible to a telescope (or detector) at any one time. See also **field of regard**.

figure The exact shape of the surface of a mirror or other optical component.

figuring The process of grinding and polishing a mirror blank in order to give it a specific geometric shape.

filter An optical device which removes portions of the spectrum of an incident beam of light. Colored-glass filters work by selective absorption and transmission. Interference filters work by selective reflection and transmission within thin coating layers.

fine guiding sensor An instrument in the focal plane of a space telescope used for fine tracking by centroiding on guide stars in the telescope's field of view. A device used for coarse tracking which has its own optical system is called a star tracker (q.v.).

firm fixed price A type of contract in which the contractor receives for his efforts a fixed price, which is agreed upon in advance. See also **cost plus fee**.

FIRST Far Infrared and Submillimetre Telescope (renamed the Herschel Space Observatory). A 3 m ESA space observatory mission operating in the 85–900 µm range, with an anticipated launch in 2007.

flat field An image taken with a light source having a flat (uniform) energy distribution, (e.g., a uniformly illuminated screen). This is used to calibrate the response of individual pixels in a two-dimensional detector.

flight model A realization of a system using design, processes, and components in all ways identical to those of the final product and which undergoes testing in simulated space environment. The flight model may be used as a spare.

fluence In radiation effects, the total number of particles incident on a sample (i.e., integration of flux over irradiation time).

flux The rate at which energy crosses a unit area of a surface in a transverse direction.

focal ratio (f-ratio) The ratio of the effective focal length of an optical system to the diameter of the aperture.

fold-flat mirror A flat mirror used to change the direction of an optical beam (e.g., in a coudé configuration, or to reduce the overall dimensions of an instrument by folding the beam).

Foucault test Also called "knife-edge test." A test developed by Jean Bernard Foucault for the qualitative evaluation of figure errors in a mirror. It consists of using a straight edge (knife edge) to block parts of the rays converging near focus. Figure errors appear on the illuminated mirror as areas of variable intensity.

Fourier transform A mathematical operation which, when applied to a function $f(x)$ of the variable x, generates the function $F(u)$ with $u = 1/x$. For a time-dependent signal, u is a frequency, and thus $F(u)$ represents the distribution of frequencies present in the signal. The transformation can be generalized to a two-dimensional function, $f(x,y)$. If x and y are space dimensions, $f(x,y)$ describes a surface and u and v are in units of spatial frequencies. The Fourier transform of F is f. An efficient digital implementation of the transform is the "fast-Fourier transform," or FFT (q.v.).

Fourier transform spectrometer (FTS) A Michelson interferometer with a movable mirror. By scanning the movable mirror over some distance, an interference pattern is produced that encodes the spectrum of the source (it is its Fourier transform). Fourier transform spectrometers offer a flexible choice of resolution and a multiplex advantage over grating spectrometers since they cover the total spectrum in a single data acquisition, but have a multiplex disadvantage for photon noise.

FOV Field of View (q.v.).

FPA Focal Plane Assembly. In an instrument, the assembly containing the detector and associated elements (window, cooling finger, connectors, etc.).

f-ratio See **focal ratio**.

frequency The number of cycles over a specified time period during which an event occurs; normally expressed in Hertz, or cycles per second.

frequency domain When the Fourier transform is applied to a time-dependent signal, $f(t)$, or to the distribution of a signal on a surface, $f(x,y)$, the resulting function, F, is transposed in frequency. Since the variables, t or x,y, are only defined on a given domain, the frequency variables are only defined over a limited range called the "frequency domain" of the signal. The two representations, F and f, carry the same information expressed in different ways.

frequency response The frequency-dependent characteristic that determines the phase and amplitude relationship between a system's sinusoidal input and output.

Fried parameter (r_0) Also called "Fried length" or "coherence length" (q.v.).

fringe The light and dark bands caused by interference of light waves.

FSM Fine Steering Mirror. A small mirror near the focus to correct for line-of-sight jitter and drift. Also called "Fast Steering Mirror" in some military applications, where the mirror is driven at up to kHz rate.

Full width at half-maximum (FWHM) The diameter of the image of a point source at half the peak intensity. Used as a measure of image quality.

FUSE Far-Ultraviolet Spectroscopic Explorer. A NASA satellite for high-resolution observations at far-UV wavelengths; launched in 1999.

fused quartz A type of glass made by melting quartz sand, cooling it rapidly to the annealing temperature, maintaining it there for hours, and then cooling it slowly to avoid devitrification. See also **fused silica**.

fused silica An amorphous silica glass made by flame hydrolysis. Fused silica and fused quartz are essentially the same material, but fused quartz has some short-range order — a residual of the original crystal structure. The impurities are also slightly different: fused quartz has metallic impurities that cause UV absorption and some fluorescence, whereas fused silica has hydroxyl ions (a by-product of flame hydrolysis) that cause infrared absorption. Physical properties (e.g., CTE) are virtually the same.

FWHM Full Width at Half-Maximum (q.v.).

gain For a linear control system or element, the ratio of the amplitude of a steady-state sinusoidal output relative to a causal input.

Galileo Galilei Italian physicist and astronomer who, in 1609, was the first to develop and use the telescope for astronomical purposes.

gamma-rays Photons with the highest energies, in excess of 10^5 eV, and highest frequencies, above 10^{20} Hz.

Gantt A scheduling chart developed by Henry Gantt in 1916, where project activities are plotted against a time line. These charts are used for planning, scheduling, and then recording progress.

Gaussian distribution The Gaussian (or "normal") distribution describes the behavior of a continuous random variable. The probability density is

$$P(x) = \frac{1}{\sigma\sqrt{2\pi}} e^{-(x-\mu)^2/2\sigma^2},$$

where μ is the mean and σ is the standard deviation. In this distribution, 68.3% of the events fall in the range of $\mu \pm \sigma$, 95.4% fall within $\mu \pm 2\sigma$, and 99.7% fall within $\mu \pm 3\sigma$.

gegenschein A diffuse glowing area on the ecliptic in the direction opposite the Sun, caused by sunlight backscattering from zodiacal dust.

Gemini observatory An international partnership that operates twin 8 m telescopes, one on Hawaii's Mauna Kea and the other on Chile's Cerro Pachon. The partners include the United States, the United Kingdom, Canada, Chile, Australia, Brazil, and Argentina. AURA manages Gemini under the auspices of an international board and the U.S. National Science Foundation.

GEO Geostationary Earth Orbit. See **geostationary orbit**.

geostationary orbit A geosynchronous orbit which is circular and has zero inclination. The geostationary orbit is at an altitude of 36 000 km. A geostationary satellite remains stationary over the same location on Earth.

geosynchronous orbit Any orbit about the Earth which has a period of rotation equal to that of the Earth, and in the same sense. This orbit can be circular or elliptic.

glass ceramic A material composed of a glassy matrix within which microscopic crystals have precipitated. Such a material is made by fusion and cooling to an amorphous solid, which is subsequently heated to develop the crystal phase. Nucleating agents are used to control the degree of crystallization.

glow discharge A method for cleaning optical surfaces prior to coating, which consists of bombarding that surface with ions.

GMT Greenwich Mean Time. Identical to universal time (q.v.).

GO Guest Observer. Guest observers are astronomers who use instruments on spacecraft to make scientific observations, but who are not part of the original team that planned and built the spacecraft and instruments. See **PI**.

Golay configuration A particular configuration of a diluted aperture system where the subapertures are located so as to provide near-uniform and nonredundant coverage of the *uv* plane (q.v.). Such configurations with, for example, 6, 9, or 12 subapertures are referred to as Golay 6, Golay 9, or Golay 12. Named after M.J.E. Golay, who studied such systems (see Ref. [17] in Chapter 1).

GPS Global Positioning System. A set of 24 U.S. Air Force satellites used for determining position and altitude on or near Earth with an accuracy of about 10 m or better. The system also provides time with nanosecond accuracy.

Gran Telescopio Canarias See **GTC**.

gravity gradient Refers to the gradient in the approximately spherical gravity field around a celestial body. Unless counteracted, the gravity gradient around the Earth forces spacecraft to align themselves with their principal inertia axis along the local vertical.

gray time See **dark time**.

Greenwood frequency The inverse of the coherence length (q.v.).

Gregorian A two-mirror telescope combination with a concave secondary mirror.

GrEp Graphite Epoxy. See **CFRP**.

grism Contraction of grating + prism. A dispersing device composed of a transmission grating ruled or glued onto the surface of a prism. The prism deviation compensates for the grating dispersion angle, such that the output beam remains aligned with the input beam.

GSFC Goddard Space Flight Center. A NASA center in Greenbelt, Maryland.

GTC Gran Telescopio Canarias. A 10.4 m aperture telescope being built at the Roque de los Muchachos Observatory, La Palma, Canary Islands.

guide star catalog See **Hubble Space Telescope guide star catalog**. See also **Hipparcos catalog** and **Tycho catalog**.

gyroscope A rapidly spinning wheel which responds to an impressed torque by changing its angular momentum in magnitude and direction; used to sense direction changes.

Hale Telescope A 5 m telescope of the Palomar Observatory, California, owned and operated by the California Institute of Technology. This optical telescope, designed in the 1930s and completed in 1949, incorporated a number of technological innovations including a lightweighted, low-thermal-expansion primary mirror, a Serrurier truss, a horseshoe mount, hydrostatic bearings, aluminum-coated mirror, and an automated dome tracking system. Scientifically extremely successful, the Hale telescope remained the world's largest for three decades. It is named in honor of George Ellery Hale, an American astronomer who was the main force behind its construction as well as that of the pacesetting 60 inch and 100-inch telescopes on Mt. Wilson.

Harlan Smith Telescope A 2.7 m telescope of the McDonald Observatory on Mt. Locke, Texas.

health and safety For a ground observatory or a space mission, refers to the monitoring and trending of critical engineering parameters to verify that all systems are functioning properly, are within their environmental ranges, and that there is no predictable risk of failure or damage.

Hertz (Hz) The unit of frequency, defined as one cycle per second.

HET Hobby-Eberly Telescope (q.v.).

hexapod A particular mechanical system with six actuated legs used to position and orient a body in all of its six degrees of freedom. Also called a "Stewart platform" after D. Stewart who first developed it for flight simulators. The design is unpatented and in the public domain.

HgCdTe Chemical abbreviation for mercury cadmium telluride, used in infrared detectors.

HIP Hot Isostatic Pressing. A technique used in the consolidation of powdered materials, particularly beryllium.

Hipparcos An astrometry satellite of the European Space Agency which operated from 1989 to 1993. The pronunciation of the acronym, standing for HIgh-Precision PARallax COllecting Satellite, is close to Hipparchus, the name of an early Greek astronomer. This observatory permitted the measurement of the position of more than 1 million stars with an accuracy of $0.02''$ to better than $0.001''$. See also **Hipparcos catalog** and **Tycho catalog**.

Hipparcos catalog A star catalog based on data obtained by the Hipparcos satellite. It supplies the positions and photometry of about 118 000 stars with an accuracy of about 0.7 mas and 0.0015 magnitude, respectively. The limiting magnitude of the catalog is about 12.4 in V. See also **Tycho catalog**.

Hobby-Eberly Telescope A 10 m fixed-elevation telescope at the McDonald Observatory in Texas.

HOE HOlography Element. A holographic grating patch used for phasing segmented-mirror systems.

honeycomb mirror A mirror consisting of thin front and back sheets sandwiching a honeycomb structure.

Hooker Telescope A 2.5 m telescope at the Mt. Wilson Observatory, California. This pace-setting reflecting telescope, completed in 1917, was responsible for a number of advances in astronomy, including the discovery of the expansion of the universe.

hour angle Angular distance on the celestial sphere measured westward along the celestial equator from the meridian to the hour circle passing through a celestial object.

hour circle A great circle on the celestial sphere that passes through the celestial poles.

Hubble Space Telescope guide star catalog The all-sky catalog of guide stars up to magnitude 14.5 and with an accuracy of about 1 arcsecond, established for the operation of the Hubble Space Telescope.

HST Hubble Space Telescope. A 2.4 m optical space telescope developed by NASA and the European Space Agency and launched in 1990. Named after Edwin P. Hubble, the American astronomer who discovered the expansion of the universe.

Hubble Space Telescope See **HST**.

hunting An undesirable oscillation which continues for some time after an external stimulus has disappeared.

hydrostatic bearing A bearing system using oil under pressure to support heavy rotating or sliding loads with essentially no friction.

hysteresis An undesirable property of a mechanical or electrical system wherein output is dependent, not only on the value of the input, but also on the direction of the movement or current.

IAU The International Astronomical Union. The IAU is an organization founded in 1919 to promote and safeguard the science of astronomy through international cooperation. It has over 8300 individual members and 66 adhering countries.

ICD Interface Control Document. A document defining the interfaces between subsystems. A draft is typically presented at PDR and the final version at CDR. See also **IRD**.

IEEE (pronounced Eye-triple-E) Institute of Electrical and Electronic Engineers. A professional association of more than 350 000 individual members in 150 countries which organizes conferences, publishes technical documentation, and establishes standards in domains such as computer engineering, telecommunications, electric power, and aerospace and consumer electronics.

image For an optical system, a point-to-point mapping of a luminous object located in one region of space (the object space) to another region of space (the image space).

image quality A qualification of the image of a point source supplied by an optical system. Traditionally measured by the angular size of the image (e.g., FWHM of the core), the energy contained in a given diameter, or the Strehl ratio.

image space The region downstream of an optical system, where the image is formed.

incentive contract A contract of either a fixed-price or cost-reimbursement nature, with a special provision for adjustment of the fixed price or fee as a function of the performance of the contractor (schedule compliance, cost containment, technical performance, etc.).

incoherent Denotes the lack of a fixed-phase relation between two electromagnetic waves.

index of refraction For a given wavelength, the ratio of the velocity of light in a vacuum to the velocity of light in a refractive material. It is a measure of the ability of an optical material to refract light. The denser the material, the higher the index.

inertial reference frame A frame which is not accelerating. In classical mechanics, the Sun is considered nonaccelerating with respect to the fixed stars, establishing a true inertial frame.

infrared The wavelength region between the visible and the shortest radio waves (i.e., microwaves). Infrared is usually divided into three spectral regions: near, mid- and far infrared. The boundaries between these three regions are not fully agreed upon but are generally taken as 0.7–5μm, 5–25μm, and 25–500μm, respectively. The region between 500 μm and 1 mm, sometimes considered part of the infrared, is commonly referred to as "submillimetric."

462   Appendix F. Glossary

infrared cirrus Patches of interstellar dust which emit in the infrared and resemble cirrus clouds in infrared sky surveys.

Infrared Telescope Facility See **IRTF**.

InSb Indium antimonide. A material used in infrared detectors.

intensity In optics, the light power per unit area transverse to the direction of propagation.

interface control document See **ICD**.

interface requirements document See **IRD**.

interference The constructive and destructive superposition of two wavefronts with different phases. In an optical testing interferometer, the two wavefronts are produced by the reference surface and the test sample surface.

interference filter An optical filter with multilayered coatings selected to remove specific wavelength bands by destructive interference.

interferometer In optical testing, an instrument that employs the interference of light waves to measure wavefront errors. In astronomy, two or more telescopes that combine their signals from the same source to create interferences which permit the determination of direction and size of the observed object and also limited imaging. The spatial resolution of an interferometer is that of a single telescope with a diameter equal to the largest separation of the individual telescopes.

IPSRU Inertial Pseudo-Stellar Reference Unit. A device that provides an inertially stable light beam which can be tracked by an optical system in a spacecraft to maintain stable pointing.

IRAD Internal Research And Development (also IR&D). Company-funded technical research and development activity that is not strictly required in the performance of a contract.

IRAS Infrared Astronomical Satellite. A joint project of NASA, the Netherlands, and the United Kingdom. Launched in 1983, IRAS carried out an infrared survey of the entire sky for 10 months, before its liquid helium coolant became exhausted.

IRD Interface Requirements Document. A document defining the interface requirements between subsystems. A draft is normally presented for approval at PDR.

IRTF Infrared Telescope Facility. A 3 m telescope on Mauna Kea, Hawaii, operated for NASA by the Institute for Astronomy, University of Hawaii.

Isaac Newton Telescope A 2.5 m telescope at the Roque de los Muchachos Observatory on La Palma, Canary Islands.

ISO Infrared Space Observatory. A cryogenically cooled 60 cm infrared space telescope operated by ESA from November 1995 to May 1998 at wavelengths from 2.5 to 240 µm.

isoplanatic patch The angular region in which the turbulence characteristics of the atmosphere remain nearly constant. Formally, the angular distance between two beams arriving at a given aperture point, over which the rms of the phase error difference is 1 radian.

isostatic press The consolidation of a powdered material by application of pressure at ambient (cold isostatic press) or high temperature (hot isostatic press).

isotropic Having the same properties in all dimensions.

IUE International Ultraviolet Explorer. A space telescope developed jointly by NASA, ESA, and the United Kingdom for observations at UV wavelengths. IUE, which operated in geosynchronous orbit from 1978 to 1996, was one of the longest-lived satellites ever.

James Webb Space Telescope See **NGST**.

Jet Propulsion Laboratory See **JPL**.

jitter Spurious, unpredictable movement of the line of sight.

JPL Jet Propulsion Laboratory. A semiautonomous NASA center managed by the California Institute of Technology in Pasadena, California. JPL was the center of U.S. rocket development in World War II. Today, it is the focus of NASA's exploration of the planets.

J-T Joule-Thompson effect or Joule-Thompson cooler. A J-T cooler is a cryogenic cooler that employs the expansion of a gas through an orifice to produce a cooling effect.

Julian date (JD) The interval of time in days and fraction of days since January 1, 4713 B.C.E., Greenwich noon. Contrary to common belief, the name "Julian" does not refer to Julius Caesar, the Roman emperor. The system was proposed in 1582 by the Italian mathematician Joseph Scaliger, who named it in honor of his father, Julius Caesar Scaliger. The year 4713 was selected somewhat arbitrarily, but thought to be early enough to include all historical events and all precisely recorded astronomical phenomena.

Julian year A period of exactly 365.25 days which serves as a basis for the Julian calendar.

Kanigen A coating process patented by Electro-Coatings of Iowa, Inc. using a nickel-phosphorous alloy to improve corrosion resistance, polishability, hardness and coat adhesion of metals such as aluminum and beryllium.

KAO Kuiper Airborne Observatory. An infrared observatory consisting of a 90 cm Cassegrain telescope mounted in a Lockheed C-141 airplane flying at an altitude of 12 000 m. The observatory was operated by NASA from 1974 to 1995, at the rate of about 70 nights per year.

Karhunen-Loeve transformation An orthogonal representation of a wavefront or an image, similar to the Zernike decomposition.

Keck telescopes A pair of 10 m telescopes on Mauna-Kea, Hawaii. The observatory is operated by the California Institute of Technology, the University of California, and NASA. The Keck I telescope began science observations in 1993, and Keck II in 1996.

kickoff meeting The initial meeting held to discuss the organization and plans for a new phase of a project.

kinematic mount A mounting system which does not constrain more than the six rigid-body degrees of freedom of the supported body. Such a mount avoids inducing stresses in the supported body.

knife edge In a telescope, a thin member used to support the secondary mirror (also called a "vane," q.v.). Also, the blade used in a Foucault test (q.v.).

knife-edge test See **Foucault test**.

KPNO Kitt Peak National Observatory, part of the National Optical Astronomy Observatory (NOAO). KPNO operates the 4 m Mayall and 3.5 m WIYN telescopes on Kitt Peak, Arizona.

Lagrangian points Points in the space around a system of two large bodies (e.g., Sun-Earth, Earth-Moon) where a small third body will remain in a fixed position relative to the other two. Named after Joseph Louis Lagrange, who first studied these points and who showed that five exist for each such system. Two are stable (L4 and L5), the other three are metastable. In the Sun-Earth system, the two points which are relatively close to Earth are the L1 and L2. Both are on the Earth-Sun line, the L1 point at 236 Earth radii sunward of Earth and the L2 point at a similar distance on the night side.

laminar flow Flow without vortices or turbulence.

LAMOST Large sky Area Multi-Object fiber Spectroscopic Telescope. A 4 m Schmidt telescope being built at the Beijing Astronomical Observatory in Xinglong, China.

LAMP Large Active Mirror Program. A DoD-sponsored program from the late 1980s for the development of a 4-meter, actively controlled, segmented mirror for use in a space-based laser weapon.

lap A tool in the form of a disk charged with abrasive used in polishing mirrors.

lapping The operation of grinding, figuring, or polishing a mirror using a revolving circular lap supplied with an abrasive powder suspended in water.

Large Binocular Telescope See **LBT**.

Large Deployable Reflector (LDR) A NASA concept from the 1980s for a 30-meter aperture telescope dedicated to far-infrared and submillimeter observations from space.

Large Zenithal Telescope See **LZT**.

laser An acronym of Light Amplification by Stimulated Emission of Radiation. A device that produces highly amplified and coherent visible or infrared radiation.

laser star An artificial star created by a laser beam for use in the correction of atmospheric seeing.

LBT Large Binocular Telescope (formerly Columbus Project). A set of two 8 m telescopes sharing the same alt-azimuthal mount. The LBT is being built on Mt. Graham in Arizona by the Mt. Graham International Observatory, the main partners of which are the University of Arizona, Italy and the Research Corporation.

learning curve The reduction in cost per unit as more such units are produced.

LEO Low Earth Orbit. An Earth-centered orbit at an altitude of between 300 and 500 nautical miles (i.e., within the first Van Allen belt).

level of effort Effort of a general or supportive nature with no firm commitment to produce definite products or results. Often used to refer to a constant number of personnel assigned to a given program for a specified period of time.

LGS Laser Guide Star. See **laser star**.

LHe Liquid Helium. Its boiling point at 1 atmospheric pressure is 3.2 K.

libration Oscillation of a body in space about a point of equilibrium (e.g., around a Lagrange point).

Lick Observatory A University of California observatory located on Mt. Hamilton, California.

life cycle The total life span of a system, commencing with concept formulation and extending through operation and eventual retirement of the system.

life cycle cost (LCC) The total cost of a system over its complete life cycle. LCC includes the cost of development, acquisition, operation, maintenance, and, when applicable, disposal.

light bucket A slang term for a large-aperture telescope operating in a mode where geometrical aberrations and phase errors have not been minimized.

line of sight (LOS) The direction on the sky corresponding to the center of the field of the telescope. When there is no image-compensation system, this usually coincides with the optical axis. The LOS is not necessarily the same as the boresight, which is the mechanical axis of the telescope.

LN2 Liquid Nitrogen. Its boiling point at 1 atmospheric pressure is 77.4 K

LN2 temperatures Temperatures associated with the use of liquid nitrogen, generally between 72 and 82 K.

load path The region in a structure with the highest concentration of stress.

long-lead part A part, component, or subassembly with a long delivery time compared to the program's overall schedule.

LOS Line Of Sight (q.v.).

Lyot stop A stop limiting the beam at the exit pupil in such a way as to prevent the detector from seeing any surface preceding the stop other than the optics itself. Used in coronagraphs (q.v) and as a cold stop (q.v.) in infrared instruments. Named after Bernard Lyot, who first used it.

LZT Large Zenith Telescope. A 6 m telescope with a fixed vertical optical axis using a liquid mercury mirror. The telescope is being built by the University of British Columbia in Vancouver, Canada.

Magellan I and II Two 6.5 m telescopes of the Las Campanas Observatory, on Cerro Manqui, Chile. The Magellan Project is a collaboration among the Universities of Arizona and Michigan, Harvard, and MIT. The first of the two telescopes has been in operation since 2000, and the second one since 2002.

magnetic storm A large-scale disturbance of the Earth's magnetosphere, often initiated by the arrival of a plasma cloud originating in the Sun. Such storms can cause severe disturbances to spacecraft.

magnetic torquer An attitude-control or momentum-dumping device on spacecraft, in which an electromagnetic coil interacts with the Earth's magnetic field to provide torque.

magnetosphere The region surrounding Earth or another planet where the magnetic field of that planet tends to exclude the solar wind.

magnification The magnifying power of an optical system can be described in two ways: linear and angular. Linear magnification is the ratio of the size of the object to the size of the image. Angular magnification is the ratio of the angular size of the object as seen through the instrument to the angular size of the object as seen without it. In astronomical telescopes, the object is at infinity for all practical purposes, and only angular magnification is applicable.

magnitude A logarithmic unit of brightness used for stars and other celestial objects. The fainter the star, the greater the magnitude.

Maksutov-Cassegrain telescope See **Schmidt-Cassegrain telescope**.

Maréchal condition A condition proposed by André Maréchal for the practical definition of diffraction-limited optical systems, namely a Strehl ratio greater than or equal to 0.8.

mas milliarcsecond.

master schedule The master schedule for a project, showing key milestones and critical tasks over the full duration of design and implementation phases.

Max Planck Institute See **MPI**.

Mayall telescope A 3.8 m telescope at the Kitt Peak National Observatory, Arizona.

MB Abbreviation for megabyte, a data unit equal to approximately 1 million bytes (1 048 576 bytes exactly). Not to be confused with Mb, the abbreviation for megabit.

meniscus mirror A mirror which is solid (not lightweighted), very thin (e.g., in the 10–20 cm range for 8 m in diameter), and typically has a back face parallel to the front face. For ground telescopes, meniscus mirrors require a large number of active supports to maintain their shapes.

meridian A great circle passing through the celestial poles and through the local zenith.

meteorite A meteoroid that survives passage through the atmosphere and strikes the ground.

meteoroid A small rocky or metallic bodies in interplanetary space. Meteroids have velocities of several tens of kilometers per second. When they enter the Earths atmosphere, the friction of their passage produces a brief luminescent trail called a meteor (popularly called a shooting star). The great majority of meteorites are fragments of asteroids, ranging in size from millimeters up to about 15 cm (very exceptionally). Larger bodies (> 10 m) are called asteroids. See also **micrometeoroid**, **meteorite**, and **orbital debris**.

microdensitometer A device for measuring the optical density of minute areas on a photographic plate (e.g., star images).

micrometeoroid A meteoroid with a diameter of less than 0.1 mm. These are too small to cause a luminous effect when entering the Earths upper atmosphere.

microthermal fluctuations Temperature fluctuations in the atmosphere, at the origin of "seeing."

microwaves The part of the electromagnetic spectrum between the infrared and short-wave radio wavelengths, i.e., approximately 1 mm to 30 cm in wavelength. See also **infrared**.

mid-infrared The region of the infrared spectrum between 5 and $\sim$ 30 µm. See also **infrared**.

MIDEX Medium-class Explorer. A medium-sized (cost <$140M) NASA mission.

Mie scattering The scattering of light caused by particles with dimensions on the order of the wavelength of light.

mil specs Short for U.S. military specifications. Detailed specifications defining materials, processes, and test procedures for military contracts.

milestone A significant event in a project, used as a monitoring tool for assessing progress.

mirror blank See **blank**.

mirror cell The mechanical and structural assembly supporting a mirror.

mirror substrate See **blank**.

MLE Maximum Likely Earthquake. The maximum earthquake level adopted for the design of an observatory. At this level, major damage is acceptable, but not to the point where it would be uneconomical to repair rather than completely rebuild the facility. See also **OBE**.

MLI MultiLayer Insulation. A radiation-insulating blanket used in spacecraft thermal control.

MMT Multiple-Mirror Telescope. Originally composed of six 1.8 m Cassegrain telescopes working together on the same mount. The MMT was installed on Mt. Hopkins, Arizona in 1978. It was converted in 1999 to a conventional telescope with a 6.5 m primary mirror.

modulus of elasticity See **Young modulus**.

moment of inertia of an area (I) In structural analysis, the second moment of a beam's cross-section area. If dA is an elemental area and y is its distance from a given axis (e.g., neutral axis, q.v.), the moment of inertia is equal to

$$I = \int y^2 \, dA.$$

momentum dumping A procedure employed on spacecraft for discarding excess momentum acquired through the continuous action of external torques such as those due to a gravity gradient or solar pressure.

Monte Carlo analysis An analysis of a system's behavior by evaluating its response to a large number of randomly selected discrete samples of input parameters. This technique is employed in cases where exploring the complete domain of possibilities would be too time-consuming. Some stray light analysis programs use this technique.

MPI The Max Planck Institute (Germany). Also the name of a 3.5 m telescope belonging to the German-Spanish Astronomical Center and located on Calar Alto, Spain.

MSFC Marshall Space Flight Center. A NASA center in Huntsville, Alabama.

MTBF Mean Time Between Failures. A measure of technical reliability. The total functional life of a population of an item divided by the total number of failures within the population. The definition holds for time, cycles, or other measures of life units.

MTF Modulation Transfer Function. A measure of the quality of an optical system, based on Fourier analysis.

Multiple Mirror Telescope See **MMT**.

multiplex advantage In an instrument, the advantage in integration time obtained by simultaneously measuring a signal over a range of spectral (or spatial) frequencies compared to scanning single channels. A Fourier transform spectrometer is an example of an instrument possessing a multiplex advantage because all spectral frequencies are detected at once.

multiplexer (mux or MUX) A switching device that sequentially connects multiple inputs or outputs in order to process several signal channels with a single A/D or D/A converter.

MUX See **multiplexer**.

NAR NonAdvocate Review. An evaluation of a project by reviewers who are not part of the project or the users' community.

NAS National Academy of Sciences. A private, nonprofit society of scientists in the United States. Advises the federal government on scientific and technical matters. See also **NRC**.

NASA National Aeronautics and Space Administration. The civil space agency of the United States, founded in 1958.

Nasmyth A Cassegrain focus folded along the altitude axis of an alt-az telescope.

NASTRAN Short for NASA STRuctural ANalysis. A finite element structural analysis program originally developed by NASA in 1965.

natural frequency The lowest vibration frequency of a system in the absence of externally applied excitation. Also called "fundamental frequency." See also **eigenfrequencies**.

NEA Noise Equivalent Angle. The angle on the sky corresponding to the rms random error of an attitude or guiding sensor.

near infrared The portion of the electromagnetic spectrum immediately beyond the visible, extending in wavelength from 0.8 to 5 μm. See also **infrared**.

near-net-shape processing The direct shaping of a mirror blank by casting, forging, or powder consolidation of discrete parts or components in a manner requiring little, if any, subsequent removal of material to comply with final part dimensions and tolerances.

NEMA National Electrical Manufacturer's Association. A U.S. organization which sets standards for motors and other industrial electrical equipment.

neutral axis In structural analysis, the line of zero fiber stress in a given section of a beam subjected to bending.

neutral density filter A filter which reduces the intensity of light equally over the entire bandpass. The reduction is usually expressed as the logarithm of the attenuation. An "ND2" will reduce intensity by a factor of 100.

New Technology Telescope See **NTT**.

Newtonian A telescope configuration with only one powered mirror (primary). The return beam is folded by a flat mirror to locate the focus outside the tube. The Newtonian configuration is popular for amateur telescopes, but rarely used for large ones.

NGST Next Generation Space Telescope. A joint NASA-ESA-CSA project for a successor to the Hubble Space Telescope operating in the 0.5–20 µm range, with an anticipated launch in 2010. Renamed the James Webb Space Telescope, after James E. Webb, NASA's second administrator.

NOAO National Optical Astronomy Observatory. An organization which operates telescopes at Kitt Peak, Arizona and Cerro Tololo, Chile. NOAO is managed by AURA for the NSF.

node The point at which the orbit of a celestial body or spacecraft intersects some particular plane, such as an equatorial plane. If the body passes the plane from south to north, the node is called an ascending node, and from north to south, it is called a descending node.

noise Any unwanted or contaminating signal competing with the desired signal. Also used to describe the random variation in the desired signal.

NRA NASA Research Announcement. A NASA request for a science or technology proposal. NRAs generally involve basic scientific research with end products that are expected to be published in the scientific literature. These are smaller programs than AOs (q.v.) or RFPs (q.v.).

NRC National Research Council. The principal operating agency of the National Academy of Sciences of the United States.

NSF National Science Foundation. An agency of the United States which promotes scientific progress by awarding competitive grants to institutions for research and education.

NTT New Technology Telescope. A 3.5 m telescope at the European Southern Observatory in La Silla, Chile, which was the very first to have an active primary mirror.

numerical aperture (N.A.) For a telescope with the object at infinity, the numerical aperture is $1/(2\times\text{focal ratio})$.

nutation A small periodic motion of the Earth's axis ("nodding") due to the Moon, which is superimposed on precession (q.v.).

Nyquist frequency A sampling frequency twice that of the minimum required resolution. See **Nyquist theorem**.

Nyquist theorem The law that is the basis for sampling continuous information. It states that the frequency of data sampling should be at least twice the maximum frequency at which the information might vary. This condition should be observed in order to preserve patterns in the information or data, without introducing artificial, lower-frequency patterns, a phenomenon called "aliasing" (see Section 4.5.3).

OAO Orbiting Astronomical Observatory. A set of three NASA space observatories in low Earth orbit. OAO-1 failed to deploy. OAO-2, launched in 1968, made observations in the far ultraviolet with 11 telescopes in the 20–40 cm range. OAO-3, launched in 1972 and renamed "Copernicus," made observations in the UV with an 80 cm telescope and also carried out an X-ray experiment.

OBE Operational Base Earthquake. The highest earthquake level that does not affect functionality of the observatory. See also **MLE**.

object space In an optical system, the region upstream of the optical train.

occultation The obscuration of one celestial body by another, as when the Moon passes in front of a star, or when the target observed by an Earth-orbiting telescope is blocked by the Earth.

off axis Refers to a source which is not part of the field of view. Also refers to an architecture for optical systems that positions the elements away from the axis. Off-axis systems benefit from having no central obscuration, thus avoiding parasitic spikes in the image of bright sources. They can also be well baffled.

Offner relay A 1-to-1 optical relay with limited aberrations.

off-ramp technology Standard technology that can be used in place of new technology under development, should the latter not be successful.

off the shelf Said of any equipment regularly produced by a manufacturer or stocked by a supplier.

OPD Optical Path Distance (q.v.).

open loop A system in which there is no feedback. Motion is expected to faithfully follow the input command. Stepping motor systems are an example of open-loop control.

optical astronomy The study of astronomical objects using electromagnetic radiation from the ultraviolet to the far infrared (0.01–500 μm). Sometimes restricted to the visible and the immediately neighboring spectral regions (0.3–1 μm). See also **optical telescopes**.

optical depth A measure of the integrated opacity in a transparent material or in the atmosphere. In a homogeneous material, the absorption along a light path varies as $e^{-\tau}$, where τ is the optical depth. The optical depth is equal to 1 when the intensity is decreased by a factor of e.

optical path distance (OPD) In an optical system, the distance traveled by light passing between two points along the optical path.

optical telescope A telescope working in the optical spectral domain, defined not just as the visible region but also including the adjoining spectral regions where the laws of geometric optics (reflection, refraction) apply *and* diffraction effects are neither negligible nor dominant. This domain extends from the far ultraviolet (100 nm) to about 500 µm wavelengths. In the X-ray domain, optical systems are driven only by geometric effects (diffraction is negligible), whereas in the radio domain, diffraction is dominant (antenna beam theory applies). Telescopes in these two surrounding domains require designs markedly different from those in the optical domain.

optical window The part of the spectrum around the visible wavelengths where Earth's atmospheric absorption is minimum. The optical window extends from about 320 to 760 nm.

optics The science of the generation and propagation of light. Also, the physical system that captures light and transmits it to a detector.

orbital debris Discarded man-made material in near-Earth orbit that can be as large as spent rocket motors and as small as the dust particles ejected from nozzles of maneuvering thrusters. The larger objects (>10 cm) are tracked (more than 7000 of them). The average impact speed of debris on a spacecraft is 10 km/s, only half that of meteoroids, but the population of debris in near-Earth orbits is much higher than that of meteoroids, making debris the greater hazard for low-Earth-orbit spacecraft.

orbital perturbation Deviation from the regular orbit due to a disturbing force.

OSA Optical Society of America. A professional organization founded in 1916 to promote the optical sciences, pure and applied. The society has about 14 000 members from over 70 countries and publishes the *Applied Optics* journal.

OSS Office of Space Science, at NASA Headquarters. Also known as Code S.

OTA Optical Telescope Assembly. Generally refers to the telescope proper in a space observatory.

PAMELA Phased Array Mirror Extendible Large Aperture. A concept for large segmented mirrors in space composed of small ($\sim$ 10 cm) "intelligent" segments. The segments would be mass produced and come equipped with edge sensors, actuators, and their share of the distributed control system. The concept was prototyped at MSFC in the 1990s.

parallax The apparent change in the position of an object when observed from different locations, as when a star is observed from two opposite points of Earth's orbit around the Sun. The annual parallax of a star is the angle subtended at that star by the semimajor axis of the Earth's orbit.

parsec The distance at which the heliocentric parallax would be $1''$, which is 3.26 light years.

payload In aerospace astronomy, the scientific equipment with its associated space-support systems and adapter carried by an aircraft, a balloon or a launch vehicle.

payload adapter In a launch vehicle, the hardware that provides (1) the structural interface between the payload and the launch vehicle and (2) a system for separating the payload from the launch vehicle.

PDR Preliminary Design Review (q.v.).

peak up During target acquisition, a method of refining pointing by small maneuvers so as to pinpoint the direction that will maximize the target signal in an instrument aperture.

Peltier effect A thermoelectric effect wherein electric current applied to a solid/solid or a solid/liquid junction creates heating in one side and cooling in the other. Used for cooling detectors at moderately low temperature (e.g., CCD).

penumbra The portion of a planet's shadow within which part of the disk of the Sun is still visible. See also **umbra**.

perigee The point at which a body in orbit around the Earth most closely approaches the Earth.

period In orbital mechanics, the time required to complete one orbital revolution.

PERT Program Evaluation Review Technique. A project-scheduling technique similar to CPM (q.v.).

perturbation In astrodynamics, a deviation in the position and velocity of a body from its regular trajectory due to the presence of a disturbing force.

Petzval curvature The paraboloidal optical curving of an image at the focal plane caused by astigmatism.

photon A "particle of light." Although light propagates as an electromagnetic wave, it can be created or absorbed only in discrete amounts of energy known as photons. The energy of a photon is inversely proportional to wavelength ($h\nu$): smallest for radio waves, increasingly larger for microwaves, infrared radiation, visible light, and ultraviolet light. It is largest for X-rays and gamma rays.

PI Principal Investigator. A researcher who is officially designated head of a group of scientists and technical staff submitting a proposal to carry out a project (e.g., to perform an observation or build a piece of scientific equipment). The PI is responsible for leading the effort and is usually given exclusive rights to the use of the data or equipment for a specified period following acquisition of the data or completion of the equipment.

PID Proportional Integral Derivative (q.v.).

piezoelectric An effect in a solid in which application of pressure induces a voltage, or vice versa.

piezostack mirror A type of deformable mirror. See **DM**.

pitch axis A space vehicle's axis of rotation normal to the plane of the orbit. See also **yaw axis** and **roll axis**.

pixel Contraction of "picture element." The smallest optically reactive element of an array detector used for imaging.

pixel matching The matching of pixel size to the spatial resolution of the optics.

plasma A highly ionized volume of atoms capable of supporting a current.

plate scale The angle on the sky subtended by a given unit length at the focal plane: plate scale = (angular size)/(image size). Typically expressed in arcseconds per micron or millimeter.

PNAR Preliminary NonAdvocate Review. See **NAR**.

p/n junction An interface formed by two semiconductor materials, the one containing a charge carrier which is an electron donor (n-type semiconductor) and the other containing a charge carrier which is an electron acceptor (p-type semiconductor).

pointing The direction in the sky to which a telescope is pointed. Also, the act of orienting a telescope toward a particular direction in the sky. See also **tracking**.

point spread function (PSF) The variation of intensity with distance from the center of an image of a point source created by an optical system. The PSF describes the optical system's effect on the image of a light source. An image is the convolution of the true brightness distribution on the sky with the PSF of the telescope (or instrument).

Poisson distribution The distribution describing the random fluctuation in a signal with a constant average (e.g., the arrival rate of photons from a source). The probability $p(n,t)$ of n photons falling on a given area of a detector in a time t is given by
$$p(n,t) = (Nt)^n \frac{e^{-Nt}}{n},$$
where N is the average flux (photons per unit time). This distribution has the property that the rms fluctuation in the average flux N (i.e., photon noise) is simply $\sqrt{N}$. This fundamental property is used in the calculation of the signal-to-noise ratio of an observation. The Poisson distribution approaches the Gaussian distribution when N is large, but differs significantly for a small N (i.e., weak source).

Poisson noise The random fluctuation in a signal has a Poisson distribution (q.v.) and is referred to as Poisson noise. Also called "shot noise" or "photon noise."

Poisson's ratio When a piece of material is stretched, the ratio of the lateral contraction per unit breadth to the longitudinal extension per unit length.

polar axis In an equatorial telescope, the axis parallel to the Earth's rotation axis.

polarized light A light beam in which all of the electromagnetic waves are aligned.

polar orbit A low Earth orbit with an inclination near 90°. See also **Sun-synchronous orbit**.

polishing Strictly speaking, making a mirror surface smooth enough to be specular. Loosely speaking, the successive actions of grinding, figuring, and polishing a mirror by lapping.

powered mirror A jargon term for a mirror that has curvature (i.e., is not flat).

precession For the Earth, the apparent slow movement of celestial poles due to the attraction of the Sun on the Earth's equatorial bulge. For a gyroscope, the periodic swinging of the axis of rotation accompanying a torque.

precipitable water The depth of a column of water equivalent to all precipitable water in a column of the atmosphere of the same diameter.

preliminary design review (PDR) A formal examination of the design, including functional flows, requirements, flowdowns, and concepts. The design effort is usually about one-fourth to one-third complete at this point.

primary mirror The first and usually the largest mirror in a reflective optical system. It provides the light gathering and frequently sets the aperture size.

prime focus The focus of the primary mirror of a reflecting telescope.

prime meridian The meridian passing through Greenwich, U.K. adopted as the origin for longitudes on Earth (see **transit telescope**).

principal investigator See **PI**.

project life cycle See **life cycle**.

propellant The gas ejected from a rocket. Ejection may result directly from combustion or be produced by electronic expulsion.

proper motion The change in the apparent position of a star as a function of time as seen from the Sun. Expressed in angular change per year.

proportional control A control mode which generates an output correction in proportion to the system's error (i.e., the system variable's deviation from set-point).

proportional integral derivative (PID) Also referred to as a three-mode controller, combining proportional, integral, and derivative control actions.

prototype A close hardware replica of the final system. Usually at full scale and fully functional.

PSF Point Spread Function (q.v.).

pupil Any image of the entrance aperture (generally the primary mirror). The exit pupil is the last image of the entrance pupil.

Pyrex A glass with a low coefficient of thermal expansion developed by Corning Glass Works.

QE Quantum Efficiency (q.v.).

Quality Assurance The planned, systematic actions necessary to provide adequate confidence that a product will satisfy its intended performance and use.

Quantum efficiency In a detector, the ratio of detected photoelectrons to incoming photons.

radiation A broad term covering emission and propagation of both particles and true electromagnetic waves.

Rayleigh criterion A rule for determining the angular resolution of an optical system. Resolution is defined as the separation between two point sources of equal intensity when the peak of a diffraction pattern of one of the sources falls on the first dark minimum of the diffraction pattern of the other. It is equal to $1.22\lambda/D$, where λ is the wavelength and D is the aperture diameter.

Rayleigh scattering The scattering of light by particles which are small compared to the wavelength of light. See also **Mie scattering**.

Rayleigh star An artificial source created in the lower atmosphere by Rayleigh scattering of a laser beam.

reaction wheel A spinning flywheel used for controlling the attitude of space telescopes by momentum exchange.

redshift The increase in the wavelength of a spectral line from an astronomical body as compared to its value when measured in a laboratory on Earth. Redshifting of stellar spectra is usually interpreted as being due to the Doppler effect (motion away from the observer). The wavelength shift is traditionally expressed as $z = \Delta\lambda/\lambda$. At very large distances, the redshift is interpreted in many cosmologies as being due to the expansion of the universe.

reflectance The ratio of reflected to incident light. See also **albedo**

reflecting telescope A telescope whose main optics are composed of mirrors.

refracting telescope A telescope whose main optics are composed of lenses.

refractive index See **index of refraction**.

requirements The description of a system's function, constraints, and required performance. See also **specifications**.

resolution The ability to distinguish detail in an image, usually expressed in terms of the angular size of the smallest features that can be distinguished. See also **Rayleigh criterion**.

resonance A comparatively large oscillation in a system excited by a periodic input of small amplitude having a frequency close to one of the system's natural frequencies.

response time The time elapsing between the moment a command for change in a system is issued and the moment that change is obtained. When the response of the system has an exponential form, meaning that the time as defined above would be infinite, the response time is, by convention, the time required to reach $1 - 1/e$ (63%) of the commanded value (also called "time constant").

retroreflector An optical device that returns a beam of light in a direction parallel to it, regardless of the orientation of the device. Usually made up of three mutually orthogonal reflective surfaces, forming a concave corner (corner cube retroreflector).

Reynolds number A nondimensional parameter used in assessing whether a flow is laminar or turbulent. It is equal to $V/\nu d$, where V and ν are the fluid's velocity and viscosity, respectively, and d is a characteristic length (e.g., diameter of a pipe).

RFP Request for Proposals. A solicitation of proposals for services, a specific product, or a work package.

right ascension A coordinate for measuring the east-west position of a celestial body; the angle measured eastward along the celestial equator from the vernal equinox to the hour circle passing through a body.

rms Root mean square. The square root of the arithmetic mean of the squares of a set of numbers: $\sqrt{\frac{1}{n}\sum x_i^2}$, where x_i is a series of n values. See also **standard deviation**.

roll axis A space vehicle's axis of rotation along the tangent to the orbit and in the direction of motion. Forms a right-handed coordinate system with the pitch and yaw axes (q.v.).

ROM Rough Order of Magnitude. Term denoting a coarse estimate or best guess for a value (especially cost).

roughness A measure of the smoothness of a surface, usually expressed as the rms of the surface variation.

rpm Revolutions per minute.

rps Revolutions per second.

rss Root sum of squares. The square root of the sum of the squares of a set of numbers (note the difference with **rms**). Used to combine errors of uncorrelated contributing factors (e.g., in an error budget). Must not be employed when the errors are partially correlated (in which case the errors may have to be added arithmetically).

RWA Reaction Wheel Assembly. See **reaction wheel**.

SAA South Atlantic Anomaly. A dip in the lower Van Allen belt created by a reduced magnetic field above Brazil. Passage through the SAA significantly perturbs the operation of electronics and detectors of low-Earth-orbit space telescopes.

SALT South African Large Telescope. A 9.5 m telescope planned at the South Africa Astronomical Observatory in Sutherland, South Africa.

S-band The microwave band near 10 cm wavelength used for satellite communication.

scale height The vertical distance in an atmosphere at which the density drops by the factor e ($\rho/\rho_0 = e^{-h/h_0}$, where h is the altitude, ρ is the density, and h_0 is the scale height). For the Earth's atmosphere, the scale height is about 7 km.

scattering The process by which light is deflected by reflection and diffraction or absorbed and reemitted at a different wavelength.

Schmidt-Cassegrain telescope A telescope system composed of a spherical primary mirror with a corrector plate to correct for its spherical aberration (as in a Schmidt telescope) combined with a Cassegrain secondary mirror mounted on the corrector plate. The primary-mirror/corrector-plate is not a pure Schmidt arrangement, however, since the corrector plate is not mounted at the center of curvature, but in front of the primary mirror focus. The Schmidt-Cassegrain telescope combination is popular among amateur astronomers because of its compact design and large aperture and because the optics are completely enclosed.

A variation of the Schmidt-Cassegrain arrangement is the Maksutov-Cassegrain system in which the corrector plate is replaced by a thick meniscus correcting lens with a strong curvature.

Schmidt telescope A telescope optical combination with a very wide field invented by Bernhard Schmidt in 1930. A Schmidt telescope is composed of a spherical mirror whose spherical aberration is corrected by an aspheric corrector plate located at the mirror's center of curvature. Such a system produces excellent images over a field of several degrees, but the focal plane is curved. Schmidt telescopes are generally used for sky surveys.

Scientific Advisory Committee (SAC) A committee of scientists external to a project whose role is to advise the project office and funding agencies on the scientific goals and priorities of the project. Although the SAC is most active during the definition phase, it also makes recommendations during the construction and commissioning phases concerning the scientific impact of proposed technical changes or workarounds. The project scientist is normally the chairman of the committee. Such a committee is referred to as the "Science Working Group" (SWG) when it acts in a pro-active manner (e.g., early in the project for the definition of the observatory's requirements).

secondary mirror The first powered mirror after the primary mirror of a telescope.

SEE Single-Event Effect (q.v.).

seeing Disturbance in a telescope image due to atmospheric turbulence. Ordinarily expressed as the angular size in arcseconds of a point source (star) seen through the atmosphere, assuming perfect optics.

segmented mirror A mirror composed of individual, close-packed mirror elements.

sensor A device that detects a variable, usually receiving the information in one form (e.g., displacement) and converting it into another (e.g., volts).

Serrurier truss A particular telescope tube structure that maintains optical collimation by parallelogram action. Named after its inventor, Marc Serrurier, who developed this design for the Hale telescope.

settling time The time required for a parameter to stop oscillating or ringing and reach its final value. When the amplitude decay is exponential, settling time is by convention defined as the time required to reach $1/e$ (37%) of the initial amplitude.

SEU Single-Event Upset (q.v).

Shack-Hartmann A type of wavefront-error sensor.

Shane telescope A 3 m telescope at the University of California's Lick Observatory, on Mt. Hamilton, California.

sharpness In telescope optics, a figure of merit for the detection of point sources in background-limited mode introduced by Christopher Burrows. It is the second moment of the pixelized image. See Section 4.4 and Appendix D.

shroud The upper part of a rocket that contains the payload. Also called "fairing."

SI (1) Science Instrument. (2) Système International d'Unités (q.v.).

sidereal time The measure of time defined by the apparent diurnal motion of the stars, hence a measure of the rotation of the Earth with respect to the stars rather than to the Sun.

signal A variable that carries information about another variable that it represents.

signal-to-noise ratio The ratio of signal amplitude to the rms amplitude of the background fluctuation. A measure of the detectability of a signal.

single-event effect (SEE) An electronic dysfunction caused by the lone strike of a charged particle. The effect on the part can be temporary or permanent. An example of a temporary effect is a single-event upset (SEU). A permanent effect is a single-event burnout (SEB), a condition that can cause device destruction in a power transistor due to high-current.

single-event upset (SEU) A radiation-induced, nondestructive error in a microelectronic circuit caused when a charge particle loses energy by ionizing the medium through which it passes, leaving behind a wake of electron-hole pairs. A reset or rewriting of the device results in normal device behavior.

sintering A thermal process in which powdered material is consolidated by heating without melting.

SIRTF Space Infrared Telescope Facility. A NASA 85 cm cryogenic telescope for infrared (3–180 μm) observations. Scheduled for launch in 2003.

skunk works A separate program operation established to operate outside the normal process, either to expedite the program or because of high-security classification.

slew The action of repointing a telescope.

Sloane Digital Sky Survey (SDSS) A survey program using a 2.5 m telescope at the Apache Point Observatory on Sacramento Peak, New Mexico.

slow optics An optical system with a large f-ratio (imported from the terminology of photography).

SNR Also S/N. Signal-to-noise ratio (q.v.).

SOAR telescope SOuthern Astrophysical Research telescope. A 4.2 m telescope under construction on Cerro Pachón, Chile. The SOAR project is being developed through a partnership between NOAO, Brazil, Michigan State University and the University of North Carolina.

sodar SOund Detection And Ranging. See **acoustic sounder**.

sodium laser star An artificial source created in the upper atmosphere by a laser beam scattering off the sodium layer at 80 km altitude.

SOFIA Stratospheric Observatory For Infrared Astronomy. An observatory operated by NASA and DLR (German Aerospace Center) using a 2.5 m telescope carried by a Boeing 747 airplane flying at an elevation of 13 000 m. The observatory is to go into operation in 2005.

solar constant The flux of solar radiation at the Earth's distance, but outside of the atmosphere. It is equal to 1358 W/m^2.

solar flare A strong, temporary emission of hard X-rays and charged particles originating primarily in sunspots. Can perturb the operation of high-orbit space observatories.

solar pressure The pressure created on a surface by sunlight photons. This is the dominant external disturbance for high-orbit space observatories.

solar wind A fairly continuous stream of low-energy charged particles (mostly protons of 100 keV or less) from the Sun. Not a major source of disturbance for space observatories except during solar flares (q.v.). Not to be confused with solar pressure (q.v.).

sole source Refers to a procurement contract that is entered into after soliciting and negotiating with only one potential source.

South African Large Telescope. See **SALT**.

SOW Statement Of Work. Also "scope of work." A detailed description of the efforts and tasks required of a contractor. Usually coupled with a requirements document.

space debris See **orbital debris**.

Sparrow criterion A rule for determining angular resolution of an optical system. It is defined as λ/D, where λ is the wavelength and D is the aperture diameter.

sparse aperture An observing interferometer where the number and distribution of individual apertures is not adequate to fully sample the *uv* plane (q.v) in a single exposure. The dilution factor of a sparse aperture is typically 5% or less. See also **diluted aperture**.

specifications A precise, detailed description of a system's functionality and constraints. Specifications are more precise and elaborate than "requirements" (q.v.) and typically put conditions on material, manufacturing, and testing procedures.

specific heat The amount of energy absorbed by a material that is required to raise its temperature by one unit (expressed in joules per kilogram and degree C).

specific impulse In a rocket motor, the total thrust attainable by a propellant divided by its burning rate. See also **thrust**.

speckle The broken-up pattern caused by atmospheric turbulence in a short-exposure image of a point source. Individual speckles are created by the regions of coherence (r_0) in the incoming beam and have an angular size of $\sim \lambda/D$ (where D is the diameter of the telescope aperture), whereas the size of the full image is $\sim \lambda/r_0$. The speckle pattern in repeated short-time exposures can be exploited to obtain diffraction-limited stellar images.

spectrometer/spectrograph An instrument used to record the spectrum of a celestial object.

specular Having the qualities of a mirror, that is, where a narrow incident beam of light is reflected in one direction only as opposed to being scattered (diffused). Specular reflection occurs when the irregularities of the reflecting surface are small compared to the wavelength of light. In practical optics parlance, "reflection" is synonymous with "specular reflection." See **speculum**.

speculum "Mirror" in Latin (from "speculor," to observe, to watch). Also, the name of a copper-tin alloy used by Isaac Newton and others to make early telescope mirrors.

spider The structure made of thin members (vanes) that holds the secondary mirror at the center of a two-mirror on-axis telescope.

SPIE Society of Photoelectric Instrumentation Engineers; now The International Society for Optical Engineering. A U.S.-based international professional society that organizes yearly technical meetings on optical engineering, including telescopes and instruments. The society has about 15 000 members worldwide and publishes the *Optical Engineering* journal.

SSM Space Support Module. The module of a space observatory that houses support systems such as communications, attitude control, and propulsion. Also referred to as "the bus."

stand-alone system A system requiring little or no assistance from interfacing systems to perform its functions.

standard deviation A measure of the average dispersion of a data distribution (usually denoted as σ). The standard deviation is the rms of the deviations from the mean: $\sigma = \sqrt{\frac{1}{n}\sum(x_i - \mu)^2}$, where x_i is a series of n values with a mean of μ.

standards Established or accepted rules, measures, or criteria against which comparisons are made.

star tracker An attitude determination device for spacecraft in which the locations of stars within the tracker's field of view are measured and compared with the coordinates from a star catalog. After supplying attitude determination, star trackers can also be used to correct gyroscope drift. Star trackers have their own optical systems. Guiding systems that use the main optics of a space telescope are called "fine guiding sensors" (q.v.).

station keeping Orbital maneuvers to maintain a spacecraft in a given orbit.

steady state A characteristic of a condition, such as value, rate, period, or amplitude, exhibiting only negligible change over an arbitrary, extended period of time.

Stewart platform See **hexapod**.

stick-slip Noncontinuous motion due to friction effects at low speed, characterized by successive starts and stops.

strain See **deformation**.

strain energy The mechanical energy stored in a stressed material.

Stratoscope A 30 cm telescope carried by a balloon up to an altitude of 25 km in 1957. This first quasi "space telescope" was used for observation of the Sun in the visible.

Stratospheric Observatory For Infrared Astronomy See **SOFIA**.

stray light Unwanted light from an off-axis source. Light that leaks into a system from outside the field of view.

Strehl ratio A measure of the quality of an optical instrument, equal to the ratio of the amplitude of the point spread function to that of an equivalent, ideal instrument.

STScI Space Telescope Science Institute. An institute that carries out the scientific mission of the Hubble Space Telescope from Baltimore, Maryland. AURA manages STScI under contract with NASA.

Subaru An 8.2 meter optical-infrared telescope on Mauna Kea, Hawaii, operated by the National Astronomical Observatory of Japan.

submillimetric Refers to the portion of the electromagnetic spectrum immediately below microwaves. See also **infrared**.

sunshade A "forebaffle" or shield extending forward from the front end of a telescope to block radiation from the Sun.

sunshield A surface used to block sunlight.

Sun-synchronous orbit A near-polar orbit around the Earth which precesses at the rate of 1 day per year in the eastward direction, thus keeping its plane at a fixed angle with respect to the Sun.

Système International d'Unités International system of units established in 1960 by the General Conference of Weights and Measures, to which most industrialized countries adhere, including the United States. A coherent system (derived from the metric system) with seven base units: meter (m), kilogram (kg), second (s), ampere (A), kelvin (K), mole (mol), and candela (cd). Unit definitions and usage can be found on the National Institute of Standards and Technology's website. Unit abbreviations are capitalized only when derived from proper names (e.g., "V" for volt, "A" for ampere, but "m" for meter). Spelled-out units are not capitalized (e.g., "ampere," "kelvin", "hertz"). Prefixes of 1000 or lower are lowercase (e.g., "kg," not Kg).

systems engineering An engineering technique for large, complex systems that controls the total life-cycle process and results in the definition, development, implementation, and operation of a system that is reliable, cost-effective, and responsive to users' needs. The discipline was formalized in the 1950s and 1960s by NASA and the aerospace and defense industries.

TDRSS Tracking and Data Relay Satellite System. A set of three NASA communications satellites in geostationary orbit used to communicate with the Space Shuttle, HST, and other low-Earth-orbit satellites. The spacecraft constellation is distributed to provide global coverage. TDRSS is operated by the Goddard Space Flight Center.

TEC ThermoElectric Cooler (q.v.).

telemetry Radio signals from a spacecraft used to encode and transmit data to a ground station.

Telescopio Infrarrojo Mexicano See **TIM**.

Telescopio Nazionale Galileo See **TNG**.

terminator The boundary between the illuminated and dark areas of the apparent disk of a planet or planetary satellite.

tertiary mirror The third powered mirror in an optical train.

testbed A system consisting of software simulators or actual hardware used to develop and validate a new technology.

thermal conductivity The ability of a material to transmit heat. Expressed in units of energy per unit area transverse to the direction of energy flow per unit thickness in the direction of flow per unit time per unit of temperature difference across the unit thickness.

thermal diffusivity Thermal conductivity divided by the product of density and specific heat.

thermal inertia The reluctance of a body to change its temperature.

thermoelectric cooler A solid state cooler that cools one side by employing the Peltier effect (q.v.). These devices, although highly reliable, typically have an efficiency of 1% or less.

thrust The propelling force of a rocket engine. It is equal to the propellant mass flow rate multiplied by the propellant exhaust velocity. The impulse is the time-integrated thrust, whereas the specific impulse is the total thrust attainable by a propellant divided by its burning rate.

TID Total Ionizing Dose. A long-term degradation of electronics due to cumulative radiation damage.

TIM Telescopio Infrarrojo Mexicano. A 7 m telescope being built at San Pedro Martir, in Baja California, Mexico.

time constant The amount of time a system requires during a transitory behavior to rise to $1 - 1/e$ (approximately 63%) of its peak final value.

TNG Telescopio Nazionale Galileo. A 3.5 m Italian telescope located at the Roque de los Muchachos observatory on the island of La Palma, Canary Islands.

tolerance A measure of the acceptable range in the dimensions of a part or in the characteristics of an assembly or function.

torr A unit of pressure equal to 1 mm of mercury. An atmosphere is equal to 760 torrs. Named after Evangelista Torricelli.

tracking The action of rotating a telescope so as to follow a target, compensating for Earth's rotation on the ground or drift in a space telescope, or to correct for the proper motion of a target, if a planet. See also **pointing** and **guiding**.

trade study An analysis conducted to compare a technique, architecture, or component to others of its class over all pertinent attributes.

trajectory The dynamical path followed by an object under the influence of gravity or other forces.

transducer A device that receives information in one form and converts it into another. See also **sensor**.

transfer function A mathematical, graphical, or tabular statement of the influence which a system or element has on a signal or action when compared at the input and output terminals.

transient The behavior of a variable during the transition between two steady states.

transit The passage of a celestial body across the local meridian due to the Earth's rotation.

transit telescope A telescope mounted on a fixed east-west axis, which allows it to swing only along the local meridian. Generally used for timing the passages of stars across the meridian. Transit instruments were once the basis of all practical timekeeping. The location of the Airy transit instrument at the Greenwich observatory in the United Kingdom constitutes the zero for all longitudes on the Earth.

TRL Technology Readiness Level. A measure of progress in the development of a new technology. NASA has produced a formal TRL description which is widely accepted.

troposphere The lowest part of the Earths atmosphere where most weather occurs. Its height varies from about 8 km at the poles to about 18 km at the equator. Above it lies the stratosphere, and then the ionosphere and the exosphere.

twilight The time between full night and sunrise or between sunset and full night. Civil twilight is when the Sun is less than 12° below the horizon. Astronomical twilight is when the Sun is between 12° and 18° below the horizon.

Tycho catalog A star catalog based on data obtained by the Hipparcos satellite. The catalog supplies the positions and photometry of about 1 million stars with an accuracy of about 25 mas and 0.07 magnitude, respectively. The limiting magnitude of the catalog is about 11.5 in V. See also **Hipparcos catalog**.

UBV system A photometric system which consist of measuring stellar flux through three color filters: the ultraviolet, U, centered on 360 nm, the blue, B, centered on 420 nm, and the "visible", V, in the green-yellow centered on 540 nm.

UKIRT United Kingdom Infrared Telescope. A 3.8 m telescope on Mauna Kea, Hawaii, owned by the United Kingdom Particle Physics and Astronomy Research Council.

ULE UltraLow Expansion. A type of glass produced by Corning which is 92.5% SiO_2 and 7.5% TiO_2 and has an expansion coefficient of $0 \pm 30 \times 10^{-9}/°C$ over the 5 to 35 °C interval.

ultraviolet (UV) Photons of wavelengths shorter than those of visible light, in the range of 100–400 nm.

umbra The central, completely dark part of a planet's shadow. See also **penumbra**.

United Kingdom Infrared Telescope See **UKIRT**.

universal time (UT) The local mean time of the prime meridian (see **transit telescope**). Identical with Greenwich mean time.

UT Universal time (q.v.).

UV Ultraviolet (q.v.).

***uv* plane** In an observing interferometer, the mathematical plane used to represent the baselines of each pair of subapertures, projected perpendicular to the line of sight. The *uv* plane is so named after the commonly used coordinates u and v. It is also referred to as the "Fourier plane." A single pair of subapertures only provides information on the spatial frequency content of the source corresponding to the angular resolution of that subaperture pair. Satisfactory imaging of an extended source requires measurements with a sufficient number of such pairs, a condition referred to as good *uv*-plane coverage. With a limited number of subapertures, coverage of the *uv* plane can be significantly improved by repeating measurements after either moving the subapertures or rotating them about a common axis. On the ground, the Earth's rotation provides this subaperture rotation "gratis."

valence band The energy band in a semiconductor that is filled with electrons at 0 K. Electrons cannot conduct in the valence band.

validation The process by which it is proved that a system accomplishes its purpose. Validation can only occur at the system level, whereas verification (q.v.) is accomplished at the subsystem level.

value engineering An engineering function which examines proposed designs, methods, and processes with the object of identifying lower-cost techniques or processes to produce an item more economically, yet still consistent with the requirements for performance, reliability, quality, and maintainability.

Van Allen belts Zones of intense radiation surrounding Earth in broad bands parallel to the geomagnetic equator, caused by charged particles trapped in Earth's magnetic field. There are two main zones: the outer belt centered at approximately 12 000 km altitude, made up chiefly of electrons, and the inner belt, centered at approximately 4000 km altitude, made up chiefly of protons.

vane In a two-mirror telescope, one of the thin members holding the secondary mirror at the center of the tube. Also called a "knife edge." See also **spider**. In an optical baffle, one of the concentric rings used to reduce stray light.

verification The process by which it is proved that a subsystem complies with its requirements and intended purpose. Verification can be accomplished by inspection, analysis, or test (in ascending order of confidence). Differs from validation (q.v.), which is done at the system level.

vernal equinox The ascending node of the ecliptic on the celestial equator. Also, the time at which the apparent longitude of the Sun is 0° (around March 21).

vertex The point on an axisymmetric surface at the intersection of its axis of symmetry.

vignetting Gradual fading near the edge of the field of an optical system due to partial obstruction by intermediate optical components.

VIS or V Visible (q.v.).

viscosity The property of a fluid measured by its resistance to deformation under the influence of shear.

visible The portion of the electromagnetic spectrum detectable by the human eye. It extends from about 0.35 to 0.7 μm. The transmission of visible filters ("V") are formally defined as part of the UBV photometric system (q.v.).

VLT Very Large Telescope. A set of four 8 m telescopes operated by the European Southern Observatory and located on Cerro Paranal, Chile. The individual telescopes are referred to as Unit Telescopes (UT) 1 to 4, or by their South American Indian names, Antu, Kueyen, Melipal, and Yepun, respectively.

VLTI An interferometric telescope array composed of the four VLTs (q.v.) and four auxiliary 2 m telescopes.

wave Short for "wavelength" when referring to wavefront errors. For example, "half a wave" is a $\lambda/2$ wavefront error where λ is assumed to be a specific optical test wavelength.

wavefront An imaginary surface of constant phase in a propagating electromagnetic wave. For a source at infinity, the wavefront is a plane. Near the focus of a perfect optical system, the wavefront becomes a portion of a sphere centered at the focus.

wavefront error Departure from an ideal wavefront surface (plane or spherical). This departure can be due to geometric differences in the optical path length or to chromatic effects in refractive systems.

wavelength The distance between two successive recurring features in a periodic sequence, such as the crests or troughs in a wave. For electromagnetic waves, wavelength (λ) is equal to c/ν, where c is the velocity of light and ν is the frequency.

wave number The number of complete wave cycles of electromagnetic radiation that exist in one unit of length. It is simply the reciprocal of the wavelength, expressed, for example, in reciprocal meters (m^{-1}). In a vacuum, the wave number w

(in reciprocal meters) is related to the frequency ν (in hertz) according to $w = \nu/c$, where c is the speed of light (in m/s).

WBS Work Breakdown Structure. A technique used to decompose a project into elements that can be readily priced and managed.

well A term used to describe the holding capacity of a unit cell (pixel) in a focal plane array.

WHT William Herschel Telescope. A U.K. 4.2 m telescope at the Roque de los Muchachos Observatory on the island of La Palma, Canary Islands.

William Herschel Telescope See **WHT**.

WIYN A 3.5 m telescope located on Kitt Peak, Arizona, and operated by the Wisconsin, Indiana, and Yale universities and NOAO (hence the name).

work package An element of a project that can be the subject of a separate contract.

X rays Photons of energies greater than those of ultraviolet rays and smaller than those of gamma rays, with frequencies from 10^{17} to 10^{20} Hz.

yaw axis A space-vehicle axis of rotation in the plane of the orbit normal to the orbit. See also **pitch axis** and **roll axis**.

yield point The point at which strain increases without significant increase in stress.

Young modulus The ratio of the change in stress applied to a body to the change in strain. When the relation is linear, the modulus is the slope of the stress-strain relation. Also called the modulus of elasticity.

Zemax A ray-tracing program for optical design, by Focus Software, Inc.

zenith distance The angular distance on the celestial sphere measured along the great circle from the local zenith to a body on the celestial sphere. Zenith distance is 90° minus altitude.

Zernike polynomials Orthogonal polynomials used in the description of wavefront errors.

Zerodur An ultralow-expansion glass ceramic manufactured by Schott, Germany.

zodiacal emission A faint glow caused by sunlight being reflected and scattered off interplanetary dust near the plane of the ecliptic.

Index

100-inch telescope, *see* Hooker telescope
200-inch telescope, *see* Hale telescope

AAT, 429, 439
AB magnitude, 6
aberration of starlight, 33, 439
    annual, 33
    diurnal, 33
aberrations (optical)
    astigmatism, 111, 442
    coma, 111, 121
    distortion, 111
    field curvature, 111
    general, 110
    off-axis (or oblique), 108
    spherical, 110
acceleration of gravity, 427
acoustic modes (in domes), 376
acoustic sounder, 402
acquisition (of targets), 35, 440
active isolation, 305
active lap, *see* figuring
active optics, 147, 440
    architecture of, 313
    control for, 331
    defined, 312
    VLT, 332
actuators
    force, 222, 327
    moment, 327

adaptive optics, 440
    architecture of, 313
    defined, 312
    multi-conjugate, 342
    with laser stars, 340
    with natural guide stars, 340
aerodynamic torque (space telescopes), 296
Aeroglaze, 195
air mass, 11, 440
aircraft (as observatory platform), 392
airglow, 11
Airy disk, 116, 440
ALADDIN (detector), 55
albedo, 190, 440
alignment of telescopes, 242
ALOT, 152, 329, 440
alt-alt mount, 236
alt-az mount, 30, 235
altitude (angle), 30, 31, 440
altitude axis, 441
altitude-azimuth mount, 30, 235
aluminizing tank, 176
aluminum
    as coating, 174
    for mirror blanks, 146
    material properties, 139, 211
Angel, J.R.P., 142
Anglo-Australian telescope, *see* AAT
Ångström, 441
angular resolution, 37, 120

annealing, 441
annual aberration, 33
Antarctica, 10, 198, 399
APART, 192
aperture
    dilution, 38
    numerical –, 469
    of instruments, 36
    size (of telescopes)
        defined, 441
        evolution of, 2
        influence on scientific performance, 68, 76
        influence on signal-to-noise ratio, 26
    synthesis, 38
aperture stop, *see* stops
aplanatic telescope, 123
Apollonius theorem, 106
ARC telescope, 429, 442
Arecibo telescope, 236, 442
arsenic-doped silicon (Si:As), *see* detectors
artificial stars, *see* laser stars
ASAP, 192
aspect ratio (in mirror blanks), 147, 451
astatic levers, 220
astigmatism, *see* aberrations
Astro-Sitall, 143
astrometry, 442
Astronomical Almanac, 442
astronomical unit (AU), 427, 442
athermalization, 212, 350
atmosphere
    adiabatic lapse rate, 13, 440
    cloud cover, 393, 398, 401
    dispersion by, 13
    gravity waves, 397
    layers, *see* boundary layers
    water vapor, *see* precipitable water
atmospheric background, *see* backgrounds
atmospheric boundary layer, *see* boundary layers
atmospheric drag, 296, 409
atmospheric emission, 11
atmospheric extinction, 9
atmospheric refraction, 12
atmospheric turbulence, 13, 17
    coherence length, 14, 18, 334, 448, 457
    coherence time, 15, 448
    isoplanatic angle, 15
    Kolmogorov spectrum, 17
    structure function, 17
atmospheric windows, 9
attenuation
    of structural modes, 265
    of wind disturbances, 372
    factor
        aerodynamic, 291
        stray light, 197–198
    in neutral density filters, 469
AU, *see* astronomical unit
AURA, 443
autocollimation, 443
autocorrelation (of aperture function), 435
auxiliary optics, 126
available observing time, 84
AXAF, *see* Chandra, 443
azimuth
    angle, 31, 443
    axis, 443
    bearings, 241

Babcock, H., 339
back focal length (or distance), 124, 443
background-limited observations, 27, 443
backgrounds
    cosmic rays, *see* cosmic rays
    due to city lights, 394
    from a detector, 22
    from the atmosphere, 22
    galactic, 19
    moonshine, 22
    noise, 24
    sources, 19
    zodiacal, 19, 485
backlash (mechanical), 207, 272, 282, 443
baffle (stray light)
    design, 187–188
    HST, 197
    in Gregorian telescopes, 122
    in space telescopes, 410
    role of, 183, 443
    scatter from, 189
bake out, 212, 443
balancing systems, 245
balloons (as observatory platforms), 10, 390, 392, 429
band gap, 50, 54, 443
bandpass, 443
bar chart, *see* Gantt chart
bathtub curve, *see* reliability
beam walk, 42, 326, 444
beamsplitter, 444
bearings (of telescopes), 237–241

alignment, 242
  carrying load from, 213, 241
  earthquake loading on, 217, 249
  friction, 256, 271
  hydrostatic
    advantages for pointing control, 273
    design of, 238
    effect of low damping, 261
    effect on seeing, 354
    oil for, 240
  modeling of, 259
  roller, 238
  selection of, 210
beryllium
  for mirror blanks, 145
  material properties, 139, 211
bidirectional reflection distribution function, see BRDF
bidirectional transmission distribution function (BTDF), 191
bimetallic effects
  in mirrors, 146
  in structures, 351
bimorph, see deformable mirrors
Blanco telescope, 429
blank (mirror), 444
  aluminum, 146
  beryllium, 145
  borosilicate glass, 142
  boules for, see boule (of glass)
  diameter-to-thickness ratio, 147, 451
  fused silica, 143
  glass ceramics, 143
  honeycomb, 142, 149, 150, 348
  lightweighting, 142, 147, 150
  material properties, 139
  materials, 139–146
  meniscus, 142, 149, 257, 359, 466
  segmented, 149
  silicon carbide, 144
  specific stiffness, 142
  thermal effects in, 140
  thermal figures of merit, 141
  ULE, 143
blind offset, 36, 253
blind pointing, 35
blind spot (alt-az mount), 236
Bol'shoi Teleskop Azimultal'nyl (BTA), 429, 445
Boller & Chivens, 91
Boltzmann constant, 57, 427
borosilicate glass, 142, 445

material properties, 139
boule (of glass), 143, 154, 445
boundary layers, 402
  atmospheric, 396, 403, 445
  planetary, 395, 445
  surface, 375, 382, 395, 445
brakes, 217, 247
braking load, 217
BRDF, 190, 193
  of mirrors, 193
bright time, 22
Brush-Wellman, 146
BTA, see Bol'shoi Teleskop Azimultal'nyl (BTA)
BTDF, 191
budgeting
  of errors, see error budgets
  of projects, 100, 103

C_3 coefficient, 446
C_n^2 coefficient
  defined, 18, 446
  determined from radiosondes, 405
  modeling, 406
C_T^2 coefficient
  defined, 18
  from acoustic sounder measurements, 402
  inside domes, 352
cable twist, see cable wrap
cable wrap, 245, 256, 271, 272, 276
CAIV, see cost as a design variable
calibration (of instruments), 85
cameras, 41, 42
  in spectrometers, 45
  infrared, 186
  infrared (for seeing investigation), 354
  stray light in, 86
  WFPC, 118
Canada-France-Hawaii telescope, see CFHT
capital cost, distribution, 102
Carbon-fiber-reinforced plastic, see CFRP
Cassegrain focus
  f-ratio, 136, 139, 312
  cage, 288
  instruments, 225, 387
  defined, 135, 446
Cassegrain telescopes
  alignment tolerances, 215
  as relay optics, 60
  baffling, 187, 197
  cable routing, 232

comatic aberration, 111
deformation geometry, 214
emission from secondary mirror, 200
invention of, 122
optical configuration, 107, 122
secondary mirror spider, 230
top unit, 206
PSF, 116
Cassegrain, N., 122
Cauchy formula, 17
caustic zone, 316, 446
CCD, 52
advantages, 49
dark current, 57
quantum efficiency, 53
radiation effects in, 419
readout noise, 58
reason for use of silicon, 50
spectral response, 53
celestial backgrounds, see backgrounds
celestial equator, 29
CELT (edge sensors), 321
central intensity ratio (CIR), 132, 133
central obscuration (defined), 447
Cer-Vit, 143, 447
Cerenkov radiation, 21, 447
Cerro Pachon, 400
Cerro Tololo, 400
CFD, 375
CFHT, 429, 447
balancing, 245
completion time, 98
dome shape, 382
efficiency, 85
mirror cell, 233
mirror cover, 246
secondary mirror, 221
seeing at, 377
seeing investigation, 355
site testing, 403
CFRP, 447
material properties, 211
moisture sensitivity, 212
near zero CTE, 211
CGH, 171
Chandra, 447
completion time, 98
isolators, 303
channel stops, 52
charge coupled device, see CCD
chopping
defined, 23, 447
effect on mirror support, 231

minimizing disturbance due to, 302
typical characteristics, 243
use of beryllium mirrors, 145
use of SiC mirrors, 144
CIR, 132, 133
citation statistics, 82
city lights (effect on sky darkness), 393
clean room, 447
classes of, 447
cost, 103
for space telescopes, 176, 194
cleaning (of mirrors), 174
cleanliness level, 193, 199, 448
$v.$ dust coverage, 195
defined, 194
closed-cycle coolers, see cryocoolers
cloud cover, 398
CO_2 snow cleaning, 175, 176
coating plant, 176
COBE, 11, 21, 448
CodeV (optics design program), 448
coefficient of thermal expansion, see CTE
cogging (in motors), 276
coherence length, see atm. turbulence
coherence time, see atm. turbulence
coherent light (in interferometers), 169
cold biased (thermal control), 347
cold head (cryocoolers), 60
collimated beam (defined), 448
collimation
control, 348, 349
defined, 205, 448
devices, 242
minimizing decollimation, 205
collimator, 171, 448
column of precipitable water, see precipitable water
coma, see aberrations
comparison spectrum, 47
compensation
in chopping mirrors, 243
of atmospheric turbulence, 330, 338, 340
of deflection, 205
of gravity, 35, 147
in mirrors, 147, 149, 220
in space telescopes (gravity release), 164, 223
of jitter, 314
of thermal deformation, 35, 213
of vibrations, 302
of wavefront errors, 35, 325, 327
computational fluid dynamics, 375

computer-controlled lapping (polishing), 160
computer-generated hologram, 171
conduction band, 50, 449
conductivity (thermal), *see* thermal properties of materials
cone effect (in laser stars), 341
configuration management, 95
conic constant, 108
conic surfaces (as mirrors), 108
constants (physical and astronomical), 427
control law, 264, 265
control system (for telescopes), *see* pointing
controller, 263
conversion of units, *see* unit conversions
cooling floor, 353, 376
cooling pipes, 354
coordinate systems, 29
cophasing of segmented mirrors, *see* segmented mirrors
Corning, 142, 143, 445, 474, 483
coronagraph, 186, 449
cosmic rays, 21, 449
    effect on electronics, 419
cosmological window, 20, 390, 449
cost
    as a design variable, 79
    effectiveness, 80, 84
    estimates, 100, 101
    models, 78
    of observatories (breakdown), 102
    scaling laws, 78
coudé focus, 126, 135, 449
Couder, A., 147
countertorque preloading, 278
counterweight lever systems, 221
CPM, 96
critical objects (stray light), 184
critical path, 97, 449
critical path methods, 96
critical sampling, *see* Nyquist sampling
cryo-null figuring, *see* figuring
cryocoolers, 60, 302
cryogenic systems, 60
cryostat, 60, 302, 450
C/SiC, 144
CTE
    defined, 450
    of mirror blanks, 141
    of structural materials, 139
CTIO, 450
curvature sensing, 168, 316, 450

CVD process, 144, 450

Dahl model (friction), 272
damping, 256, 260
    at cryogenic temperatures, 261
    coefficient, 260
    for vibration control, 302
    in earthquake response, 217
    Keck telescopes, 256
    matrix, 259
    telescope foundations, 386
    telescope pier, 216, 385
dark current, *see* detectors, 56
dark sky, 393
dark time, 22
Davenport spectrum (wind), 290
DEC, 29, 450
Decadal Survey, 450
declination, 29, 450
deformable mirrors, 330, 339
    bimorph, 330
    control of, 314, 332
    field limitation due to, 331
    in MCAO systems, 342
    location (at pupil), 109, 125, 326
    piezostack, 331
depletion region (or zone), 52
depth of focus, 121
derotator (prism), 127
descope (of projects), 64, 66, 69, 88
design, 62
    forgiveness, 75
    margin, 66
    optical, 106
    phases, 65
    reference program (or mission), 67
    requirements
        mechanical, 214
        pointing, 253
    reviews, *see* reviews
    testability, 75
detector-noise-limited observations, 26
detectors, 49–55
    arsenic-doped silicon (Si:As), 54
    background, *see* background
    CCD, *see* CCD
    dark current, 25, 55, 56
    HgCdTe, *see* mercury–cadmium–telluride
    infrared, 51, 53
    InSb, *see* indium–antimonide
    linearity, 24
    optical, 51, *see* CCD
    pixel size, 55

quantum efficiency, 2, 53, 55
  readout noise, 55, 57
  readout time, 55
  well depth, 55
devitrification, 451
dewar, see cryostat
Dewar, J., 451
DGT, 204, 205
diameter-to-thickness ratio (in mirror blanks), 147, 451
dichroic, 59, 451
Dierickx, P., 132
differential image motion monitor, 406
differential refraction, 12
differential velocity aberration, 34, 286
diffraction, 114, 451
  effects in spectrographs, 46
  grating, 44, 451
  in image formation, 116
  in radio telescopes, 4
  limit, 76, 451
  spikes (due to vanes), 231
diffraction-limited
  observations, 27
  optics, 119, 452
diffuse surfaces (scatter from), 195
diffusivity (thermal), 141
digital tachometers, 283
diluted aperture, 38, 452
DIMM, 406
DIRBE, 21, 452
direct drives, 217, 228, 275–277
dispersion, 452
  by the atmosphere, 13
  in spectrometers, 44
displacement damage dose (DDD), 419
distortion, see aberrations
disturbances
  correction of, 268, 313
  due to appendages, 300
  due to cable wraps, 246
  due to cryocoolers, 302
  due to fuel slosh, 301
  due to magnetic torque, 297
  due to mechanism motion, 301
  due to reaction wheels, 298
  due to solar radiation pressure, 296
  due to stick-slip, 208
  due to wind, 124, 289–294
  effect of reduction ratio, 277
  in equation of motion, 257
  in ground-based telescopes, 288–294
  in NGST, 269
  in space, 294–302
  linear, 271
  microlurch, 208
  modeling, 258
  nonlinear, 272
  rejection, 270, 271
  vector of, 259, 262
dithering
  mechanical, 273, 452
  to reduce pixelization, 23, 138, 452
diurnal aberration, 33
DM, see deformable mirrors
dome, see enclosures
doping (semiconductors), 51
Dove prism derotator, 127
drag
  atmospheric (space telescopes), see atmospheric drag
  coefficient, 289, 296
Draper Laboratories, 324
drift-away orbits, 414
DRM, see design reference program
DRP, see design reference program
Duran 50, 142
dust (celestial)
  galactic, 19
  in solar system, 19, 346
  infrared cirrus, 19, 462
  zodiacal disk, 20
dust (in optical systems)
  cleaning, see mirror, washing
  coverage v. cleanliness level, 194
  effect on PSF, 116, 119
  impact on infrared background, 188
  on HST mirror, 119, 194
  on mirrors, 115, 199
  protection by mirror cover, 246
  scatter by, 127, 188, 193–195
dynamic range
  defined, 453
  in astronomical sources, 25, 56
dynamical models, see structural models

E6, 453
Earth
  gravity gradient, 295
  magnetic field, 297
  magnetosphere, 418
  mass, 427
  mean radius, 427
  radiation, 297
earthquake
  analysis, 216

492 Index

loading, 216, 249, 381
    maximum likely (MLE), 216
    operational base (OBE), 216
    response spectrum, 217
    restraints, 249
échelle grating, 46, 453
échelle spectrograph, 47, 453
edge sensors, 319–322
    CELT, 321
    Keck telescopes, 320
    number of, 153
effective focal length, 107, 453
eigenfrequencies, 259, 453
eigenvalues, 259
electromagnetic compatibility, 453
electromagnetic spectrum, 9
elemental exposures, 56
EMC, 453
encircled energy (image), 117, 131
enclosures
    acoustic modes, 376
    bogies, 383
    configuration, 369
    corotating, 370
    drive, 383
    floor (effect on seeing), 353
    flow modeling, 375
    functions, 369
    handling equipment, 369, 386
    heat sources, 354
    height above ground, 372
    louvers, 370, 374
        control, 307, 358
        to avoid internal whirl, 370
    relative cost, 102
    requirements, 369
    retractable, 371
    rolling hangar, 371
    seals, 384
    seeing, see seeing, local (dome)
    shape, 381
    shutter, 382, 383
    shutter downflow, 380
    skin emissivity, 380
    snow load, 381
    structural and mechanical design, 380
    thermal control, 352
    thermal design, 376
    vorticity, 370
    wind and water tunnel studies, 375
    wind flushing, 372
    windscreen, 374
        control, 358

energy transfer to high frequencies, 372, 379
encoders, 280
end stops, 248
end-to-end model, 74, 365
English mount, 234
entrance pupil, see pupil
environmental testing, 363
EOST, 91
epoch, 32, 454
equations of motions, 436
equatorial mount, 233
equinox, 32, 454
    vernal, 29, 32
equivalent focal length, 107
error budgets, 72
    image quality, 72
    optical, 127
    pointing, 73
ESA
    launch site, 409
    member states, 454
    yearly budget, 101
escape velocity, 454
ESO, 455
    3.6 m telescope, 429
    member states, 455
    NTT, see NTT
    VLT, see VLT
    yearly budget, 101
étendue, 455
European Southern Observatory, see ESO
European Space Agency, see ESA
exit pupil, see pupil
extinction (atmospheric), 9
extremely large telescopes
    adaptive optics for, 342
    sites for, 397

f-ratio, see focal ratio
Fabry lens, 43
Fabry-Perot interferometer, 455
Fabry-Perot spectrometer, 47
fast Fourier transform (FFT), 116, 455
fast steering mirror, see fine steering mirror (FSM)
fatigue (of materials), 455
fault tree (in reliability analysis), 93
feedback control, 252, 455
    for disturbance rejection, 302, 313
    for mirror support, 222
FEM, see finite element models
field

    curvature, 111
    derotator, 126
    of regard, 456
    of view, 456
    rotation, 235
    stabilization, 244
    stop, *see* stops
figure of merit
    image quality, 129
    mirror thermal behavior, 141, 142
figuring, 159, 456
    of Schmidt plates, 161
    active lap, 159, 162
    by mechanical-deformation, 164
    cost of, 155
    cryo-null, 156, 449
    difficulty v. f-ratio, 160
    ion beam, 157, 162, 163
    of aspheric surfaces, 161
    of Keck segments, 164
    of off-axis mirror segments, 152
    print-through, 151
    stressed mirror, 160–162
    ultraprecision machining, 162
    zero-pressure, 151
filtering (of sensor noise), 270
fine guiding sensor, 456
fine steering mirror (FSM), 244
    distortion, 326
    field rotation, 327
    for jitter compensation, 269
    for seeing compensation, 339
    location (at pupil), 326
finite element models, 254, 259
FIRST, 414, 429, 456
fixed cost (of a project), 103
flat field, 456
flight software, 307
FLIP, 170
flow visualization, 404
flushing (of enclosures), 372–375, 378
    Gemini telescopes, 358, 382
    MMT design, 370
    role of louvers, 370
    traditional domes, 370
focal configurations, 134
focal ratio
    defined, 107
    for Nyquist sampling, 56
    for photographical plates, 139
    of relay optics, 59
    selection of, 136
focus

    (of a telescope), *see* prime, Newtonian, Cassegrain, Nasmyth or coudé focus
    anisoplanatism, 341
    control, 205, 348–350
    depth of, 121
    selection of, 134
focusing, 242
force actuators, *see* actuators
fork mount, 234
formulation phase, 64
Foucault test, 456
four-mirror telescopes, 124
Fourier transform, 116, 457
Fourier transform spectrometer, 48, 457
Fowler sampling, 58
Fowler, A., 58
free atmosphere, 13, 396, 402
frequency, *see* natural frequency
frequency domain, 457
friction drives, 274
Fried length, *see* atm. turbulence
Froude number, 357
FSM, *see* fine steering mirror
FTS, *see* Fourier transform spectrometer
fuel slosh, 301
full width at half-maximum, 131
functional testing, 363
FUSE, 457
fused silica, 457
FWHM, 131

galactic background, 19
Gantt chart, 96, 98, 458
gas thruster, 280
Gaussian distribution, 458
Gaussian noise, 270
gear backlash, *see* backlash
gegenschein, 20
Gemini telescopes, 458
    completion time, 98
    efficiency, 85
    flushing of, 382
    image quality, 72
    mirror, 149
        cooling of, 359
        seeing, 358
        surface heating of, 359
    roller drive, 275
    schedule, 97, 98
    secondary mirror, 244
    sensitivity, 391
    shutter, 383

494   Index

wavefront error PSD, 128
GEO, see orbits, geosynchronous
geometrical configuration factor (GCF), 191
geostationary orbit, see orbits
geosynchronous orbit, see orbits
German mount, 234
glass ceramics, 143, 458
Golay configuration, 38, 459
gold coating, 174
GPS, 308, 459
Gran Telescopio Canarias, see GTC
grating spectrometers, 44
gravitational constant, 427
gravity gradient, 295, 459
gravity waves (atmospheric), 397
gray time, 22
Greenwood frequency, 459
Gregorian telescopes, 107, 122
grinding, 157
grism, 45
ground v. space comparison, 389
ground layer, see boundary layers
ground software, 307
Grubb-Parsons, 91
GTC, 459
    direct drive, 275, 276
    mirror support, 164
guide stars, 35
    (natural) for adaptive optics, 340
    artificial, see laser stars
    catalogs, 36
    density, 287
    effect of parallax, 32
    field limitations, 268
    need for reacquisition (HST), 410
    probability of finding, 288
guiding, 36, 308
    accuracy limitations, 285
    correction of refraction, 35
    correction of starlight aberration, 33
    correction of velocity aberration, 34
    effects of fine steering mirrors, 326
    implementation, 285
    loop, 267
    need for two guide stars, 285
    noise equivalent angle, see noise equivalent angle
    of HST, 268
    star density, 287
gyroscopes, 265, 283
    blending with guiding signal, 268
    drift correction by star trackers, 280

HST, 268
precession, 473

Hale telescope, 1, 98, 206, 368, 429, 459
Hale, George, 1, 459
halo orbits, see orbits, halo
handling equipment, 369, 386
Harlan Smith telescope, 459
Hartmann test, 167
Harvey-Shack law, 193
Hatheway actuator, 223
HAWAII (detector), 55
HDF, 8
heat pipes, 347
heaters, 346, 347, 350
heliocentric orbits, 415
hemispherical reflectance, 190
HET, see Hobby-Eberly telescope
hexapod, 204, 242, 243, 304, 460
HgCdTe, see mercury–cadmium–telluride
high-Earth orbits (HEO), see orbits
Hindle sphere, 172
Hindle-Simpson test, 172
Hipparchus, 5
Hipparcos, 83, 460
Hipparcos Star Catalog, 37
Hobby-Eberly telescope, 429, 460
    mirror, 152, 164
    mount, 236
hologram (computer generated), 171
holographic patches, 322
Hooker telescope, 460
Horn d'Arturo, G., 152
horseshoe mount, 234
hour angle, 30, 460
HST, 2, 84, 429, 460
    aperture door, 247
    baffles, 189, 197
    breathing, 351
    completion time, 98
    differential velocity aberration, 34
    efficiency, 85
    guide stars, 288
    guiding, 268
    gyros, 283
    isolators, 303
    lightshield, 410
    metering truss, 351
    mirror
        actuators, 164
        characteristics, 149, 151
        dust on, 119, 194
        figuring error, 165, 171

microroughness, 119
MTF, 131
operations, 306
phase retrieval, 319
pointing error, 73
PSF, 118
reaction wheel disturbances, 300
resolution, 340
safemode, 247
scientific productivity, 83
star selector system, 285
star tracker, 284
target distribution, 8
thermal control, 349, 351
zero wheel speed crossing, 273
Hubble Deep Field, 8
Hubble Space Telescope, see HST
Hubble Space Telescope Guide Star Catalog, 36
Hubble, E.P., 460
hunting (in control systems), 278, 461
hydrostatic bearings, see bearings

I & V, 361
ICD, 95
ice load, 381
illuminated objects (stray light), 184
image (defined), 461
image quality
    advantage of space, 390
    criteria for, 129–134
    effect of cooling floor on, 377
    effect of phase errors on, 334
    effect of wind on, 291, 373
    for NGST, 134
    from space mirrors, 349
    modeling, 258
    observatory level test, 366
    of KAO and SOFIA, 392
    site testing for, 401
    thermal effects on, 212
image space, 109, 461
IMOS, 74
implementation phase, 64
incremental verification, 363
index of refraction structure coefficient,
    see C_n^2
indium–antimonide (InSb), 51, 54, 55
    dark current in, 57
    quantum efficiency of, 55
infrared cameras (for seeing investigation), 354
infrared cirrus, 19, 462

infrared interferometry, see mirror, testing
infrared telescopes
    chopping, 243
    ideal location, 416
    in Antarctica, 400
    in space, 410
    secondary mirror, 200
    thermal background, 199
InSb, see indium–antimonide
instruments, 41–60
    alignment of, 242
    apertures, 36
    calibration, 85
    comparison spectrum for, 47
    handling equipment for, 386
    kinematic mounting of, 204
    load paths, 208
    radiation effects in, 419
    relative cost, 102
    relay optics for, 59
    stray light in, 183
integrated model, 254, 257
integrated product team, 88
integration and verification, 361
integration time
    calculation of, 25
    dependence on aperture size, 76
    factors in, 85
    reduction by going to space, 391
interface control documents, 95
interface requirements document, 95
interference filter, 462
interferences, 44, 462
interferometers, 462
    $v.$ telescopes, 37
    on the Moon, 415
    Dyson, 323
    Fabry-Perot, 455
    for optical testing, 169
    infrared (for optical testing), 170
    Keck, 3
    LUPI, 169
    Michelson, 48
    Palomar, 337
    shearing, 406
    space, 416
    Twyman-Green, 169
    VLTI, 3
internal metrology, 313, 315, 319, 324
intrinsic photoconductors, 51
inversion layer, 13, 396, 399
ion sputtering, 177

ion-beam figuring, *see* figuring
IPSRU, 324
IPT, 88
IRAS, 198, 411, 429, 462
IRD, 95
IRTF, 429, 462
Isaac Newton telescope, 462
ISO, 429, 462
isolation
    active, 305
    passive, 303
isoplanatic angle, 15, 462
IUE, 412, 429, 461, 463

James Webb Space Telescope, *see* NGST
Jansky, 6
jet stream (effect on seeing), 13
jitter, 314, 463
Joule-Thompson cooler, 463
JPL, 463
Julian date, 29, 463
Jupiter
    brightness, 7
    Lagrange point 2 (as an observatory site), 416

Kalman filter, 270
Kanigen, 145
KAO, 392, 429, 463
    image quality, 392
Karhunen-Loeve modes, 338
Keck telescopes, 429, 463
    azimuth bearings, 241
    completion time, 98
    cophasing, 334
    damping, 256
    dome shape, 382
    edge sensors, 320
    efficiency, 85
    enclosure, 368
    lumped-mass model, 256
    mirror, 149, 152, 161, 164
        actuators, 223
        radial support, 223
        support system, 220
    roller drive, 274
    scientific productivity, 83
    secondary mirror vanes, 232
    warping harness, 164
    whiffletree, 220
kinematic mount, 203, 219, 464
    for bearings, 241
    for instruments, 204

    for mirrors, 220, 222
Kitt Peak, 400
knife edge (mirror support), *see* vanes
knife-edge test, 456, 464
Kolmogorov spectrum, 17
KPNO, 464
Kuiper Airborne Observatory, *see* KAO

L2, *see* orbits, Lagrange points
La Palma, 400
La Silla, 397, 400, 455
Lagrange equations, 256, 436
Lagrange points, *see* orbits
Lambert's law, 190
LAMOST, 429
LAMP, 152, 464
Laplacian, 317
lapping, 157, 158, 464
Large Binocular Telescope, *see* LBT
Large Zenithal Telescope, *see* LZT
largest telescopes (list), 429
Las Campanas, 400
laser metrology, 324
laser stars
    Rayleigh, 341
    sodium, 341
laser, use in interferometers, 169
launch costs, 423
launch load, 217
launchers, 422
LBT, 149, 222, 429, 464
learning curve, 102
Lemaitre, G., 161
life cycle
    cost, 81
    of a project, 63, 465
light pollution, 86, 394
Livermore National Laboratory, 163
load limiter, 208
load paths, 208, 465
loadings
    earthquake, 216
    emergency braking, 217
    launch, 217
    snow and ice, 381
    wind, *see* wind, static effects
local metrology, *see* internal metrology
locking devices, 248
louvers, *see* enclosures, louvers
low Earth orbit (LEO), *see* orbits
lumped-mass model, 254, 255
LUPI, 169
Lyot stop, *see* stops

Lyot, B., 376, 449
LZT, 429, 465

MACOS, 75, 118
Magellan telescope, 222, 429, 465
magnetic torquers, 280
magnification
 effects on deformable mirror performance, 331
 of a telescope, 466
 of secondary mirror, 124
magnitude (of stars), 5
maintainability, 72
manufacture (of mirrors), *see* mirror
MAP, 414
Maréchal rule, 119, 466
marching army syndrome, 103
marginal rays, 110
mass of telescopes *v.* aperture size, 235
matching plate scale to pixel size, *see* pixel, matching to resolution
materials properties
 structural, 210
 thermal, 347
Mauna Kea
 atmospheric refraction, 12
 basic site data, 400
 infrared background, 11
 seeing, 16
 surface layer, 395
 temperature variation, 378
 upper wind, 397
Maxorb, 356
Mayall telescope, 429, 466
MCAO, 342
MCT, *see* mercury–cadmium–telluride
mean time between failures, *see* MTBF
mechanical requirements, 214
mechanism motion (disturbance from), 301
Meinel, A., 124, 163
meniscus mirror, *see* mirror, meniscus
mercury–cadmium–telluride (HgCdTe), 51, 54, 55
 dark current, 57
meso-scale meteorological modes, 405
meteorite, 466
meteoroids, 414, 466
Michelson interferometer, 48
microlurch, 208, 210
micrometeoroid, 467
microroughness, *see* mirror
microthermal sensors, 402
Mie scattering, 9

mirror
 actuators, 223
 blank, *see* blank (mirror)
 cell, 233
 cleaning, 174
 cleanliness, *see* cleanliness level
 coating, 174, 176
 computer controlled lapping, 160
 conic constant, 108
 cooling, 358
 cover, 246
 deflection, 146
 dust coverage, 199
 dust scattering, 119, 188, 194
 effect of wind on, 293
 figure control, 348
 figuring, *see* figuring
 grinding, 157
 handling, 387
 microroughness, 115, 119, 127, 193
 mid-spatial frequencies, 127
 mount, *see* mount for mirrors
 natural frequency, 146
 pads, 213
 polishing, *see* figuring
 print-through, 151
 production, 157
 reflectivity, 86
 scatter, 193
 seeing due to, 356
 segmented, *see* segmented mirrors
 silvering, 176
 structural design, 146
 support vanes, *see* vanes
 surface heating, 359
 testing, *see* optical testing, 167
 thermal effects in, 140, 155, 348
 ventilation, 348, 357, 358
 washing, 174, 175
mission design review (MDR), 64
MLI, 347
MMT, 222, 467
 enclosure, 368, 370
 secondary mirror, 205
 thermal design, 378
modal control, 328
modal density plot, 259, 260
moment actuators, *see* actuators
moment of inertia (structural), 467
momentum dumping, 280, 297, 467
monocoque (design), 213
Moon
 as an observatory site, 415

498 Index

baffling against, 189, 198
brightness, 7
scatter (bright time), 22
mount for mirrors, 219–224
    back support, 220
    for segmented mirrors, 223
    in space telescopes, 222
    radial support, 222
    reactionless, 243
    whiffletree, 220
mounts for telescopes, 233–237
    alt-alt, 236
    alt-az, 30, 235
    drive rates, 235
    English, 234
    equatorial, 233
    field rotation due to, 235
    fixed, 236
    fork mount, 234
    German, 234
    horseshoe type, 234
MPI telescope, 429
Mt. Fowlkes, 400
Mt. Graham, 400
Mt. Locke, 400
Mt. Palomar, 400
Mt. Palomar telescope, *see* Hale telescope
MTBF, 93, 468
MTF, 129
multi-layer insulation, 347
multiconjugate adaptive optics, *see* MCAO
multimirror telescope, *see* MMT
multiplexer, *see* MUX
MUX, 54, 468

NASA, 468
    launch site, 409
    project phases, 63
    space transportation system (STS), 392
    technology readiness levels, 92
Nasmyth focus, 135, 468
NASTRAN, 468
National Aeronautics and Space Administration, *see* NASA
National Optical Astronomy Observatory, *see* NOAO
National Science Foundation, *see* NSF
natural frequency
    of mirrors, 146
    of telescope tubes, 227
NDI, 192
NEA, *see* noise equivalent angle

near infrared, 468
near net shape, 150, 469
New Technology Telescope, *see* NTT
Newton, I., 122
Newtonian focus, 469
Next Generation Space Telescope, *see* NGST
NGST, 429, 469
    cophasing, 319, 336
    detector critical sampling, 138
    disturbances in, 269
    efficiency, 85
    elemental exposure, 56
    image quality, 134
    mirror, 149
    mirror actuators, 223
    mirror testing, 166, 173
    optical configuration, 125
    orbit, 414
    phase retrieval, 336
    segmented mirror support, 224
    thermal emission, 198
    verification flow, 364
    wavefront error correctability, 328
    WBS, 89
NOAO, 469
    yearly budget, 101
node regression, 409
noise (Gaussian), 270
noise equivalent angle, 268, 286, 287
nonadvocate review (NAR), 65
nonlinear disturbances, *see* disturbances
nonstationarity, 401
normalized detector irradiance, 192
notch filters, 269
NSF, 469
NTT, 293, 332, 370, 429, 469
null corrector, 168, 171
numerical aperture, 469
numerical modeling, 375
nutation, 32
Nyquist sampling, 137, 138
Nyquist theorem, 470

OAO, 429, 470
object space, 109
observation overhead, 85
observatory
    completion time, 97
    control software, 306
    cost (typical apportionment), 102
    efficiency, 85
    sites, 389, 398
        Antarctica, 399

coastal, 399
compared, 400
desirable characteristics of, 393
inland, 399
islands, 398
stratosphere, 392
testing, *see* site testing
validation, 366
observatory-level tests, 366
off-axis aberration, 108, 110
off-axis instruments, 123
off-axis mirror segments
  fabrication, 152, 159
  surface formulas, 108
off-axis optical design, 124
off-axis sources
  defined, 183, 470
  stray light from, *see* stray light
Offner relay, 59, 126
offsetting, 36
Ohara, 453
OPD, 37, 470
Opteon, 161
optical depth, 470
optical astronomy (defined), 470
optical design, 106
optical path distance (OPD), 470
optical performance, 76
optical sensitivities, 262
optical telescope (defined), 4, 471
optical testing, 165–173
  at cryogenic temperatures, 173
  CGH, 171
  effect of gravity, 173
  Hartmann test, 167
  Hindle sphere, 172
  infrared interferometry, 170
  interferometric, 169
  null corrector, 171
  primary mirrors, 171
  radius of curvature, 173
  secondary mirror, 172
  spherometer, 167
  surface finish, 170
optics, *see* mirror
optics (relative cost of), 102
orbital debris, 471
orbits
  drift-away, 414
  geostationary, 411
    defined, 458
  geosynchronous, 411
    v. geostationary, 458

altitude of, 427
  radiation effects in, 418, 420
halo, 413, 421
heliocentric, 415
high Earth, 412
Lagrange points
  defined, 464
  Sun-Earth L2, 412, 421, 427
  Sun-Jupiter L2, 416
low Earth (LEO), 408, 419, 465
Moon, 415
polar, 411
Sun-synchronous, 410
OTF, 130
overrun (in projects), 65
Owens-Illinois, 143
OWL, 126

p/n junction, 51, 52, 473
Pamela, 152
parallactic angle, 30
parallax, 32, 471
Paranal, 455
  basic site data, 400
  flow visualization, 404
  seismic characteristics, 216
  site layout, 396
  surface layer, 395
  upper wind, 397
  VLT site, 484
paraxial rays, 110
PDB, *see* project, data base
peak-up (target acquisition), 36
Peltier effect, 472
performance
  budget, 72
  metric
    scientific, 82
    technical, 84
PERT, 96
Petzval curvature, 472
phase diversity, 319
phase retrieval, 318
phase-shift intererometer, 169
photocathode devices, 49
photographic plate, 2, 49, 139
photometers, 43
photon noise
  in guiding systems, 268, 286
  of a source, 26
PI controller, 264
Pic du Midi telescope, 377
PID controller, 264

500   Index

pier (of telescope), 369, 385
    cable routing, 245
    earthquake loading, 216, 249
    effect of wind, 293
    foundation, 386
    in alt-az telescopes, 241
    insulation, 354, 379
    stiffness, 217
    telescope alignment, 242
    vibrations, 325
piezoelectric deformable mirrors, 330
piezoelectric effect, 472
pixel, 472
    channel stops, 52
    critical sampling, 56, 137
    dithering to reduce pixelization, 23
    effect of cosmic rays, 21, 56
    matching to resolution, 27, 56
    sharpness (scaling with), 433
    typical sizes, 55
Planck constant, 427
planetary boundary layer, *see* boundary layers
plate scale, 107, 137, 472
point spread function (PSF)
    by Fourier transform, 116
    characterization, 129
    defined, 116, 473
    modeling, 118
    of a perfect system, 117
    of actual systems, 118
    of HST, 118
    of Keck telescopes, 335
    of NGST, 134
    relation to MTF, 130
pointing
    control law, 263–270
    control software, 306
    corrections, 31
    disturbances, *see* disturbances
    error budget, 73
    guiding, *see* guiding
    guiding loop, 267
    of HST, 73
    position loop, 266, 267, 269, 270, 272, 282
    procedure, 35
    requirements, 253
    servo system, 263–270
    structural filters, 269
    system modeling, 253
    target acquisition, *see* acquisition
    velocity loop, 266, 267, 272, 282

Poisson distribution, 473
Poisson equation, 317
Poisson noise, 24, 473
Poisson ratio, 473
    of blank materials, 139
    of structural materials, 211
polar orbits, 410, 473
polarimeters, 44
polishing, 157, 159
    defined (strict sense), 473
position loop, *see* pointing
postfiguring deformation, 164
potted optics, 213
precession, 32
precipitable water (column of)
    as a function of altitude, 10
    at major observatory sites, 400
    defined, 10, 473
    in Antarctica, 400
preload, 207, 231, 278
Preston law, 160
prestressing, *see* preload
primary mirror
    defined, 107, 474
prime contractor, 90
prime focus, 134
print-through (in mirrors), 151
procurement strategy, 90
profilometer, *see* spherometer
project
    completion time, 98
    data base, 90
    life cycle, 63
    management, 62, 86
    manager, 88
    organization, 88
    overrun, 65
    phases, 63, 64
    scheduling, 96
proper motion, 32
PSF, *see* point spread function
pupil, 109, 474
Pyrex, 142, 474

quality assurance (QA), 94
quantum efficiency (of detectors), *see* detectors
quilting, *see* mirror, print-through

RA, 29
radiation
    effects of, 418
    in space, 417–421

radiators (for space telescopes), 348
radiosondes, 404
radius of curvature (measure of), *see* mirror testing
Rayleigh criterion (resolution), 120, 474
Rayleigh scattering, 9
Rayleigh stars, 341
Rayleigh's quarter wavelength rule, 119
RC, *see* Ritchey-Chrétien
reaction bonded process, 144
reaction wheels, 279
    disturbances due to, 298
reactionless mount, *see* mirror, mount
readout noise, *see* detectors, 57
redshift, 475
reduction ratio, 277
reference star, 35
reflecting telescope (defined), 475
refracting telescope (defined), 475
rejection ratio (stray light), 197
relay optics, 59
reliability, 72, 93
REOSC, 161, 166, 170
requirements, 475
    design, *see* design, requirements
    flowdown, 69
    levels, 69
    matrix, 364
    operational, 71
    safety, 71
resolution (angular), 76, 120
resolution (spectral), 45
resolving power (spectrometer), 45
resonance (excitation of), 207, 266, 279, 290, 292, 299, 301–303, 331, 373
resource planing, 103
response time
    defined, 475
    pointing system, 264
retroreflector, 322, 475
reviews
    critical design (CDR), 66
    design, 451
    mission design (MDR), 64
    nonadvocate (NAR), 65
    observatory design, 64
    preliminary design (PDR), 65
    systems requirements (SRR), 65
Reynolds number, 289, 475
Richardson number, 405
right ascension, 29, 475
risk analysis, 99

Ritchey, George W., v
Ritchey-Chrétien telescope, 123–125
rms (defined), 475
Roddier, F., 168
roll axis, 475
roller bearings, *see* bearings
roller drives, 274
root mean square (defined), 475

S-band, 476
SAA, *see* South Atlantic Anomaly
safe mode, 307
sagittal plane, 111
sampling, *see* Nyquist sampling
scale height (defined), 476
scaling laws, 76, 77
scattering, 188, 189
    from diffuse surfaces, 195
    Mie, 9
    off mirrors, *see* mirror
    Rayleigh, 9
schedule slack (float), 97
scheduling of projects, 96
Schmidt telescopes, 350
Schmidt, B., 161
Schott, 142, 143, 485
Schwarzschild theorem, 123
science verification, 366
science working group (SWG), 477
scientific advisory committee (SAC), 477
scientific productivity, 82
scintillation, 15
secondary mirror, 477
    alignment devices, 242, 243
    baffling, 187
    chopping, 231, 243, 356
    defined, 107
    diffraction effects, 116, 118
    emission from, 200
    hexapod support, 205
    mirror mount, 221
    parameters, 124
    seeing, 356
    spider, 206, 230, 232
    support, 228
    testing, 172
    thermal effects, 356
    tolerances, 123, 215
seeing
    correction, 338
    defined, 13, 477
    due to floor, 353
    due to heat sources in domes, 354

due to hydrostatic pads, 354
from mirrors, 356
local (dome), 351
monitors, 406
origin of, 394
use of infrared cameras, 354
segmented mirrors, 151–155
cophasing, 333, 336
geometry, 152
mount for, 223
thermal control, 349
seism, seismic, see earthquake
self emission, 199
semiconductors, 50
sensing noise, filtering, 270
sensitivity, 76
Serrurier truss, see tube
service life, 71
settling time (defined), 477
Shack-Hartmann sensor, 168, 315, 333, 335
Shane telescope, 429
sharpness, 132, 433
shot noise (in detectors), 57
shutters (for enclosures), see enclosures
sidereal time, 28, 477
Siding Springs, 400
signal-to-noise ratio, 24, 433, 477
silicon carbide, 139, 144
silver coating, 174
single-event effect (SEE), 418, 477
single-event upset (SEU), 418, 478
SIRTF, 429, 478
detectors, 55
efficiency, 85
mirror, 149
temperature control, 350
Sitall, see Astro-Sitall
site testing, 401–404
acoustic sounder, 402
flow visualization, 404
image quality, 401
microthermal sensors, 402
numerical modeling, 405
radiosondes, 404
seeing monitors, 406
sites (of observatories), see observatory, sites
sky brightness, 22, 86
sky-limited observations, 27
slip joints, 245
Sloane telescope, 478
enclosure, 371

focus control, 350
snow cleaning, see CO_2 snow cleaning
snow cleaning (of mirrors), 175
snow load, 381
snow on enclosures, 381, 383
SNR, see signal-to-noise ratio
SOAR, 429
sodium stars, 341
SOFIA, 391, 392, 429, 478
solar activity, 417
solar arrays, 300
solar constant, 427, 478
solar cycle, 417
solar luminosity, 427
solar mass, 427
solar radiation pressure, 280, 296, 478
solar wind, 478
solid state detectors, see detectors
source noise, 26
South Atlantic Anomaly (SAA), 21, 418, 476
South Pole, see Antarctica
space environment, see radiation
space frame, see truss
Space Infrared Telescope Facility, see SIRTF
space orbits, see orbits
Space Telescope Science Institute, 480
space-based facilities, advantages, 390
Sparrow criterion, 120
sparse aperture, 479
specific impulse, 479
specific stiffness, 140, 142, 211
speckle, 479
spectral resolution, 45
spectrographs, see spectrometers
spectrometers
dispersing, 44
echelle, 47
Fabry-Perot, 47
Fourier transform, 48
grating, 44
resolving power, 45
specular reflection, 157–159, 190, 193, 473, 479
speculum, 479
speed of light, 427
spherical aberration, see aberrations
spherical mirrors (telescopes with), 126, 236
spherometer, 167, 173
spider, 206, 230
spoke wheel, 210
spring equinox, 29

SPSI, 169
spur gear, 274
star density, 287
star in a box, 324
star selector systems, 285
star tracker, 284, 480
star trails, 406
state-space representation, 262
steel (material properties), 211
Stefan-Boltzmann constant, 348, 427
Stewart platform, see hexapod
stick-slip, 208, 272
stiffness matrix, 259
stigmatism, 106
stops
    aperture, 108, 185
    cold, 448
    field, 108, 186
    Lyot, 186, 465
    pupil, 108
Stratoscope, 429, 480
stratosphere (observing from the), 392
Stratospheric Observatory For Infrared Astronomy, see SOFIA
stray light
    analysis, 189
    critical objects, 184
    defined, 183
    direct paths, 184
    from off-axis sources, 183
    illuminated objects, 184
    thermal, 183, 198
Strehl ratio, 119, 131, 339, 480
Strehl seeing angle, 340
stressed mirror figuring, see figuring
Strouhal number, 290
structural design, 213, 380
structural filters, 269
structural model modal density plot, 260
structural models, 217, 254, 258, 455
structure coefficients, see C_n^2 and C_T^2
structure function (of turbulence), 17
STScI, 480
Subaru telescope, 480
    completion time, 98
    direct drive, 275
    enclosure, 370
    mirror, 149
    secondary mirror, 242
    thermal control, 356
submillimetric telescopes, 4
Sun
    baffling against, 197, 247
    brightness, 7
Sun sensor, 284
Sun-Earth L2, 198, 412–414
Sun-Jupiter L2, 416
Sun-synchronous orbit, see orbits
sunshield, 198
surface layer, see boundary layers
Système International d'Unités, 481
systems engineering, 62, 66
systems requirements review (SRR), 65

tachometers, 266, 272, 282–283
    motor-mounted, 274
    quantization effect, 272
tangential plane, 111
target acquisition, 35
TDRSS, 481
technical problems, time lost due to, 85
technology
    development, 91
    enabling, 92
    enhancing, 92
    off-ramp, 92
    readiness level, see TRL
telescope
    alignment, 242
    balancing, 245
    Cassegrain, see Cassegrain telescopes
    collecting power, see aperture size
    control, 307
    cost, see cost models
    diameter, see aperture size
    end stops, 248
    focus, see prime, Newtonian, Cassegrain, Nasmyth or coudé focus
    focusing, 242
    four-mirror, 124
    infrared, see infrared telescopes
    list of the largest, 429
    locking device, 248
    mass (v. aperture size), 235
    materials, 210
    mounts, see mounts for telescopes
    pier, see pier (of telescope)
    pointing, see pointing
    reflecting (defined), 475
    refracting (defined), 475
    sensitivity, 1, 76
    single-mirror, 121
    structure (relative cost of), 102
    three-mirror, 124
    throughput, 86, 200
    tracking, see pointing

504   Index

tube, *see* tube (of telescope)
two-mirror, *see* Casssegrain telescopes
    with spherical mirror, 126, 236
Telescopio Nazionale Galileo, *see* TNG
temperature control, 346
temperature structure coefficient, *see* C_T^2
Tenerife, 400
testing (of mirrors), *see* mirror, testing
thermal background, 199
thermal control, 345–360
thermal diffusivity, 141
thermal environment, 346
thermal models, 261
thermal properties of materials, 347
thermal snaps, 294
thermographs (for seeing investigation), 355
threading (of baffles), 196
three-mirror telescopes, 124
throughput, 200
thrust, 481
TIM (Telescopio Infrarrojo Mexicano), 482
TIM image modeling software, 118
time
    from GPS, 308
    Julian, 29
    sidereal, 28
time constant
    defined, 475
    thermal, 149
Tiny TIM, 118
tip-tilt mirror, *see* fine steering mirror
TIS, *see* total integrated scatter
TNG, 429, 482
tolerances
    of a Cassegrain telescope, 215
    of mirror support systems, 221
    of the secondary mirror, 123
torr, 176
total integrated scatter (TIS), 195
total ionizing dose (TID), 419
total systems authority, 90
transit telescope, 482
TRL, 92, 482
troposphere, 482
    turbulence in, 13
troughput, 86
truss design, 214
tube (of telescope), 224–233
    multibay truss, 226
    natural frequency, 227
    Serrurier truss, 206, 225
    thermal effects, 229

tower-type, 228
tripod-type, 228
twilight (astronomical), 84, 482
Twyman-Green interferometer, 169
Tycho Star Catalog, 37

UBV system, 6, 483
UKIRT, 429, 483
ULE fused silica, 143, 483
    material properties, 139
ultraprecision machining, *see* figuring
unit conversions, 427
universal joint, 210
up-the-ramp sampling, 58
upper winds, 397
uv-plane, 483

validation, 94, 483
value engineering, 483
Van Allen belts, 21, 409–412, 417, 419, 476, 484
Vandenberg Air Force Base, 411
vanes for secondary mirror support, 115, 118, 124, 230, 484
    arrangement, 231
    diffraction from, *see* diffraction
    emission from, 200
    for hiding cables, 232
    in Keck telescopes, 232
    prestressing, 230
vanes in baffles, 187, 196, 197
VanZandt model, 405
velocity loop, *see* pointing
ventilation
    for cooling dome floors, 354
    of enclosures, 370, 374
    of mirrors, *see* mirror, ventilation
verification, 94, 484
    flow, 364
    incremental, 363
    matrix, 364
    methods, 363
    with end-to-end model, 365
Very Large Telescopes (ESO), *see* VLT
vibration
    control, 302
    damping of, 260
    during launch, 422
    from cryocoolers, 60
    from machinery, 385
    isolation of, 303
    modes (of mirrors), 328
vignetting, 109, 484

VISTA, 429
VLT, 484
    absolute encoder, 281
    active optics, 329, 332
    azimuth bearings, 241
    cable wrap, 246
    completion time, 98
    direct drive, 275
    earthquake restraint, 249
    efficiency, 85
    enclosure, 368
    enclosure louvers, 374
    locked rotor frequency, 276
    lumped-mass model, 257
    mirror cell, 233
    mirror cooling, 359
    mirror cover, 246
    mirror figuring, 161
    mirror seeing, 358
    open air concept, 371
    position loop transfer gain, 270
    secondary mirror, 244
    site layout, 396
    tachometer, 282
    top end, 232
    wind feed-forward loop, 271
    wind torque PSD, 292
VLTI, 484
Von Karman spectrum, 290
Von Karman vortices, 290
vortex shedding, 290, 292

warping harness, 164
washing (of mirrors), 174
water vapor in the atmosphere, see precipitable water
water-tunnel studies, 375
wave (as a unit of measure of wavefront error), 112, 484
wave number, 485
wavefront, 484
wavefront error
    as a metric for image quality, 131
    correction, 325, 328
    defined, 111
    due to aberrations, 113
    sensing, 315
    vs. Strehl ratio, 76
WBS, 89, 90, 96, 101, 103, 485
weather
    monitoring, 308
    time lost due to, 85
weather seals, 384

WFPC, 118
whiffletree (mirror mount), 220
WHT, 429, 485
    efficiency, 85
William Herschel Telescope, see WHT
Wilson, R.N., 123, 327
wind
    disturbance due to, 124, 289
    effect on mirrors, 293
    effect on telescope pier, 293
    flushing of enclosures, 372
    power spectrum, 290, 293, 373
    static effects due to, 289
    telescope shake due to, 370
wind buffeting
    correction by active optics, 312
    correction with fine steering mirror, 325
    in extremely large telescopes, 398
    tube top end, 232
wind-tunnel studies, 375, 404
window plate (glass)
    emission of, 200
    scatter of, 191
windows
    atmospheric, 9
    cosmological, 20, 390, 449
    infrared, 9
    optical, 471
windscreen, 374
WIYN telescope, 429, 464, 485
work breakdown structure, see WBS
worm gear, 273

Young modulus, 140, 211, 485
    of blank materials, 139
    of structural materials, 211

Zeiss, 91, 161
Zener theory, 261
zenith distance, 31, 485
Zernike polynomials
    defined, 112, 114
    wavefront correction, 325, 328, 331
    wavefront error representation, 118
zero-pressure figuring, see figuring
Zerodur, 143, 485
    material properties, 139
zodiacal light, 19

ASTRONOMY AND ASTROPHYSICS LIBRARY

Continued from page ii

Modern Astrometry
By J. Kovalevsky

Astrophysical Formulae 3rd Edition (2 volumes)
Volume 1: Radiation, Gas Processes, and
 High Energy Astrophysics
Volume 2: Space, Time, Matter, and Cosmology
By K.R. Lang

Observational Astrophysics 2nd Edition
By P. Lena, F. Lebrun, and F. Mignard

Astrophysics of Neutron Stars
Editors: V.M. Lipunov and G. Börner

Galaxy Formation
By M.S. Longair

Supernovae
Editor: A.G. Petschek

General Relativity, Astrophysics, and Cosmology
By A.K. Raychaudhuri, S. Banerji, and A. Banerjee

Tools of Radio Astronomy 3rd Edition
By K. Rohlfs and T.L. Wilson

Atoms in Strong Magnetic Fields
Quantum Mechanical Treatment and Applications in Astrophysics and Quantum Chaos
By H. Ruder, G. Wunner, H. Herold, and F. Geyer

The Stars
By E.L. Schatzman and F. Praderie

Physics of the Galaxy and Interstellar Matter
By H. Scheffler and H. Elsässer

Gravitational Lenses
By P. Schneider, J. Ehlers, and E.E. Falco

Relativity in Astrometry, Celestial Mechanics, and Geodesy
By M.H. Soffel

The Sun
An Introduction
By M. Stix

Galactic and Extragalactic Radio Astronomy 2nd Edition
Editors: G.L. Verschuur and K.I. Kellermann

Reflecting Telescope Optics (2 volumes)
Volume I: Basic Design Theory and Its Historical Development
Volume II: Manufacture, Testing, Alignment, Modern Techniques
By R.N. Wilson

Tools of Radio Astronomy
Problems and Solutions
By T.L. Wilson and S. Hüttemeister